Parallel Computational Fluid Dynamics

Theory and Applications

PARALLEL COMPUTATIONAL FLUID DYNAMICS

THEORY AND APPLICATIONS

Proceedings of the Parallel CFD 2005 Conference
College Park, MD, U.S.A. (May 24-27, 2005)

Edited by

ANIL DEANE
Conference Chair
University of Maryland
College Park, MD, U.S.A.

AKIN ECER
IUPUI
U.S.A.

JAMES MCDONOUGH
University of Kentucky
U.S.A.

NOBUYUKI SATOFUKA
University of Shiga Prefecture
Shiga, Japan

GUNTHER BRENNER
Technical University of Clausthal
Germany

DAVID R. EMERSON
Daresbury Laboratory
U.K.

JACQUES PERIAUX
Dassault-Aviation
Saint-Cloud
France

DAMIEN TROMEUR-DERVOUT
Université Claude Bernard Lyon I
France

ELSEVIER

Amsterdam – Boston – Heidelberg – London – New York – Oxford – Paris
San Diego – San Francisco – Singapore – Sydney – Tokyo

Elsevier
Radarweg 29, PO Box 211, 1000 AE Amsterdam, The Netherlands
The Boulevard, Langford Lane, Kidlington, Oxford OX5 1GB, UK

First edition 2006

Copyright © 2006 Elsevier B.V. All rights reserved

No part of this publication may be reproduced, stored in a retrieval system
or transmitted in any form or by any means electronic, mechanical, photocopying,
recording or otherwise without the prior written permission of the publisher

Permissions may be sought directly from Elsevier's Science & Technology Rights
Department in Oxford, UK: phone (+44) (0) 1865 843830; fax (+44) (0) 1865 853333;
email: permissions@elsevier.com. Alternatively you can submit your request online by
visiting the Elsevier web site at http://elsevier.com/locate/permissions, and selecting
Obtaining permission to use Elsevier material

Notice
No responsibility is assumed by the publisher for any injury and/or damage to persons
or property as a matter of products liability, negligence or otherwise, or from any use
or operation of any methods, products, instructions or ideas contained in the material
herein. Because of rapid advances in the medical sciences, in particular, independent
verification of diagnoses and drug dosages should be made

Library of Congress Cataloging-in-Publication Data
A catalog record for this book is available from the Library of Congress

British Library Cataloguing in Publication Data
A catalogue record for this book is available from the British Library

ISBN-13: 978-0-444-52206-1
ISBN-10: 0-444-52206-9
ISSN: 1570-9426

For information on all Elsevier publications
visit our website at books.elsevier.com

Printed and bound in The Netherlands

06 07 08 09 10   10 9 8 7 6 5 4 3 2 1

Preface

Computational Fluid Dynamics, broadly encompassing fluid flows in engineering, atmospheric and ocean sciences, geophysics, physics and astrophysics, has seen calculations of great fidelity performed routinely that were only imagined at the inception of this conference series in 1989. These advances have not only been made with the available horsepower of increasingly more powerful parallel computers, but in tandem with advances in implementations of core and of entirely new algorithms, including multi-scale and adaptive methods.

Far from being a done deal however, the complexity of the logical and physical hardware continues to exert itself in the creation of algorithms that seek excellence in performance. Fluid Dynamics is sufficiently challenging that virtually all known algorithms need to be mapped to parallel architectures as our applications include more and more complex physics and elucidate ever more complex phenomena.

In Parallel CFD 2005, the traditional emphases of the Parallel CFD meetings was included - parallel algorithms, CFD applications, and experiences with contemporary architectures: parallel unstructured and block-structured solvers, parallel linear solvers for implicit fluid problems (domain decomposition, multigrid, Krylov methods) and adaptive schemes; unsteady flows, turbulence, complex physics, reactive flows, industrial applications, and multidisciplinary applications; developments in software tools and environments, parallel performance monitoring and evaluation of computer architectures, - as well as newer areas such as grid computing, and software frameworks and component architectures.

Parallel CFD 2005 was held on the campus of the University of University of Maryland College Park, at the Inn and Conference Center, May 24-27, 2005. Eight invited talks, two special session talks, and one hundred and ten registered attendees have led to this volume of sixty-two technical papers divided into seventeen sections that show the enduring diversity of the conference and its international participation.

The Editors

Acknowledgements

Parallel CFD 2005 gratefully acknowledges support by:

- Institute for Physical Science and Technology (IPST), University of Maryland, College Park.

- Center for Scientific Computation and Mathematical Modeling (CSCAMM), University of Maryland, College Park.

- Burgers Program in Fluid Dynamics, University of Maryland College Park.

- Army Research Laboratory, Aberdeen, Maryland.

- National Aeronautics and Space Administration (NASA).

- Army High Performance Computing Research Center, Minneapolis, Minnesota.

- Intel Supercomputer Users' Group.

- IBM, Armonk, New York.

- Silicon Graphics, Inc. (SGI), Mountain View, California.

- James River Technical, Glen Allen, Virginia.

- International Association for Mathematics and Computers in Simulation (IMACS), Universite libre de Bruxelles, Belgium & Rutgers University, New Brunswick, New Jersey (IMACS endorsed meeting).

- American Physical Society (APS endorsed meeting).

The local organizing committee members (*viz.* Anil Deane, Jan Sengers and Eitan Tadmor) also thank Susan Warren and Lisa Press, UMCP Conference and Visitors Services and Leyla Lugonjic, UMUC Inn and Conference Center, for making the meeting a success, and extend a special thanks to Teri Deane for serving as Conference Secretary, and Alice Ashton, Aaron Lott and Dongwook Lee, IPST, for assisting with the conference.

TABLE OF CONTENTS

Preface   v
Acknowledgements   vi

1. Invited Speakers and Special Sessions

C.B. Allen
Parallel Simulation of Lifting Rotor Wakes   1

Roland Glowinski, J. Hao and T.W. Pan
*A Distributed Lagrange Multipliers Based Fictitious Domain Method for the
Numerical Simulation of Particulate Flow and its Parallel Implementation*   11

William D. Henshaw
Solving Fluid Flow Problems on Moving and Adaptive Overlapping Grids   21

G. Wellein, P. Lammers, G. Hager, S. Donath and T. Zeiser
*Towards Optimal Performance for Lattice Boltzmann Applications
on Terascale Computers*   31

Rainald Löhner, Chi Yang, Juan R. Cebral, Fernando F. Camelli,
Fumiya Togashi, Joseph D. Baum, Hong Luo, Eric L. Mestreau and
Orlando A. Soto
*Moore's Law, the Life Cycle of Scientific Computing Codes and
the Diminishing Importance of Parallel Computing*   41

David Keyes
"Letting physicists be physicists," and Other Goals of Scalable Solver Research   51

2. Turbulence

Jin Xu
Investigation of Different Parallel Models for DNS of Turbulent Channel Flow   77

O. Frederich, E. Wassen and F. Thiele
*Flow Simulation Around a Finite Cylinder on Massively Parallel
Computer Architecture*   85

Paul R. Woodward, David A. Porter, Sarah E. Anderson, B. Kevin Edgar,
Amitkumar Puthenveetil and Tyler Fuchs
*Parallel Computation of Turbulent Fluid Flows with
the Piecewise-Parabolic Method*   93

Jianming Yang and Elias Balaras
Parallel Large-Eddy Simulations of Turbulent Flows with Complex Moving Boundaries on Fixed Cartesian Grids   101

F.X. Trias, M. Soria, A. Oliva and C.D. Pérez-Segarra
Direct Numerical Simulation of Turbulent Flows on a Low Cost PC Cluster   109

Wei Lo and Chao-An Lin
Large Eddy Simulation of Turbulent Couette-Poiseuille Flows in a Square Duct   117

R. Giammanco and J.M. Buchlin
Development of a Framework for a Parallel Incompressible LES Solver Based on Free and Open Source Software   125

3. Grid Computing

M. Garbey and H. Ltaief
On a Fault Tolerant Algorithm for a Parallel CFD Application   133

R.U. Payli, H.U. Akay, A.S. Baddi, A. Ecer, E. Yilmaz and E. Oktay
Computational Fluid Dynamics Applications on TeraGrid   141

Christophe Picard, Marc Garbey and Venkat Subramaniam
Mapping LSE Method on a Grid: Software Architecture and Performance Gains   149

D. Tromeur-Dervout and Y. Vassilevsky
Acceleration of Fully Implicit Navier-Stokes Solvers with Proper Orthogonal Decomposition on GRID Architecture   157

4. Software Frameworks and Component Architectures

Craig E. Rasmussen, Matthew J. Sottile, Christopher D. Rickett and Benjamin A. Allan
A Gentle Migration Path to Component-Based Programming   165

Michael Tobis
PyNSol: A Framework for Interactive Development of High Performance Continuum Models   171

V. Balaji, Jeff Anderson, Isaac Held, Michael Winton, Jeff Durachta, Sergey Malyshev and Ronald J. Stouffer
The Exchange Grid: A Mechanism for Data Exchange Between Earth System Components on Independent Grids   179

Shujia Zhou and Joseph Spahr
A Generic Coupler for Earth System Models — 187

Kum Won Cho, Soon-Heum Ko, Young Gyun Kim, Jeong-su Na,
Young Duk Song and Chongam Kim
CFD Analyses on Cactus PSE — 195

Benjamin A. Allan, S. Lefantzi and Jaideep Ray
*The Scalability Impact of a Component-Based Software Engineering
Framework on a Growing SAMR Toolkit: a Case Study* — 203

5. BioFluids

I.D. Dissanayake and P. Dimitrakopoulos
*Dynamics of Biological and Synthetic Polymers Through Large-Scale
Parallel Computations* — 211

Paul Fischer, Francis Loth, Sang-Wook Lee, David Smith, Henry Tufo
and Hisham Bassiouny
Parallel Simulation of High Reynolds Number Vascular Flows — 219

6. Multiphysics and MHD

Ding Li, Guoping Xia and Charles L. Merkle
*Large-Scale Multidisciplinary Computational Physics Simulations Using
Parallel Multi-Zone Methods* — 227

Ying Xu, J.M. McDonough and K.A. Tagavi
*Parallelization of Phase-Field Model for Phase Transformation Problems
in a Flow Field* — 235

Dongwook Lee and Anil E. Deane
A Parallel Unsplit Staggered Mesh Algorithm for Magnetohydrodynamics — 243

V.A. Gasilov, S.V. D'yachenko, O.G. Olkhovskaya, O.V. Diyankov,
S.V. Kotegov and V.Yu. Pravilnikov
*Coupled Magnetogasdynamics – Radiative Transfer Parallel Computing
Using Unstructured Meshes* — 251

7. Aerodynamics

A.G. Sunderland, D.R. Emerson and C.B. Allen
*Parallel Performance of a UKAAC Helicopter Code on HPCx and Other
Large-Scale Facilities* — 261

Jubaraj Sahu
Parallel Computations of Unsteady Aerodynamics and Flight Dynamics of Projectiles — 269

S. Bhowmick, D. Kaushik, L. McInnes, B. Norris and P. Raghavan
Parallel Adaptive Solvers in Compressible PETSc-FUN3D Simulations — 277

Gabriel Winter, Begoña González, Blas Galván and Esteban Benítez
Numerical Simulation of Transonic Flows by a Double Loop Flexible Evolution — 285

B.N. Chetverushkin, S.V. Polyakov, T.A. Kudryashova, A. Kononov and A. Sverdlin
Numerical Simulation of 2D Radiation Heat Transfer for Reentry Vehicles — 293

8. Parallel Algorithms and Solvers

D. Guibert and D. Tromeur-Dervout
Parallelism Results on Time Domain Decomposition for Stiff ODEs Systems — 301

R. Steijl, P. Nayyar, M.A. Woodgate, K.J. Badcock and G.N. Barakos
Application of an Implicit Dual-Time Stepping Multi-Block Solver to 3D Unsteady Flows — 309

B. Hadri, M. Garbey and W. Shyy
Improving the Resolution of an Elliptic Solver for CFD Problems on the Grid — 317

9. Structured AMR

K.B. Antypas, A.C. Calder, A. Dubey, J.B. Gallagher, J. Joshi, D.Q. Lamb, T. Linde, E.L. Lusk, O.E.B. Messer, A. Mignone, H. Pan, M. Papka, F. Peng, T. Plewa, K.M. Riley, P.M. Ricker, D. Sheeler, A. Siegel, N. Taylor, J.W. Truran, N. Vladimirova, G. Weirs, D. Yu and J. Zhang
FLASH: Applications and Future — 325

R. Deiterding
An Adaptive Cartesian Detonation Solver for Fluid-Structure Interaction Simulation on Distributed Memory Computers — 333

Kevin Olson
PARAMESH: A Parallel, Adaptive Grid Tool — 341

10. General Fluid Dynamics

Andrei V. Smirnov, Gusheng Hu and Ismail Celik
Embarrassingly Parallel Computations of Bubbly Wakes   349

S.C. Kramer and G.S. Stelling
Parallelisation of Inundation Simulations   357

X.J. Gu, D.R. Emerson, R.W. Barber and Y.H. Zhang
Towards Numerical Modelling of Surface Tension of Microdroplets   365

I. Abalakin, A. Alexandrov, V. Bobkov and T. Kozubskaya
DNS Simulation of Sound Suppression in a Resonator with Upstream Flows   373

A. Hamed, D. Basu, K. Tomko and Q. Liu
Performance Characterization and Scalability Analysis of a Chimera Based Parallel Navier-Stokes Solver on Commodity Clusters   381

I. Nompelis, T.W. Drayna and G.V. Candler
A Parallel Unstructured Implicit Solver for Hypersonic Reacting Flow Simulation   389

11. Boundary Methods

M. Garbey and F. Pacull
Toward A MatlabMPI Parallelized Immersed Boundary Method   397

Yechun Wang, Walter R. Dodson and P. Dimitrakopoulos
Dynamics of Multiphase Flows via Spectral Boundary Elements and Parallel Computations   405

12. Parallel Tools and Load Balancing

Stanley Y. Chien, Lionel Giavelli, Akin Ecer and Hasan U. Akay
SDLB - Scheduler with Dynamic Load Balancing for Heterogeneous Computers   413

Sameer Shende, Allen D. Malony, Alan Morris, Steven Parker and J. Davison de St. Germain
Performance Evaluation of Adaptive Scientific Applications using TAU   421

B. Norris, L. McInnes and I. Veljkovic
Computational Quality of Service in Parallel CFD   429

13. Combustion

G. Eggenspieler and S. Menon
Parallel Numerical Simulation of Flame Extinction and Flame Lift-Off   437

R.Z. Szasz, M. Mihaescu and L. Fuchs
Parallel Computation of the Flow- and Acoustic Fields in a Gas Turbine Combustion Chamber   445

14. Discrete Methods and Particles

George Vahala, Jonathan Carter, Min Soe, Jeffrey Yepez, Linda Vahala and Angus Macnab
Performance of Lattice Boltzmann Codes for Navier-Stokes and MHD Turbulence on High-End Computer Architectures   453

Dan Martin, Phil Colella and Noel Keen
An Incompressible Navier-Stokes with Particles Algorithm and Parallel Implementation   461

G. Brenner, A. Al-Zoubi, H. Schwarze and S. Swoboda
Determination of Lubrication Characteristics of Bearings Using the Lattice Boltzmann Method   469

Masaaki Terai and Teruo Matsuzawa
MPI-OpenMP Hybrid Parallel Computation in Continuous-Velocity Lattice-Gas Model   477

15. High Order Methods and Domain Decomposition

Amik St-Cyr and Stephen J. Thomas
Parallel Atmospheric Modeling with High-Order Continuous and Discontinuous Galerkin Methods   485

A. Frullone and D. Tromeur-Dervout
A New Formulation of NUDFT Applied to Aitken-Schwarz DDM on Nonuniform Meshes   493

16. Unstructured Grid Methods

B. Chetverushkin, V. Gasilov, M. Iakobovski, S. Polyakov, E. Kartasheva, A. Boldarev, I. Abalakin and A. Minkin
Unstructured Mesh Processing in Parallel CFD Project GIMM   501

Thomas Alrutz
Investigation of the Parallel Performance of the Unstructured DLR-TAU-Code on Distributed Computing Systems 509

17. Visualization

Marina A. Kornilina, Mikhail V. Iakobovski, Peter S. Krinov, Sergey V. Muravyov, Ivan A. Nesterov and Sergey A. Sukov
Parallel Visualization of CFD Data on Distributed Systems 517

Parallel Simulation of Lifting Rotor Wakes

C. B. Allen[a*]

[a]Department of Aerospace Engineering,
University of Bristol,
Bristol, BS8 1TR, U.K.

Lifting rotors in both hover and forward flight are considered. These flows are extremely expensive to simulate using finite-volume compressible CFD codes due to the requirement to capture the vortical wake, and its influence on the following blades, over many turns. To capture the wake to any reasonable accuracy thus requires very fine meshes, and so an efficient parallel flow solver has been developed and is presented here. The code is an implicit unsteady, multiblock, multigrid, upwind finite-volume solver, and has been paralellised using MPI. An efficient structured multiblock grid generator has also been developed to allow generation of high quality fine meshes. Results of four-bladed rotors in hover and forward flight, in terms of global loads, wake capturing, and more detailed flow features, are presented and analysed, using meshes of up to 32 million points, and excellent parallel performance on upto 1024 CPUs demonstrated. In hover it is shown that fine meshes can capture unsteady wakes, as have been shown from low diffusion schemes, for example free-vortex or vorticity transport methods, and theoretical stability analyses. In forward-flight it is shown that if global loads are required coarse(ish) grids can be used, but for more detailed flow features, for example blade-vortex interaction, grid convergence is not approached even with 32 million points.

1. INTRODUCTION

It is well known that numerical diffusion inherent in all CFD codes severely compromises the resolution of high flow gradients. This is a serious problem for rotor flow simulation, where the vortical wake capture is essential. Hover simulation requires the capture of several turns of the tip vortices to compute accurate blade loads, resulting in the requirement for fine meshes away from the surface, and a long numerical integration time for this wake to develop. Forward flight simulation also requires accurate capture of the vortical wake but, depending on the advance ratio, fewer turns need to be captured, as the wake is swept downstream. However, not only does the entire domain need to be solved, rather than the single blade for hover, but the wake is now unsteady, and so an unsteady solver must be used, which is not only more expensive than the steady solver used for hover, but can easily result in even higher numerical diffusion of the wake. Hence, it is extremely expensive to simulate these flows, and rotor simulations using conventional finite-volume compressible CFD codes are normally run with meshes that are

---

[*]Reader in Computational Aerodynamics. Email: c.b.allen@bris.ac.uk

not fine enough to capture enough physics to give anything of interest except global blade loads. This problem is considered here. Of particular interest is the development and capture of the unsteady vortical wake, and the effect of this wake on global loads and more detailed flow physics, for example blade-vortex interaction, which is a major design consideration for rotor blades. Hence, a grid dependence study has been performed, using a finite-volume, implicit unsteady, upwind, multiblock, multigrid solver, which has been parallelised using MPI to allow the use of very fine meshes.

In this paper the unsteady formulation of the flow-solver will be presented, along with parallelisation details, followed by grid generation aspects for structured grids for rotors in hover and forward flight. Previous work has shown [1,2] that multiblock grids result in much better wake capturing than single block for hovering rotors, and since multiblock grids are essential in forward flight, a multiblock structured grid generator has been developed. Numerical solutions for lifting rotors are then presented and examined, for meshes of upto 32 million points. Grid dependence of the vorticity capturing, and the influence of the vortical wake capturing on the loading of the following blades, is considered. The question is: how much of the physics can be captured, how much needs to be captured when considering global loads for example, and how expensive is it ? Parallel performance of the code is also presented.

2. FLOW SOLVER

The Euler equations in integral form in a blade-fixed rotating coordinate system are:

$$\frac{d}{dt}\int_{V_r} \mathbf{U}_r dV_r + \int_{\partial V_r} \mathbf{F}_r . \mathbf{n}_r dS_r + \int_{V_r} \mathbf{G}_r dV_r = 0 \tag{1}$$

where

$$\mathbf{U}_r = \begin{bmatrix} \rho \\ \rho u_r \\ \rho v_r \\ \rho w_r \\ E \end{bmatrix}, \quad \mathbf{F}_r = \begin{bmatrix} \rho[\mathbf{q_r} - (\omega \times \mathbf{r})] \\ \rho u_r[\mathbf{q_r} - (\omega \times \mathbf{r})] + P\mathbf{i_r} \\ \rho v_r[\mathbf{q_r} - (\omega \times \mathbf{r})] + P\mathbf{j_r} \\ \rho w_r[\mathbf{q_r} - (\omega \times \mathbf{r})] + P\mathbf{k_r} \\ E[\mathbf{q_r} - (\omega \times \mathbf{r})] + P\mathbf{q_r} \end{bmatrix}, \quad \mathbf{G}_r = \begin{bmatrix} 0 \\ \rho(\omega \times \mathbf{r}).\mathbf{i_r} \\ \rho(\omega \times \mathbf{r}).\mathbf{j_r} \\ \rho(\omega \times \mathbf{r}).\mathbf{k_r} \\ 0 \end{bmatrix}. \tag{2}$$

$\mathbf{G_r}$ is the source term resulting from the transformation from fixed to rotating coordinates. This form is used with a steady solver for hover calculations. For forward flight, a fixed axis system is used with an unsteady solver. In this case $\mathbf{G_r} = \underline{0}$, and the coordinate vector $\mathbf{r}$ is now time dependent, i.e.

$$\mathbf{r}(t) = [R(t)]\mathbf{r}(0) \tag{3}$$

where $[R(t)]$ is the time-dependent rotation matrix.

The equation set is closed in both cases by

$$P = (\gamma - 1)[E - \frac{\rho}{2}\mathbf{q}^2]. \tag{4}$$

An unsteady finite-volume upwind scheme is used to solve the integral form of the Euler equations. The code has been developed for structured multiblock meshes, and incorporates an implicit temporal derivative, following Jameson [3], with an explicit-type scheme within each real time-step. A third-order upwind spatial stencil is adopted, using the flux vector splitting of van-Leer [4,5] along with a continuously differentiable limiter [6]. Multigrid acceleration is also adopted [7–9].

2.1. Parallelisation

The code has been parallelised using MPI, to allow the use of fine meshes. However, there is a balance to be struck here, since multigrid is implemented: more CPU's means smaller blocks and, hence, faster solution time per global time-step, but smaller blocks mean fewer multigrid levels can be used, resulting in slower convergence. A pre-processing algorithm has been developed which reads in the mesh in its coarsest (in terms of number of blocks) form, and splits it into more blocks, if required, while attempting to balance the number of points on each CPU, and maintaining the maximum number of multigrid levels.

The solver is coded such that each block boundary simply has a boundary condition tag, a neighbouring block number, and an orientation flag. The only restriction applied is that a boundary can only have one type of tag. Hence, each block only requires NI, NJ, NK, the $NI \times NJ \times NK$ physical coordinates, then six lines, of three integers each, defining the boundary conditions. The spatial stencil is five points in each direction, and so for parallel send/receives, at each internal block boundary, two planes of solution are simply packed and sent to the appropriate processor, and received boundary data unpacked according to the orientation flag. All communications are thus non-blocking. This mesh approach has the advantage that no halo or multigrid data is required in the grid file and, hence, there is no 'preprocessing' stage. This is important as it avoids the mesh having to be processed on a single CPU, which can result in a mesh size limit, or the development of a parallel grid generator/preprocessor. A separate file can be written for each block, and a header file created which defines the block-CPU relationship and the name of each block file. This also means that if the spatial stencil is changed, for example to a higher-order scheme, the number of levels of solution sent and received can simply be increased.

The code has been written to require no global flow or geometry data storage. Only arrays of block dimensions and connectivity are required globally, and this vastly reduces the memory requirements of the code. The 32 million point mesh cases shown later can be run with five levels of multigrid on 16 dual-CPU nodes, with only 1GByte RAM per node, i.e. less than 0.5GByte/million points.

The preprocessing algorithm discussed above has been linked to the grid generator, and so can be run before generating the mesh. This then optimises the block dimensions to ensure load balancing, rather than generating a mesh and then trying to optimise the block-CPU distribution.

3. GRID GENERATION

A multiblock grid generation tool has been developed which is applicable to fixed- and rotary-wing cases, but is particularly effective for rotor flows. A transfinite interpolation based scheme [11,12,10] is employed within each block, but with significantly improved surface normal weighting and orthogonality, along with a periodic transformation [10] to ensure high quality meshes.

As the flow-solver has been parallelised there is, in theory, no limit on the mesh size that can be used and, hence, it is important that the grid generator does not place a limit this. The software developed has been coded to be extremely efficient in terms of memory requirement and, as an example, a 64 million point, 408 block, mesh can be generated in around 30 minutes on a P4 machine running Linux, and requires less than 2GBytes RAM.

A wake grid dependence study has been performed, considering the ONERA 7A blade [13]. A typical mesh density for a simulation may be one million points, and so to perform the grid dependence study, this density was doubled five times, i.e. meshes of density 1, 2, 4, 8, 16, and 32 million points were generated, for hover and forward flight.

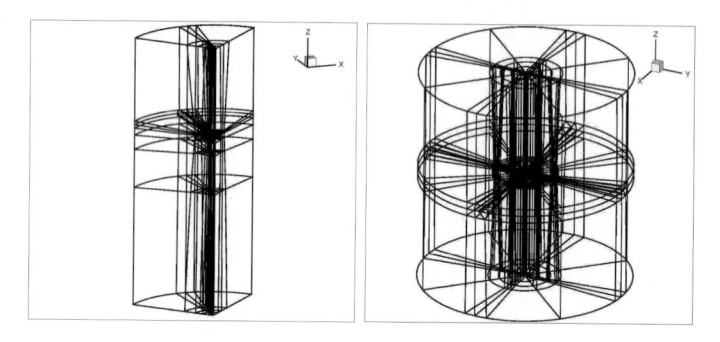

Fig.1:*Multiblock domain and block boundaries. Hover and forward flight.*

Figure 1 shows the computational domain and block boundaries, for the 7A four-bladed case, left is hover and right is forward flight. The farfield for the forward flight mesh is set at 50 chords spanwise and 50 chords vertically, i.e. a cylinder of radius 50 chords, height 100 chords, and for hover the spanwise boundary is 100 chords, and lower is 150 chords. The grid density above and below the rotor disk in forward flight is much smaller than used for hover, since the wake is swept downstream rather than downwards, and so the grid density there is less significant.

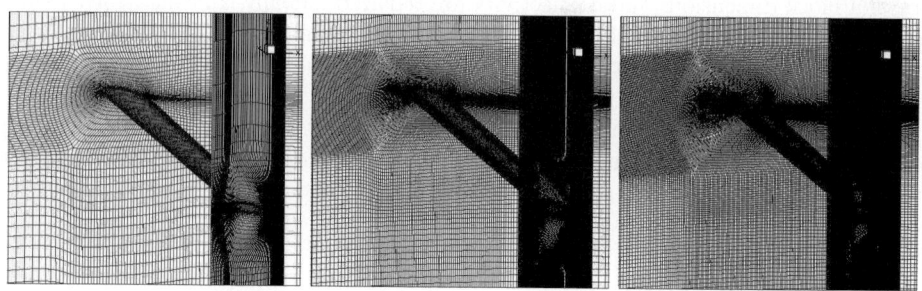

Fig.2:*Hover mesh in selected planes,* $1, 8, 32 \times 10^6$ *points.*

These meshes were all individually generated, i.e. coarser meshes have not been generated by removing points from finer ones. This was done to ensure the highest possible grid quality and also, in the forward flight case, to maintain the maximum proportion of points in the rotor disk area, as this is the region of interest.

Figure 2 shows selected grid planes for the hover meshes, and figure 3 part of the grid in the rotor disk for forward fight, in both cases for 1, 8, and 32 million point grid densities.

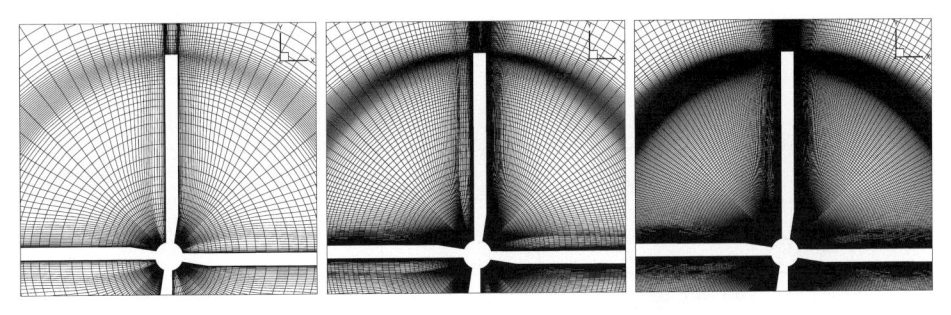

Fig.3: *Multiblock forward flight mesh. Rotor disk grid, $1, 8, 32 \times 10^6$ points.*

4. 7A ROTOR LIFTING CASES

4.1. Hover Case

The hover test case is $M_{Tip} = 0.661$, with a thrust coefficient of around 0.08, and the solver was run in steady mode, in a blade-fixed rotating coordinate system. The case corresponds to a 7.5^o effective incidence at 75% span. Figure 4 shows vorticity contours in the downstream periodic plane, i.e. 45° behind the blade, for 1, 8, and 32 million points. The effect of grid density on vorticity capturing is clear; upto ten turns of the tip vortices are captured with 32 million points, but only one with one million.

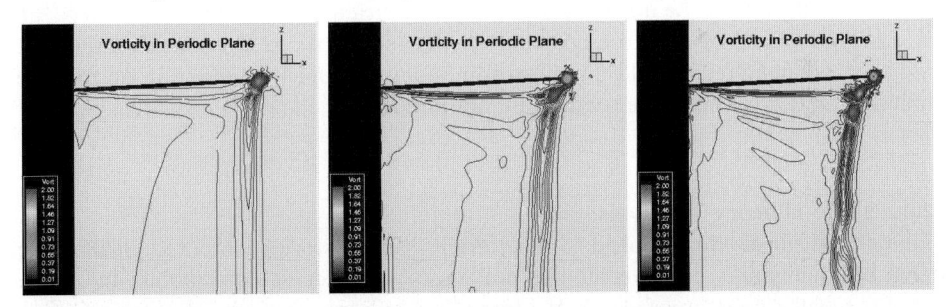

Fig.4: *Vorticity contours in periodic plane, $1, 8, 32 \times 10^6$ points.*

Figure 5 shows convergence histories of the simulations. It is clear that an oscillatory wake has been captured for 16 million points and upwards.

Previous work has shown that hover is not actually a steady case, but little, if any, has been presented on simulation of unsteady hover flows using finite-volume CFD methods. Recently, wake breakdown in forward flight [14], and unsteady hovering wakes have been simulated [15], using the low diffusion vorticity transport formulation. Bhagwat and Leishman [16] have also shown unsteady wakes using a free-vortex method, and a stability analysis shows that the wake should actually be in periodic equlibrium, but is unstable to all perturbations. Hence, any experimental or numerical result is almost certain to become unstable, due to any perturbation. Hence, it is stated in the above paper, the fact that similar unsteady results are shown in experiment and simulation is "fortuitous".

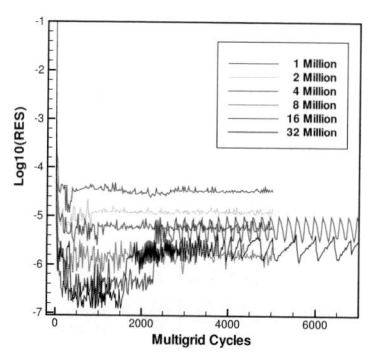

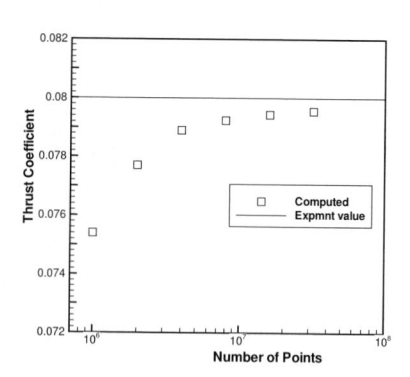

Fig.5: 7A convergence history.　　　　　　Fig.6: 7A thrust coefficient.

The experimental single blade load coefficient for this case is 0.16, corresponding to a normalised thrust coefficient ($\frac{C_T}{\sigma}$) of 0.08. The thrust coefficient grid dependence is shown in figure 6. Considering figures 4 to 6 it is clear that increasing the grid density significantly improves the wake capturing but this, even capturing an unsteady wake, has little effect on the rotor load.

4.2. Unsteady Simulation

This case was then run as an unsteady case. The final 'steady' solution was used as the initial solution, and the unsteady solver started with a few very small time-steps, and this was gradually increased to a constant 90 time-steps per revolution. Two questions could then be answered: would any unsteadiness in the solution remain in an unsteady simulation, and if it did, is the solution in fact periodic ?

4.2.1. Time History Analysis

Fourier transforms were applied to the vorticity time histories computed on the downstream periodic plane. However, it was not sensible to process every point, and so just four of the block boundary faces in the periodic plane were considered. Figure 7 shows the block boundaries in the region of interest, and the four planes considered. To reduce the amount of data, each plane was split into 12 sub-blocks, and the total vorticity in each sub-block summed at each time step. Fourier transforms of this quantity were then computed over the entire simulation (the first five revolutions were ignored in case of any transients).

Figure 8 (left) shows the magnitude of the Fourier coefficients for the frequency range, relative to the rotation frequency, of 0 to 100, for plane 1, and right is the lower frequency range only for this plane. The sub-block index is 1 to 12. Hence, there are peaks around 1, 4, and 8 times the rotation frequency and a strong peak at 90 time rotation frequency. This is due to the time-stepping input, as there are 90 time steps per revolution. Similar behaviour was found in blocks 2, 3, and 4. (The block planes below planes 1 and 2 were also considered but showed no unsteadiness.)

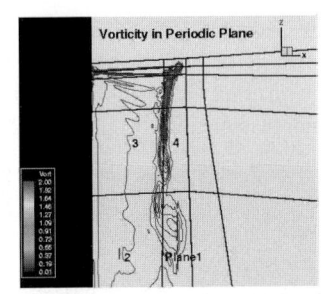

Fig.7: *Block boundaries in region of interest.*

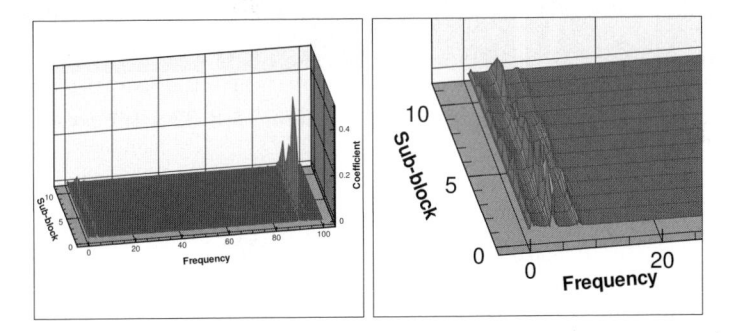

Fig.8: *Fourier transform of total vorticity time-history in sub-blocks in plane 1.*

4.3. Forward Flight Case

The forward flight case corresponds to datapoint Dpt0165, where the M_{Tip} is 0.618, and the advance ratio, μ, set to 0.214. The case also has a shaft inclination of -3.72 degrees, i.e. backward, so has significant BVI effects. This test case in fact has a time-dependent blade pitch, but that is not considered here. The inclusion of blade motion is an important stage of the development, but rigid pitch leads to stronger vortices on the advancing side of the disk, and this is more useful for the wake capturing analysis considered here.

The case was run as an unsteady simulation, using 180 real time-steps per revolution, by spinning the entire rotor mesh at Ω_z with a uniform freestream onflow velocity of

$$\mathbf{q}_f = (\mu M_{Tip} a_\infty \cos\Theta_{inc}, 0, -\mu M_{Tip} a_\infty \sin\Theta_{inc})^T. \qquad (5)$$

where Θ_{inc} is the rotor shaft inclination.

Figure 9 shows the vorticity field on selected cutaway planes in the rotor disc, for the 1, 8 and 32 million grid densities (V_{FF} shows the forward flight direction of the rotor, and the scale is the same in all pictures). This shows both the tip vortex path and the effect of numerical diffusion. The grid dependence is clear, with even the 32 million point simulation exhibiting significant diffusion. This is as expected.

Figure 10 shows total blade load coefficient around the azimuth, for the six grid densities. Hence, if only blade loads are required, the grid density is not as significant as if more detailed quantities are of interest. Figure 11 shows normal force coefficient variation around the azimuth at r/R = 0.5 (left), and r/R = 0.82 (right). This clearly shows the blade-vortex interactions, which are particularly strong on the advancing side. Figure 11 demonstrates how, with less than 16 million points there is little, if any, evidence of blade-vortex interaction being captured. However, it is also clear that the solution has not converged to a grid independent solution, even with 32million points.

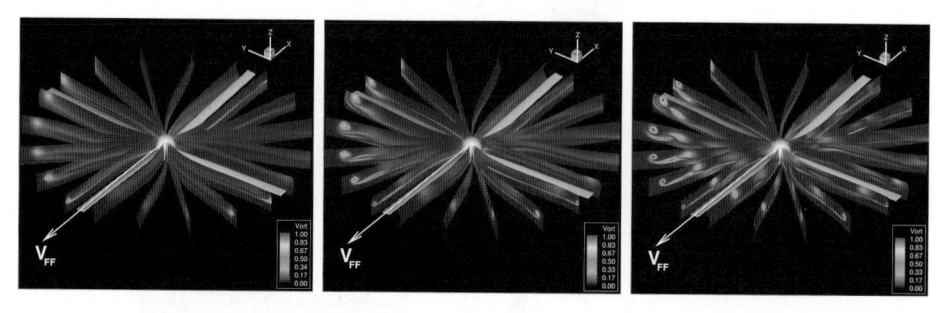

Fig.9: *7A forward flight vorticity shading, $1, 8, 32 \times 10^6$ points.*

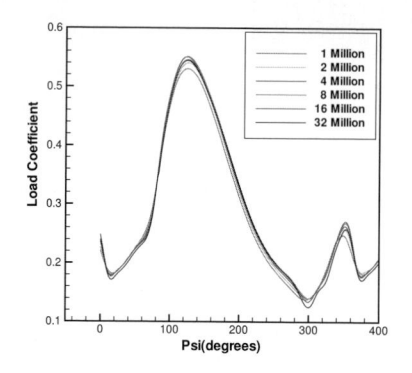

Fig.10: *7A forward flight blade loading.*

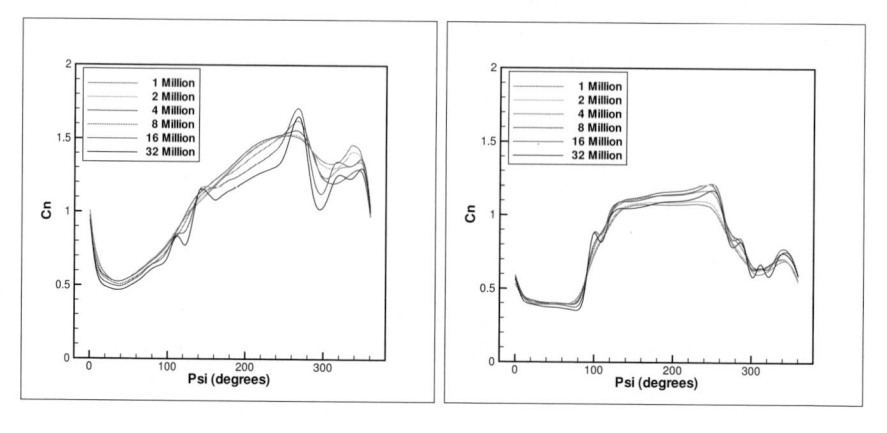

Fig.11: *7A Section Normal Force Variation, $r/R = 0.50$ (left), $r/R = 0.82$ (right).*

5. COMPUTATIONAL PERFORMANCE

Smaller cases were run on the Beowulf cluster in the LACMS (Laboratory for Advanced Computation in the Mathematical Sciences), at the Department of Mathematics at Bristol. This consists of 80 nodes, each comprising two 1GHz P3's with 1GByte RAM, and these cases were run on upto 48 CPU's. The 16 and 32 million point forward flight cases were run on the national HPCx 1600 CPU machine. This consists of 50 nodes, each comprising 32 1.7GHz IBM CPU's with 32GBytes of shared memory.

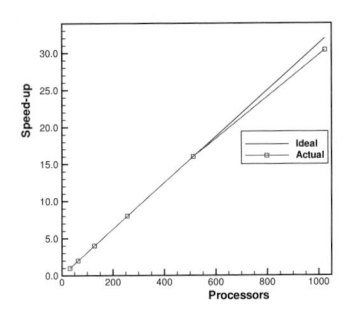

Fig.12: *Parallel performance.*

Figure 12 shows the parallel performance of the code[2]. The case was not run on under 32 CPU's, but a speed-up factor of one was set for this case, and subsequent cases scaled from this. Hence, excellent scaling has been achieved, and the code has been awarded a 'Gold Star' for performance by the National Supercomputing Centre.

6. CONCLUSIONS

Numerical simulation of multi-bladed lifting rotor flows in hover and forward flight has been presented. An implicit, upwind, unsteady, multiblock, multigrid scheme has been developed, and used to compute lifting test cases. The code has been parallelised to allow use of very fine meshes, and excellent parallel performance has been demonstrated.

Hovering simulations performed using 16 million and 32 million points do not converge to a steady state solution, instead exhibiting oscillatory wakes. This agrees with other published simulation results using less diffusive formulations, theoretical stability analysis, and also some experiments where vortex pairing is thought to occur. Unsteady simulation of the same case has also demonstrated unsteady behaviour, and detailed analysis of the time-accurate wake history has shown there appear to be three distinct unsteady modes present, with frequencies of 1, 4, and 8 times the rotational frequency.

For forward flight, it has been demonstrated that the numerical dissipation inherent in the solver limits the accuracy of wake capturing. Even the solutions computed using 32 million points exhibits significant wake diffusion resulting in the vorticity diffusing quickly behind each blade. If only total blade loads are of interest this is not such a huge problem, but if more detailed data is required, for example BVI effects are to be analised, it does not seem sensible to attempt to capture all the physics with a standard CFD code.

---

[2] These timings have been certified by the Terascaling Team at the National Supercomputing Centre

Acknowledgements

Thanks must go to Dr's Ian Stewart and Jason Hogan-O'Neill, and Professor Steve Wiggins, of the University of Bristol for granting access to the LACMS Beowulf cluster. The HPCx computer time was provided through the UK Applied Aerodynamics Consortium under EPSRC grant EP/S91130/01.

REFERENCES

1. Allen, C.B., *"Time-Stepping and Grid Effects on Convergence of Inviscid Rotor Flows"*, AIAA paper 2002-2817, proceedings 20th Applied Aerodynamics Conference, St Louis, June 2002.
2. Allen, C.B., *"Multigrid Multiblock Hovering Rotor Solutions"*, Aeronautical Journal, Vol. 108, No. 1083, pp255-262, 2004.
3. Jameson, A., "Time Dependent Calculations Using Multigrid, with Applications to Unsteady Flows Past Airfoils and Wings", AIAA Paper 91-1596.
4. Van-Leer, B., *"Flux-Vector Splitting for the Euler Equations"*, Lecture Notes in Physics, Vol. 170, 1982, pp. 507-512.
5. Parpia, I. H., *"Van-Leer Flux-Vector Splitting in Moving Coordinates"*, AIAA Journal, Vol. 26, January 1988, pp. 113-115.
6. Anderson, W. K. Thomas, J. L. and Van-Leer, B., *"Comparison of Finite Volume Flux Vector Splittings for the Euler Equations"*, AIAA Journal, Vol. 24, September 1986, pp. 1453-1460.
7. Allen, C.B., *"Multigrid Acceleration of an Upwind Euler Code for Hovering Rotor Flows"*, The Aeronautical Journal, Vol. 105, No. 1051, September 2001, pp517-524.
8. Allen, C.B., *"Multigrid Convergence of Inviscid Fixed- and Rotary-Wing Flows"*, International Journal for Numerical Methods in Fluids, Vol. 39, No. 2, 2002, pp121-140.
9. Allen, C.B., *"An Unsteady Multiblock Multigrid Scheme for Lifting Forward Flight Simulation"*, International Journal for Numerical Methods in Fluids, Vol. 45, No. 7, 2004.
10. Allen, C. B., *"CHIMERA Volume Grid Generation within the EROS Code"*, I. Mech. E. Journal of Aerospace Engineering, Part G, 2000.
11. Gordon, W. J. and Hall, C. A., *"Construction of Curvilinear Coordinate Systems and Applications of Mesh Generation"*, International Journal of Numerical Methods in Engineering, Vol. 7, 1973, pp. 461-477.
12. Eriksson, L. E., *"Generation of Boundary-Conforming Grids Around Wing-Body Configurations Using Transfinite Interpolation"*, AIAA Journal, Vol. 20, No. 10, 1982, pp. 1313-1320.
13. Schultz, K.-J., Splettstoesser, W., Junker, B., Wagner, W., Scheoll, E., Arnauld, G., Mercker, E., Fertis, D., *"A Parametric Wind Tunnel Test on Rotorcraft Aerodynamics and Aeroacoutics (HELISHAPE) - Test Documentation and Representative Results"*, 22nd European Rotorcraft Forum, Brighton, U.K., 1996.
14. Brown, R.E., Line, A.J. and Ahlin, G.A., *"Fuselage and Tail-Rotor Interference Effects on Helicopter Wake Development in Descending Flight"*, Proceedings 60th American Helicopter Society Annual Forum, Baltimore, Maryland, June 2004.
15. Line, A.J. and Brown, R.E., *"Efficient High-Resolution Wake Modelling using the Vorticity Transport Equation"*, Proceedings 60th American Helicopter Society Annual Forum, Baltimore, Maryland, June 2004.
16. Bhagwat, M.J., Leishman, J.G., *"On the Aerodynamic Stability of Helicopter Rotor Wakes"*, Proceedings of the 56th Annual AHS Forum, Virginia Beach, VA, May 2000.

…

A distributed Lagrange multipliers based fictitious domain method for the numerical simulation of particulate flow and its parallel implementation

R. Glowinski[a], J. Hao[a] and T.-W. Pan[a]

[a]Department of Mathematics, University of Houston, Houston, TX 77204, USA

The numerical simulation of particulate flow, such as mixtures of *incompressible viscous fluids* and thousands of *rigid particles*, is computational expensive and parallelism often appears as the only way towards large scale of simulations even we have a fast Navier-Stokes solver. The method we advocate here combines *distributed Lagrange multipliers* based *fictitious domain methods*, which allows the use of *fixed structured finite element grids* on a simple shape auxiliary domain containing the actual one for the fluid flow computations, with *time discretizations* by *operator splitting* to decouple the various computational difficulties associated to the simulation. This method offers an alternative to the *ALE* methods investigated in [1]-[6] and can be easily parallelized via OpenMP due to the use of uniform structured grids and no need to generate mesh at each time step right after finding the new position of the rigid particles. Numerical results of particulate flow obtained on a SUN SMP cluster are presented.

1. A model problem

For simplicity, we consider only the *one-particle* case (see [7] and [8] (Chapter 8) for the *multi-particle case*). Let Ω be a bounded, connected and open region of $\mathbb{R}^d$ ($d = 2$ or 3 in applications); the boundary of Ω is denoted by Γ. We suppose that Ω contains:

   (i) A Newtonian incompressible viscous fluid of density ρ_f and viscosity μ_f ; ρ_f and μ_f are both positive constants.

   (ii) A rigid body B of boundary ∂B, mass M, center of mass G, and inertia $\mathbf{I}$ at the center of mass (see Figure 1, below, for additional details).

The fluid occupies the region $\Omega \setminus \bar{B}$ and we suppose that $distance\,(\partial B(0), \Gamma) > 0$. From now on, $\mathbf{x} = \{x_i\}_{i=1}^d$ will denote the generic point of $\mathbb{R}^d$, $d\mathbf{x} = dx_1 \ldots dx_d$, while $\phi(t)$ will denote the function $\mathbf{x} \to \phi(\mathbf{x}, t)$. Assuming that the only external force is *gravity*, the *fluid flow-rigid body motion* coupling is modeled by

$$\rho_f \left(\frac{\partial \mathbf{u}}{\partial t} + (\mathbf{u} \cdot \boldsymbol{\nabla})\mathbf{u} \right) - \mu_f \Delta \mathbf{u} + \boldsymbol{\nabla} p = \rho_f \mathbf{g} \;\; in \;\; \{(\mathbf{x},t) | \mathbf{x} \in \Omega \setminus \bar{B}(t), t \in (0,T)\}, \quad (1)$$

$$\boldsymbol{\nabla} \cdot \mathbf{u}(t) = 0 \; in \; \Omega \setminus \bar{B}(t), \forall t \in (0,T), \quad (2)$$

$$\mathbf{u}(t) = \mathbf{u}_\Gamma(t) \; on \; \Gamma, \forall t \in (0,T), \;\; with \;\; \int_\Gamma \mathbf{u}_\Gamma(t) \cdot \mathbf{n} \, d\Gamma = 0, \quad (3)$$

$$\mathbf{u}(0) = \mathbf{u}_0 \; in \; \Omega \setminus \bar{B}(0) \;\; with \;\; \boldsymbol{\nabla} \cdot \mathbf{u}_0 = 0, \quad (4)$$

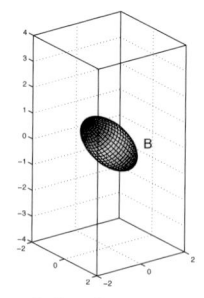

Figure 1: Visualization of the flow region and of the rigid body

and
$$\frac{dG}{dt} = \mathbf{V}, \tag{5}$$
$$\mathbf{M}\frac{d\mathbf{V}}{dt} = M\mathbf{g} + \mathbf{R}_H, \tag{6}$$
$$\frac{d(\mathbf{I}\boldsymbol{\omega})}{dt} = \mathbf{T}_H, \tag{7}$$
$$G(0) = G_0, \ \mathbf{V}(0) = \mathbf{V}_0, \ \boldsymbol{\omega}(0) = \boldsymbol{\omega}_0, \ B(0) = B_0. \tag{8}$$

In relations (1)-(8), $\mathbf{u} = \{u_i\}_{i=1}^d$ is the *fluid (flow) velocity* and p is the *pressure*, $\mathbf{u}_0$ and $\mathbf{u}_\Gamma$ are given functions, $\mathbf{V}$ is the velocity of the center of mass of body B, while $\boldsymbol{\omega}$ is the angular velocity; $\mathbf{R}_H$ and $\mathbf{T}_H$ denote, respectively, the *resultant* and the *torque* of the *hydrodynamical forces*, namely the forces that the fluid exerts on B; we have, actually,

$$\mathbf{R}_H = \int_{\partial B} \boldsymbol{\sigma}\mathbf{n}\,d\gamma \ \text{ and } \ \mathbf{T}_H = \int_{\partial B} \overrightarrow{G\mathbf{x}} \times \boldsymbol{\sigma}\mathbf{n}\,d\gamma. \tag{9}$$

In (9) the *stress-tensor* $\boldsymbol{\sigma}$ is defined by $\boldsymbol{\sigma} = 2\mu_f D(\mathbf{u}) - p\mathbf{I}_d$, with $D(\mathbf{v}) = \frac{1}{2}(\boldsymbol{\nabla}\mathbf{v} + (\boldsymbol{\nabla}\mathbf{v})^t)$, while $\mathbf{n}$ is a unit normal vector at ∂B and $\mathbf{I}_d$ is the *identity tensor*.

Concerning the compatibility conditions on ∂B we have: (i) the forces exerted by the fluid on the solid body *cancel* those exerted by the solid body on the fluid, and we shall assume that: (ii) on ∂B the *no-slip boundary condition* holds, namely

$$\mathbf{u}(\mathbf{x}, t) = \mathbf{V}(t) + \boldsymbol{\omega}(t) \times \overrightarrow{G(t)\mathbf{x}}, \ \forall \mathbf{x} \in \partial B(t). \tag{10}$$

Remark 1. *System (1)-(4) (resp., (5)-(8)) is of the incompressible Navier-Stokes (resp., Euler-Newton) type. Also, the above model can be generalized to multiple-particles situations and/or non-Newtonian incompressible viscous fluids.*

The (local in time) *existence of weak solutions* for above problem (1)–(10) has been proved in [9], assuming that, at $t = 0$, the particles do not touch each other and do not touch Γ (see also [10], [11]). Concerning the numerical solution of (1)-(4) and (5)-(8) completed by the above interface conditions we can divide them, roughly, in two classes, namely: (i) The *Arbitrary Lagrange-Euler* (**ALE**) methods; these methods, which rely on *moving meshes*, are discussed in, e.g., [1]-[6]. (ii) The non-boundary fitted *fictitious domain* methods; these methods rely on fixed meshes and are discussed in, e.g., [7] and [8] (Chapter 8) (see also the references therein). These last methods seem to enjoy a growing popularity, justifying thus the (brief) discussion hereafter.

Remark 2. *Even if theory suggests that collisions may never take place in finite time (if we assume that the flow is still modeled by the Navier-Stokes equations as long as the particles do not touch each others, or the boundary), near-collisions take place, and, after discretization, "real" collisions may occur. These phenomena can be avoided by introducing well-chosen short range repulsion potentials reminiscent of those encountered in Molecular Dynamics (see [7] and [8] (Chapter 8) for details).*

2. A fictitious domain formulation

Considering the fluid-rigid body mixture as a unique medium we are going to derive a fictitious domain based variational formulation. The principle of this derivation is pretty simple; it relies on the following steps (see, e.g., [7] and [8] for details):

a. Start from the following *global weak* formulation (of the *virtual power* type):

$$\begin{cases} \rho_f \int_{\Omega \setminus \bar{B}(t)} \left[\frac{\partial \mathbf{u}}{\partial t} + (\mathbf{u} \cdot \nabla)\mathbf{u} \right] \cdot \mathbf{v}\, d\mathbf{x} + 2\mu_f \int_{\Omega \setminus \bar{B}(t)} \mathbf{D}(\mathbf{u}) : \mathbf{D}(\mathbf{v})\, d\mathbf{x} \\ - \int_{\Omega \setminus \bar{B}(t)} p \nabla \cdot \mathbf{v}\, d\mathbf{x} + \mathbf{M} \frac{d\mathbf{V}}{dt} \cdot \mathbf{Y} + \frac{d(\mathbf{I}\boldsymbol{\omega})}{dt} \cdot \boldsymbol{\theta} = \rho_f \int_{\Omega \setminus \bar{B}(t)} \mathbf{g} \cdot \mathbf{v}\, d\mathbf{x} + \mathbf{Mg} \cdot \mathbf{Y}, \\ \forall \{\mathbf{v}, \mathbf{Y}, \boldsymbol{\theta}\} \in (H^1(\Omega \setminus \bar{B}(t)))^d \times \mathbb{R}^d \times \Theta \ \text{and verifying}\ \mathbf{v} = 0\ \text{on}\ \Gamma, \\ \mathbf{v}(\mathbf{x}) = \mathbf{Y} + \boldsymbol{\theta} \times \overrightarrow{G(t)\mathbf{x}},\ \forall \mathbf{x} \in \partial B(t),\ t \in (0, T), \\ (\text{with}\ \Theta = \{(0, 0, \theta)\ |\ \theta \in \mathbb{R}\}\ \text{if}\ d = 2,\ \Theta = \mathbb{R}^3\ \text{if}\ d = 3), \end{cases} \quad (11)$$

$$\int_{\Omega \setminus \bar{B}(t)} q \nabla \cdot \mathbf{u}(t)\, d\mathbf{x} = 0,\ \forall q \in L^2(\Omega \setminus \bar{B}(t)),\ t \in (0, T), \quad (12)$$

$$\mathbf{u}(t) = \mathbf{u}_\Gamma(t)\ \text{on}\ \Gamma, t \in (0, T), \quad (13)$$

$$\mathbf{u}(\mathbf{x}, t) = \mathbf{V}(t) + \boldsymbol{\omega}(t) \times \overrightarrow{G(t)\mathbf{x}},\ \forall \mathbf{x} \in \partial B(t),\ t \in (0, T), \quad (14)$$

$$\frac{dG}{dt} = \mathbf{V}, \quad (15)$$

$$\mathbf{u}(\mathbf{x}, 0) = \mathbf{u}_0(\mathbf{x}), \forall \mathbf{x} \in \Omega \setminus \bar{B}(0), \quad (16)$$

$$G(0) = G_0,\ \mathbf{V}(0) = \mathbf{V}_0,\ \boldsymbol{\omega}(0) = \boldsymbol{\omega}_0,\ B(0) = B_0. \quad (17)$$

b. Fill B with the surrounding fluid.

c. Impose a rigid body motion to the fluid inside B.

d. Modify the global weak formulation (11)-(17) accordingly, taking advantage of the fact that if $\mathbf{v}$ is a rigid body motion velocity field, then $\nabla \cdot \mathbf{v} = 0$ and $\mathbf{D}(\mathbf{v}) = \mathbf{0}$.

e. Use a *Lagrange multiplier* defined over B to force the rigid body motion inside B.

Assuming that B is made of a homogeneous material of density ρ_s, the above "program" leads to:

$$\begin{cases} \rho_f \int_\Omega \left[\frac{\partial \mathbf{u}}{\partial t} + (\mathbf{u} \cdot \nabla)\mathbf{u} \right] \cdot \mathbf{v}\, d\mathbf{x} + 2\mu_f \int_\Omega \mathbf{D}(\mathbf{u}) : \mathbf{D}(\mathbf{v})\, d\mathbf{x} - \int_\Omega p \nabla \cdot \mathbf{v}\, d\mathbf{x} \\ +(1 - \rho_f/\rho_s) \left[\mathbf{M} \frac{d\mathbf{V}}{dt} \cdot \mathbf{Y} + \frac{d(\mathbf{I}\boldsymbol{\omega})}{dt} \cdot \boldsymbol{\theta} \right] + <\boldsymbol{\lambda}, \mathbf{v} - \mathbf{Y} - \boldsymbol{\theta} \times \overrightarrow{G(t)\mathbf{x}} >_{B(t)} \\ = \rho_f \int_\Omega \mathbf{g} \cdot \mathbf{v}\, d\mathbf{x} + (1 - \rho_f/\rho_s)\mathbf{M}\mathbf{g} \cdot \mathbf{Y},\ \forall \{\mathbf{v}, \mathbf{Y}, \boldsymbol{\theta}\} \in (H^1(\Omega))^d \times \mathbb{R}^d \times \Theta, \\ t \in (0, T),\ with\ \Theta = \mathbb{R}^3\ if\ d = 3,\ \Theta = \{(0, 0, \theta) \mid \theta \in \mathbb{R}\}\ if\ d = 2, \end{cases} \quad (18)$$

$$\int_\Omega q \nabla \cdot \mathbf{u}(t)\, d\mathbf{x} = 0, \forall q \in L^2(\Omega),\ t \in (0, T), \quad (19)$$

$$\mathbf{u}(t) = \mathbf{u}_\Gamma(t)\ on\ \Gamma,\ t \in (0, T), \quad (20)$$

$$<\boldsymbol{\mu}, \mathbf{u}(\mathbf{x}, t) - \mathbf{V}(t) - \boldsymbol{\omega}(t) \times \overrightarrow{G(t)\mathbf{x}} >_{B(t)} = 0, \quad (21)$$
$$\forall \boldsymbol{\mu} \in \Lambda(t)\ (= (H^1(B(t)))^d),\ t \in (0, T),$$

$$\frac{dG}{dt} = \mathbf{V}, \quad (22)$$

$$\mathbf{u}(\mathbf{x}, 0) = \mathbf{u}_0(\mathbf{x}), \forall \mathbf{x} \in \Omega \setminus \bar{B}_0, \quad (23)$$

$$G(0) = G_0,\ \mathbf{V}(0) = \mathbf{V}_0,\ \boldsymbol{\omega}(0) = \boldsymbol{\omega}_0,\ B(0) = B_0, \quad (24)$$

$$\mathbf{u}(\mathbf{x}, 0) = \begin{cases} \mathbf{u}_0(\mathbf{x}), \forall \mathbf{x} \in \Omega \setminus \bar{B}_0, \\ \mathbf{V}_0 + \boldsymbol{\omega}_0 \times \overrightarrow{G_0 \mathbf{x}}, \forall \mathbf{x} \in \bar{B}_0. \end{cases} \quad (25)$$

From a theoretical point of view, a natural choice for $<\cdot, \cdot>_{B(t)}$ is provided by, e.g.,

$$<\boldsymbol{\mu}, \mathbf{v}>_{B(t)} = \int_{B(t)} [\boldsymbol{\mu} \cdot \mathbf{v} + l^2 \mathbf{D}(\boldsymbol{\mu}) : \mathbf{D}(\mathbf{v})]\, d\mathbf{x}; \quad (26)$$

in (26), l is a characteristic length, the diameter of B, for example. From a practical point of view, when it come to space discretization, a simple and efficient strategy is the following one (cf. [7] and [8] (Chapter 8)): "approximate" $\Lambda(t)$ by

$$\Lambda_h(t) = \{\boldsymbol{\mu} \mid \boldsymbol{\mu} = \sum_{j=1}^N \boldsymbol{\mu}_j \delta(\mathbf{x} - \mathbf{x}_j),\ with\ \boldsymbol{\mu}_j \in \mathbb{R}^d,\ \forall j = 1, .., N\}, \quad (27)$$

and the above pairing by

$$<\boldsymbol{\mu}, \mathbf{v}>_{(B(t),h)} = \sum_{j=1}^N \boldsymbol{\mu}_j \cdot \mathbf{v}(\mathbf{x}_j). \quad (28)$$

In (27), (28), $\mathbf{x} \to \delta(\mathbf{x} - \mathbf{x}_j)$ is the Dirac measure at $\mathbf{x}_j$, and the set $\{\mathbf{x}_j\}_{j=1}^N$ is the union of two subsets, namely: (i) The set of the points of the velocity grid contained in $B(t)$ and whose distance at $\partial B(t)$ is $\geq ch$, h being a space discretization step and c a constant ≈ 1. (ii) A set of control points located on $\partial B(t)$ and forming a mesh whose step size is of the order of h. It is clear that, using the approach above, one forces the rigid body motion inside the particle by *collocation*.

3. Solving problem (18)-(25) by operator-splitting

Let us first consider the following autonomous initial value problem:

$$\frac{d\phi}{dt} + A(\phi) = 0 \text{ on } (0,T), \quad \phi(0) = \phi_0 \tag{29}$$

with $0 < T \leq +\infty$. Operator A maps the vector space V into itself and we suppose that $\phi_0 \in V$. We suppose also that A has a *non-trivial decomposition* such as $A = \sum_{j=1}^{J} A_j$ with $J \geq 2$ (by *non-trivial* we mean that the operators A_j are individually simpler than A). It has been known for a long time that many schemes have been designed to take advantage of the decomposition of A when solving (29), one of them will be briefly discussed in the following, namely the *Lie's scheme*.

Let $\tau(> 0)$ be a *time-discretization step* (we suppose τ uniform, for simplicity); we denote $n\tau$ by t^n. With ϕ^n denoting an approximation of $\phi(t^n)$, the *Lie's scheme* reads as follows (for its derivation see, e.g., [8] (Chapter 6)):

$$\phi^0 = \phi_0; \tag{30}$$

then, for $n \geq 0$, assuming that ϕ^n is known, compute ϕ^{n+1} via

$$\begin{cases} \dfrac{d\phi}{dt} + A_j(\phi) = 0 \text{ on } (t^n, t^{n+1}), \\ \phi(t^n) = \phi^{n+(j-1)/J}; \phi^{n+j/J} = \phi(t^{n+1}), \end{cases} \tag{31}$$

for $j = 1, \ldots, J$.

If (29) is taking place in a finite dimensional space and if the operators A_j are smooth enough, then $\|\phi(t^n) - \phi^n\| = O(\tau)$, function ϕ being the solution of (29).

Remark 3. *The above scheme applies also for multivalued operators (such as the subgradient of proper l.s.c. convex functionals) but in such a case first order accuracy is not guaranteed anymore.*

Remark 4. *The above scheme is easy to generalize to non-autonomous problems by observing that*

$$\frac{d\phi}{dt} + A(\phi, t) = 0, \quad \phi(0) = \phi_0, \quad \Leftrightarrow \quad \begin{cases} \dfrac{d\phi}{dt} + A(\phi, \theta) = 0, \quad \phi(0) = \phi_0, \\ \dfrac{d\theta}{dt} - 1 = 0, \quad \theta(0) = 0. \end{cases}$$

For the problem (18)-(25), if we do not consider *collisions*, after (formal) elimination of p and $\boldsymbol{\lambda}$, it is reduced to a *dynamical system* of the following form

$$\frac{d\mathbf{X}}{dt} + \sum_{j=1}^{J} A_j(\mathbf{X}, t) = \mathbf{0} \text{ on } (0,T), \quad X(0) = X_0, \tag{32}$$

where $\mathbf{X} = \{\mathbf{u}, \mathbf{V}, \boldsymbol{\omega}, G\}$ (or $\{\mathbf{u}, \mathbf{V}, \mathbf{I}\boldsymbol{\omega}, G\}$). A typical situation will be the one where, with $J = 4$, operator A_1 will be associated to incompressibility, A_2 to advection, A_3 to diffusion, A_4 to fictitious domain and body motion; other decompositions are possible as shown in, e.g., [7] and [8] (Chapter 8). The Lie's scheme applies "beautifully" to the solution of the formulation (32) of problem (18)-(25). The resulting method is quite modular implying that different space and time approximations can be used to treat the various steps; the only constraint is that two successive steps have to communicate (by projection in general).

4. Numerical Implementation and its Parallelization

The methods described (quite briefly) in the above paragraphs have been validated by numerous experiments (see, e.g., [7], [8] (Chapters 8 and 9), and [12]). All the flow computations have been done using the *Bercovier-Pironneau finite element approximation*; namely (see [8] (Chapters 5, 8 and 9) for details) we used a globally continuous piecewise affine approximation of the velocity (resp., the pressure) associated to a uniform structured triangulation (in 2-D) or tetrahedral partition (in 3-D) $\mathcal{T}_h$ (resp., $\mathcal{T}_{2h}$) of Ω, h being a space discretization step. The pressure mesh is thus *twice coarser* than the velocity one. All our calculations have been done using *uniform partitions* $\mathcal{T}_h$ and $\mathcal{T}_{2h}$. The saddle point problem associated with the velocity field and pressure has been solved by an Uzawa/preconditioned conjugated gradient algorithm as in [13]. At the preconditioned step, the discrete equivalent of $-\Delta$ for the homogeneous Neumann boundary condition has been solved by a red-black SOR iterative method parallelized via OpenMP.

The advection problem for the velocity field is solved by a wave-like equation method as in [14] and [15], which is an explicit time stepping method with its own local time steps (so the stability condition can be easily satisfied) and can be parallelized easily. The diffusion problem, which is a classical discrete elliptic problem, is solved by the parallel Jacobi iterative method.

Finally, to solve the problem of the rigid-body motion projection via the fictitious domain technique, which is a classical *saddle-point* problem, we have applied an *Uzawa/conjugate gradient* algorithm (in which there is no need to solve any elliptic problems); such an algorithm is described in [16].

Due to the fact that distributed Lagrange multiplier method uses *uniform meshes* on a rectangular domain, we have been able to use OpenMP to implement the parallelization within the Fortran codes. OpenMP is a set of standards and interfaces for parallelizing programs in a shared memory environment. The parallel jobs were run on a Sun SMP cluster (104 processors, 750MHz).

5. Numerical results

The parallelized codes have been used to simulate the motions of thousands of particels in the following two test problems. The first test problem that we considered concerns the simulation of the motion of 6400 sedimenting circular disks in the closed cavity $\Omega = (0,8) \times (0,12)$. The diameter d of the disks is $1/12$ and the position of the disks at time $t = 0$ is shown in Figure 2. The solid fraction in this test case is 34.9%. The disks and the fluid are at rest at $t = 0$. The density of the fluid is $\rho_f = 1$ and the density of the disks is $\rho_s = 1.1$. The viscosity of the fluid is $\mu_f = 0.01$. The time step is $\Delta t = 0.001$. The mesh size for the velocity field is $h_\Omega = 1/160$ (resp., $1/192$) and the velocity triangulation thus has about 2.46×10^6 (resp., 3.54×10^6) vertices while the pressure mesh size is $h_p = 2h_\Omega$, implying approximately 6×10^5 (resp., 8.87×10^5) vertices for the pressure triangulation. The optimal value of ω for the SOR method is 1.99. In Figure 2 the snapshots of the disk position and velocity field at time $t = 0.6$, 1 and 3.2 are shown. In Table 1, we have provided the averaged elapsed time per time step and speed-up for 6400 sedimenting disks on various numbers of processors. The overall speed-up on 16 processors compared with one processor is 10.47 (resp., 12.37) when the mesh size is $h_\Omega = 1/160$ (resp., $1/192$).

Table 1:
The elapsed time (min.) per time step and speed-up for 6400 sedimenting disks.

processors	$h_\Omega = 1/160$	speed-up	$h_\Omega = 1/192$	speed-up
1	11.52	—	16.20	—
2	6.12	1.88	8.44	1.92
4	3.09	3.73	4.37	3.71
8	1.68	6.86	2.29	7.07
16	1.10	10.47	1.31	12.37

The second test problem we considered is the simulation of the motion of 1200 neutrally buoyant particles in a pressure-driven Poiseuille flow. The computational domain is $\Omega = (0, 42) \times (0, 12)$. The flow velocity is periodic in the horizontal direction and zero at the top and bottom of the domain. A fixed horizontal force is given so that the flow moves from the left to the right (and the maximum horizontal speed is 25 when there is no particle). The initial flow velocity and particle velocities are at rest. The densities of the fluid and the particles are 1 and the viscosity of the fluid is $\mu_f = 1$. The diameter of the particles is 0.45. The initial position of the particles is arranged like a rectangular lattice shown in Figure 3. The time step is $\Delta t = 0.001$. The mesh size for the velocity field is $h_v = 1/24$ (resp., $1/48$) and the velocity triangulation thus has about 2.9×10^5 (resp., 1.16×10^6) vertices while the pressure mesh size is $h_p = 2h_v$, implying approximately 7.3×10^4 (resp., 2.9×10^5) vertices for the pressure triangulation. The optimal value of ω for the SOR method is 1.40. In Figure 3 the snapshots of the disk position and velocity field at time $t = 30$, 100 and 176 are shown. The averaged elapsed time per time step and speed-up for 1200 neutrally buoyant particles on various numbers of processors is listed in Table 2. The overall speed-up on 8 processors compared with one processor is 5.34 (resp., 5.97) when the mesh size is $h_v = 1/24$ (resp., $1/48$).

Table 2:
The elapsed time (sec.) per time step and speed-up for 1200 neutrally buoyant particles.

processors	$h_v = 1/24$	speed-up	$h_v = 1/48$	speed-up
1	21.57	—	83.26	—
2	11.25	1.92	41.82	1.99
4	6.14	3.51	22.49	3.70
8	4.04	5.34	13.95	5.97

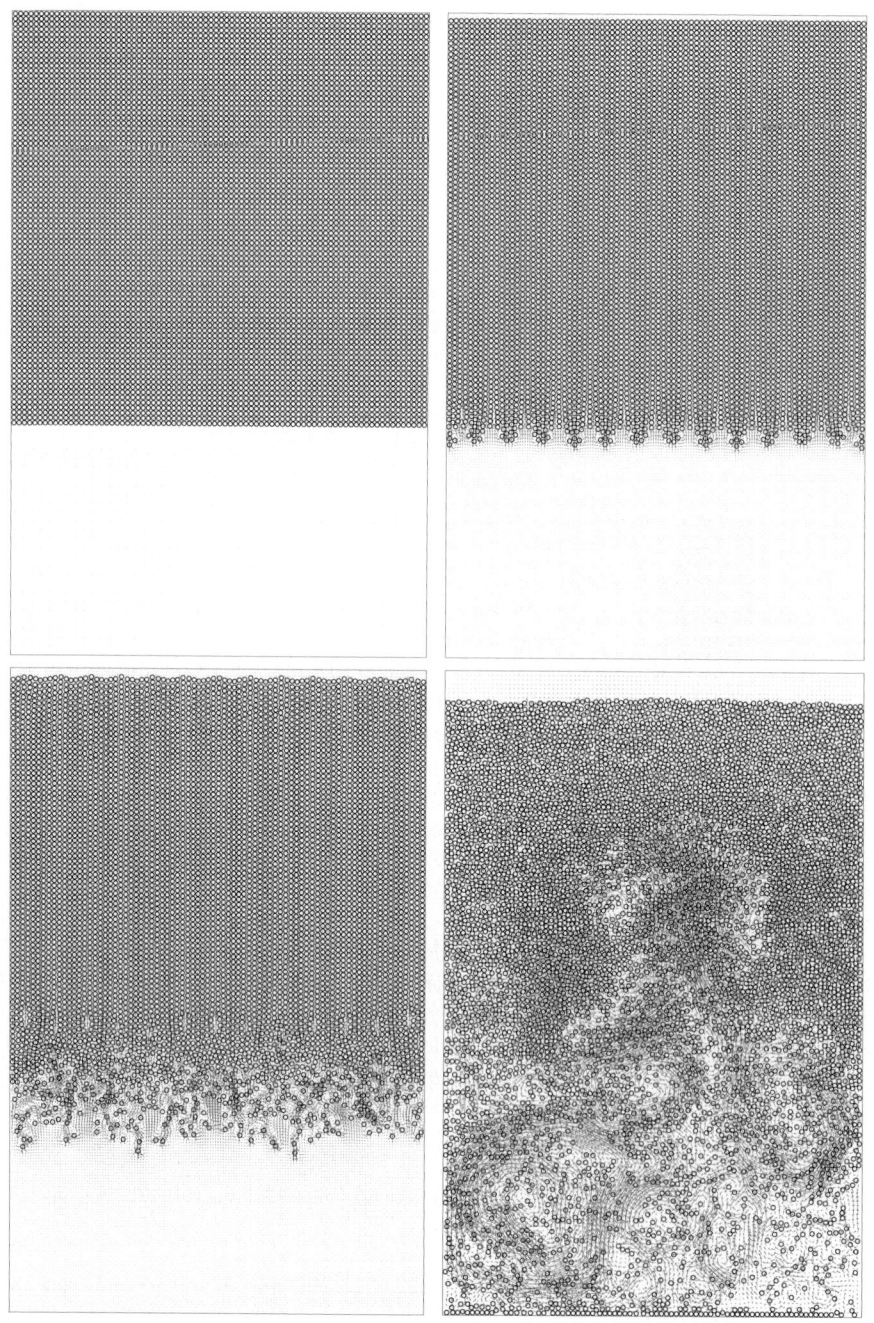

Figure 2: Sedimentation of 6400 particles: positions at $t =$0, 0.6, 1.0 and 3.2 (from left to right and from top to bottom) for mesh size $h_\Omega = 1/192$ (computed on 16 processors).

6. Conclusion

We have presented in this article a parallel implementation of a distributed Lagrange multiplier based fictitious domain method for the direct simulation of particulate flow with thousands of particles. We also compared it with a variant using fast solvers, such as FISHPAK. For both test cases, our codes run in sequential mode are slightly slower (about 10%) than the ones with Fishpak. Considering the amount of speed-up after parallelization, this difference in computing time is insignificant.

Acknowledgments: This parallel work was performed using the computational resources of the Sun Microsystems Center of Excellence in the Geosciences at the University of Houston. We acknowledge the support of the DOE/LASCI (grant R71700K-292-000-99) and the NSF (grants ECS-9527123, CTS-9873236, DMS-9973318, CCR-9902035, DMS-0209066, DMS-0443826).

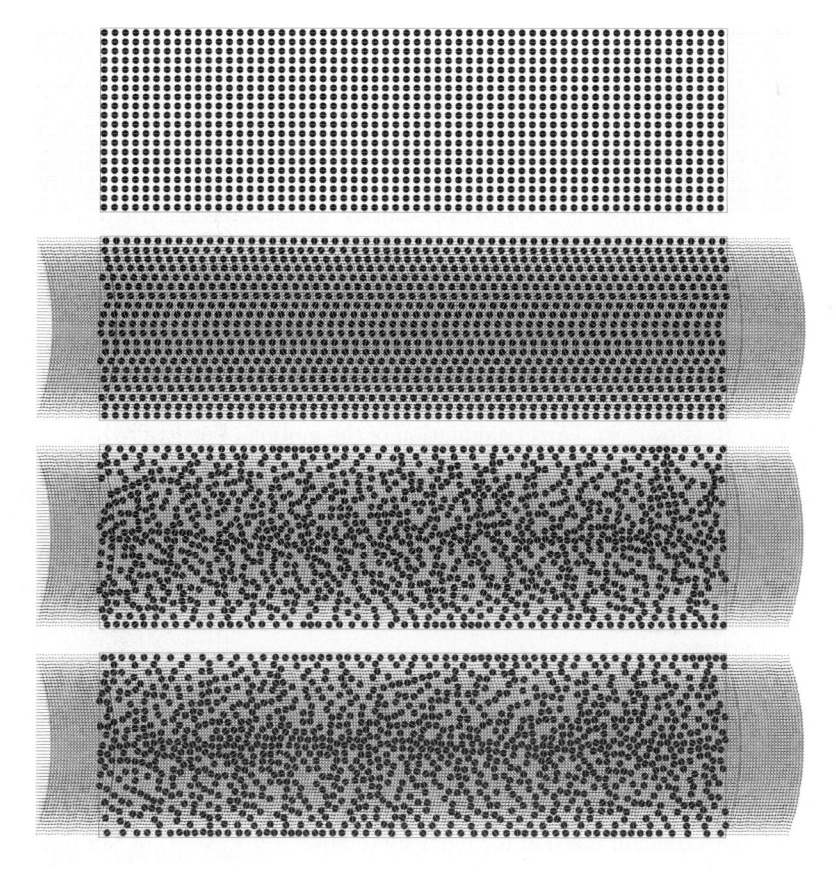

Figure 3: The 1200 neutrally buoyant particles at t=0, 30, 100 and 176 (from top to bottom) for mesh size $h_v = 1/24$ (computed on 8 processors).

REFERENCES

1. O. Pironneau, J. Liou and T. Tezduyar, Characteristic-Galerkin and Galerkin least squares space-time formulations for advection-diffusion equations with time-dependent domains, Comp. Meth. Appl. Mech. Eng., 16 (1992) 117.
2. H.H. Hu, Direct simulation of flows of solid-liquid mixtures, Internat. J. Multiphase Flow, 22 (1996) 335.
3. B. Maury and R. Glowinski, Fluid particle flow: a symmetric formulation, C.R. Acad. Sci., Paris, t. 324, Série I (1997) 1079.
4. A.A. Johnson, T. Tezduyar, 3-D simulations of fluid-particle interactions with the number of particles reaching 100, Comp. Methods Appl. Mech. Engrg., 145 (1997) 301.
5. B. Maury, Direct simulation of 2-D fluid-particle flows in bi-periodic domains, J. Comp. Phys., 156 (1999) 325.
6. H.H. Hu., N.A. Patankar, M.Y. Zhu, Direct numerical simulation of fluid-solid systems using arbitrary Lagrangian-Eulerian techniques, J. Comp. Phys., 169 (2001) 427.
7. R. Glowinski, T.W. Pan, T.I. Hesla, D.D. Joseph, J. Périaux, A fictitious domain approach to the direct numerical simulation of incompressible viscous fluid flow past moving rigid bodies: Application to particulate flow, J. Comp. Phys., 169 (2001) 363.
8. R. Glowinski, Finite Element Methods for Incompressible Viscous Flow, in Handbook of Numerical Analysis, Vol. IX, P.G. Ciarlet and J.L. Lions eds., North-Holland, Amsterdam, 2003, 3-1176.
9. B. Desjardin, M.J. Esteban, On weak solution for fluid-rigid structure interaction: compressible and incompressible models, Arch. Ration. Mech. Anal., 146 (1999) 59.
10. C. Grandmont, Y. Maday, Existence for an unsteady fluid-structure interaction problem, Math. Model. Num. Anal., 34 (2000) 609.
11. J.A. San Martin, V. Starovoitov, M. Tucsnak, Global weak solutions for the two-dimensional motion of several rigid bodies in an incompressible viscous fluid, Arch. Ration. Mech. Anal., 161 (2002) 113.
12. T.W. Pan, R. Glowinski, Direct simulation of the motion of neutrally buoyant circular cylinders in plane Poiseuille flow, J. Comp. Phys. , 181 (2002) 260.
13. R. Glowinski, T.W. Pan, J. Périaux, Distributed Lagrange multiplier methods for incompressible flow around moving rigid bodies, Comput. Methods Appl. Mech. Engrg., 151 (1998) 181.
14. E.J. Dean, R. Glowinski, A wave equation approach to the numerical solution of the Navier-Stokes equations for incompressible viscous flow, C.R. Acad. Sc. Paris, Série 1, 325 (1997) 783.
15. E.J. Dean, R. Glowinski, T.W. Pan, A wave equation approach to the numerical simulation of incompressible viscous fluid flow modeled by the Navier-Stokes equations, in Mathematical and Numerical Aspects of Wave Propagation, J.A. De Santo ed., SIAM, Philadelphia, 1998, 65.
16. R. Glowinski, T.W. Pan, T.I. Hesla, D.D. Joseph, A distributed Lagrange multiplier/fictitious domain method for particulate flows, Internat. J. of Multiphase Flow, 25 (1999) 755.

Solving Fluid Flow Problems on Moving and Adaptive Overlapping Grids

William D. Henshaw[a]*

[a]Centre for Applied Scientific Computing,
Lawrence Livermore National Laboratory,
Livermore, CA, USA 94551.

The solution of fluid dynamics problems on overlapping grids will be discussed. An overlapping grid consists of a set of structured component grids that cover a domain and overlap where they meet. Overlapping grids provide an effective approach for developing efficient and accurate approximations for complex, possibly moving geometry. Topics to be addressed include the reactive Euler equations, the incompressible Navier-Stokes equations and elliptic equations solved with a multigrid algorithm. Recent developments coupling moving grids and adaptive mesh refinement and preliminary parallel results will also be presented.

1. Introduction

The are many interesting problems that involve the solution of fluid dynamics problems on domains that evolve in time. Examples include the motion of valves in a car engine and the movement of embedded particles in a flow. The numerical solution of these problems is difficult since the discrete equations being solved change as the domain evolves. The problems can be especially hard when there are fine scale features in the flow such as shocks and detonations.

In this paper an approach will be described that uses composite overlapping grids to resolve complex geometry, moving component grids to track dynamically evolving surfaces and block structured adaptive mesh refinement (AMR) to efficiently resolve fine scale features. The numerical method uses composite overlapping grids to represent the problem domain as a collection of structured curvilinear grids. This method, as discussed in Chesshire and Henshaw [1], allows complex domains to be represented with smooth grids that can be aligned with the boundaries. The use of smooth grids is particularly attractive for problems where the solution is sensitive to any grid induced numerical artifacts. This approach can also take advantage of the large regions typically covered by Cartesian grids. These Cartesian grids can be treated with efficient approximations leading to fast methods with low memory usage. Overlapping grids have been used successfully for the numerical solution of a variety of problems involving inviscid and viscous flows, see the references in [2,3] for example. The use of adaptive mesh refinement in combination with

---

*This research was supported under the auspices of the U.S. Department of Energy by the University of California, Lawrence Livermore National Laboratory under contract No. W-7405-Eng-48.

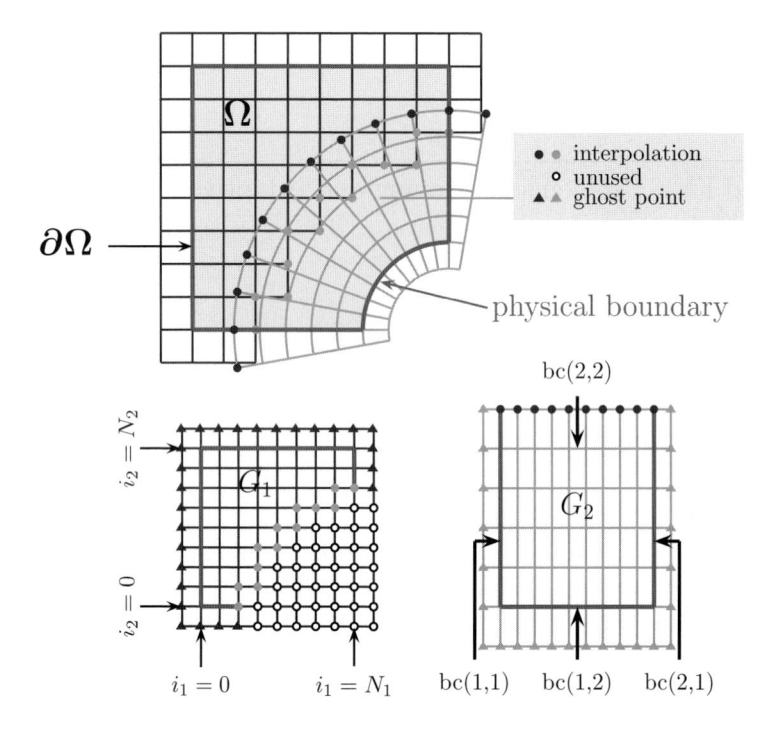

Figure 1. The top view shows an overlapping grid consisting of two structured curvilinear component grids. The bottom views show the component grids in the unit square parameter space. Grid points are classified as discretization points, interpolation points or unused points. Ghost points are used to apply boundary conditions.

overlapping grids has been considered by Brislawn, Brown, Chesshire and Saltzman[4], Boden and Toro[5], and Meakin[6].

Figure 1 shows a simple overlapping grid consisting of two component grids, an annular grid and a background Cartesian grid. The top view shows the overlapping grid while the bottom view shows each grid in parameter space. In this example the annular grid cuts a hole in the Cartesian grid so that the latter grid has a number of unused points that are marked as open circles. The other points on the component grid are marked as discretization points (where the PDE or boundary conditions are discretized) and interpolation points. Solution values at interpolation points are generally determined by a tensor-product Lagrange interpolant in the parameter space of the donor grid. Ghost points are used to simplify the discretization of boundary conditions. In a moving grid computation one or more of the component grids will move, following the boundaries as they evolve. As the grids move the overlapping connectivity information, such as the location of interpolation points, will be recomputed. In our work the grid generation is performed by the Ogen grid generator which has a specialized algorithm to treat moving grid problems efficiently. When adaptive mesh refinement is used on an overlapping grid, a hierarchy of refinement grids is added to the parameter space of each component grid. The

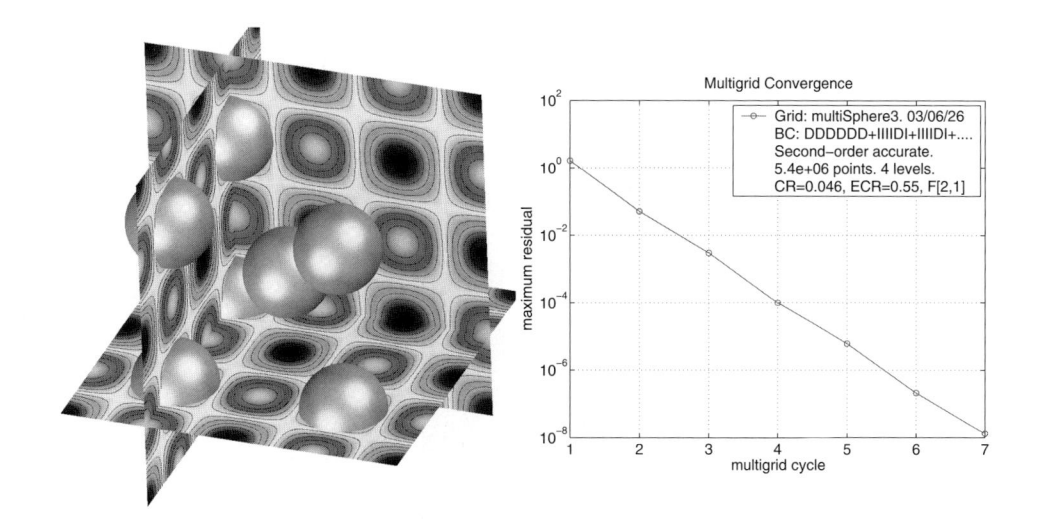

Figure 2. Solution of Poisson's equation by the multigrid algorithm for a domain containing some spheres. The average convergence rate per F-cycle with 2 pre-smooths and 1-post smooth was about .046 .

locations of these refinement patches are determined by an appropriate error estimate. The software that we develop, collectively known as the Overture framework, is freely available in source form [7].

2. Multigrid

A fast multigrid algorithm has been devised for solving elliptic boundary value problems on overlapping grids [3]. This method can be used to solve the implicit time-stepping equations and pressure equation in an incompressible Navier-Stokes solver, for example. In moving grid applications it is particularly important that the elliptic equation solver have a low startup cost since the equations will be changing at each time step. The Ogmg multigrid solver was developed to solve elliptic boundary value problems, in two and three space dimensions, of the form

$$Lu = f \quad \mathbf{x} \in \Omega ,$$
$$Bu = g \quad \mathbf{x} \in \partial\Omega ,$$

where L is chosen to be a second-order, linear, variable-coefficient operator and B is chosen to define a Dirichlet, Neumann or mixed boundary condition. The key aspects of the multigrid scheme for overlapping grids are an automatic coarse grid generation algorithm, an adaptive smoothing technique for adjusting residuals on different component grids, and the use of local smoothing near interpolation boundaries. Other important features include optimizations for Cartesian component grids, the use of over-relaxed Red-Black smoothers and the generation of coarse grid operators through Galerkin averaging.

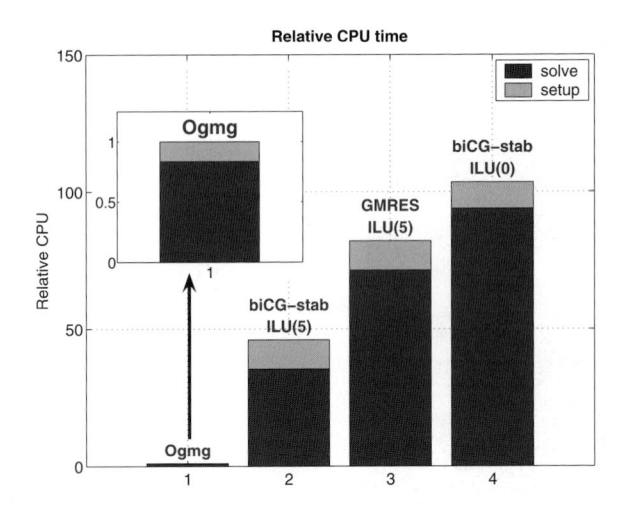

Figure 3. In comparison to Krylov solvers the Ogmg multigrid solver is an order of magnitude faster and uses an order of magnitude less storage. These results are for the solution of a two-dimensional Poisson problem, for a cylinder in a square domain, using an overlapping grid with about 1 million grid points.

Numerical results in two and three dimensions show that very good multigrid convergence rates can be obtained for both Dirichlet and Neumann/mixed boundary conditions.

Figure 2 shows the solution and convergence rates when solving Poisson's equation on a region containing some spheres. The convergence rates are similar to the *text-book* convergence rates that one can obtain on single Cartesian grids. Figure 3 presents a comparison of the multigrid solver to some Krylov based solvers for a two-dimensional problem with about 1 million grid points. The results show that that the multigrid solver can be much faster (over 45 times faster in this case) and also that the multigrid scheme has a low startup cost. Moreover, the multigrid solver uses about 10 times less memory in this case.

3. Solution of the reactive Euler Equations

The reactive Euler equations are solved on a domain $\Omega(t)$ whose boundaries, $\partial\Omega(t)$ may evolve in time. In two space dimensions the initial boundary value problem for the solution $\mathbf{u} = \mathbf{u}(\mathbf{x}, t)$ is

$$\mathbf{u}_t + \mathbf{F}(\mathbf{u})_x + \mathbf{G}(\mathbf{u})_y = \mathbf{H}(\mathbf{u}), \quad \mathbf{x} \in \Omega(t) ,$$
$$B(\mathbf{u}) = 0, \quad \mathbf{x} \in \partial\Omega(t) ,$$
$$\mathbf{u}(\mathbf{x}, 0) = \mathbf{u}_0(\mathbf{x}),$$

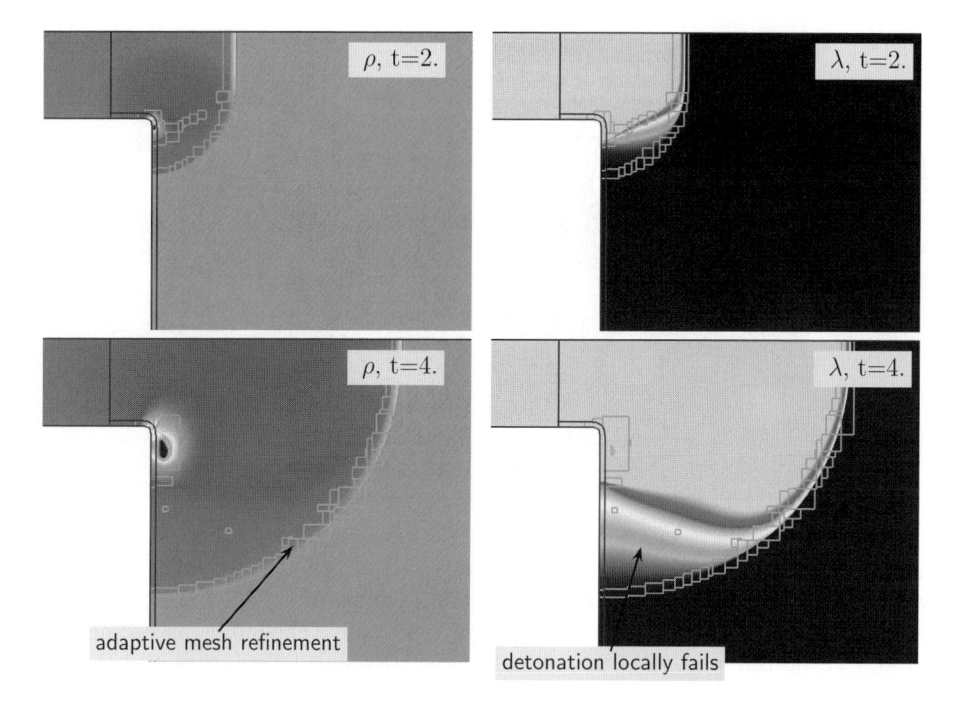

Figure 4. Diffraction of a detonation by a corner. The computations were performed with an ignition and growth model and a JWL equation of state. The density and reaction progress variable are shown. The boundaries of the component base grids and of the AMR grids are also displayed.

where

$$\mathbf{u} = \begin{bmatrix} \rho \\ \rho u \\ \rho v \\ E \\ \rho \mathbf{Y} \end{bmatrix}, \quad \mathbf{F} = \begin{bmatrix} \rho u \\ \rho u^2 + p \\ \rho u v \\ u(E+p) \\ \rho u \mathbf{Y} \end{bmatrix}, \quad \mathbf{G} = \begin{bmatrix} \rho v \\ \rho v u \\ \rho v^2 + p \\ v(E+p) \\ \rho v \mathbf{Y} \end{bmatrix}, \quad \mathbf{H} = \begin{bmatrix} 0 \\ 0 \\ 0 \\ 0 \\ \rho \mathbf{R} \end{bmatrix}.$$

The state of the flow depends on the position $\mathbf{x} = (x, y) = (x_1, x_2)$ and the time t and is described by its density ρ, velocity $\mathbf{v} = (u, v)$, pressure p and total energy E. The flow is a mixture of m_r reacting species whose mass fractions are given by $\mathbf{Y}$. The source term models the chemical reactions and is described by a set of m_r rates of species production given by $\mathbf{R}$. The total energy is taken to be

$$E = \frac{p}{\gamma - 1} + \frac{1}{2}\rho\left(u^2 + v^2\right) + \rho q,$$

where γ is the ratio of specific heats and q represents the heat energy due to chemical reaction.

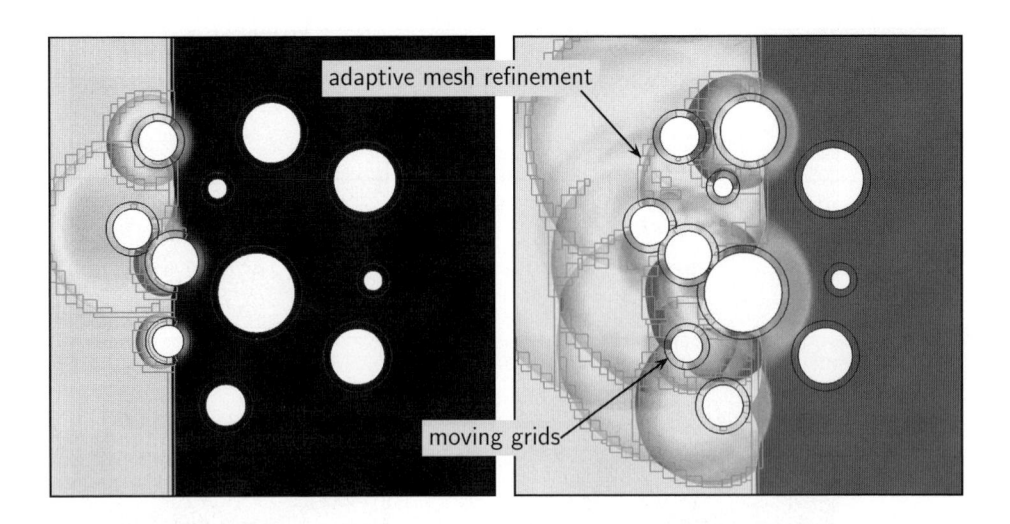

Figure 5. Computation of a shock hitting a collection of cylinders, the density is shown at two different times. This computation illustrates the use of moving grids and adaptive mesh refinement. Each cylinder is treated as a rigid body that moves according to the forces exerted by the fluid. The boundaries of the component base grids and the AMR grids are also shown.

These equations are discretized, as part of the OverBlown solver, with a high-order accurate Godunov scheme coupled to an adaptive Runge-Kutta time stepper for the stiff source terms that model the chemistry [2]. Figure 4 shows results of a computation of a detonation diffracting around a corner. The detonation locally fails in the expansion region.

The motion of a rigid body $\mathcal{B}$ embedded in the flow is governed by the Newton-Euler equations. Let M^b be the mass of the body, $\mathbf{x}^b(t)$, and $\mathbf{v}^b(t)$ the position and velocity of the center of mass, $\mathcal{I}_i$ the moments of inertia, ω_i the angular velocities about the principal axes of inertial, $\mathbf{e}_i$, $\mathbf{F}^b(t)$ the resultant force, and $\mathbf{G}^b(t)$ the resultant torque about $\mathbf{x}^b(t)$. The Newton-Euler equations are then

$$\frac{d\mathbf{x}^b}{dt} = \mathbf{v}^b, \qquad M^b \frac{d\mathbf{v}^b}{dt} = \mathbf{F}^b,$$
$$\mathcal{I}_i \dot{\omega}_i - (\mathcal{I}_{i+1} - \mathcal{I}_{i+2})\omega_{i+1}\omega_{i+2} = \mathbf{G}^b \cdot \mathbf{e}_i, \quad \dot{\mathbf{e}}_i = \boldsymbol{\omega} \times \mathbf{e}_i \qquad i = 1, 2, 3,$$

where the subscripts on $\mathcal{I}_i$ and ω_i are to be taken modulo 3 in the sense $\mathcal{I}_{i+1} := \mathcal{I}_{(i \bmod 3)+1}$. Define the force, $\mathbf{F}^b$, and the torque, $\mathbf{G}^b$, on the body by

$$\mathbf{F}^b = \mathbf{B}^b + \int_{\partial \mathcal{B}} \boldsymbol{\tau} \cdot \mathbf{n} \, dS, \qquad \mathbf{G}^b = \int_{\partial \mathcal{B}} (\mathbf{r} - \mathbf{x}^b) \times \boldsymbol{\tau} \cdot \mathbf{n} \, dS.$$

Here the integrals are over the surface of the rigid body, $\partial \mathcal{B}$. The force on the body will be a sum of body forces, $\mathbf{B}^b$, such as those arising from buoyancy, plus hydrodynamic forces on the boundary of the body, exerted by the fluid stresses. In the case of the Euler-equations, the stress tensor is simply $\boldsymbol{\tau} = -p\mathbf{I}$; the effects of viscosity are assumed to be negligible.

Figure 5 shows a computation of a shock hitting a collection of cylinders. The cylinders are rigid bodies that move due to the hydrodynamic forces. Adaptive mesh refinement is used in combination with moving grids. The grids around each cylinder move at each time step. The refinement grids move with their underlying base grid. The locations of all refinement grids are recomputed every few time steps. More details on this approach will be available in a forthcoming article.

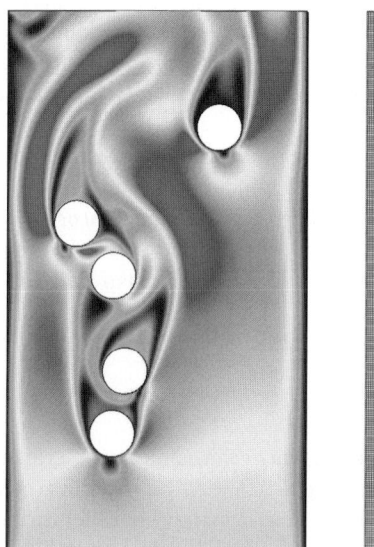

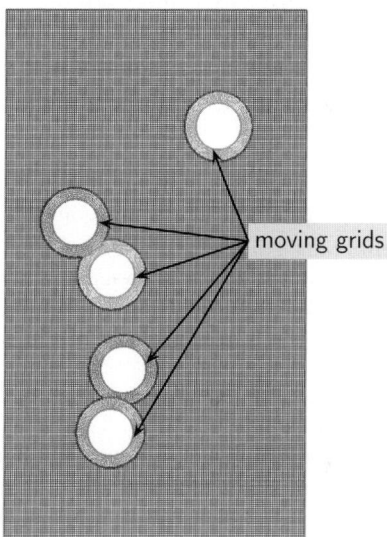

Figure 6. Falling cylinders in an incompressible flow. Left: a contour plot of the magnitude of the velocity. Right: A coarsened version of the overlapping grid. The cylinders have a mass and move under the forces of gravity and the forces exerted by the viscous fluid. The annular grids are moved at each time step. A Poisson equation for the pressure is solved using the overlapping grid multigrid solver Ogmg.

4. Incompressible Flow

The incompressible Navier-Stokes equations are solved using the velocity-pressure formulation,

$$\left. \begin{array}{rcl} \mathbf{u}_t + (\mathbf{u} \cdot \nabla)\mathbf{u} + \nabla p & = & \nu \Delta \mathbf{u} \\ \Delta p + \nabla \cdot (\mathbf{u} \cdot \nabla \mathbf{u}) & = & \alpha(\mathbf{x}) \nabla \cdot \mathbf{u} \end{array} \right\} \quad \mathbf{x} \in \Omega \,,$$

with boundary conditions

$$\left.\begin{array}{rcl}B(\mathbf{u},p) & = & 0 \\ \nabla \cdot \mathbf{u} & = & 0\end{array}\right\} \quad \mathbf{x} \in \partial\Omega \;,$$

and initial conditions

$$\mathbf{u}(\mathbf{x},0) = \mathbf{u}_0(\mathbf{x}) \quad \text{at } t = 0 \;.$$

The term $\alpha(\mathbf{x})\nabla \cdot \mathbf{u}$ is used to damp the dilatation. The boundary condition $\nabla \cdot \mathbf{u} = 0$ is the additional *pressure boundary condition* needed for this formulation to ensure that the dilatation is zero everywhere. The numerical scheme is a split-step formulation. The velocity is advanced first in an explicit or implicit manner. The pressure is then determined. The scheme has been implemented using second-order and fourth-order accurate approximations using a predictor-corrector time stepping scheme. The discretization of the boundary conditions on no-slip walls is an important element of the scheme. This solution algorithm is also implemented in the OverBlown code. See [8,9] for further details.

Figure 6 shows results of a computation of some rigid cylinders falling through a channel containing a viscous fluid. The grids around each cylinder move at each time step according to the Newton-Euler equations of motion. The Ogen grid generator is used to update the overlapping grid connectivity information. The multigrid solver is used to solve the pressure equation. The Poisson equation changes at each time step but this equation can be treated efficiently with the multigrid solver. More details on the approach used to solve the incompressible equations with moving grids will appear in a future publication.

5. Parallel computations

In a distributed parallel computing environment, each component grid-function (representing the solution variables such as ρ, $\mathbf{u}$, p, etc.) can be distributed across one or more processors. The grid functions are implemented using a parallel distributed arrays from the P++ array class library[10]. Each P++ array can be independently distributed across the available processors. The distributed array consists of a set of serial arrays, one serial array for each processor. Each serial array is a multi-dimensional array that can be operated on using array operations. The serial array can also be passed to a Fortran function, for example. The serial arrays contain extra ghost lines that hold copies of the data from the serial arrays on neighbouring processors. P++ is built on top of the Multiblock PARTI parallel communication library [11] which is used for ghost boundary updates and copying blocks of data between arrays with possibly different distributions. Figure 7 presents a section of C++ code showing the use of P++ arrays.

A special parallel overlapping grid interpolation routine has been developed for updating the points on grids that interpolate from other grids, see Figure 1. Overlapping grid interpolation is based on a multi-dimensional tensor product Lagrange interpolant. In parallel, the Lagrange formula is evaluated on the processor that owns the data in the stencil (the donor points), the resulting sums are collected into a message and then sent to the processor that owns the interpolation point.

Figure 8 shows some preliminary parallel results from solving the Euler equations on an overlapping grid with the parallel version of OverBlown. The speed-up running on up

```
Partitioning_Type partition; // object that defines the parallel distribution
partition.SpecifyInternalGhostBoundaryWidths(1,1);

realDistributedArray u(100,100,partition); // build a distributed array
Range I(1,98), J(1,98);

// Parallel array operation with automatic communication:
u(I,J)=.25*( u(I+1,J) + u(I-1,J) + u(I,J+1) + u(I,J-1) ) + sin(u(I,J))/3.;

// Access local serial arrays and call a Fortran routine:
realSerialArray & uLocal = u.getLocalArray(); // access the local array
myFortranRoutine(*uLocal.getDataPointer(),...);
u.updateGhostBoundaries(); // update ghost boundaries on distributed arrays
```

Figure 7. The P++ class library provides parallel multi-dimensional arrays. The class supports array operations with automatic communication. It is also possible to use Fortran or C kernels to operate on each local serial array.

to 32 processors is shown for a problem with a fixed number of grid points. The scaling is quite reasonable. The computations were performed on a Linux cluster.

6. Acknowledgments

The development of the reactive Euler equation solver was performed in collaboration with Professor Don Schwendeman. Thanks also to Kyle Chand and Jeffrey Banks for contributions to this work.

REFERENCES

1. G. Chesshire, W. Henshaw, Composite overlapping meshes for the solution of partial differential equations, J. Comp. Phys. 90 (1) (1990) 1–64.
2. W. D. Henshaw, D. W. Schwendeman, An adaptive numerical scheme for high-speed reactive flow on overlapping grids, J. Comp. Phys. 191 (2003) 420–447.
3. W. D. Henshaw, On multigrid for overlapping grids, SIAM Journal of Scientific Computing 26 (5) (2005) 1547–1572.
4. K. Brislawn, D. L. Brown, G. Chesshire, J. Saltzman, Adaptively-refined overlapping grids for the numerical solution of hyperbolic systems of conservation laws, report LA-UR-95-257, Los Alamos National Laboratory (1995).
5. E. P. Boden, E. F. Toro, A combined Chimera-AMR technique for computing hyperbolic PDEs, in: Djilali (Ed.), Proceedings of the Fifth Annual Conference of the CFD Society of Canada, 1997, pp. 5.13–5.18.

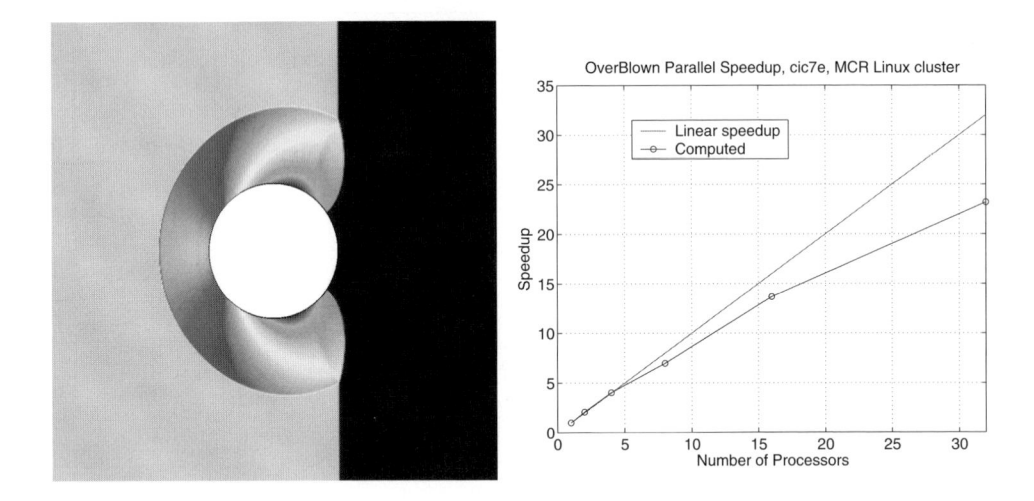

Figure 8. Left: the computation of a shock hitting a cylinder (density). Right: parallel speedup for this problem, keeping the problem size fixed (4 million grid points), on a linux cluster (Xeon processors).

6. R. L. Meakin, Composite overset structured grids, in: J. F. Thompson, B. K. Soni, N. P. Weatherill (Eds.), Handbook of Grid Generation, CRC Press, 1999, Ch. 11, pp. 1–20.
7. D. Brown, K. Chand, P. Fast, W. Henshaw, A. Petersson, D. Quinlan, Overture, Tech. Rep. http://www.llnl.gov/CASC/Overture.html, Lawrence Livermore National Laboratory (2003).
8. W. Henshaw, A fourth-order accurate method for the incompressible Navier-Stokes equations on overlapping grids, J. Comp. Phys. 113 (1) (1994) 13–25.
9. W. D. Henshaw, N. A. Petersson, A split-step scheme for the incompressible Navier-Stokes equations, in: M. Hafez (Ed.), Numerical Simulation of Incompressible Flows, World Scientific, 2003, pp. 108–125.
10. D. Quinlan, A++/P++ class libraries, Research Report LA-UR-95-3273, Los Alamos National Laboratory (1995).
11. A. Sussman, G. Agrawal, J. Saltz, A manual for the Multiblock PARTI runtime primitives, revision 4.1, Technical Report CS-TR-3070.1, University of Maryland, Department of Computer Science (1993).

Towards Optimal Performance for Lattice Boltzmann Applications on Terascale Computers

G. Wellein[a], P. Lammers[b], G. Hager[a], S. Donath[a], and T. Zeiser[a]

[a]Regionales Rechenzentrum Erlangen (RRZE)
Martensstraße 1, D-91058 Erlangen, Germany

[b]Höchstleistungsrechenzentrum Stuttgart (HLRS)
Allmandring 30, D-70550 Stuttgart, Germany

On popular cluster computers the performance of many CFD applications can fall far short of the impressive peak performance numbers. Using a large scale LBM application, we demonstrate the different performance characteristics of modern supercomputers. Classical vector systems (NEC SX8) still combine excellent performance with a well established optimization approach and can break the TFlop/s application performance barrier at a very low processor count. Although, modern microprocessors offer impressive peak performance numbers and extensive code tuning has been performed, they only achieve 10% or less of the vector performance for the LBM application. Clusters try to fill this gap by massive parallelization but network capabilities can impose severe performance restrictions if the problem size is not arbitrarily scalable. Tailored HPC servers like the SGI Altix series can relieve these restrictions.

1. Introduction

In the past decade the lattice Boltzmann method (LBM) [1-3] has been established as an alternative for the numerical simulation of (time-dependent) incompressible flows. One major reason for the success of LBM is that the simplicity of its core algorithm allows both easy adaption to complex application scenarios as well as extension to additional physical or chemical effects. Since LBM is a direct method, the use of extensive computer resources is often mandatory. Thus, LBM has attracted a lot of attention in the High Performance Computing community [4-6]. An important feature of many LBM codes is that the core algorithm can be reduced to a few manageable subroutines, facilitating deep performance analysis followed by precise code and data layout optimization [6-8]. Extensive performance evaluation and optimization of LBM for terascale computer architectures is required to find an appropriate computer architecture for the various LBM applications and to make best use of computer resources. It also allows us to better understand the performance characteristics of modern computer architectures and to develop appropriate optimization strategies which can also be applied to other applications.

In this report we discuss LBM performance on commodity "off-the-shelf" (COTS) clusters (GBit, Infiniband) with Intel Xeon processors, tailored HPC systems (Cray XD1,

SGI Altix) and a NEC SX8 vector system. We briefly describe the main architectural differences and comment on single processor performance as well as optimization strategies. Finally, we evaluate and present the parallel performance of a large scale simulation running on up to 2000 processors, providing 2 TFlop/s of sustained performance. For those studies we use the LBM solver $\mathcal{BEST}$ (Boltzmann Equation Solver Tool), which is written in FORTRAN90 and parallelized with MPI using domain decomposition [9]. As a test case we run simulations of flow in a long channel with square cross-section, which is a typical application in turbulence research.

2. Basics of the Lattice Boltzmann Method

The widely used class of lattice Boltzmann models with BGK approximation of the collision process [1–3] is based on the evolution equation

$$f_i(\vec{x}+\vec{e}_i\delta t,\, t+\delta t) = f_i(\vec{x},\, t) - \frac{1}{\tau}\left[f_i(\vec{x},\, t) - f_i^{\text{eq}}(\rho, \vec{u})\right], \qquad i = 0\ldots N. \tag{1}$$

Here, f_i denotes the particle distribution function which represents the fraction of particles located in timestep t at position $\vec{x}$ and moving with the microscopic velocity $\vec{e}_i$. The relaxation time τ determines the rate of approach to local equilibrium and is related to the kinematic viscosity of the fluid. The equilibrium state f_i^{eq} itself is a low Mach number approximation of the Maxwell-Boltzmann equilibrium distribution function. It depends only on the macroscopic values of the fluid density ρ and the flow velocity $\vec{u}$. Both can be easily obtained as the first moments of the particle distribution function.

The discrete velocity vectors $\vec{e}_i$ arise from the $N+1$ chosen collocation points of the velocity-discrete Boltzmann equation and determine the basic structure of the numerical grid. We chose the D3Q19 model [1] for the discretization in 3-D, which uses 19 discrete velocities (collocation points) and provides a computational domain with equidistant Cartesian cells (voxels). Each timestep ($t \to t + \delta t$) consists of the following steps which are repeated for all cells:

- Calculation of the local macroscopic flow quantities ρ and $\vec{u}$ from the distribution functions, $\rho = \sum_{i=0}^{N} f_i$ and $\vec{u} = \frac{1}{\rho}\sum_{i=0}^{N} f_i \vec{e}_i$.

- Calculation of the equilibrium distribution f_i^{eq} from the macroscopic flow quantities (see [1] for the equation and parameters) and execution of the "collision" (relaxation) process, $f_i^*(\vec{x},\, t^*) = f_i(\vec{x},\, t) - \frac{1}{\tau}[f_i(\vec{x},\, t) - f_i^{\text{eq}}(\rho, \vec{u})]$, where the superscript * denotes the post-collision state.

- "Propagation" of the $i = 0\ldots N$ post-collision states $f_i^*(\vec{x},\, t^*)$ to the appropriate neighboring cells according to the direction of $\vec{e}_i$, resulting in $f_i(\vec{x}+\vec{e}_i\delta t,\, t+\delta t)$, i.e. the values of the next timestep.

The first two steps are computationally intensive but involve only values of the local node while the third step is just a direction-dependent uniform shift of data in memory. A fourth step, the so called "bounce-back" rule [2,3], is incorporated as an additional routine and "reflects" the distribution functions at the interface between fluid and solid cells, resulting in an approximate no-slip boundary condition at walls.

3. Implementation and Optimization Strategies

The basic implementation of our code follows the guidelines described in [7]. For a better understanding of the subsequent performance evaluation we briefly highlight some important aspects:

- Removing contributions with zero components of $\vec{e}_i$ and precomputing common subexpressions results in roughly $F = 200$ floating point operations (Flops) per lattice site update. The actual number can vary because of compiler optimizations.

- Collision and propagation steps are done in the same loop ("collision-propagation"), which reduces the data transfer between main memory and processor for one time step to one read of the whole distribution function at time t and one store of the whole distribution function at time t^*. Thus, $B = 2 \times 19 \times 8$ Bytes have to be transfered per lattice site update, at first sight.

- Two arrays are used which hold data of successive timesteps t and $t+1$. The data layout was chosen as $f(x, y, z, i, t)$ with the first index addressing consecutive memory locations due to Fortran's column major order making spatial blocking techniques dispensable [7].

The achieved performance for this implementation on IA32 and IA64 systems was, however, only 25–45% of the theoretical maximum set by memory bandwidth. Inspection of compiler reports and performance counters has revealed two additional bottlenecks, which have not been addressed by the optimizations discussed so far:

- The large loop body prevents the Intel IA64 compiler from software pipelining, which is essential for Itanium 2, and causes massive register spills on all architectures.

- Concurrently writing to 19 different cache lines interferes with the limited number of write combine buffers on IA32 architectures (6 for Intel Xeon/Nocona).

In order to remove these problems the innermost loop (in x-direction) of our "collision-propagate" step is split up into 3 loops of length N_x, which equals the domain size in x-direction. However, additional auxiliary arrays of length N_x are required which hold intermediate results. As long as N_x is not too large ($N_x < 1000$), these arrays can be kept in the on-chip caches and accessed quickly. This implementation increases the number of floating point operations per lattice site update ($B \approx 230-250$) but usually provides best performance (see next section). Note that for vector systems no additional optimizations besides vectorization, are required and the number of floating point operations per lattice site update is rather small ($B \approx 170$).

4. Single Node Specification and Performance

Most up-to-date HPC systems are configured as shared-memory nodes (e.g. dual Xeon) connected by an high speed interconnect (e.g. GBit or Infiniband) or a proprietary network (e.g. NUMALink in SGI systems). In order to clearly separate the contribution of processor and interconnect to the total performance of a terascale system, we first turn to the single node performance characteristics.

Table 1
Single processor specifications. Peak performance, maximum memory bandwidth and sizes of the largest on-chip cache levels are given in columns 2-4. Columns 5,6 show the maximum MLUPS rate as limited by peak performance or memory bandwidth. The last two colums contain performance achieved for original and tuned implementations at a domain size of 128^3.

Platform	CPU specs Peak GFlop/s	CPU specs MemBW GB/s	CPU specs Cache MB	Limits Peak MLUPS	Limits MemBW MLUPS	Versions Orig. MLUPS	Versions Tuned MLUPS
Intel Xeon DP (3.4 GHz)	6.8	5.3	1.0 (L2)	34.0	11.8	4.5	4.9
AMD Opteron848 (2.2 GHz)	4.4	6.4	1.0 (L2)	22.0	14.0	2.9	4.7
Intel Itanium 2 (1.4 GHz)	5.6	6.4	1.5 (L3)	28.0	14.0	7.9	8.5
NEC SX8 (2 GHz)	16.0	64.0	–	80.0	210.0	68.0	–

Since a lot of optimization was done at the CPU level, we have built a benchmark kernel including the routines for "collision-propagation" and "bounce-back" which mainly determine the performance of the LBM step. In order to test for limitations of the single node performance, a shared-memory parallelization using OpenMP was implemented as well.

4.1. Architectural Specifications

In the left part of Table 1 we briefly sketch the most important single processor specifications[1] of the architectures examined. Concerning the memory architecture of COTS systems, we find a clear tendency towards large on-chip caches which run at processor speed, providing high bandwidth and low latency. The NEC vector system incorporates a different memory hierarchy and achieves substantially higher single processor peak performance and memory bandwidth.

Intel Xeon/Nocona and AMD Opteron

The Intel Xeon/Nocona and the AMD Opteron processors used in our benchmarks are 64-bit enabled versions of the well-known Intel Xeon and AMD Athlon designs, respectively, but maintain full IA32 compatibility. Both are capable of performing a maximum of two double precision floating point (FP) operations (one multiply and one add) per cycle. The most important difference is that in standard multi-processor configurations (2- or 4-way are in common use) all processors of the Intel based designs have to share one memory bus while in AMD based systems the aggregate memory bandwidth scales with processor count. The benchmarks were compiled using the Intel Fortran Compiler for Intel EM64T-based applications in version 8.1.023 and were run on a 2-way Intel and 4-way AMD system. The parallel benchmarks were run on a cluster with Intel Xeon nodes featuring both GBit and Infiniband interconnect. As an Opteron based parallel system

---

[1] The Intel Xeon processor supports a maximum bandwidth of 6.4 GB/s; the benchmark system was equipped with DDR-333 memory providing a bandwidth of 5.3 GB/s only.

we have chosen the Cray XD1 which uses a proprietary network (RapidArray) to connect 2-way Opteron nodes.

Intel Itanium 2

The Intel Itanium 2 processor is a superscalar 64-bit CPU using the Explicitly Parallel Instruction Computing (EPIC) paradigm. The Itanium concept does not require any out-of-order execution hardware support but demands high quality compilers to identify instruction level parallelism at compile time. Today clock frequencies of up to 1.6 GHz and on-chip L3 cache sizes from 1.5 to 9 MB are available. Two Multiply-Add units are fed by a large set of 128 floating point (FP) registers, which is another important difference to standard microprocessors with typically 32 FP registers. The basic building blocks of systems used in scientific computing are 2-way nodes (e.g. SGI Altix, HP rx2600) sharing one bus with 6.4 GByte/s memory bandwidth.

The system of choice in this section is a 2-way HP zx6000 with 1.4 GHz CPUs. For compilation we used Intel's Fortran Itanium Compiler V9.0. The parallel benchmarks were run on SGI Altix systems with NUMALink interconnect and 1.5 GHz or 1.6 GHz CPUs but with the same memory bandwidth per CPU.

NEC SX8

From a programmers' view, the NEC SX8 is a traditional vector processor with 4-track vector pipes running at 2 GHz. One multiply and one add instruction per cycle can be executed by the arithmetic pipes delivering a theoretical peak performance of 16 GFlop/s. The memory bandwidth of 64 GByte/s allows for one load or store per multiply-add instruction, providing a balance of 0.5 Word/Flop. The processor features 64 vector registers, each holding 256 64-bit words. Basic changes compared to its predecessor systems, as used e.g. in the Earth Simulator, are a separate hardware square root/divide unit and a "memory cache" which lifts stride 2 memory access patterns to the same performance as contiguous memory access. An SMP node comprises eight processors and provides a total memory bandwidth of 512 GByte/s, i.e. the aggregated single processor bandwidths can be saturated.

The benchmark results presented in this paper were measured on a 72-node NEC SX8 at the High Performance Computing Center Stuttgart (HLRS).

4.2. Single Node Performance: Theory and Reality

Before looking at the "pure" performance measurements an estimation of the maximum achievable performance numbers for the platforms under consideration may be helpful to understand the "experiments". All performance numbers are given in MLUPS (**M**ega **L**attice **S**ite **U**pdates **p**er **S**econd), which is a handy unit for measuring the performance of LBM.[2] In this section we report the update rate of the fluid cells only, i.e. we do not count obstacle cells at the boundaries where no computation is being done. Based on characteristic quantities of the benchmarked architectures (cf. Table 1), an estimate for a theoretical performance limit can be given. Performance is either limited by available memory bandwidth or peak performance. Therefore, the attainable maximum performance is either given as

---

[2] Note that 5 MLUPS are roughly equal to 1 GFlop/s of sustained performance.

Table 2
Single node performance for optimized LBM implementation using OpenMP parallelization and a fixed domain size of 128^3. Arrays include one additional ghost layer in each direction, making cache trashing effects [7] negligible. For AMD, performance for naive and NUMA-aware OpenMP implementations are shown.

Platform	MLUPS 1 CPU	2 CPUs	4 CPUs
AMD Opteron - naive	4.6	5.8	5.8
AMD Opteron - NUMA	4.6	9.2	17.6
Intel Xeon	4.8	5.5	—
Intel Itanium 2	8.4	8.9	—

$$P = \frac{\text{MemBW}}{B} \quad \text{or} \quad P = \frac{\text{Peak Perf.}}{F} \qquad (2)$$

The values of P are given in units of MLUPS, since B is the number of bytes per lattice site update to be transferred between CPU and main memory and F is the number of floating point operations to be performed per lattice site update. For both quantities, a typical number arising from the LBM implementation itself has been given in Section 3. As a peculiarity of most cache based architectures, read for ownership (RFO) transfers on write misses must be taken into account. In our implementation, this occurs when the updated values are propagated to adjacent cells. The corresponding cache line must be loaded from main memory before it can be modified and written back. Consequently, the value of B increases by 50%, further reducing the effective available bandwidth. The performance limits imposed by hardware and the performance numbers measured both for original and optimized implementations (additional inner loop splitting as described in Section 3) are presented in Table 1 for a computational domain size of 128^3. A substantial effect of the additional optimization can be seen on the AMD system which is lifted from a poor performance level to the same performance as its Intel competitor.

The striking features of the vector machine are extremely high performance and very efficient use of the processor, e.g. our LBM code achieves 85% of the algorithm's theoretical limit. Note that this limit is not set by the memory subsystem but by the peak performance of the processor.

The scalability of our simple LBM kernel using OpenMP within one node is depicted in Table 2. Although the Opteron architecture provides a separate path to the main memory for each processor, a naive implementation with "collision-propagation" and "bounce-back" routines being the only parallelized parts, shows poor scalability. The Opteron design is not a symmetric multiprocessing (SMP) node with flat memory (as the Intel systems are), but implements a Cache Coherent Non Uniform Memory (ccNUMA) architecture. Each processor has its own local memory and can access data on remote nodes through a network at lower bandwidth and higher latencies. For that reason, data placement becomes a major issue on ccNUMA machines. Since memory pages are mapped on the node of the processor which initializes the data (first-touch placement), all data is put into the memory of a single processor if the initialization routine is not parallelized. A

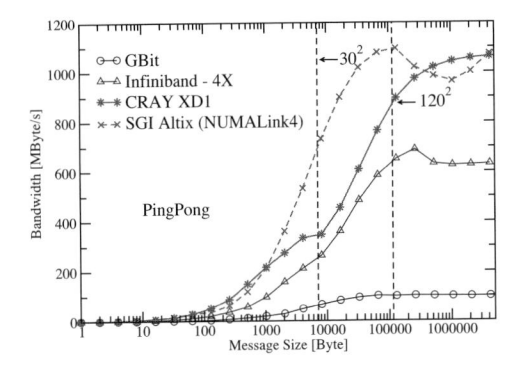

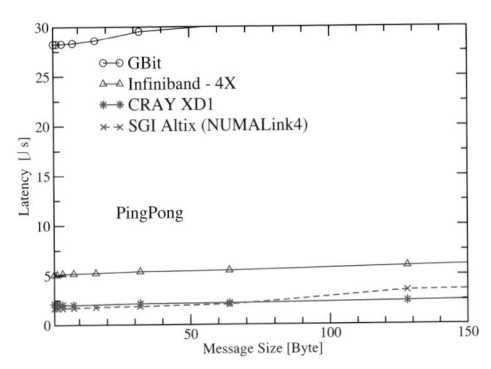

Figure 1. MPI bandwidth measured with PMB. Vertical lines denote message sizes with 30^2 and 120^2 double words.

Figure 2. MPI latency (total time for message transfer) at small message sizes as measured with PMB.

NUMA-aware implementation uses parallelized data initialization in a manner compatible with the computational kernels and achieves very good scalability on the Opteron system, leading to a parallel efficiency of roughly 96% on 4 processors.[3]

Intel systems, on the other hand, do not benefit from parallel initialization as the memory model is flat. Consequently, scalability is poor due to the memory bottleneck, and gets even worse on 4-way SMP systems which we did not include in our evaluation.

5. Scalability and Parallel Performance

The lessons learned from [7] and Section 3 have been used to optimize the MPI parallel LBM application $\mathcal{BEST}$. Because of additional program overhead, $\mathcal{BEST}$ performance may of course slightly deviate from the numbers presented so far.

Scalability beyond one node is determined by the ability of the network to meet the application's communication requirements. In our scalability analysis of $\mathcal{BEST}$, we use several very different interconnects ranging from standard GBit to the NEC IXS. A good insight into basic network capabilities is provided by the Pallas MPI Benchmark (PMB, available as part of the Intel Cluster Toolkit). We have chosen the PingPong test to measure round trip times for sending messages between two arbitrary nodes. Bandwidth and latency (one way time) as a function of the message length are depicted in Figure 1 and Figure 2, respectively. Obviously, rather large message lengths of roughly 100 KBytes are required to get maximum bandwidth on all interconnects presented. Typical message sizes arising from simple domain decompositions of LBM on clusters (cf. vertical lines in Figure 1) can be too short to get maximum bandwidth. While GBit is not competitive in both categories, the Infiniband (IB) solution does much better but is still a factor of 2 behind the two proprietary interconnects used in the SGI Altix (NUMALink4) and the Cray XD1 (RapidArray). Although, the Cray XD1 technology seems to be related to IB there is a substantial technical difference which limits the capabilities of standard IB solutions as used in most clusters: The IB card is connected via an external PCI-X

---
[3]Parallel efficiency is defined as $P(n)/(n \times P(1))$, where $P(n)$ is the performance on n CPUs.

Figure 3. Strong scaling performance and parallel efficiency (inset) at a domain size of 256×128^2. Horizontal bold line denotes NEC SX8 single CPU performance.

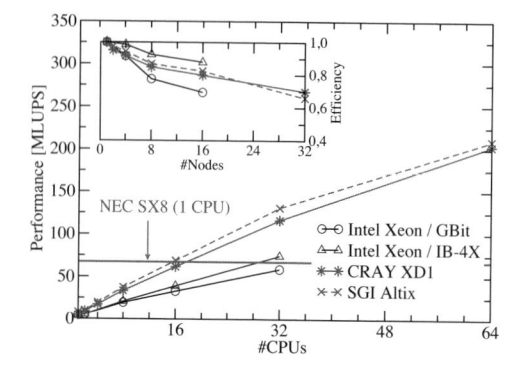

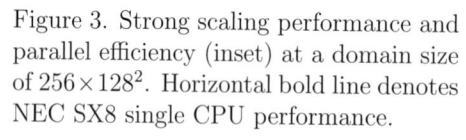

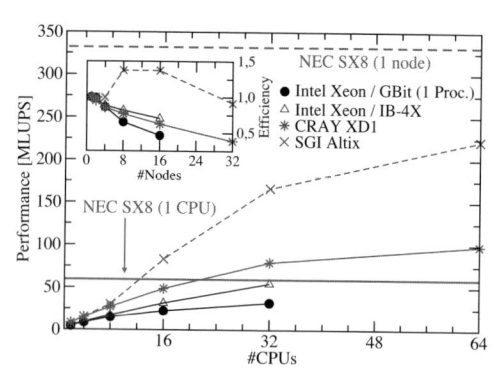

Figure 4. Strong scaling performance and parallel efficiency (inset) with a domain size of 64^3. Horizontal bold (dashed) line denotes NEC SX8 single CPU (node) performance.

interface to the compute node. The RapidArray chip, on the other hand, is directly connected to the HyperTransport channel of one Opteron processor, avoiding additional overhead and limitations of PCI-X. For comparison, the IXS interconnect of the NEC SX8 achieves up to 28.75 GByte/s bandwidth and 3-7 μs small message latency between two nodes [10].

For a final scalability analysis one must differentiate between *weak scaling* and *strong scaling* scenarios. In *weak-scaling* experiments the total problem size per processor is kept constant, i.e. total domain size increases linearly with the number of CPUs used. Consequently, the ratio of communication and computation remains constant and one can achieve a perfectly linear increase in total performance even on GBit [5] provided the domain size per node is large enough and the network capabilities scale with processor count. A typical area for the weak scaling LBM scenario is basic turbulence research, where the total domain size should be as large as possible.

In *strong scaling* experiments the total domain size remains constant, independent of CPU number. Here, scalability is dominated by interconnect features and two effects must be considered: First, with increasing processor count the domain size per processor decreases, raising the ratio of communication ($\propto$ local domain surface) versus computation ($\propto$ volume of the local domain). Second, smaller surface size is equivalent to smaller message size leading to substantially worse network bandwidth for small domains (cf. discussion of Figure 1). For large processor counts runtime will thus mainly be governed by communication and performance saturates, i.e. scalability breaks down. Strong scaling is a major issue if turnaround time for a fixed problem is important, e.g. for practical engineering problems or if computations would run for months.

In Figures 3 and 4 we present strong scaling experiments for two typical domain sizes with 4.2×10^6 and 0.26×10^6 lattice sites. Note that the baseline for parallel efficiency as given in the insets is the one-node performance $P(2)$ in order to eliminate the different intranode performance characteristics. The deficiencies of GBit for decreasing message

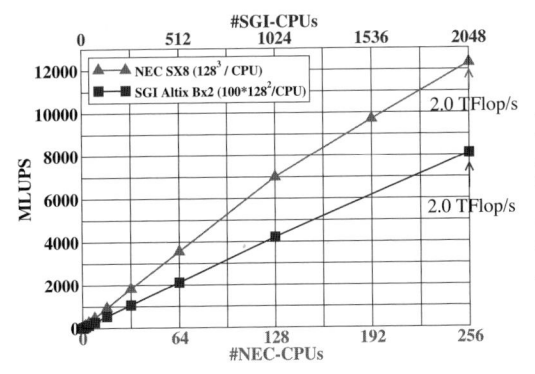

Figure 5. Weak scaling performance of $\mathcal{BEST}$ on SGI Altix Bx2 and NEC SX8. The number of NEC SX8 (SGI Altix Bx2) CPUs is given on the lower (upper) x-axis. The total domain size for the largest run was 0.5×10^9 (3.3×10^9) on NEC SX8 (SGI Altix Bx2).

sizes become quite obvious. In both cases the GBit cluster with 32 CPUs (16 nodes) could not match the performance of a single vector CPU. Even more, in Figure 4 we only show results for one CPU per node because total performance drops significantly if both CPUs are used. The IB network provides higher total performance and better parallel efficiency, the latter even larger than on the Cray XD1, which is at first sight an inconsistency with our findings for the PMB benchmark. However, the substantially lower single node performance of the Intel Xeon node (≈ 5.5 MLUPS as compared to ≈ 9.2 MLUPS for 2 Opteron processors) reduces the fraction of total time spent for communication and thus allows the slower network to "scale better", albeit at lower total performance. A peculiarity of systems with large caches is presented in Figure 4 for an SGI Altix with Itanium 2 processors having 6 MB of L3 cache each. At and beyond 16 CPUs, the aggregate cache size is large enough to hold the whole data set, speeding up the computation part substantially as compared to the baseline result on two processors. This effect in combination with the fast NUMALink network allows us to achieve a parallel efficiency much larger than 1 and to get an even higher performance on 64 CPUs than for the larger domain used in Figure 3.

Concerning single CPU and single node performance, the NEC SX8 system is a class of its own for both domain sizes. Note that the NEC SX8 single node performance for Figure 3 is 435 MLUPS which eceeds the maximum y-scale of this figure by more than 50%.

In a final weak scaling experiment we have chosen NEC SX8 and SGI Altix Bx2 systems in Figure 5 to demonstrate that the $\mathcal{BEST}$ application is able to break the TFlop/s barrier on modern HPC systems. Both systems scale very well to a maximum sustained performance of roughly 2 TFlop/s for the largest simulations. However, the equivalent MLUPS rate, which is the important performance number for LBM applications, is 50% higher on the vector machine due to its computationally less intensive implementation (cf. discussion in Section 3). The striking feature of the vector system is, of course, that less than a tenth of the processors is required to reach the same performance level as on the SGI system. On clusters this ratio will be even worse. With only 128 processors the NEC machine achieves more than one TFlop/s of sustained application performance which is more than 50% of its peak.

6. Conclusion

The main focus of our report was to discuss single processor optimization strategies and scalability issues for large scale LBM applications. Although a lot of work was put into optimization, cache based microprocessors achieve only a fraction of the performance of a modern NEC SX8 vector CPU. In contrast to the microprocessors, LBM performance is not limited by the memory subsystem on vector systems. For our turbulence application code $\mathcal{BEST}$ we achieved a sustained performance of 2 TFlop/s on 256 NEC vector CPUs. Even though the SGI Altix system is best in the microprocessor class we require more than 2000 processors on this architecture to reach a comparable performance level.

Acknowledgments

Part of this work is financially supported by the Competence Network for Technical, Scientific High Performance Computing in Bavaria KONWIHR and the High performance computer competence center Baden-Wuerttemberg. We gratefully acknowledge the support of R. Wolff (SGI), who provided access to the SGI Altix Bx2 and gave many helpful hints to make best use of the system. Benchmarks were also run on the SGI Altix system at CSAR Manchester.

REFERENCES

1. Y. H. Qian, D. d'Humières, P. Lallemand, Lattice BGK models for Navier-Stokes equation, Europhys. Lett. 17 (6) (1992) 479–484.
2. D. A. Wolf-Gladrow, Lattice-Gas Cellular Automata and Lattice Boltzmann Models, Vol. 1725 of Lecture Notes in Mathematics, Springer, Berlin, 2000.
3. S. Succi, The Lattice Boltzmann Equation – For Fluid Dynamics and Beyond, Clarendon Press, 2001.
4. L. Oliker, J. C. A. Canning, J. Shalf, S. Ethier, Scientific computations on modern parallel vector systems, in: Proceedings of SC2004, CD-ROM, 2004.
5. T. Pohl, F. Deserno, N. Thürey, U. Rüde, P. Lammers, G. Wellein, T. Zeiser, Performance evaluation of parallel large-scale lattice Boltzmann applications on three supercomputing architectures, in: Proceedings of SC2004, CD-ROM, 2004.
6. F. Massaioli, G. Amati, Achieving high performance in a LBM code using OpenMP, in: EWOMP'02, Roma, Italy, 2002.
7. G. Wellein, T. Zeiser, S. Donath, G. Hager, On the single processor performance of simple lattice Boltzmann kernels, Computers & Fluids.
8. T. Pohl, M. Kowarschik, J. Wilke, K. Iglberger, U. Rüde, Optimization and profiling of the cache performance of parallel lattice Boltzmann codes, Par. Proc. Lett. 13 (4) (2003) 549–560.
9. P. Lammers, K. Beronov, T. Zeiser, F. Durst, Testing of closure assumptions for fully developed turbulence channel flow with the aid of a lattice Boltzmann simulation, in: S. Wagner, W. Hanke, A. Bode, F. Durst (Eds.), High Performance Computing in Science and Engineering, Munich 2004. Transactions of the Second Joint HLRB and KONWIHR Result and Reviewing Workshop, Springer Verlag, 2004, pp. 77–92.
10. http://icl.cs.utk.edu/hpcc/hpcc_results_lat_band.cgi?display=combo.

Moore's Law, the Life Cycle of Scientific Computing Codes and the Diminishing Importance of Parallel Computing

Rainald Löhner[a], Chi Yang[a], Juan R. Cebral[a], Fernando F. Camelli[a], Fumiya Togashi[a], Joseph D. Baum[b], Hong Luo[b], Eric L. Mestreau[b] and Orlando A. Soto[b]

[a]School of Computational Sciences
M.S. 4C7, George Mason University, Fairfax, VA 22030-4444, USA

[b]Advanced Concepts Business Unit
Science Applications International Corp., McLean, VA

A description is given of the typical life cycle of scientific computing codes. Particular relevance is placed on the number of users, their concerns, the machines on which the codes operate as they mature, as well as the relative importance of parallel computing. It is seen that parallel computing achieves the highest importance in the early phases of code development, acting as an enabling technology without which new scientific codes could not develop. Given the typical times new applications tend to run at their inception, Moore's law itself is perhaps the biggest incentive for new scientific computing codes. Without it, computing time would not decrease in the future, and the range of applications would soon be exhausted.

1. MOORE'S LAW

One of the most remarkable constants in a rapidly changing world has been the rate of growth for the number of transistors that are packaged onto a square inch. This rate, commonly known as Moore's law, is approximately a factor of 2 every 18 months, which translates into a factor of 10 every 5 years [27], [28]. As one can see from Figure 1 this rate, which governs the increase in computing speed and memory, has held constant for more than 3 decades, and there is no end in sight for the foreseeable future [29].

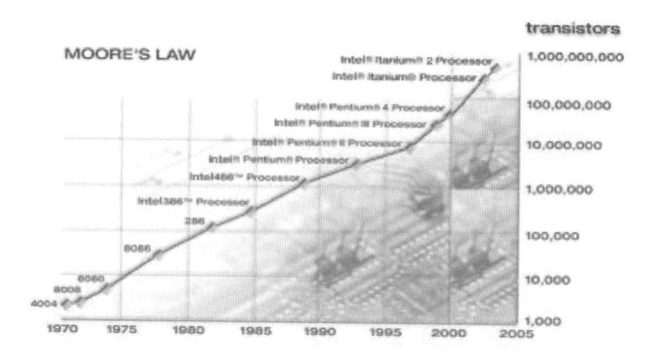

Figure 1: Evolution of Transistor Density

One may argue that the raw number of transistors does not translate into CPU performance. However, more transistors translate into more registers and more cache, both important elements to achieve higher throughput. At the same time, clockrates have increased, and pre-fetching and branch prediction have improved. Compiler development has also not stood still. Programmers have become conscious of the added cost of memory access, cache misses and dirty cache lines, employing techniques such as the use of edge-based data structures [23], chunky loops, and many renumbering strategies [15], [18], [21]. The net effect, reflected in all current projections, is that CPU performance is going to continue advancing at a rate comparable to Moore's law.

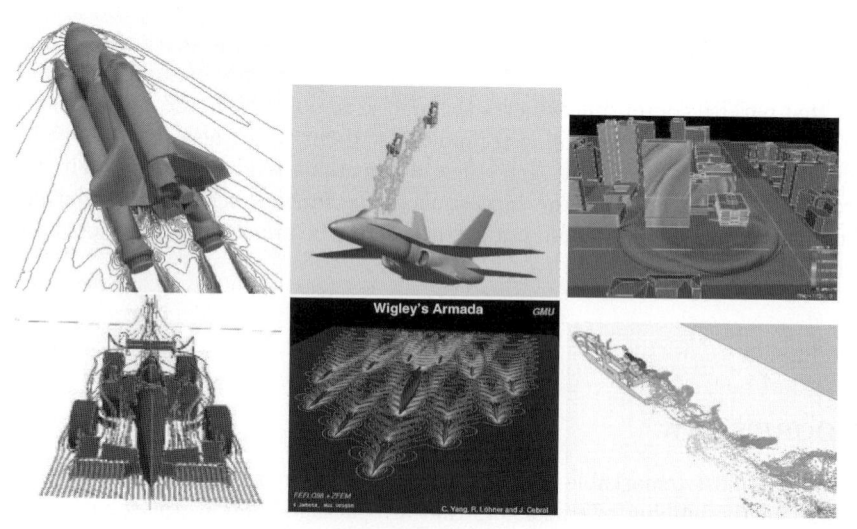

Figure 2: Snapshots of Large-Scale Problems Solved on Parallel Machines by the Authors

The present group of authors has used parallel computers for a large variety of applications over the course of two decades. Figure 2 shows a small list of examples taken from different problem classes (for a detailed description, see [6] for the shuttle, [30] for the pilot ejection, [7] for the blast in Nairobi, [13] for the indy-car, [19] for the armada of Wigley hulls and [9] for the dispersion calculation of the TAKE-1 ship). The largest of these problem have exceeded $5 \cdot 10^8$ elements and were run on shared memory machines using 128 processors. Given the experience with vector [17], [21] distributed [16], [31], and shared memory [18], [30] parallel machines, and a perspective of two decades of high-end computing, it was felt that it would be interesting to take a look at the consequences Moore's law has had on the application classes considered.

2. THE LIFE CYCLE OF SCIENTIFIC COMPUTING CODES

Let us consider the effects of Moore's law on the lifecycle of typical large-scale scientific computing codes. The lifecycle of these codes may be subdivided into the following stages:

- Conception;

- Demonstration/Proof of Concept;

- Production Code;

- Widespread Use and Acceptance;

- Commodity Tool;

- Embedding.

In the **conceptual** stage, the basic purpose of the code is defined, the physics to be simulated are identified, proper algorithms are selected and coded. The many possible algorithms are compared, and the best is kept. A run during this stage may take weeks or months to complete. A few of these runs may even form the core of a Ph.D. thesis.

The **demonstration** stage consists of several large-scale runs that are compared to experiments or analytical solutions. As before, a run during this stage may take weeks or months to complete. Typically, during this stage the relevant time-consuming parts of the code are optimized for speed.

Once the basic code is shown to be useful, the code may be adopted for **production** runs. This implies extensive benchmarking for relevant applications, quality assurance, bookkeeping of versions, manuals, seminars, etc. For commercial software, this phase is also referred to as **industrialization** of a code. It is typically driven by highly specialized projects that qualify the code for a particular class of simulations, e.g. air conditioning or external aerodynamics of cars.

If the code is successful and can provide a simulation capability not offered by competitors, the fourth phase, i.e. **widespread** use and acceptance, will follow naturally. An important shift is observed: the 'missionary phase' (why do we need this capability ?) suddenly transitions into a 'business as usual phase' (how could we ever design anything without this capability ?). The code becomes an indispensable tool in industrial research, development, design and analysis. It forms part of the widely accepted body of 'best practices' and is regarded as commercial off the shelf (COTS) technology.

One can envision a fifth phase, where the code is **embedded** into a larger module, e.g. a control device that 'calculates on the fly' based on measurement input. The technology embodied by the code has then become part of the common knowledge and the source is freely available.

The time from conception to widespread use can span more than two decades. During this time, computing power will have increased by a factor of 1:10,000 (Moore's law). Moreover, during a decade, algorithmic advances and better coding will improve performance by at least another factor of 1:10. Let us consider the role of parallel computing in light of these advances.

During the **demonstration** stage, runs may take weeks or months to complete on the largest machine available at the time. This places heavy emphasis on parallelization. Given that optimal performance is key, and massive parallelism seems the only possible way of solving the problem, **distributed memory parallelism** on $O(10^3)$ processors is perhaps the only possible choice. The figure of $O(10^3)$ processors is bourne out of experience: even as high-end users with sometimes highly visible projects we have not been able to obtain a larger number of processors with consistent availability in the last two decades, and do not envision any improvement in this direction. The main reason is the usage dynamics of large-scale computers: once online, a large audience requests time on it, thereby limiting the maximum number of processors available on a regular basis for production runs.

Once the code reaches **production** status, a shift in emphasis becomes apparent. More and more 'options' are demanded, and these have to be implemented in a timely manner. Another five years have passed and by this time, processors have become faster (and memory has increased) by a further factor of 1:10, implying that the same run that used to take $O(10^3)$ processors can now be run on $O(10^2)$ processors. Given this relatively small number of processors, and the time constraints for new options/variants, **shared memory parallelism** becomes the most attractive option.

The **widespread** acceptance of a successful code will only accentuate the emphasis on quick implementation of options and user-specific demands. Widespread acceptance also implies that the code will no longer run exclusively on supercomputers, but will migrate to high end servers and ultimately personal computers. The code has now been in production for at least 5 years, implying that computing power has increased again by another factor of 1:10. The same run that used to take $O(10^3)$ processors in the demonstration stage can now be run using $O(10^1)$ processors, and soon will be within reach of $O(1)$ processors. Given that user-specific demands dominate this stage, and that the developers are now catering to a large user base working mostly on low-end machines, **parallelization diminishes in importance**, even to the point of completely disappearing as an issue. As parallelization implies extra time devoted to coding, hindering fast code development, it may be removed from consideration at this stage.

One could consider a 5th phase, 20 years into the life of the code. The code has become an indispensable commodity tool in the design and analysis process, and is run thousands of times per day. Each of these runs is part of a stochastic analysis or optimization loop, and is performed on a commodity chip-based, uni-processor machine. Moore's law has effectively **removed parallelism** from the code.

Figure 3 summarizes the life cycle of typical scientific computing codes.

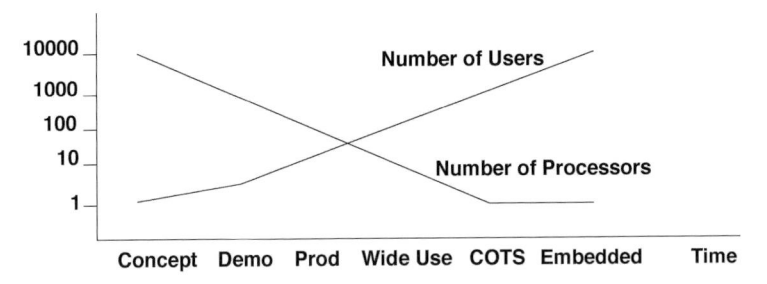

Figure 3: Life Cycle of Scientific Computing Codes

3. EXAMPLES

Let us show three examples where the life cycle of codes described above has become apparent.

3.1 Crash Simulations: The first example considers crash simulation codes. Worldwide, the car industry produces more than 300 new models or variations thereof a year. For each of these, a crash safety demonstration is required. Any prototype test is expensive, and this has given rise to the so-called 'crash analysis market'.

Figure 4: Typical Crash Simulations

A crash simulation will require approximately $O(10^5 - 10^6)$ elements, many material models, contact, and numerous specific options such as spot welds. Crash codes grew out of the DYNA3D impact-code legacy of Lawrence Livermore National Labs [11], [8] The first demonstration/feasibility studies took place in the early 1980's. At that time, it took the fastest machine (CRAY-XMP) a night to compute a car crash. Several commercial codes were soon adopted in the car industry [12] and found their way into a large segment of consumer products (for droptests) by the mid 1990's. At present (mid-2005), crash simulations can be accomplished on a PC in a matter of hours, and are carried out by the hundreds on a daily basis in the car industry for stochastic analysis on PC clusters.

Figure 4 shows an example.

2. <u>External Missile Aerodynamics</u>: The second example considers aerodynamic force and moment predictions for missiles. Worldwide, approximately 100 new missiles or variations thereof appear every year. In order to assess their flight characteristics, the complete force and moment data for the expected flight envelope must be obtaied. Simulations of this type based on the Euler equations require approximately $O(10^6 - 10^7)$ elements, special limiters for supersonic flows, semi-empirical estimation of viscous effects, and numerous specific options such as transpiration boundary conditions, modeling of control surfaces, etc. The first demonstration/ feasibility studies took place in the early 1980's. At that time, it took the fastest machine (CRAY-XMP) a night to compute such flows. The codes used were based on structured grids [10] as the available memory was small compared to the number of gridpoints. The increase of memory, together with the development of codes based on unstructured [25], [23] or adaptive Cartesian grids [26], [1] as well as faster, more robust solvers [24] allowed for a high degree of automation. Presently, external missile aerodynamics can be accomplished on a PC in less than an hour, and runs are carried out daily by the thousands for envelope scoping and simulator input on PC clusters [32]. Figure 5 shows an example.

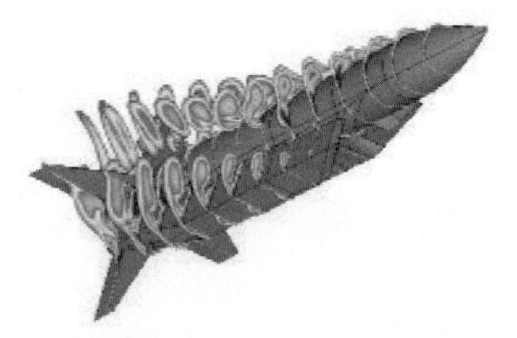

Figure 5: External Missile Aerodynamics

3. <u>Blast Simulations</u>: The third example considers pressure loading predictions for blasts. Simulations of this type based on the Euler equations require approximately $O(10^6 - 10^8)$ elements, special limiters for transient shocks, and numerous specific options such as links to damage prediction post-processors. The first demonstration/ feasibility studies took place in the early 1990's [2], [3], [4], [5]. At that time, it took the fastest machine (CRAY-C90 with special memory) several days to compute such flows. The increase of processing power via shared memory machines during the last decade has allowed for a considerable increase in problem size, physical realism via coupled CFD/CSD runs [17], [7], and a high degree of automation. Presently, blast predictions with $O(2 \cdot 10^6)$ elements can be carried out on a PC in a matter of hours [22], and runs are carried out daily by the hundreds for maximum possible damage assessment on networks of PCs. Figure 6 shows the results of such a prediction based on genetic algorithms for a typical city environment [33]. Each dot

represents an end-to-end run (grid generation of approximately 1.5 Mtets, blast simulation with advanced CFD solver, damage evaluation), which takes approximately 4 hours on a high-end PC. This run was done on a network of PCs and is typical of the migration of high-end applications to PCs due to Moore's law.

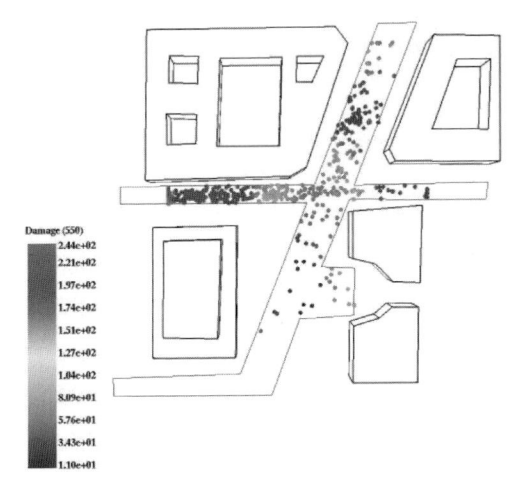

Figure 6: Maximum Possible Damage Assessment for Inner City

4. DISCUSSION

The statement that parallel computing diminishes in importance as codes mature is predicated on two assumptions:

- The doubling of computing power every 18 months will continue;
- The total number of operations required to solve the class of problems the code was designed for has an asymptotic (finite) value.

The second assumption may seem the most difficult to accept. After all, a natural side-effect of increased computing power has been the increase in problem size (grid points, material models, time of integration, etc.). However, for any class of problem there is an intrinsic limit for the problem size, given by the physical approximation employed. Beyond a certain point, the physical approximation does not yield any more information. Therefore, we may have to accept that parallel computing diminishes in importance as a code matures.

This last conclusion does not in any way diminish the overall significance of parallel computing. Parallel computing is an **enabling** technology of vital importance for the development of new high-end applications. Without it, innovation would seriously suffer. On the other hand, without Moore's law many new code developments would appear as unjustified. If computing time does not decrease in the future, the range of applications would soon be exhausted.

REFERENCES

1. M.J. Aftosmis, M.J. Berger and G. Adomavicius - A Parallel Multilevel Method for Adaptively Refined Cartesian Grids with Embedded Boundaries; *AIAA*-00-0808 (2000).
2. J.D. Baum and R. Löhner - Numerical Simulation of Shock Interaction with a Modern Main Battlefield Tank; *AIAA*-91-1666 (1991).
3. J.D. Baum. H. Luo and R. Löhner - Numerical Simulation of a Blast Inside a Boeing 747; *AIAA*-93-3091 (1993).
4. J.D. Baum, H. Luo and R. Löhner - Numerical Simulation of Blast in the World Trade Center; *AIAA*-95-0085 (1995).
5. J.D. Baum, H. Luo, R. Löhner, C. Yang, D. Pelessone and C. Charman - A Coupled Fluid/Structure Modeling of Shock Interaction with a Truck; *AIAA*-96-0795 (1996).
6. J.D. Baum, E. Mestreau, H. Luo, D. Sharov, J. Fragola and R. Löhner - CFD Applications in Support of the Space Shuttle Risk Assessment; *JANNAF* (2000).
7. J.D. Baum, E. Mestreau, H. Luo, R. Löhner, D. Pelessone and Ch. Charman - Modeling Structural Response to Blast Loading Using a Coupled CFD/CSD Methodology; *Proc. Des. An. Prot. Struct. Impact/ Impulsive/ Shock Loads (DAPSIL)*, Tokyo, Japan, December (2003).
8. T. Belytchko and T.J.R. Hughes (eds.) - *Computer Methods for Transient Problems*, North Holland, Dordrecht (1983).
9. F. Camelli, R. Löhner, W.C. Sandberg and R. Ramamurti - VLES Study of Ship Stack Gas Dynamics; *AIAA*-04-0072 (2004).
10. S.R. Chakravarthy and K.Y. Szema - Euler Solver for Three-Dimensional Supersonic Flows with Subsonic Pockets; *J. Aircraft* 24, 2, 73-83 (1987).
11. G.L. Goudreau and J.O. Hallquist - Recent Developments in Large-Scale Finite Element Lagrangean Hydrocode Technology; *Comp. Meth. Appl. Mech. Eng.* 33, 725-757 (1982).
12. E. Haug, H. Charlier, J. Clinckemaillie, E. DiPasquale, O. Fort, D. Lasry, G. Milcent, X. Ni, A.K. Pickett and R. Hoffmann - Recent Trends and Developments of Crashworthiness Simulation Methodologies and their Integration into the Industrial Vehicle Design Cycle; *Proc. Third European Cars/Trucks Simulation Symposium (ASIMUTH)*, Oct. 28-30 (1991).
13. J. Katz, H. Luo, E. Mestreau, J.D. Baum and R. Löhner - Viscous-Flow Solution of an Open-Wheel Race Car; SAE 983041 (1998).
14. R. Löhner, K. Morgan and O.C. Zienkiewicz - Effective Programming of Finite Element Methods for CFD on Supercomputers; pp.117-125 in The Efficient Use of Vector Computers with Emphasis on CFD (W. Schönauer and W. Gentzsch eds.), Vieweg Notes on Numerical Fluid Mechanics, Vol 9, Vieweg Verlag (1985).
15. R. Löhner - Some Useful Renumbering Strategies for Unstructured Grids; *Int. J. Num. Meth. Eng.* 36, 3259-3270 (1993).
16. R. Löhner, A. Shostko and R. Ramamurti - Parallel Unstructured Grid Generation and Implicit Flow Solvers; paper presented at the *Parallel CFD'94 Conf.* , Kyoto, Japan, May (1994).
17. R. Löhner, C. Yang, J. Cebral, J.D. Baum, H. Luo, D. Pelessone and C. Charman

- Fluid-Structure Interaction Using a Loose Coupling Algorithm and Adaptive Unstructured Grids; *AIAA*-95-2259 [Invited] (1995).
18. R. Löhner - Renumbering Strategies for Unstructured-Grid Solvers Operating on Shared-Memory, Cache-Based Parallel Machines; *Comp. Meth. Appl. Mech. Eng.* 163, 95-109 (1998).
19. R. Löhner, C. Yang and E. Oñate - Viscous Free Surface Hydrodynamics Using Unstructured Grids; *Proc. 22nd Symp. Naval Hydrodynamics*, Washington, D.C., August (1998).
20. R. Löhner - A Parallel Advancing Front Grid Generation Scheme; *Int. J. Num. Meth. Eng.* 51, 663-678 (2001).
21. R. Löhner and M. Galle - Minimization of Indirect Addressing for Edge-Based Field Solvers; *Comm. Num. Meth. Eng.* 18, 335-343 (2002).
22. R. Löhner, J.D. Baum and D. Rice - Comparison of Coarse and Fine Mesh 3-D Euler Predictions for Blast Loads on Generic Building Configurations; *Proc. MABS-18 Conf.*, Bad Reichenhall, Germany, September (2004).
23. H. Luo, J.D. Baum and R. Löhner - Edge-Based Finite Element Scheme for the Euler Equations; *AIAA J.* 32, 6, 1183-1190 (1994).
24. H. Luo, J.D. Baum and R. Löhner - A Fast, Matrix-Free Implicit Method for Compressible Flows on Unstructured Grids; *J. Comp. Phys.* 146, 664-690 (1998).
25. D.J. Mavriplis - Three-Dimensional Unstructured Multigrid for the Euler Equations; *AIAA*-91-1549-CP (1991).
26. J.E. Melton, M.J. Berger and M.J. Aftosmis - 3-D Applications of a Cartesian Grid Euler Method; *AIAA*-93-0853-CP (1993).
27. G.E. Moore - Cramming More Components Onto Integrated Circuits; *Electronics* 38, 8 (1965).
28. G.E. Moore - A Pioneer Looks Back at Semiconductors; *IEEE Design & Test of Computers* 16, 2, 8-14 (1999).
29. G.E. Moore - No Exponential is Forever...but We Can Delay 'Forever'; paper presented at the *Int. Solid State Circuits Conference (ISSCC)*, February 10 (2003).
30. D. Sharov, H. Luo, J.D. Baum and R. Löhner - Time-Accurate Implicit ALE Algorithm for Shared-Memory Parallel Computers; pp. 387-392 in *Proc. First Int. Conf. on CFD* (N. Satofuka ed., Springer Verlag), Kyoto, Japan, July 10-14 (2000).
31. R. Ramamurti and R. Löhner - A Parallel Implicit Incompressible Flow Solver Using Unstructured Meshes; *Computers and Fluids* 25, 2, 119-132 (1996).
32. M.A. Robinson - Modifying an Unstructured Cartesian Grid Code to Analyze Offensive Missiles; *Rep. Pan 31941-00*, US Army Missile and Space Int. Comm., Huntsville, AL (2002).
33. F. Togashi, R. Löhner, J.D. Baum, H. Luo and S. Jeong - Comparison of Search Algorithms for Assessing Airblast Effects; *AIAA*-05-4985 (2005).

"Letting physicists be physicists," and other goals of scalable solver research

David E. Keyes
Columbia University
Department of Applied Physics and Applied Mathematics, MC 4701
New York, NY 10027

INTRODUCTION AND MOTIVATION

Computational fluid dynamicists and other scientists and engineers (henceforth collectively "physicists") who make extensive use of large-scale simulation inherit the accumulation of centuries or decades of developments in mathematical models, numerical algorithms, computer architecture, and software engineering, whose recent fusion can be identified with the beginning of a new era of simulation. As a symbolic date for the modern beginning of the mathematization of natural law, few would likely quibble with the publication of Newton's *Principia* in 1687 [1]. Similarly, we can associate with Von Neumann the modern thrust to approximate the solutions of mathematical conservation laws, especially nonlinear conservation laws, on digital computers. We date the beginning of this enterprise at 1947, when Von Neumann and Goldstine published the famous treatise that demonstrated that Gaussian elimination could be expected to preserve a certain level of accuracy in the presence of floating point roundoff error [2], a fact that we take for granted today as we commit trillions of floating point errors per second in our CFD simulations, but which is far from trivial to establish. The contemporary drive to improve the speed and storage capacity of computer hardware, which allows us to resolve more of the natural scales of fluid dynamical and other phenomena, we can associate with the genius of Cray, whose first vector machine was delivered in 1976 [3]. The confluence of these three streams of scientific and engineering endeavor alone three decades ago powered an incredibly inventive and fruitful period of computational fluid dynamics, which phrase seems first to have appeared in 1971 [4]. However, CFD would be destined to remain the province of a relatively small number of abundantly talented visionaries and not a tool in the box of practicing engineers, were it not for the emergence of software engineering.

The subject of this chapter is contemporary developments in scalable solver software and their transforming effect by enabling "physicists" to function at the frontier of algorithmic technology while concentrating primarily on their application – that is, refining their model and understanding and employing its results – rather than, for instance, debugging split-phase communication transactions. This is possible only through a division of labor, or, as a computer scientist would say, a "separation of concerns" [5], since the sophistication of solution algorithms and their implementation on massively parallel distributed hierarchical memory computers has outstripped the understanding even of any individual mathematician and computer scientist who builds a career on this task alone. Modern scientific software engineering means many things to many people, but in its cluster of distinguishing characteristics are: abstract interface definition, object orientation (including encapsulation and polymorphism), componentization, self-description, self-error-checking, self-performance-monitoring, and design for performance, portability, reusability and extensibility. As an exemplar of scientific software engineering, we mention PETSc, the *Portable, Extensible Toolkit for Scientific Computing* [6], first released in May 1992, and employed in a variety of community and commercial CFD codes, among many other applications.

Historically, the challenge for software designers, who were once primarily the physicists who understood the entire problem in a vertically integrated way, was to *increase the functionality and capability for a small number of users expert in the domain of the software*. A new challenge for software designers is to *increase the ease of use (including both correctness and efficiency) for a large number of users expert in something else*. A simple but crucial design principle that has evolved to accommodate these challenges is that of multiple levels.

At the top level, which is seen by all users, is an abstract interface that features the language of the application domain, itself, and hides implementation details and call options with conservative parameter defaults. An example of this within the realm of linear solvers is MATLAB's **x=A\b**, which requires the user to know nothing about how the matrix **A** is represented or indeed about any of its solution-salient properties (such as symmetry, definiteness, sparsity, etc.) or how the mathematical objects, vectors and matrices, are distributed across storage. The top level description is intended to offer ease of use, correctness, and robustness. However, it might not be particularly efficient to solve problems through the top-layer interface, and the user does not at this level exploit significant knowledge about the system being solved.

The middle level is intended to be accessed by experienced users. Through it is offered a rich collection of state-of-the-art methods and specialized data structures beyond the default(s). The parameters of these methods are exposed upon demand and the methods are typically highly configurable. Accessing the software at this level gives the user control of algorithmic efficiency (complexity), and enables her to extend the software by registering new methods and data structures that interoperate with those provided. At this level, the user may also expose built-in performance and resource monitors, in the pursuit of comprehending performance. For example, in the realm of linear solvers, a user might elect a domain-decomposed multigrid-preconditioned Krylov iterative method. All associated parameters would have robust defaults, but he could specify the type of Krylov accelerator, its termination criteria, and whether any useful by-products, such as spectral estimates of the preconditioned operator are desired. For the multigrid preconditioner, typical user-controllable parameters include the number of successively coarsened grids, the type of smoother to be used at each level and the number of smoothing sweeps, the prolongation and restriction operators, and the means of forming coarsened operators from their predecessors. For the (domain-decomposed) smoothers, one may have many further parameters. Flexible composability is possible; e.g., the smoother could be another Krylov method. Extensibility is possible, e.g., the coarsening method may be something supplied by the user, or something that comes from an externally linked third-party package.

The bottom level is for implementers. It provides support for a variety of execution environments. It is accessed for portability and implementation efficiency. For instance, different blocking parameters may be appropriate for machines with different cache structures. Different subdomain partitions and message granularities may be appropriate to exploit locally shared memories, if present.

Though this introduction skims over issues of philosophical depth to computer scientists while simultaneously dredging up details that physicists may consider unsightly, its purpose is to convey that the era of high-performance simulation for "everyday" scientists and engineers depends as critically upon the emergence software engineering as it does on mathematical modeling, numerical analysis, and computer architecture. Solvers make up just one class of the enabling technologies required for today's large-scale simulation portfolio. Sixteen chapters describing different enabling technologies for large-scale simulation are described in the latter sections of the two-volume "SCaLeS" report [7]. We focus on solvers because they are relatively mature compared to some of the other technologies and need for them is ubiquitous.

DEMAND FOR SCALABLE SOLVERS

The principal demand for scalable solvers arises from multiscale applications. Here we give operational definitions of "multiscale" in space and time, see why scalable solvers are needed, and mention some of their recent successes.

Multiple spatial scales exist when there are hard to resolve features, such as interfaces, fronts, and layers, typically thin and clustered around a surface of co-dimension one relative to the computational domain, whose width is small relative to the length of the domain. An example would be a thin sheet embedded in three-dimensional space, though features of greater co-dimension may also exist, e.g., a thin tube or a small "point" source. The computational domain is usually taken as small as possible subject to the constraint that one must be able to specify boundary conditions that do not interfere with the phenomena of interest. Boundary layers, combustion fronts, hydrodynamic shocks, current sheets, cracks, and material discontinuities are typical multiscale phenomena in this sense. Multiple spatial scales are also present in isotropic phenomena, such as homogeneous turbulence or scattering, when the wavelength is small relative to the target. Multiple spatial scales demand fine mesh spacing relative to the domain length, either adaptively or uniformly.

Multiple temporal scales exist when there are fast waves in a system in which only slower phenomena, such as material convection, diffusion, or other waves, are of interest. The physicist must isolate the dynamics of interest from the multitude of dynamics present in the system, and model the rest of the system in a way that permits discretization over a computably modest range of scales. She may rely on physical assumptions or mathematical closures. Often, assumptions of quasi-equilibrium or filtration of (presumed energetically unimportant) fast modes are invoked since it is infeasible for an explicit integrator to resolve the fastest transients while respecting Courant-like stability limits. For example, fast reactions may be assumed to be in instantaneous equilibrium relative to slow. Fast waves (e.g., acoustic waves in aerodynamics, surface gravity waves in physical oceanography, and magnetosonic waves in plasma dynamics) may be projected out. The dynamics is, in practice, often reduced to a computable manifold by enforcing algebraic or elliptic constraints. The discrete algebraic systems that must be solved to enforce these constraints at every iteration to keep the solution on the manifold are typically ill-conditioned (due to the spatial multiscale nature of the problem) and of extremely high algebraic dimension (millions or billions of degrees of freedom on terascale hardware).

To enforce the constraints, scalable implicit solvers must be called at every step. In this context, "scalable" refers both to complexity of operations and storage, and to parallel implementation on many processors. Classical implicit solvers, such as Gaussian elimination and Krylov iteration are not generally, in and of themselves, sufficient for this task. Both have operation count complexities that grow superlinearly in the discrete dimension of the problem. Depending upon the problem and the particular method, both may have storage complexity that grows superlinearly as well. This implies that even if one adds processors in linear proportion to the discrete problem size (in so-called "weak" or "memory constrained scaling"), the execution time grows without bound. Only hierarchical algorithms, such as multigrid, multipole, and FFT, are algorithmically scalable in this context – their operation counts and storage typically grow only as fast as $O(N \log N)$. FFTs are widely exploited in Poisson projections for structured problems. However, they are being replaced by multigrid, which is less fragile, in many contexts. In particular, algebraic multigrid methods extend $O(N \log N)$ optimal complexity to many unstructured problems. Fortunately, multigrid methods often have enough parameters to be tuned to be nearly scalable in a parallel sense as well. Since multigrid methods typically use other direct and iterative linear methods as components, a well-stocked scalable solver toolkit has a wide range of composable solvers available.

Recent Gordon Bell Prizes in the "special" category [8,9,10] illustrate that implicit solvers have scaled to massive degrees of parallelism once deemed highly challenging for unstructured problems. The 1999 Gordon Bell "special" prize, which was introduced that year (alongside the traditional "peak" and "price performance" prizes and the intermittently awarded "compiler-based parallelism" prize), went for a steady-state CFD simulation: external Euler flow on a wing-conforming tetrahedral mesh containing 11M degrees of freedom (see Figure 1). Nearly 0.25 Teraflop/s were sustained on over 3000 dual processors of the Intel ASCI Red machine on a version of a code employed in analysis and design missions by NASA. The 2003 Gordon Bell "special" prize was awarded for a transient inverse problem simulation: seismic inversion of the modulus field of the Los Angeles basin on a tetrahedral, adaptively refined mesh containing 17M degrees of freedom. Approximately 1 Tflop/s was sustained on over 2000 processors of the HP Lemieux machine on a code employed in NSF-funded seismic research. The 2004 Gordon Bell "special" prize was given for an incrementally loaded large-deformation analysis of human trabecular bone on a porous domain hexahedrally meshed with nearly 0.5B degrees of freedom. Approximately 0.5 Tflop/s were sustained on over 4000 processors of the IBM ASCI White machine on a code employed in NIH-funded bone research. Table 1 summarizes these three submissions.

Each of these prize-winning simulations employed PETSc [6], each in a different way. The 2003 simulation employed PETSc at a low level of functionality: as an application programmer interface to MPI-distributed data structures. The 2004 simulation employed PETSc at an intermediate level of functionality: Adams introduced his own version of smoothed-aggregation algebraic multigrid as a preconditioner, but he used PETSc for distributed data structures and Krylov iterative infrastructure. The 1999 simulation, which was joined by PETSc developers and stimulated the extension of PETSc to unstructured distributed mesh and matrix objects, used PETSc for all aspects, outward beyond linear solvers to its Jacobian-free Newton-Krylov iteration and pseudo-transient continuation timestepping modules.

Solver software is among the least exciting of technical subjects for many users. *Ax=b* has been long-regarded as a simple blackbox presenting a relatively obvious interface. Nevertheless, this unpretentious module is asymptotically the bottleneck to massively parallel scaling in any implicit or semi-implicit CFD code, and in many others. To see why this is the case, consider that most physics applications discretized on a mesh of N degrees of freedom require $O(N)$ work for a basic work step, e.g., evaluating the residual of the conservation laws over the mesh for an implicit problem, or evaluating the right-hand side of a transient problem for the instantaneous rate of change of the state variables. The constant of the $O(N)$ may be large, to accommodate a variety of physical models and constitutive laws, but weak memory-constrained scaling of the problem to large sizes is routine. If carried out by domain decomposition, and if the operator is local, excellent communication-to-computation ratios follow the surface-to-volume scaling of the subdomains [11]. On the other hand, conventional solvers suffer from an $O(N^p)$ scaling for some $p>1$. Suppose $p=1.5$, which would be typical of diagonally preconditioned conjugate gradients for a 2D problem. If the computation is well balanced on 64 processors, with the physics phase and the solver phase each costing 50% of the execution time, and if the problem is weakly scaled up to 64K processors, preserving all constants and exponents (Figure 2), the solver will cost 97% of the execution time! We note that many contemporary users will be scaling in the coming decade from 64 nodes in a departmental cluster to 64K nodes, which are available already on the IBM BG/L machine at LLNL, as of 2005 [12]. We also note that many contemporary implicit and semi-implicit large-scale simulations have reached the point in which 90% of the execution time is spent in *Ax=b* on problem sizes that require a number of processors far less than 64K. These applications require an optimal $O(N \log N)$ scalable solver now.

Progress in arriving at today's optimal solvers from the starting point of the highly nonoptimal Gaussian elimination, first considered for floating point hardware in [2], is charted in Table 2. Observe the reduction in operation complexity from $O(n^7)$ to $O(n^3)$ and the reduction in storage complexity from $O(n^5)$ to $O(n^3)$ in a span of approximately 36 years, based on publication dates of the corresponding papers. For a cube modestly resolved with $n=64$, this is reduction of floating point operators of $(64)^4 \approx 16M$. Given that there are approximately 31M seconds in one year, this is like reducing a 6-month calculation to one second, on a constant computer. Of course, computers are not constant. Over the 36-year period of this algorithmic inventiveness, Moore's Law [13], which states that transistor density doubles every 18 months, also dictates a factor of $2^{24} \approx 16M$ in transistor density for integrated circuits, which metric translates (imperfectly) into performance. Therefore, algorithmic improvements across the abstract **Ax=b** interface have contributed an equal factor as hardware improvements over a long period (Figure 3). Of course, the scaling laws are different and coincide only for $n=64$. For n taking on larger values in multiscale problems, the improvement from algorithms is much more impressive. This recalls the statement of Phillipe Tointe, "I would much rather have today's algorithms on yesterday's computers than vice versa."

DETAILED REQUIREMENTS SPECIFICATION

A requirements specification is part of the software engineering process, whereby developers identify requirements a user demands of the product. Though often simple, it is a vital part of the process, since failing to identify requirements makes it unlikely for the finished software to meet the users' needs.

In this section, we spell out a number of ideals concerning solver software. Not all of them are perfectly met in any solver software of which we are aware, and not all of them have been explicitly requested by most computational physicists. However, as simulation begins to fulfill its potential at the high end and penetrate the scientific and engineering community at the low end, we believe that these truths will become self-evident.

Application developers wish to accomplish certain abstract "blackbox" mathematical (and other) tasks without having to "make bets" about particular algorithms or (more importantly) the specialized data structures that algorithms require. In general, they prefer to stay agnostic about particular methods and implementations. They also desire portability, from personal laptops, to departmental clusters, to national supercomputing resources. They require each of the following by the time they get to production use, but they need them in order, or their journey will terminate at the first one missing: usability and robustness, portability, and algorithmic and implementation efficiency.

The author's Terascale Optimal PDE Simulations (TOPS) scalable solvers project [14] was created to serve such users, and intends to equip its users to:
1) Understand the full range of algorithmic options available, with trade-offs (e.g., memory versus time, computation versus communication, iteration loop work versus number of iterations),
2) Try all reasonable options "easily" (e.g., at runtime, without recoding of extensive recompilation),
3) Know how their solvers are performing, with access to detailed profiling information,
4) Intelligently drive solver research, as measured by the ultimate metric of publishing joint research papers with solver developers,
5) Simulate truly new physics, free of artifactual solver limitations, including finer meshes, complex coupling, full nonlinearity, and
6) Go beyond "one-off" solutions to the ultimate reason that users compute, including probing the stability and sensitivity of solutions, and performing PDE-constrained optimization.

Regarding applications, TOPS holds that the solution of a PDE is rarely an end in itself. The actual goal is often characterization of a response surface, or design, or control. Solving the PDE is just one forward map (from inputs to outputs) in this process. Together with analysis, sensitivities and stability are often desired. Solver toolkits for PDEs should support these follow-on desires, which are just other instances of solvers. For instance, stability may be a matter of doing an eigenanalysis of a linear perturbation to the base nonlinear solution. Sensitivity may be a matter of solving an augmented set of equations for the principal unknowns and their partial derivatives with respect to uncertain parameters. No general purpose PDE solver can anticipate all needs. Extensibility is important. A solver software library improves with user feedback and user contributions.

Regarding users, TOPS acknowledges that solvers are used by people of varying numerical sophistication. Some expect MATLAB-like defaults; others wish complete control. Hence, the multilevel design philosophy described above. TOPS is motivated by the assumption that user demand for resolution in multiscale PDE simulations is unquenchable. Relieving resolution requirements with superior multiscale models (e.g., turbulence closures, homogenization) only defers the demand for resolution to the next level of modeling. Furthermore, validating such closures requires simulations of the first-principles model at very high resolution. Therefore, algorithmic optimality and implementation scalability are both critical.

Regarding legacy code, it is not retired easily, and TOPS holds that it need not be. Porting a CFD code to a scalable parallel framework through domain decomposition does not mean starting from scratch. High-value physics routines associated with evaluating the right-hand side of an integration routine or the residual of an implicit problem can be substantially preserved, in their original language. The overhead of partitioning, reordering, and mapping onto distributed data structures (that a solver infrastructure, such as PETSc, may provide) adds code but little runtime overhead. For ease of use, solver libraries should contain code samples exemplifying this separation of concerns. Whereas legacy physics routines may be substantially preserved, legacy solvers generally must be replaced. Though the best available at the time the code was assembled, with time they likely impose limits to resolution, accuracy, and generality of the simulation overall. Replacing the solver with something more robust and efficient may "solve" several of these issues. However, segments of the old solver may have value as part of the preconditioner for a modern preconditioned Newton-Krylov solver. Scope does not permit a launch into technical details, but see the review article [15] for examples of "physics-based preconditioning" that use routines previously regarded a solvers as preconditioners in an enlarged context.

Regarding solvers, TOPS understands that they may be employed many different ways over the life-cycle of a code. During development, robustness and verbose diagnostics are important. During production, however, solvers must be streamlined for performance and expensive robustness features must be deactivated in regimes in which the dangers against which they guard are unlikely. TOPS acknowledges further, that in the great "chain of being" solvers are subordinate to the code that calls them, and are often not the only library to be linked. The same solver may be called in multiple, even nested places, and may make callbacks. Solver threads must therefore not interfere with other component threads, including other active instances of themselves. This is a nontrivial requirement for general message-passing or shared-memory programming. The "communicator" feature of MPI, however, provides for such compositional complexity.

Regarding numerical software, TOPS understands that a continuous operator may appear in a discrete code in many different instances. Its use in residual evaluation and Jacobian estimation may be to different approximations. It may exist in the same approximation on each of a hierarchical sequence of grids. The majority of progress towards a highly resolved, high fidelity result is often achieved through cost-effective, parallel-efficient, low resolution, low fidelity stages. TOPS also understands that hardware may change many times over the life-cycle of a code. Portability is critical, and performance portability is important after a new architecture achieves critical mass.

SOLVER TOOLCHAIN

A central concept in the TOPS project is the solver toolchain, a group of inter-related tools presenting functional abstractions and leveraging common distributed data structure implementations, which ultimately inherit the structure of the mesh on which they are based. The solver tool chain is part of a longer toolchain that reaches upwards through discretization to the application and downwards through performance tools to the hardware.

Within the solver part of the toolchain, the solution of square nonsingular linear systems is the base. Literally dozens of methods – most of them iterative – through combinatorial composition of accelerators and preconditioners, are provided to fulfill this function. Linear solvers are, in turn, called by eigensolvers, mainly during the shift-and-invert phase of recovering eigencomponents other than that of the maximum eigenvalue.

Similarly, linear solvers are called by nonlinear solvers, during each iteration of Newton-like methods. Nonlinear solvers, in turn, are called by stiff implicit integrators on each time step. Stiff integrators are also called by optimizers to solve inverse problems or compute sensitivities. The TOPS toolchain is depicted in Figure 4. It does not include every type of solver that physicists may want; for instance, there is no development of least-squares methods in TOPS. This constellation of software was assembled to respond to particular needs of users with large-scale mesh-based finite-discretization PDE simulations.

ILLUSTRATIONS FROM THE TOPS SOLVER PROJECT

To illustrate this chapter with some "stories from the trenches" from the early years of TOPS project collaborations with the Scientific Discovery through Advance Computing (SciDAC) initiative [16], we consider three different collaborations through which TOPS has engaged magnetohydrodynamicists modeling magnetically confined fusion plasmas.

Magnetohydrodynamics models the dynamics of a plasma as a continuous conducting fluid, coupled with the continuous electromagnetic field equations of Maxwell. The fluid motion induces currents, which in turn produce Lorentz forces on the fluid, leading to nonlinear coupling via cross products of the magnetic field and the fluid velocity between the Navier Stokes and Maxwell equations. Many regimes are relevant to the modeling of magnetically confined fusion plasmas that arise in tokamaks and stellarators, ranging from ideal MHD, in which viscosity and magnetic resistivity are neglected and all plasma components are in local thermodynamic equilibrium, to more general situations, including resistive and extended MHD, in which the Ohm's law term includes multifluid effects.

Nimrod [17] is a toroidal geometry MHD code discretized via Fourier expansion in the periodic toroidal direction and high-order finite elements in each identically meshed poloidal crossplane. At each time step, several complex nonsymmetric linear systems must be solved in each crossplane, with, for typical contemporary resolutions, 10K to 100K unknowns. These sparse linear systems were solved with diagonally preconditioned Krylov iteration, consuming approximately 90% of the execution time. High-order discretizations lead to linear systems that typically lack diagonally dominance and convergence was slow and not very robust. Nimrod presented straightforward symptoms and a straightforward *Ax=b* interface. A natural trial was to replace the Krylov solver with a parallel direct solver, SuperLU_DIST [18], a supernodal Gaussian elimination code supported by the TOPS project. The concurrency available in SuperLU is ultimately limited, as it is in any direct solver, by sparsity, and parallel orchestration is disturbed by the dynamic pivoting required in Gaussian elimination on nonsymmetric systems, which gives rise to sophisticated heuristics for pre-factorization scaling and ordering, as well as iterative refinement. Despite the challenges of parallel direct factorization and backsolving, SuperLU has achieved parallel speedups in the hundreds for finite element-type problems with algebraic dimension in the millions, with hundreds of millions of nonzeros in the system matrix. Incorporated into Nimrod to replace the iterative linear solver with no other reformulation, SuperLU achieved solver time improvements of up to two orders of magnitude in the individual crossplane problems, leading to reduced overall running times of up to a factor of 5 relative to the slowly converging Krylov iteration [19]. This is in line with expectations from Amdahl's Law [20] for improving the solver execution and shifting the bottleneck to the remaining nonsolver portions of the code. It is well-known that direct methods are competitive with iterative methods on sparse, two-dimensional problems; therefore simulation codes with heavy requirements for such kernels should offer run-time swapable linear solvers including members of both categories.

M3D [21] is an MHD code with greater geometrical generality than Nimrod in that toroidal symmetry is not presumed in the domain or in the poloidal crossplane discretization. As a result, M3D can model configurations such as the experimental stellerator geometry sometimes employed to control the stability of fusion plasmas. Fourier decomposition does not generally apply in the toroidal direction, so M3D employs finite differences in this direction, and finite elements on the triangulated poloidal crossplane meshes, either low-order C0 elements in the standard release or higher-order C1 elements. To achieve parallelism, M3D employs domain decomposition in both toroidal and poloidal directions. The velocity and magnetic fields in M3D are expressed via a Hodge decomposition, through scalar potentials and streamfunctions. As in Nimrod, there are many linear solves per crossplane per timestep, in which Laplacian-like or Helmholtz-like operators are inverted to advance the potentials and streamfunctions. M3D employs PETSc [6] for these linear solves, and at the outset of the TOPS project, it relied on one-level additive Schwarz preconditioned GMRES iteration for each such system, with incomplete factorization being employed within each subdomain of each crossplane. One-level additive Schwarz is capable of achieving optimal conditioning for diagonally dominant Helmholtz problems, but is known to degrade in condition number and therefore number of iterations to convergence for pure Laplace problems. Moreover, some of the Poisson problems in each M3D crossplane at each timestep are of Neumann type, so there is a nontrivial nullspace. Presented with the symptom of nonscalable linear solves, requiring 70-90% of the execution time, depending upon the type of problem being run, TOPS collaborators assisted in several improvements. First, they noted that the only asymmetry in the system matrices came from the implementation of boundary conditions, which were modified to achieve symmetry. Second, they implemented a nullspace projection method for the Neumann Poisson problems. Third, they hooked a variety of algebraic multigrid solvers from Hypre [22] and Prometheus [10] into PETSc, to provide dynamic selection of linear solvers. Fourth, they selected three different iterative linear solver combinations, one for each of three different classes of Poisson problems, that produced the lowest running times on problem-architecture combinations typical of current M3D operations. The result of this collaboration, most of which was accomplished underneath the

PETSc interface, is an M3D that runs 4-5 times faster relative to a pre-collaboration baseline on a typical production problem. Greater savings ratios are expected as problems scale to the sizes required for first-principles simulations of the International Thermonuclear Experimental Reactor (ITER).

The preceding two illustrations of scalable solver enhancements to fluids codes, and collaborations in other science and engineering domains in the TOPS project, are at the *Ax=b* level. Potentially, there is much to be gained by a tighter coupling of solver research and CFD application development, as illustrated by a collaboration on a standard test case in MHD modeling known as the GEM challenge magnetic reconnection problem [23]. Magnetic reconnection refers to the breaking and reconnecting of oppositely directed magnetic field lines in a plasma, or a fluid of charged particles. In this process, the energy stored in the magnetic field is released into kinetic and thermal energy. The GEM problem is designed to bring out the requirement for adaptive meshing in the limit, but it can be run on a fixed mesh and provides a stiff nonlinear system of boundary value problems for the fluid momenta and the magnetic field. A standard case is two-dimensional with periodic boundary conditions – a *bona fide* idealized temporal multiscale problem. TOPS collaborators worked with a plasma physicist [24] to compare a nonlinearly implicit method with an explicit method on a fixed-mesh problem. The explicit code respected a CFL stability limit, and therefore ran at a fine timestep. The nonlinearly implicit code ran at a timestep up to two orders of magnitude higher than the CFL limit and showed no loss of accuracy relative to the explicit code. Even after surrendering more than an order of magnitude of this timestep size advantage due to the necessity of solving nonlinear systems on every timestep, the implicit method was approximately a factor of five faster per unit physical time simulated, "out of the box." Prospects exist for preconditionings that could substantially boost the implicit advantage. Similar opportunities for scalable implicit solvers exist throughout computational physics, from magnetohydrodynamics to aerodynamics, from geophysics to combustion.

FUTURE OF SOLVER SOFTWARE

Applied computational scientists and engineers should always be "on the lookout" for new and better solution algorithms. A lesson of 60 years of history of floating point algorithms is to be humble about present capabilities. Alongside the punctuated discovery of new solution algorithms, however, are other more predictable research thrusts. We mention five such thrusts, each of which complements in an important way the discovery of new solution algorithms: interface standardization, solver interoperability, vertical integration with other enabling technologies, automated architecture-adaptive performance optimization, and automated problem-adaptive performance optimization.

Interface Standardization. In the recent past, there have been several attempts to establish standards for invoking linear solvers. Examples include the Finite Element Interface, the Equation Solver Interface, Hypre's "conceptual interfaces," and PETSc's linear solver interface, which has become the *de facto* TOPS interface through the process of absorbing valuable features from all of the others. Fortunately, it is possible to support many independent interfaces over the same core algorithmic functionality, just as it possible to provide many independent solver libraries beneath a common interface. For ***Ax=b***, the TOPS Solver Component (TSC) [25] interface has the following object model: A *solver* applies the inverse action of an *operator* (typically representing the discretization of a differential operator) to a *vector* (typically representing field data) to produce another vector. The *layout* of the data that make up the operator and the vectors in memory and across processors need not be specified. However, since some users require it, TOPS has *viewers* that provide access to the layout and to the actual data. Available views include the classic linear algebra view (indices into $\mathbf{R}^n$ or $\mathbf{C}^n$), as well as several that preserve aspects of the spatial origin of vectors as fields: a structured grid view on "boxes" of data in a Cartesian index space, an unassembled finite element view, a hierarchically structured grid view, etc. The TSC compatible with the Common Component Architecture (CCA) definition for scientific code interfaces.

Solver interoperability. Because there is no solver that is universally best for all problems or all hardware configurations, users and developers like to mix and match solver components. TOPS has made it possible for many popular solvers to be called interoperably – in place of or as parts of each other. However, only a finite number of well published solver components can ever be included in this manner. TOPS also allows users to "register" custom solver components that support a minimal object model for callback. This feature is especially useful in allowing legacy code to become a component of a TOPS preconditioner.

Vertical Integration. The data structures required by solvers are often complex and memory intensive, and the information they contain (for instance, the number of distinct physical fields in a single mesh cell, or the neighbor list for an unstructured grid entity) is potentially valuable in other computational phases, such as mesh refinement, discretization adaptivity, solution error estimation, and visualization, where it might otherwise have to be recomputed. For example, the coarsening proceedure in a multilevel solution algorithm may be directly useful in systematically downgrading detail in a discrete field for graphical display. TOPS intends to work with the providers of other components, such as meshers and visualizers, to amortize such work and reduce the total number of copies of what is essentially the same metadata.

Architecture-adaptive Performance Optimization. It is well known as a result of the ATLAS project [26] that arithmetically neutral data layout issues (blocking and ordering) can have an enormous impact in performance in dense linear algebra. The performance of sparse linear algebra subroutines is, if anything, even more sensitive to such issues, and an additional issue of padding by explicitly stored zeros to obtain greater uniformity of addressing is added to the mix. TOPS is working directly with the developers of OSKI [27] to obtain the benefit of data layout optimizations demonstrated in that project, which for some sparse linear algebra kernels can approach an order of magnitude.

Application-adaptive Performance Optimization. Because optimal iterative methods tend to have many (literally up to dozens) of parameters that can be tuned to a specific application, and even to a specific application-architecture combination, users are frequently overwhelmed by parameter choices and run with safe but inefficient defaults. Examples of such tuning parameters include the number of levels in a multigrid method, the number of vectors in a subspace in a Krylov method, the level of fill in an incomplete factorization, and the degree of overlap in a domain decomposition method. Machine learning can be employed to assist users in the choice of these methods, on the basis of previous runs of similar problems, whose features (such as a measure of symmetry or diagonal dominance or number of nonzeros per row) are stored together with performance outcomes. Using the AnaMod test harness [28] and the AdaBOOST machine learner [29], we have demonstrated in a preliminary way the effectiveness of machine learning on recommending solvers for matrices coming from M3D [21].

CONCLUSIONS

Computational scientists and engineers possess tremendous opportunities to extend the scope and fidelity of simulation and allow it to interact as an equal partner with theory and experiment. However, no single person can any longer be the master all aspects of the vertically integrated scientific instrument of a validated, verified, algorithmically optimized, performance-optimized simulation code. Modern scientific software engineering conveys the fruits of expertise at lower levels of this vertically integrated software structure to physicists working at higher levels. There is no more fitting a way to end this chapter than with the statement of the 2005 Abel Prize winner, Professor Peter Lax: "Quality software renormalizes the difficulty of doing computation."

ACKNOWLEDGMENTS

The author has had the inestimable privilege of working during the past four years of the TOPS project with extraordinarily talented individuals from many disciplines, mathematicians and computer scientists within the multi-institutional project and collaborators from engineering and physics. For the illustrations in this chapter, contributions from Jin Chen, Rob Falgout, Steve Jardin, Sherry Li, Dan Reynolds, Ravi Samtaney, Barry Smith, Carl Sovinec, Carol Woodward, Ulrike Yang were vital.

This work was partially supported by the National Science Foundation under CCF-03-52334 and by the U.S. Department of Energy under DE-FC02-04ER25595.

BIBLIOGRAPHY

1 I. Newton, *Philosophiae Naturalis Principia Mathematica*, 1687.
2 J. von Neumann and H.H. Goldstine, *Numerical inverting of matrices of high order*, Bull. Amer. Math. Soc., 53(11):1021-1099, 1947.
3 R. M. Russell, *The Cray-1 Computer System*, Comm. ACM 21(1):63-72, 1978.
4 C. K. Chu (ed.), *Computational Fluid Dynamics*, AIAA Series of Selected Reprints, vol. 4, 1971.
5 E. W. Dijkstra, *On the role of scientific thought*, in "Selected Writings on Computing: A Personal Perspective," pp. 60-66, Springer, 1974.
6 S. Balay, K. Buschelman, V. Eijkhout, W. D. Gropp, D. Kaushik, M. G. Knepley, L. C. McInnes, B. F. Smith and H. Zhang, *PETSc Users Manual*, ANL 95/11 Revision 2.3.0, 2005.
7 D. E. Keyes, et al., *A Science-based Case for Large-scale Simulation*, U. S. Department of Energy, www.pnl.gov/scales, Vol. 1 (70 pp.) 2003 and Vol. 2 (254 pp.) 2004.
8 W. K. Anderson, W. D. Gropp, D. K. Kaushik, D. E. Keyes and B. F. Smith, *Achieving High Sustained Performance in an Unstructured Mesh Application*, Proceedings of SC'99, IEEE, Los Alamitos, 1999 (Gordon Bell Special Prize).
9 V. Akcelik, J. Bielak, G. Biros, I. Epanomeritakis, A. Fernandez, O. Ghattas, E. J. Kim, J. Lopez, D. O'Hallaron, T. Tu, J. Urbanic, *High-resolution forward and inverse earthquake modeling on terascale computers*, Proceedings of SC'03, IEEE, Los Alamitos, 2003 (Gordon Bell Special Prize).
10 M. F. Adams, H. Bayraktar, T. Keaveny and P. Papadopoulos, *Ultrascalable implicit finite element analyses in solid mechanics with over a half a billion degrees of freedom*, Proceedings of SC'04, IEEE, Los Alamitos, 2004 (Gordon Bell Special Prize).
11 D. E. Keyes, *How Scalable is Domain Decomposition in Practice?*, Proceedings of the 11th International Conference on Domain Decomposition Methods, C. H. Lai, et al. (eds.), pp. 282-293, ddm.org, 1997.
12 N. R. Adiga, et al., *An Overview of the BlueGene/L Supercomputer*, Proceedings of SC'02, IEEE, Los Alamitos, 2002.
13 G. Moore, *Cramming more components onto integrated circuits*, Electronics Magazine 38(8):114-117, 1965.

14 D. E. Keyes, et al., *TOPS Home Page*, www-unix.mcs.anl.gov/scidac-tops.org, 2001.

15 D. A. Knoll and D. E. Keyes, *Jacobian-free Newton Krylov Methods: A Survey of Approaches and Applications*, J. Comp. Phys. 193:357-397, 2004.

16 T. R. Dunning, Jr., *Scientific Discovery through Advanced Computing*, U.S. Department of Energy, www.sc.doe.gov/ascr/mics/scidac/SciDAC_strategy.pdf, March 2000.

17 C. R. Sovinec, A. H. Glasser, T. A. Gianakon, D. C. Barnes, R. A. Nebel, S. E. Kruger, D. D. Schnack, S. J. Plimpton, A. Tarditi, and M. S. Chu, *Nonlinear Magnetohydrodynamics Simulation Using High-order Finite Elements*, J. Comp. Phys., 195:355-386, 2004.

18 X. S. Li and James W. Demmel, *SuperLU_DIST: A Scalable Distributed-Memory Sparse Direct Solver for Unsymmetric Linear Systems*, ACM Trans. Math. Soft. 29(2):110-140, 2003.

19 C. R. Sovinec, C. C. Kim, D. D. Schnack, A. Y. Pankin, S. E. Kruger, E. D. Held, D. P. Brennan, D. C. Barnes, X. S. Li, D. K. Kaushik, S. C. Jardin, and the NIMROD Team, *Nonlinear Magnetohydrodynamics Simulations Using High-order Finite Elements,* J. Physics 16:25-34, 2005.

20 G. Amdahl, *Validity of the Single Processor Approach to Achieving Large-Scale Computing Capabilities*, AFIPS Conference Proceedings 30:483-485, 1967.

21 J. Breslau, *M3DP Users Guide*, w3.pppl.gov/~jchen/doc/jbreslau.pdf, 2002.

22 R.D. Falgout and U.M. Yang, *Hypre, A Library of High-performance Preconditioners*, in "Computational Science - ICCS 2002 Part III", P.M.A. Sloot, et al., eds., Lecture Notes in Computer Science 2331:632-641, Springer, 2002.

23 J. Birn et al., *Geospace Environmental Modeling (GEM) Magnetic Reconnection Problem*, J. Geophys. Res. 106(A3):3715-3719, 2001.

24 D. R. Reynolds, R. Samtaney, C. S. Woodward, *A Fully Implicit Numerical Method for Single-fluid Resistive Magnetohydrodynamics*, J. Comp. Phys. (to appear), 2006.

25 B. F. Smith et al., *TOPS Solver Component*, http://www-unix.mcs.anl.gov/scidac-tops/solver-components/tops.html, 2005.

26 R. Clint Whaley, Antoine Petitet and J. J. Dongarra, *Automated Empirical Optimization of Software and the ATLAS Project*, Parallel Comput. 27(1-2):3-35, 2001.
27 R. Vuduc, J. W. Demmel and K. A. Yelick, *OSKI: A library of automatically tunred sparse matrix kernels*, J. Physics 16:521-530, 2005.
28 V. Eijkhout, *AnaMod: An Analysis Modules Library*, www.tacc.utexas.edu/~eijkhout/doc/anamod/html/, 2006.
29 Y. Freund and R. E. Shapire, *A short introduction to boosting*, J. Japan. Soc. Art. Intel. 14(5):771-780, 1999.

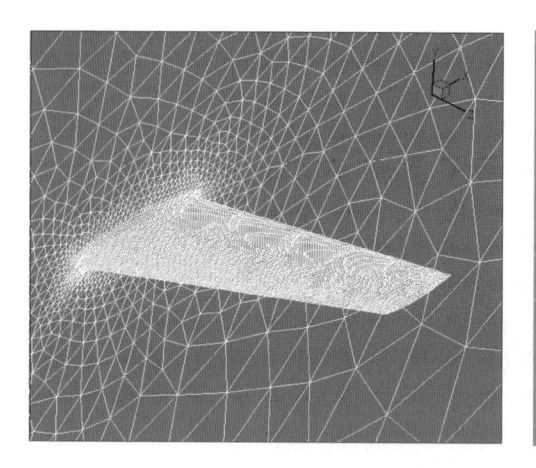

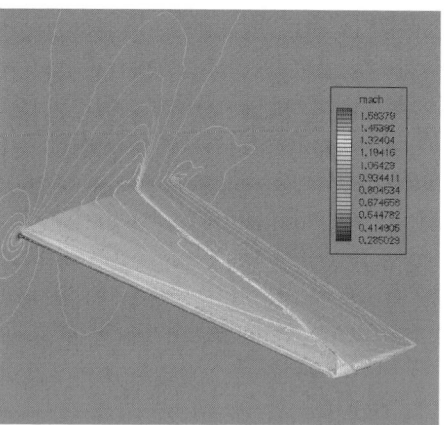

Figure 1: Footprint of tetrahedral grid on bounding surfaces and Mach contours for the lambda shock ONERA M6 test case included in the 1999 Gordon Bell "Special Prize" submission.

Year	Ref.	Application	DOFs	Procs	Flop/s	Machine
1999	[8]	Euler flow	11M	3K	0.25 Tf/s	Intel ASCI Red
2002	[9]	Seismic waves	17B	2K	1.0 Tf/s	HP Lemieux
2003	[10]	Bone deformation	0.5B	4K	0.5 Tf/s	IBM ASCI White

Table 1: Some salient parameters of Gordon Bell "Special Prize" winners employing PETSc [6]: overall number of degrees of freedom in space (and time for the 2002 submission), largest number of processors employed, sustained floating point computational rate on the largest problem, and the computational platform on which the peak execution was achieved.

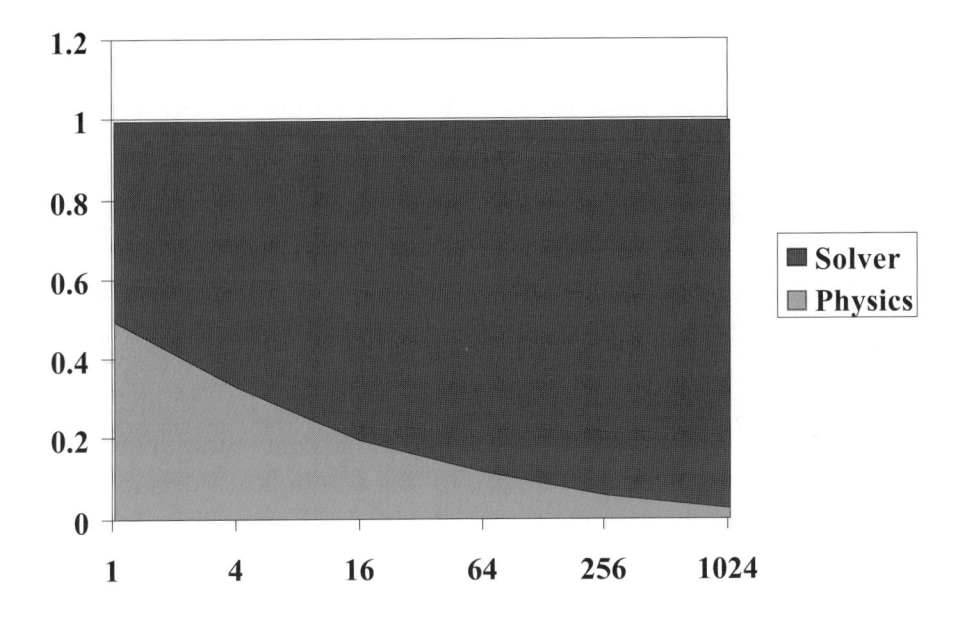

Figure 2: Fraction of time spent in the solver and in the physics phases of a typical PDE simulation as the problem is scaled by a factor of one thousand, assuming that the computation is originally well balanced on one processor and that the complexity of the solver phase grows like the 3/2ths power of the discretization size, whereas the physics phase grows linearly.

Year	Method	Reference	Storage	Flops
1947	GE (banded)	Von Neumann & Goldstine	n^5	n^7
1950	Optimal SOR	Young	n^3	$n^4 \log n$
1971	MIC-CG	Reid & others	n^3	$n^{3.5} \log n$
1984	Full MG	Brandt	n^3	n^3

Table 2: Complexity estimates in space (floating point storage) and time (floating point operations, cumulative), for the solution of Poisson's equation, in terms of the discrete dimension of a uniform cube of size n elements on a side, using different algorithms, identified by year of seminal publication (not necessarily a year in which the algorithm achieved final form or wide implementation)

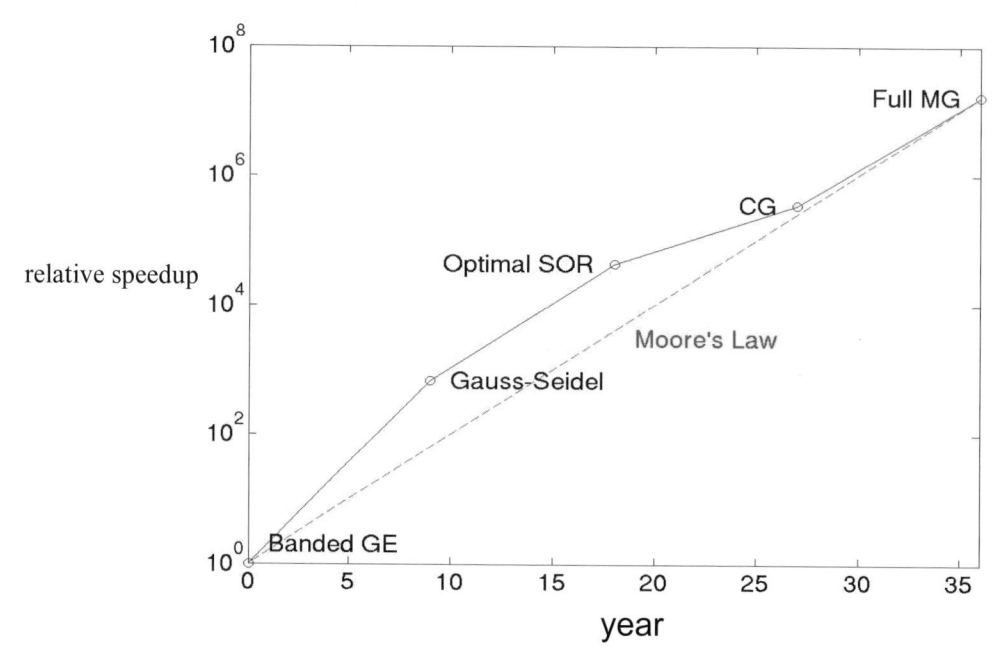

Figure 3: Stylized plot comparing the incremental progress of Moore's Law [13] and the algorithmic advances described in Table 2, for the case n=64, each over a period of 36 years. Each contributes a factor of approximately 16 million.

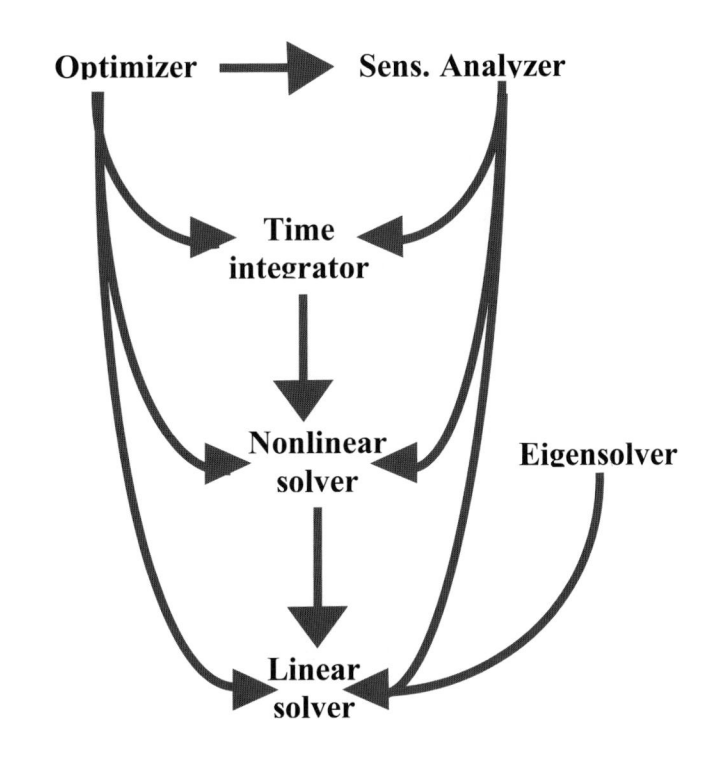

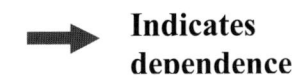

Figure 4: TOPS toolchain.

Investigation of Different Parallel Models for DNS of Turbulent Channel Flow

Jin Xu*
Division of Applied Math.,
Brown University,
Box F, 182 George Street,
Providence, RI 02912, USA

Abstract

This paper is aimed at investigating efficiency of different parallel implementations and target at DNS of high Reynolds number channel flow. It includes two parts: the first part is the investigation of different parallel implementations, including different parallel models and benchmark results. While the second part reports DNS simulation results at high Reynolds number, Re*=1000. Detailed statistics and visualization have also been given.

Keyword: DNS, Domain Decomposition, MPI, OpenMP

1. Parallel Investigation of Channel DNS Code

Turbulence remains one of the most challenging problems in physics. It is still a great challange to the scientifc comunity. Since the first channel turbulence DNS paper published in 1987, DNS has been a successful and powerful tool to investigate turbulence. Since the computer cost to do turbulence DNS is proportional to Re^3, so only relatively low Reynolds number has been achieved right now. With the rapid development of supercomputers, people are getting more and more interested in trying high Reynolds number DNS.

Because DNS at high Reynolds number is time and memory consuming, we need to optimize the code to achieve the highest possible parallel efficiency to use supercomputers. In order to obtain maximum optimization, we have implemented several different parallel models. Their performance have been benchmarked and the best model has been chosen. The first model uses domain decomposition in the stream-wise direction (model A), and it is the easiest way to implement. However, this model has potential limitation on the maximum number of processors that can be used, since we need to have at least two planes allocated to each processor in order to do de-aliasing in the nonlinear step. In order to increase the number of processors that can be used, we have also implemented another parallel model, in which the domain is decomposed in both stream-wise and span-wise

---

*Corresponding author: jin_xu@cfm.brown.edu

directions (model B). Using this parallel model, we can use thousands of processors for high Reynolds number DNS.

At very large resolutions, we found that most of the time consumed in each time loop is spent in the nonlinear step (approximately 75%). Since we need to expand the mesh 9/4 times larger than the mesh before de-aliasing, we need to exchange large mounts of data between different processors for high Reynolds number simulation. This is the major bottleneck of the code performance. In order to speed-up the nonlinear step, we also tried different ways to perform de-aliasing.

At last, we also implemented a hybrid parallel model using MPI and OpenMP. MPI has been used in stream-wise direction, while OpenMP has been used in span-wise direction (Model C).

1.1. Different Parallel Models

1. Model A: MPI in x direction

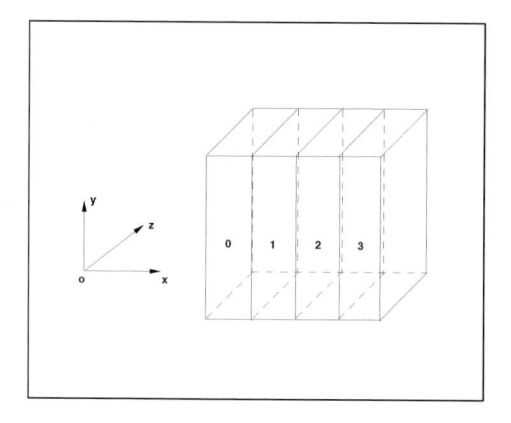

Figure 1. Sketch for Model A parallelization

Figure 1 is the sketch of model A; the domain has been decomposed only in the stream-wise direction. In the nonlinear step, the data on each processor has been shifted N times (N is the total number of processors been used) to the "following" processor. That means $0 \to 1 \to 2 \to ... \to N-1 \to 0$. During this process, each processor can hold the data which should belong to it after 3/2 expansion. Similar process apply to extrapolation.

2. Model B: MPI in x and z direction

Figure 2 is the sketch of model B, the domain has been decomposed in both stream-wise and span-wise directions.

In the nonlinear step, we have implemented Fast Fourier Transform in two different

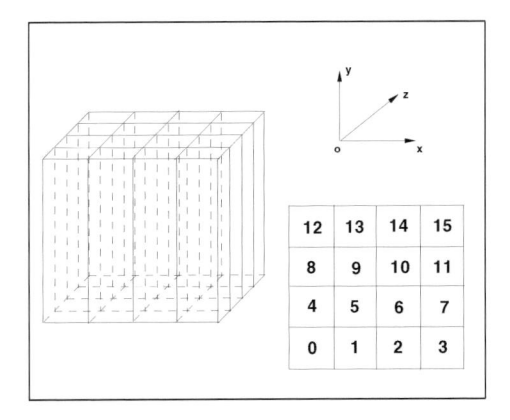

Figure 2. Sketch for Model B parallelization

ways. The first approach (referenced as model B1) is to generate two separate communication groups, one is composed of processors that have the same x planes, and the other is composed of processors which have same z planes as shown in figure 3. In the model, the FFT has been done separately in x and z directions, one after another. The second approach (model B2) is to collect data from all processors at first, and then put data on one or several planes to one processor as shown in figure 4. Then, a two-dimensional FFTW has been used, and after transformation, the data will be distributed back to original processors.

The second approach (B2) shows better performance, and it has been used in our high Reynolds number DNS simulation.

3. Model C: MPI in x and OpenMP in z direction
Model C has the same parallel structure as that of Model A, besides using MPI in the stream-wise direction, it uses OpenMP in the span-wise direction.

1.2. Benchmark Results

First we compare Model A and B on different machines, as shown in figure 5 and 6. They show that Model A and B have similar performance on different machines: SGI, SP and Linux cluster. Then we compare the parallel efficiency in two different directions for Model B and C. Figure ?? shows that Model B has same parallel efficiency in x and z directions, while figure ?? shows that Model C has different parallel efficiency in x and z directions. MPI has better performance than OpenMP as we use large number of processors. From our benchmarks, Model B2 is the most efficient, and we will use it to

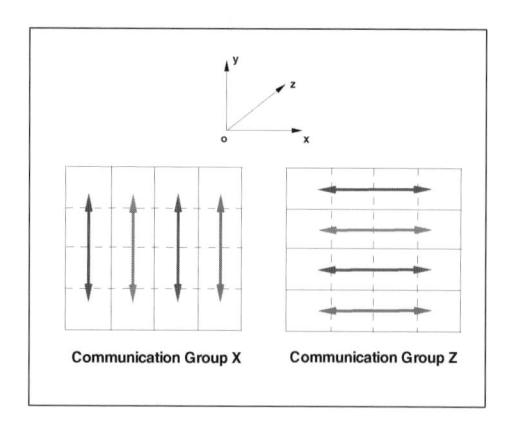

Figure 3. Sketch of the first way for Fourier transform in Model B

Case	Nx*Ny*Nz	Norm Re	Real Re	Lx	Lz	δx^+	δy^+	δz^+
Moser	384*257*384	595	587.2	2π	π	9.7	7.2	4.8
I	384*361*384	600	633	2π	π	9.81	12.9	4.91
II	768*521*768	1000	933	6π	1.5π	22.88	10.9	5.72

do high Reynolds DNS simulations.

2. Statistics and visualization of High Reynolds number simulation

In this section, we will give detail statistics of high Reynolds number simulation, and compare with other available results.

The table show the mesh used at $Re^* = 600, 1000$. The mesh we used at $Re^* = 600$ is close to Moser's mesh, and in y direction we used more points than Moser. The spacing is quite similar to his mesh. The domain for $Re^* = 1000$ is much longer than that of $Re^* = 600$, because the vortices are much longer and smaller than that of low Reynolds number. Then we show mean velocity and turbulent fluctutaion profile at $Re^* = 600$ and $Re^* = 1000$. They have been compared with other researcher's results, and show good agreement.

At last, we show the streaks at $y^+ = 5$ and Q contour of vortexes at $Re^* = 1000$. The spacing of streaks are still approximately 100 in wall units, similar to what is at low Reynolds number. But the vortices become much smaller at high Reynolds number than at the low Reynolds number. The domain size is quite different than that of low Reynolds number case, it is $6\pi \times 2 \times 1.5\pi$, while the low Reynolds number case is $2\pi \times 2 \times 2\pi$.

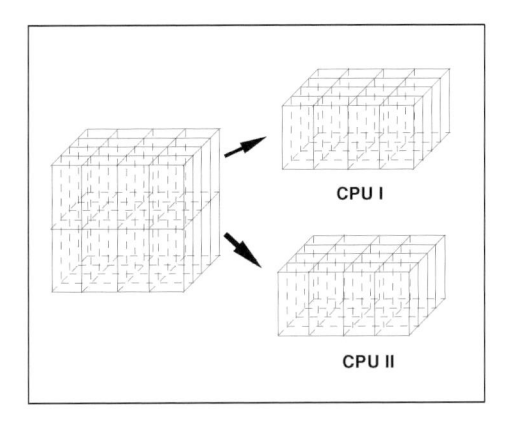

Figure 4. Sketch of the second way for Fourier transform in Model B

3. Summary and Conclusion

In this paper, we have investigated efficiency of different parallel implementations and done DNS of high Reynolds number channel flow. In the first part, different parallel implementations of channel code have been expained, and benchmark results have been reported. In the second part, DNS simulation results at high Reynolds number, Re*=1000, has been given. Detailed statistics and visualization have also been presented. In the future, even higher Reynolds number DNS will soon be realized with the fast development of supercomputers.

Acknowledgments: This work was supported by DARPA. The computations were performed mainly at Pittsburgh Supercomputer Center(PSC), and Arctic Supercomputer Center(ARSC). Since it takes large mount of computer resources, it causes lots of problem to run on these machines, I would like to give thanks to system consultants. Without their help, this work can not be done at present condition.

REFERENCES

1. I. Babuska, B.A. Szabo and I.N. Katz, The p-version of the finite element method *SIAM J. Numer. Anal.* **18**, 515 (1981).
2. J.P. Boyd, *Chebyshev and Fourier specyral methods. Springer-verlag.* New York (1989).
3. C. Canuto, M.Y. Hussaini, A. Quarteroni and T.A. Zang, Spectral Methods in Fluid Mechanics, *Springer-Verlag.* New York (1987).
4. H. Choi, P. Moin and J. Kim, Direct Numerical simulation of turbulent flow over riblets. *J. Fluid Mech.* **255**, 503 (1993).
5. J.W. Deardorff, A numerical study of three-dimensional turbulent channel flow at large Reynolds numbers. *J. Fluid Mech.* **41(2)**, 453 (1970).

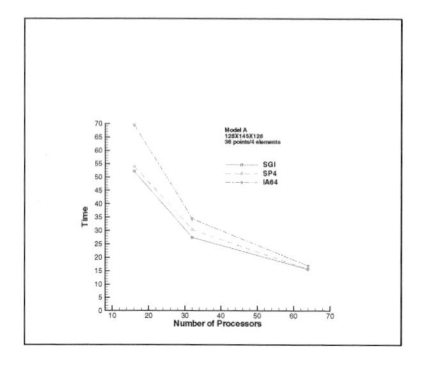

Figure 5. Model A on different machines

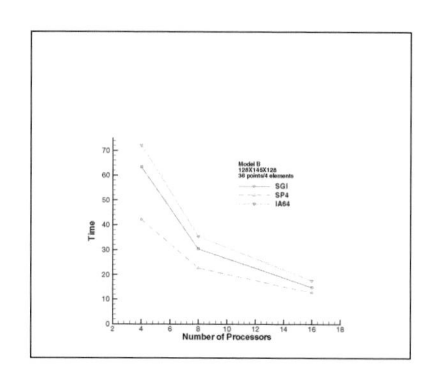

Figure 6. Model B on different machines

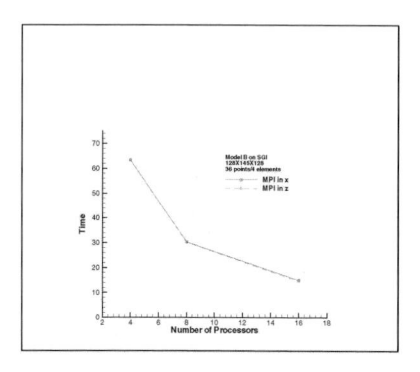

Figure 7. Model B on different directions on SGI

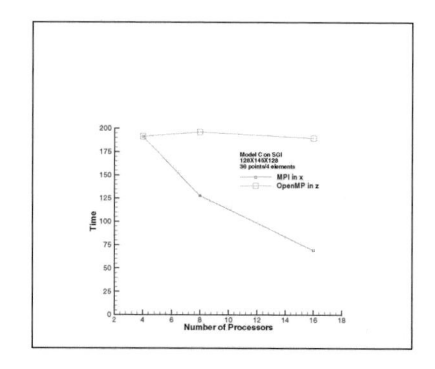

Figure 8. Model C on different directions on SGI

6. M. Dubiner, Spectral Methods on triangles and other domains. *J. Sci. Comp.* **6**, 345 (1991).
7. R.D. Henderson, *Unstructured spctral element methods: parallel algorithms and simulations*, PhD Thesis, Department of Mechanical and Aerospace Engineering, Princeton University, (1994).
8. G.E. Karniadakis, M. Isreali and S.A. Orszag, High-order splitting methods for incompressible Navier-Stokes equations, *J. Comp. Phys.* **97**, 414 (1991).
9. G.E. Karniadakis and S.J. Sherwin, Spectral/*hp* element methods for CFD, Oxford University Press, London (1999).
10. J. Kim, P. Moin and R. Moser, Turbulence statistics in fully develope channel flow at low Reynolds number, *Journal of Fluid Mechanics* **117**, 133 (1987).
11. P. Moin and J. Kim, Numerical investigation of turbulent channel flow. *J. Fluid Mech.* **188**, 341 (1982).

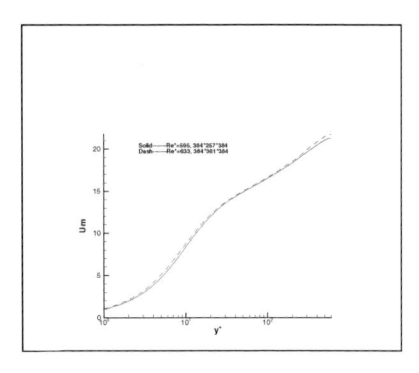

Figure 9. Mean velocity at $Re^* = 600$

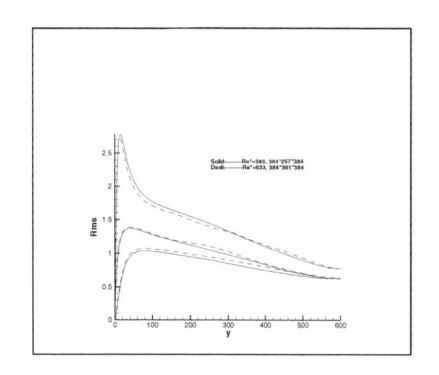

Figure 10. Turbulent fluctuation at $Re^* = 600$

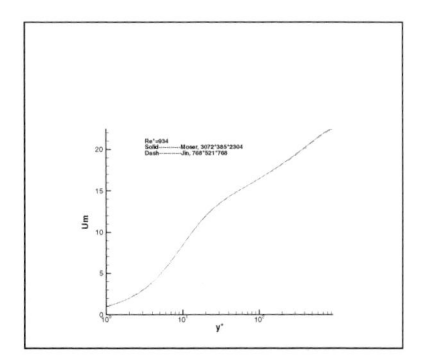

Figure 11. Mean velocity at $Re^* = 1000$

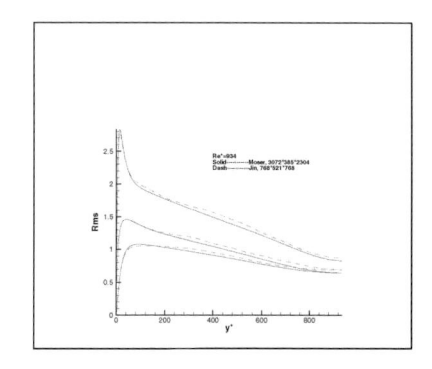

Figure 12. Turbulent fluctuation at $Re^* = 1000$

12. R. Moser, J. Kim and N. Mansour, Direct numerical simulation of turbulent channel flow up to $re_\tau = 590$. *Phys. Fluids* **11**, 943-945 (1999)

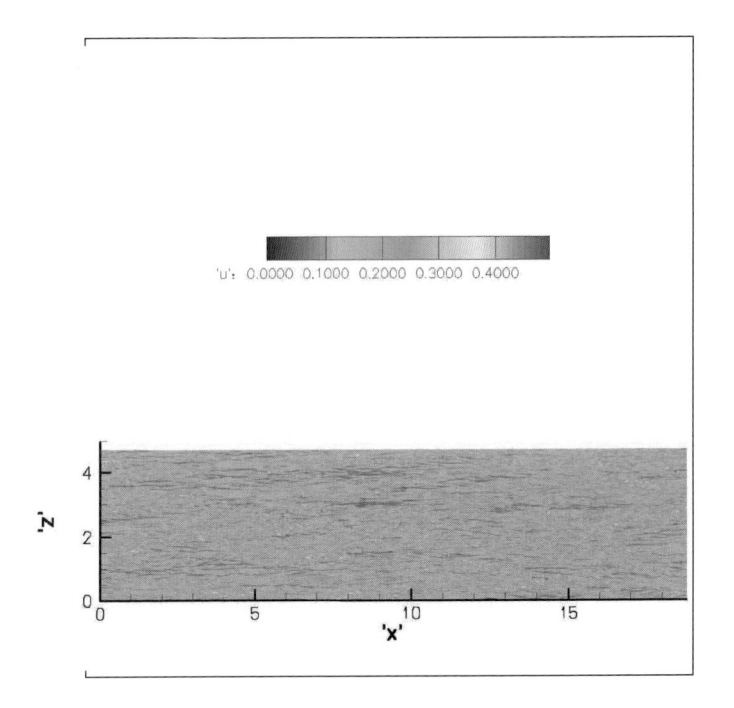

Figure 13. Streaks at $Re^* = 1000, y^+ = 5$

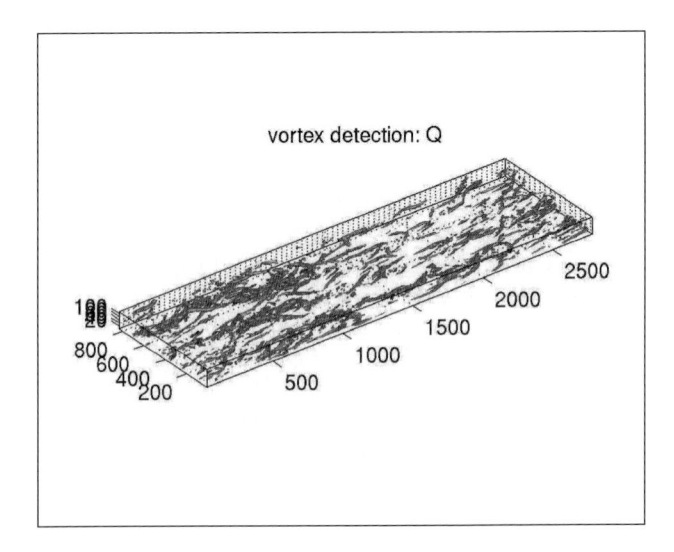

Figure 14. Vortex Q contour at $Re^* = 1000$

Flow Simulation around a Finite Cylinder on Massively Parallel Computer Architecture

O. FREDERICH[*], E. WASSEN[†] AND F. THIELE[‡]

Hermann-Föttinger-Institute of Fluid Mechanics, Berlin University of Technology, Müller-Breslau-Strasse 8, 10623 Berlin, Germany, www.cfd.tu-berlin.de

The spatio-temporal flow field around a wall-mounted finite circular cylinder is computed using Detached-Eddy Simulation and Large-Eddy Simulation on a range of grids. Therefore, the used numerical method, the suitability of parallel computers, the descretisation method and the simulation setup is described. Results are presented in comparison to experiments, showing a good prediction of expected structures and flow phenomena.

1. Introduction

The investigations presented form part of the research project "Imaging Measuring Methods for Flow Analysis" funded by the German Research Foundation (DFG). The aim of this program is to develop flow measuring techniques and to improve their performance. Because of the scarcity of experimental methods capable of producing similar results to simulations (with respect to spatial and temporal resolution), the development of improved measurement methods for the analysis of complex flows is to be furthered in the scope of this program.

The numerical simulations are to yield all flow quantities with a high resolution in time and space. In the scope of the DFG program, concurrent time-resolved experiments will be carried out by another partner. The provision of the highly spatially and temporally resolved simulated flow field together with the unsteady experimental data will form a combined database for the verification of newly developed visualisation methods and numerical turbulence models and establish a reference testcase.

The present work gives an overview of the Detached-Eddy Simulation and the Large-Eddy Simulation of the configuration described in section 2. In section 3 an overview of the approach is given. The necessity of parallel computers is explained in section 4, and the used simulation and model parameters are summarised in section 5. Obtained Results are presented in section 6 before a conclusion is given.

---

[*]Research Engineer
[†]Assistant Professor
[‡]Professor

2. Configuration and Physics

The principle configuration for the experimental and numerical investigations is the complex unsteady flow around a finite circular cylinder mounted on a plate (figure 1). The flow field is characterised by complex structures and various flow phenomena (Agui and Andreopoulus, 1992), which interact with each other.

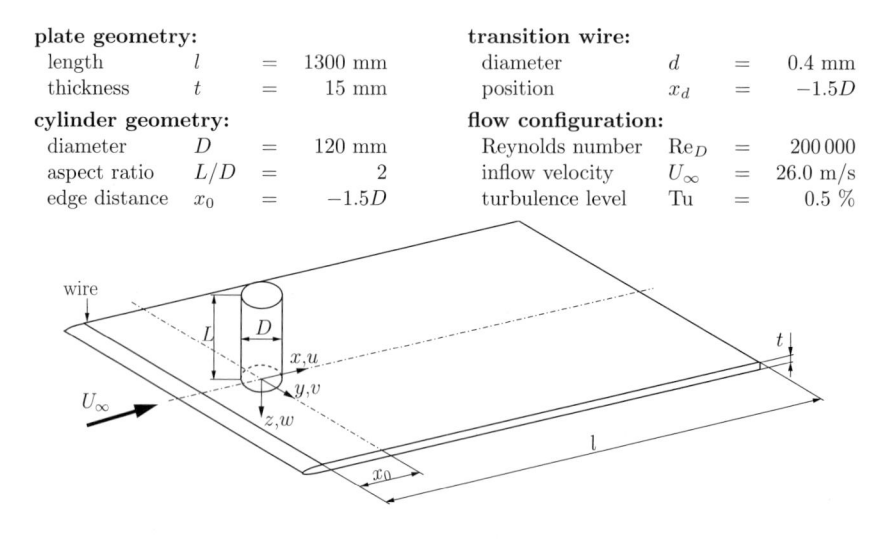

plate geometry:
- length l = 1300 mm
- thickness t = 15 mm

cylinder geometry:
- diameter D = 120 mm
- aspect ratio L/D = 2
- edge distance x_0 = $-1.5D$

transition wire:
- diameter d = 0.4 mm
- position x_d = $-1.5D$

flow configuration:
- Reynolds number Re_D = 200 000
- inflow velocity U_∞ = 26.0 m/s
- turbulence level Tu = 0.5 %

Figure 1. Configuration of experimental setup

A similar configuration was investigated by Fröhlich and Rodi (2003) using Large-Eddy Simulation at a Reynolds number $Re_D = 43\,000$ and the aspect ratio $L/D = 2.5$. They discerned a Strouhal number of about $St = 0.16$ and pointed out the long averaging times required, making the computations very expensive. The experimental results on nearly the same configuration of Leder (2003), against which the numerical results are compared, reveal that the separated flow is characterised by three superimposed vortex systems, whose axes are orthogonal to each other. These are the recirculation vortices downstream of the cylinder and on the free end (y-axis), the tip vortices separating from that end (x-axis) and the vortex structures shed from the sides (z-axis). The vortex shedding regime of the present case lies marginally between sub-critical and super-critical cylinder flow (Zdravkovich, 1997). As such, the transition is assumed to be fixed at the point of separation in order to simplify the configuration and the resolution of these phenomena.

3. Numerical Method

Due to the relatively high Reynolds number of $Re_D \approx 200\,000$ the resolution of all turbulent scales using Direct Numerical Simulation (DNS) cannot be realised on currently available high-performance computers. Thus, both Large-Eddy Simulation (LES) and

Detached-Eddy Simulation (DES) techniques are applied, which resolve the dominant energetic scales and model the smallest scales. At first, investigation using DES on a coarser grid should help to examine the prediction of expected flow phenomena, to determine their spatial position, and to provide grid and simulation parameters for the LES.

The unsteady and incompressible numerical simulations around the finite cylinder are carried out using the flow solver ELAN permanently under developement at the Hermann-Föttinger-Institute of Fluid Mechanics (HFI) of the TU Berlin. This implicit solver is of second order accuracy in time and space and is based on a fully conservative three-dimensional finite-volume approximation of the Navier-Stokes-equations in primitive variables stored in the cell centers. The spatial discretisation is semi-block-structured on the basis of general curvilinear coordinates, thereby facilitating the mapping of complex geometries and local grid refinement. All equations are solved sequentially, whereas the continuity equation is conserved by a pressure correction of the SIMPLE type. To avoid the decoupling of pressure and velocity, a generalised Rhie & Chow interpolation is used. Diffusive terms are approximated using central differences and convective ones with monotone and limited approaches of higher order. A hybrid approach, mixing both schemes for handling the convective terms as required by Detached-Eddy Simulation based on a suggestion of Travin et al. (2002), is also provided by the code.

Depending on the application of either Reynolds-Averaged Navier-Stokes-equations (RANS) or Large-Eddy Simulation, the Reynolds-averaged or spatially filtered Navier-Stokes equations are solved, respectively. A variety of RANS turbulence models and LES subgrid-scale (SGS) models are inplemented and validated in ELAN. Some of the turbulence models are also implemented as background models for DES, where the model constant has been calibrated for each model for the decaying of homogeneous isotropic turbulence (DHIT).

4. Suitability of Parallel Computers

The employed algorithm can be parallelised very efficiently through a simple domain-decomposition method. In addition, the code has the possibility to handle multiple blocks on one processor for optimal load balance. The interchange of data is realised through explicit message passing using MPI, making the code also easily portable to different platforms.

For the wall-mounted finite cylinder, a grid of approximately 12.5 million grid points has been generated. The central memory requirement of the code is about 2.6 kB per grid point, leading to a total memory requirement of about 33 GB. To obtain a sufficient statistical base for the determination of time-averaged flow variables and higher order moments, time steps numbering in the tens of thousands are necessary, whereby a single time step requires several minutes of wall-clock time using 42 processors on a IBM p690 architecture. It is clear, that such a quantity of memory and reasonable lead-times can only be achieved by a high level of parallelisation.

5. Discretisation and Simulation Parameters

For the configuration investigated, attention must be paid to three sets of walls (plate, cylinder shell and cylinder top). This difficulty leads to the compression of grid lines in all three spatial directions due to the orthogonality of the boundary layers.

Using an appropriate block topology and hanging nodes, the grid points can be concentrated in the focus regions found by the experiments (Leder, 2003) and the DES on the coarse grid (figure 2). The fixed transition on the plate was realised with a geometrical wire model equivalent to the experimental setup in the high resolution grid (figure 3). The distribution and size of cells followed the established practice for LES grids and was adapted for the present Reynolds number $\text{Re}_D = 200\,000$.

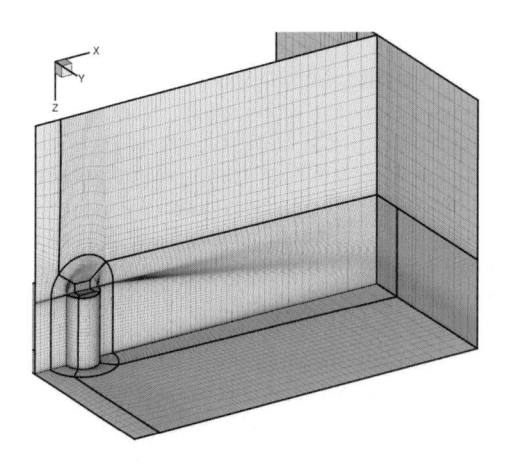

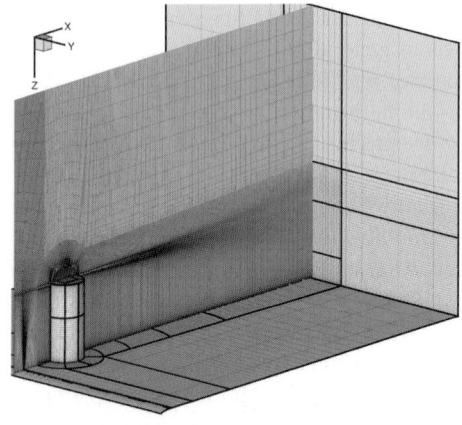

Figure 2. Coarse grid for initial investigations, 930 000 grid points

Figure 3. Grid for the high resolution simulations, 12.5 million grid points (walls and outflow show only every third grid line)

A slip condition was applied to the upper and lateral boundaries, and a convective outflow condition to the downstream x-boundary. No-slip conditions were applied to all physical walls. For the coarse grid simulations, a constant velocity inlet profile was used, whereas for the fine grid a spatially-variable velocity profile obtained from an additional simulation with an extended upstream region was implemented.

The initial DES calculation, based on the LLR k-ω turbulence model (Rung and Thiele, 1996), was conducted on the coarser grid, for which a timestep of $\Delta t = 0.025 \frac{D}{U_\infty}$ was used. From this, the required timestep of $\Delta t = 0.005 \frac{D}{U_\infty}$ could be established for all simulations on the finer grid. The LES simulations on this grid (figure 3) used a Smagorinsky constant of $C_s = 0.1$.

6. Results

6.1. Simulation on Coarse Grid

In figure 4 time-averaged and unsteady vortex structures (λ_2-isosurfaces) of the DES are shown. While the instantaneous snapshot reveals the unsteadiness of the flow, the steady visualisation presents the detected system of vortices. The recirculation vortex on the cylinder top, the side tip vortices, the shear layers of the sideways separation from the cylinder, the recirculation at the downstream side of the cylinder and the horse shoe vortex near the plate can all be seen. These structures are much harder to identify in the instantaneous snapshot.

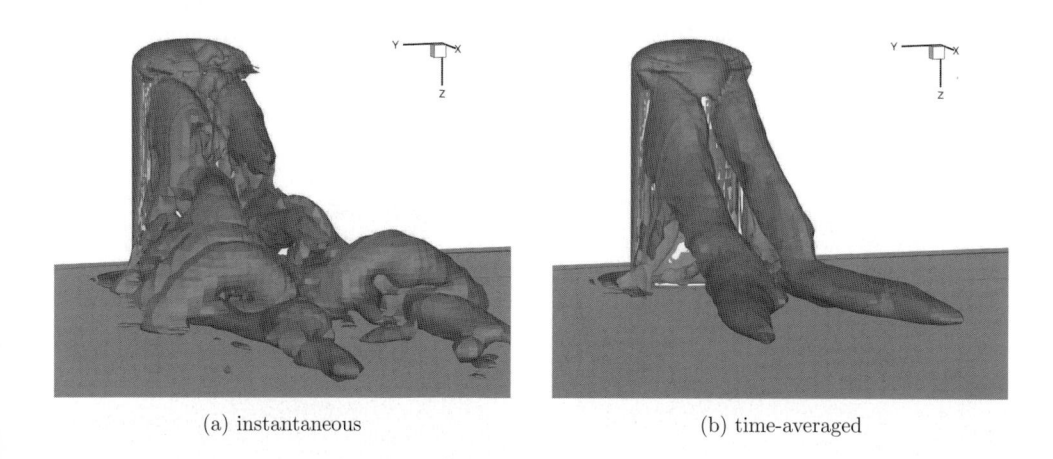

(a) instantaneous

(b) time-averaged

Figure 4. Predicted vortex structures (λ_2-isosurfaces), Detached-Eddy Simulation

Figure 5 shows the separation bubble on the cylinder top using the time-averaged mainstream velocity $\bar{u}/U_\infty$ in the symmetry plane $y = 0$. The time-averaged flow separates at the leading edge of the cylinder and reattaches at $0.84D$ downstream in the experiment (Leder, 2003) and at $0.87D$ in the Detached-Eddy Simulation. This relatively good agreement of experiment and simulation supports the usage of the hybrid approach with RANS-modelling near the walls, although there is a larger velocity above the bubble due to the used constant inflow velocity profile. This insufficiency was avoided in the high-resolution simulations by a spatially-variable inflow profile.

(a) experiment (Leder, 2003)

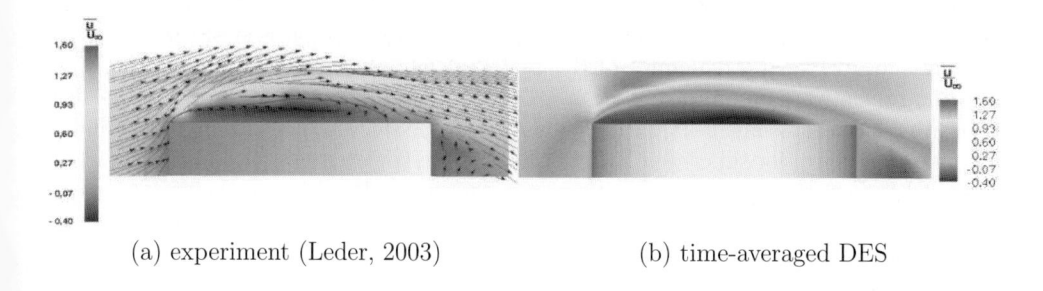

(b) time-averaged DES

Figure 5. Comparison of experiment and DES: Separation bubble on the cylinder top

6.2. Simulation on Refined Grid

Figure 6 shows time-averaged and unsteady vortex structures (λ_2-isosurfaces) predicted by the Large-Eddy Simulation on the refined grid (figure 3), and with adapted inflow velocity profile. The unsteadiness of the flow is much more highlighted by the instantaneous snapshot compared to the coarse DES. The visualisation of isosurfaces of λ_2 evaluated for the time-averaged velocity field reveals the structures also in more detail. It can be recognised that about 30 000 time steps for the time-averaging are hardly sufficient and the statistical base has to be increased.

(a) instantaneous (b) time-averaged

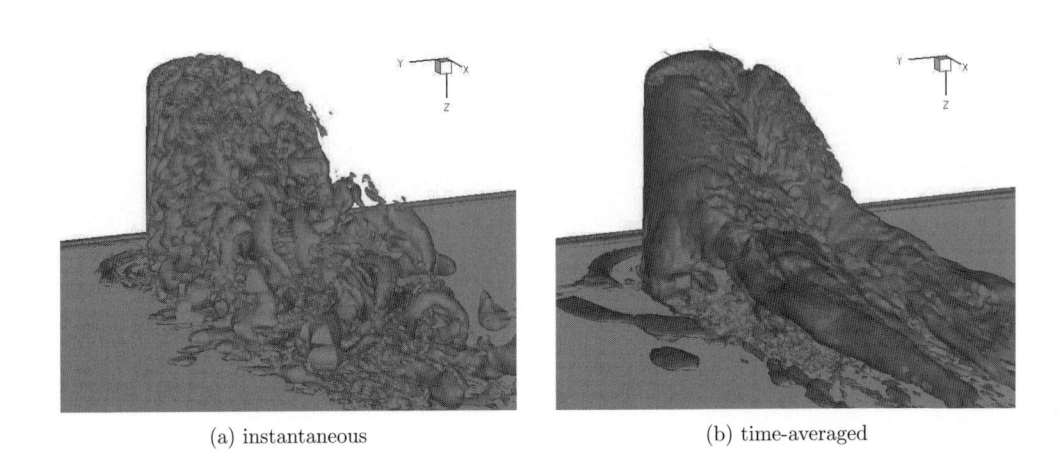

Figure 6. Predicted vortex structures (λ_2-isosurfaces), Large-Eddy Simulation

The predicted formation and decay of the side tip vortices can be depicted by the vorticity around the x-axis. Using this criterion, experiment and LES are compared with each other in selected cutting planes, shown in figure 7. A good agreement with the experiment is evident near the cylinder and in the further wake. Through the high grid density, much finer details of the flow structures are resolved. Hence an interpolation to the coarser experimental measuring grid is used for a direct comparison of the results.

Using time-averaged streamlines of the first grid line above the walls in conjunction with the respective pressure distribution, as shown in figure 8, further flow features like separation and reattachment lines can be identified. The sidewise flow around the cylinder separates, as typical for subcritical cylinder flow (Zdravkovich, 1997), at about 80° (angle measured from the x-z-plane, 0° at the stagnation point). In addition, the reattachment lines on the cylinder top as well as the flow spreading lines in front and on the top of the cylinder can be recognised clearly.

One of the aspects of the presented investigations is the time-resolved prediction of the flow. Among the temporal behaviour of field quantities, this can be characterised by

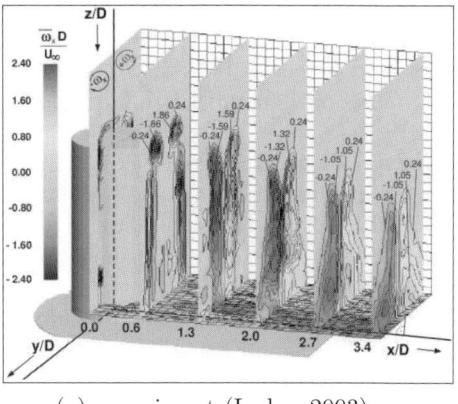

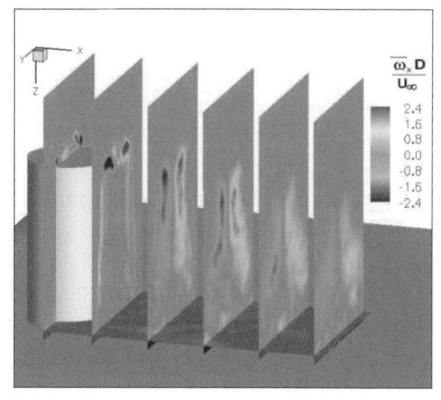

(a) experiment (Leder, 2003)   (b) LES, interpolated to measuring grid

Figure 7. Comparison of experiment and LES: Formation and decay of the side tip vortices (x-vorticity in selected cutting planes)

the Strouhal number $St = \frac{fD}{U_\infty}$. In figure 9 the temporal development of the side force coefficient is depicted. The contributions of the three main surfaces are shown as well as the global sum. Based on these data, a dominant frequency of $f = 40.45$ Hz, or $St = 0.187$, can be detected, which corresponds well to the Strouhal number of $St = 0.18$ found experimentally. Figure 9 also reveals that the flow consists of several superimposed frequencies hardly forming a cycle, and that the plate contributes a non-negligible fraction.

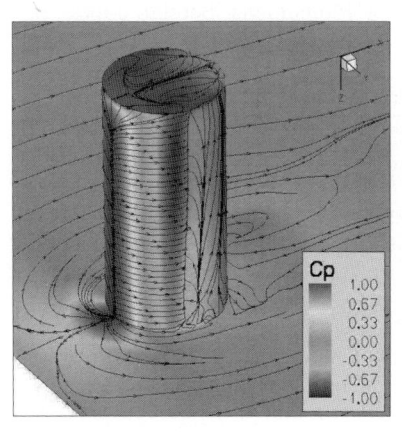

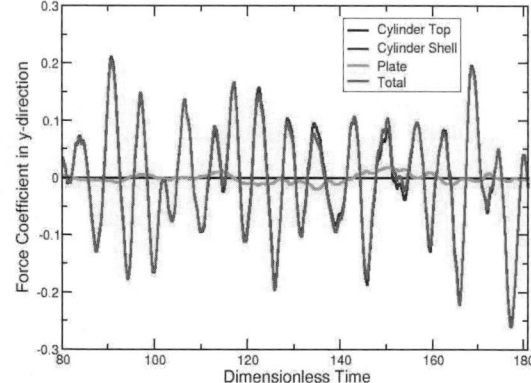

Figure 8. Time-averaged pressure and surface streamlines, LES

Figure 9. Temporal behaviour of the global side force coefficient, LES

7. Conclusion and Outlook

The main objective of the investigation presented here was the reproduction of the experimental results, whereby additional spatio-temporal flow information are provided by the simulations.

The flow around a wall-mounted finite circular cylinder was computed successfully using a coarse grid with Detached-Eddy Simulation and a finer one with Large-Eddy Simulation. The DES was used for determining the spatial position of the flow structures and model parameters for the high resolution simulation. The presented results demonstrate the ability to predict flow structures using different methods and grids, but with a large amount of ressources. When more experimental results become available a quantitative comparison of experimental and numerical results will be possible based on time-averaged velocity data, Reynolds stresses and even on triple correlations.

Acknowlegdements

The work presented here is supported by the German Research Foundation within the research project "Imaging Measuring Methods for Flow Analysis". All simulations were performed on the IBM pSeries 690 supercomputer of the North German cooperation for High-Performance Computing. We thank both organisations cordially for their support.

References

Agui, J. H., Andreopoulus, J., 1992. Experimental investigation of a three-dimensional boundary-layer flow in the vicinity of an upright-wall mounted cylinder. Journal of Fluid Mechanics 114, 566–576.

Fröhlich, J., Rodi, W., 2003. LES of the flow around a cylinder of finite height. In: Proc. of 3rd Int. Symposium on Turbulence and Shear Flow Phenomena. Sendai, Japan, pp. 899–904, june 25-27, 2003.

Leder, A., 2003. 3d-flow structures behind truncated circular cylinders. In: Proc. of 4th ASME/JSME Joint Fluid Engineering Conference. Honolulu, Hawaii, USA, July 6-10, 2003, Paper No. FEDSM2003-45083.

Rung, T., Thiele, F., June 1996. Computational modelling of complex boundary-layer flows. In: Proc. of the 9th Int. Symposium on Transport Phenomena in Thermal-Fluid Engineering. Singapore, pp. 321–326.

Travin, A., Shur, M., Strelets, M., Spalart, P. R., 2002. Physical and numerical upgrades in the Detached-Eddy Simulation of complex turbulent flows. In: Friederich, R., Rodi, W. (Eds.), Advances in LES of Complex Flows. Vol. 65 of Fluid Mechanics and its Applications. pp. 239–254.

Zdravkovich, M. M., 1997. Flow Around Circular Cylinders, Vol. 1: Fundamentals. Oxford University Press, New York.

Parallel Computation of Turbulent Fluid Flows with the Piecewise-Parabolic Method

Paul R. Woodward,[a] David H. Porter,[a] Sarah E. Anderson,[b] B. Kevin Edgar,[a] Amitkumar Puthenveetil,[a] and Tyler Fuchs[a]

[a]*Laboratory for Computational Science & Engineering, University of Minnesota, 499 Walter, 117 Pleasant St. S. E., Minneapolis, Minnesota 55455, USA*

[b]*Cray Research, Eagan, Minnesota, USA*

parallel computation; turbulence

1. Introduction

We have used the PPM gas dynamics code [1-4] on large NSF TeraGrid parallel systems as well as on SMP and cluster systems in our lab to simulate turbulent flows at grid resolutions of 1024^3 and 2048^3 cells [5-9]. We have studied both single- and two-fluid flows with statistically homogeneous and unstable shear layer initial conditions. The purpose of these simulations has been to capture detailed datasets that allow design and validation of subgrid-scale turbulence models. The parallel code implementation manages a set of shared data objects, each describing a subdomain of the problem. These data objects can be instantiated as either disk files or in-memory objects on a designated set of network nodes that serve them to computational processes over a cluster network. When implemented as disk files, large problems can be run, at no loss in performance, on systems that do not have sufficient memory to hold a complete description of the problem state. Such runs are also fault tolerant, because of the persistent disk image of the problem that is updated by the worker processes. Using either disk files or in-memory data objects, the number of participating CPUs can be dynamically adjusted during the run with no need to halt or pause the computation. This parallel computing method and the implications of our simulation results for subgrid-scale turbulence models is described below. Implications of our simulations for subgrid-scale turbulence models are also discussed.

2. New features of PPM used in this work.

Although several new features have been in use in our PPM codes that have not appeared in the published literature, a full description of the present version of the algorithm can be found on the LCSE Web site at www.lcse.umn.edu [4]. Such features that play particularly important roles in the simulations to be presented include the following: (1) estimation of the cell-averaged kinetic energy per unit mass from constrained linear distributions of all velocity components in all dimensions that are generated at the outset in each 1-D pass, and (2) estimation of the local smoothness of each variable to be interpolated with application of monotonicity constraints only when a variable is judged not to be smooth.

3. SHMOD parallel code framework.

The problem domain is decomposed into grid bricks in such a way that there are roughly 4 times as many such bricks as the maximum expected number of computing processes that will be available during the simulation. An order for updating these subdomains is established which minimizes the chance that any update process will have to wait upon the completion of previous updates in order to obtain necessary domain boundary data. This minimization is made under the assumption that updates require roughly the same amounts of time. Substantial variation in update time can occur with essentially no impact on the efficiency of code execution if there are sufficiently many subdomains. We find that 4 times the number of updating processes works well in practice, with no processes ever having to wait upon data, even if the previous time step has not yet been completed.

The subdomain updates are coordinated through a simple mySQL database. This database contains all necessary global knowledge, including the value of the time step, the status of each grid brick update, and the location of its data on the network. The database serializes access to this global data and makes it available to all participating processes, but it is otherwise passive. Processes can be started at any time. Once told the location of the database, they will check in, grab work assignments, and assist in the computation until either it is completed or they are killed. Processes can be killed at any time, in which case their assigned work is not completed. After a time-out interval, other processes take up this uncompleted work, so that the computation can proceed. If the killed process somehow wakes up, it will be aware from examination of its own clock that it has timed out, it will then abort the work it had begun, go to the database, and find new work to perform.

The key to this type of parallel computation is network bandwidth and the overlapping of data communication with computation. There are only two types of data communication used. First, of course, is database access. This is very simple. The grid brick update process simply reads the entire contents of the database, locking it, altering it, returning it, and unlocking it. Some accesses do not require locks, but the database is so small that its entire contents, a single "unstructured glob," can be transmitted over the network in a time comparable to that required to contact the database at all. Even with hundreds of participating processes, the database easily handles the traffic and is not a bottleneck. The second kind of communication is bulk data access. Here grid brick update processes either read or write back large data records (files) that describe the complete states of subdomains. This data is specially structured to permit transmission in a small number of large serial reads or writes, so that the transmission is very efficient. These grid brick data objects are implemented either in the memories of specially designated memory-server nodes or on the disks of specially designated disk server nodes (which may also participate in the computation). Large shared file systems attached to the network can also store and serve up these files very efficiently.

4. Homogeneous, Mach 1 turbulence simulation on a 2048^3 grid.

As friendly users of the TeraGrid cluster at NCSA, we used from 80 to 250 dual-CPU nodes over a 2½ month period in the fall of 2003 to simulate compressible, homogeneous turbulence on a 2048^3 grid with our PPM code (see [9] and www.lcse.umn.edu). We initialized this flow with smooth velocity disturbances centered on an energy-containing scale of half the width of our cubical, periodic problem domain. We put 10% of the initial kinetic energy into the compressional modes and the remaining 90% into the solenoidal velocity field. The initial disturbances were of course extremely well resolved. This flow was allowed to decay over a time interval equal to 2 flow-through times for the entire box, which in this Mach-1 flow was also 2 sound crossing times of the box. By the end of this time interval the decaying velocity power spectrum was no longer changing shape.

We have analyzed the data from this flow to assess the quality of a correlation we observed in earlier flows [6-8] between the rate of generation of subgrid-scale kinetic energy, which we write as F_{SGS} below, and the determinant of the rate of strain tensor for the larger-scale flow. We follow the classic Reynolds averaging approach, in which we apply a Gaussian filter with a prescribed full width at half maximum, L_f, to our simulation data to arrive at a set of filtered, or "resolved," variables and a set of "unresolved" state variables that fluctuate rapidly in space. For any particular state variable, Q, we define the filtered (or "resolved") value, $\overline{Q}$, by

$$\overline{Q}(x) = \int e^{-(k_f(x-x_1))^2} Q(x_1) d^3x \ / \ \int e^{-(k_f(x-x_1))^2} d^3x,$$

where the wavenumber of the filter, k_f, is related to the full width at half maximum, L_f, by $L_f = 1.6688/k_f$, and where the integral in the denominator is, of course, equal to $(2\pi)^{3/2}/(2k_f)^3$. The mass-weighted, or Favre, average of a state variable, Q, is denoted using a tilde. Manipulating the Euler equations using these definitions in order to arrive at the time rate of change of the kinetic energy in a frame moving with the filtered flow velocity, we get:

$$\frac{\partial k_{SGS}}{\partial t} + \partial_j(\tilde{u}_j k_{SGS}) = \frac{Dk_{SGS}}{Dt} + k_{SGS}\partial_j(\tilde{u}_j) =$$

$$\left(\overline{p\,\partial_i u_i} - \overline{p}\,\partial_i \tilde{u}_i\right) - \tau_{ij}\,\partial_j \tilde{u}_i$$

$$- \partial_j\left(\overline{u_j p} - \tilde{u}_j \overline{p} - \tilde{u}_i \tau_{ij} + \tfrac{1}{2}\overline{\rho\, u_i^2\, u_j} - \tfrac{1}{2}\overline{\rho\, u_i^2}\,\tilde{u}_j\right)$$

Here k_{SGS} is the subgrid-scale kinetic energy, D/Dt denotes the co-moving time derivative, and τ_{ij} is the subgrid-scale stress (SGS) tensor,

$$\tau_{ij} = \overline{\rho\, u_i u_j} - \overline{\rho}\,\tilde{u}_i \tilde{u}_j$$

Using our simulation data we can establish the relative importance of the various terms grouped on the right in the above expression for the time rate of increase of subgrid-scale kinetic energy per unit mass in the co-moving frame. This analysis indicates that, statistically, the divergence terms can be modeled by a diffusion of k_{SGS} and that the first terms in brackets on the right, the $p\,DV$ work terms, tend to have little effect on the average. However, the term $-\tau_{ij}\,\partial_j \tilde{u}_i$ has systematic behavior that tends to make it dominant over space and over time. We will refer to this term as the forward energy transfer to subgrid scales, or F_{SGS}. By analysis of

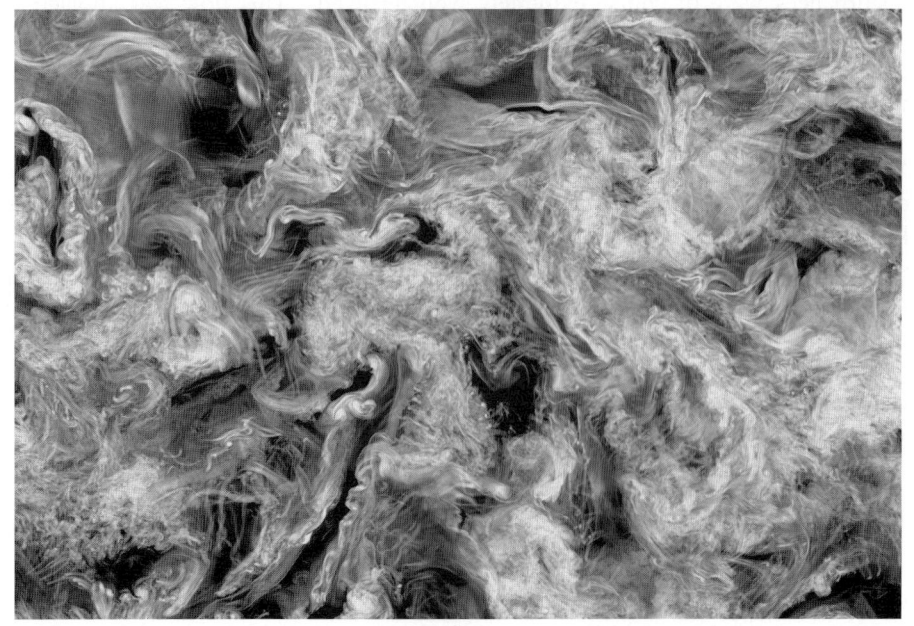

Figure 1. A diagonal slice through the computational domain with a thickness of 200 cells is shown here at time 1.25. The logarithm of the magnitude of the vorticity is visualized, with the highest values white, and then smaller values yellow, red, purple, blue, and black.

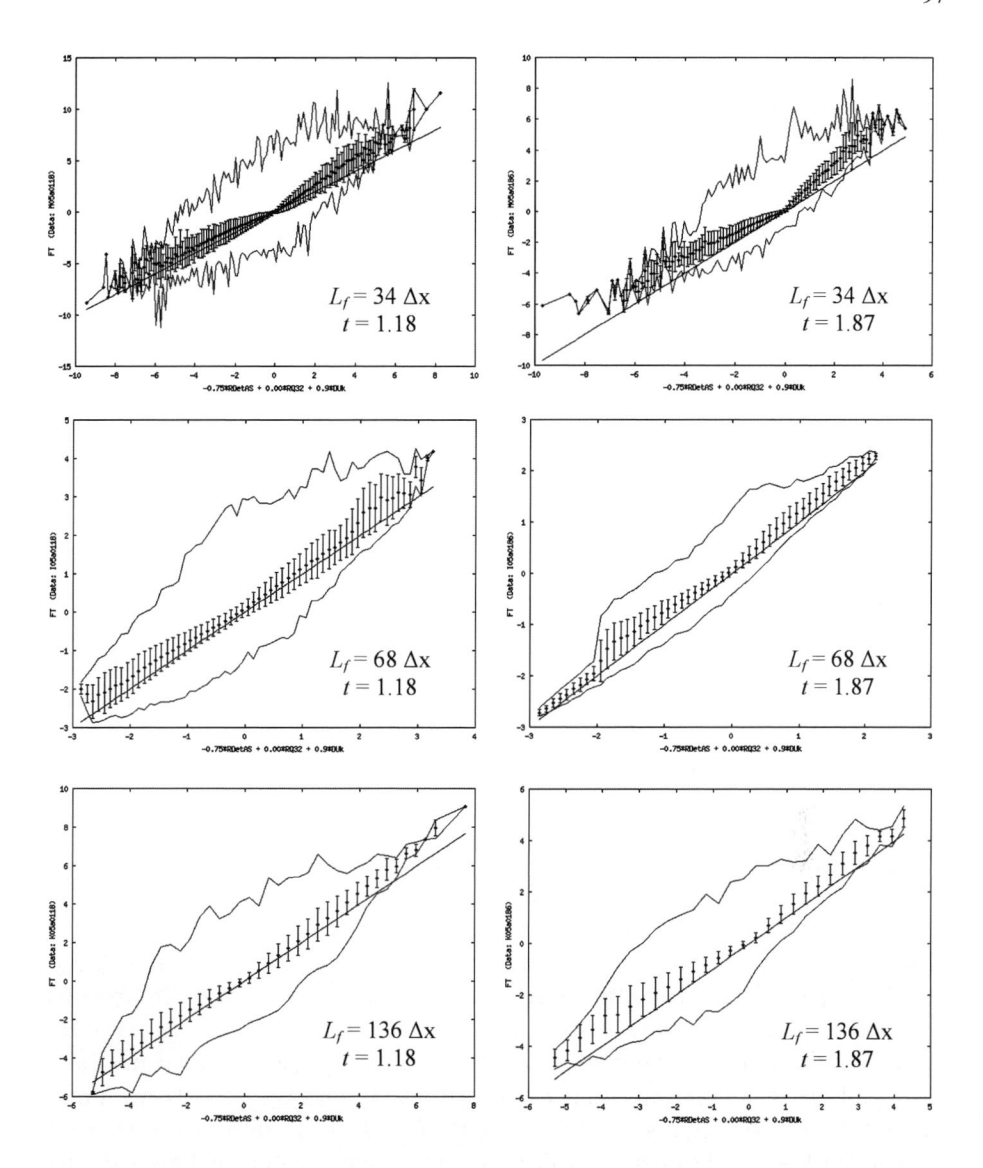

this and other detailed data sets, we have correlated this F_{SGS} term to the topology of the filtered flow field, expressed in terms of the determinant of the deviatoric symmetric rate of strain tensor, given by:

$$(S_D)_{ij} = \frac{1}{2}\left(\frac{\partial \tilde{u}_i}{\partial x_j} + \frac{\partial \tilde{u}_j}{\partial x_i} - \frac{2}{3}\delta_{ij}\nabla\cdot\tilde{u}\right)$$

There is of course also a correlation with the divergence of this velocity field.

$$FT_{MODEL} = AL_f^2\,\bar{\rho}\,\det(S_D) + Ck_{SGS}\,\nabla\cdot\tilde{u}$$

This model equation is intended for use in a large eddy simulation in which the subgrid-scale kinetic energy, k_{SGS}, is carried as an additional independent variable, so that it is available for use in the second, compressional term above. We find that outside of shock fronts the best fits for the coefficients A and C in the model are:

$$A = -0.75, \quad C = -0.67$$

The best fit coefficient for a term in the norm of the rate of strain tensor is zero.

The results of the above set of coefficients are shown on the previous page for data from our 2048^3 PPM simulation at 2 times during the run and for 3 choices of the filter width (the model values are along the x-axes and the actual ones along the y-axes). The fits are very good in all cases, so that we are encouraged to construct a subgrid-scale turbulence model using this model equation. We note that if we combine the term in FT_{MODEL} involving the divergence of the velocity with the term $k_{SGS} \nabla \cdot \tilde{u}$ on the left-hand side of the evolution equation for k_{SGS}, the combination of these terms indicates that this form of internal energy of the gas resulting from unresolved turbulent motions acts as if it had a gamma-law equation of state, with a gamma of 5/3. This is precisely the behavior under compression or expansion that results from the conservation of angular momentum in the case of an isolated line vortex, and consequently the value -0.67 for our coefficient C is no real surprise. Dissipation of this turbulent kinetic energy, k_{SGS}, into heat is not described by the Euler equations and therefore does not appear in the evolution equation for k_{SGS} derived from them. In the Euler simulation of decaying turbulence with PPM, this dissipation is provided by numerical error terms, but the rate of this dissipation is determined by the rate at which the well-resolved, inviscid turbulent cascade produces very-small-scale motions for PPM to damp. This dissipation rate can be extracted from analysis of the simulation data. Reasoning that a local value of k_{SGS} should decrease by a factor C_{decay} in one local SGS eddy turn-over time, we are led to the model that the contribution to Dk_{SGS}/Dt from this dissipation is, approximately, $-C_{decay}\,(k_{SGS}/L_f)\,(2\,k_{SGS}/\overline{\rho})^{1/2}$. A reasonably good fit to the data from the PPM simulation of decaying Mach 1 turbulence on the 2048^3 grid is then obtained using a value of 0.25 for C_{decay}. Using similar reasoning, we may suppose that local variations of k_{SGS} over one filter width should be reduced substantially by diffusion in one local SGS eddy turn-over time. This leads us to model the contribution to Dk_{SGS}/Dt from this diffusion, represented in our evolution equation for k_{SGS} by the bracketed group of divergence terms on the right, by $-C_{diffuse}\,L_f\,(2\,k_{SGS}/\overline{\rho})^{1/2}\,\nabla^2 k_{SGS}$. Our data shows that this model works fairly well with $C_{diffuse} = 0.07$ for all three of our filter widths.

To assess the value of these modeling ideas, we have implemented them in our PPM code. In an attempt to keep the model simple, we use an eddy viscosity mechanism, with a coefficient set from our model for F_{SGS} above, to bring about the exchange of energy, in either direction, between the resolved motions and the

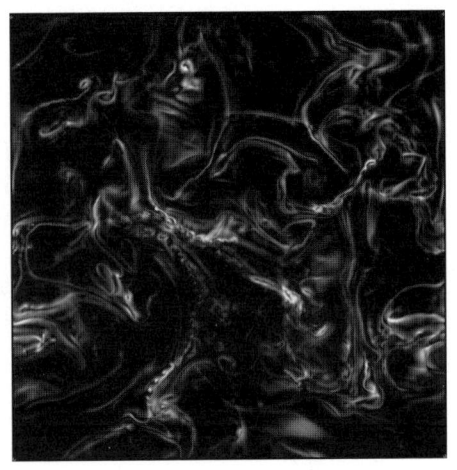

Figure 3. The distribution of k_{SGS} in a thin slice through the computational domain is shown here at time 1.10. The distribution on the left has been obtained using the turbulence model described in the text in the PPM code on a 256^3 grid. The distribution on the right comes from evaluating the differences between u^2 and $\tilde{u}^2$ in the 512^3 bricks of 4^3 cells in the PPM Euler simulation of the 2048^3 grid.

unresolved SGS turbulence. We of course conserve total energy to machine accuracy, and therefore do not allow more energy to go into the resolved motions from the unresolved turbulence than is stored locally (in space and time) in k_{SGS}. As we have briefly described it here, the model has only a single adjustable parameter, the effective filter width, L_f, of our PPM gas dynamics scheme on a given mesh. Experimentation shows that setting this to 3 or 4 times the grid cell width produces noticeable effects upon the simulated flow. At these settings, the smallest eddies produced by PPM without such a turbulence model are absent, and a distribution of k_{SGS} appears roughly in their place. Beyond this, dynamical effects of the model remain to be tested in detail.

Although a thorough analysis of the effects of the turbulence model briefly described above remains to be carried out, integration of this model's evolution equation for k_{SGS} along with the rest of the decaying turbulent flow from its initial conditions can reveal whether or not the ideas and statistical correlations that motivate this model are consistent with the data. We can determine k_{SGS} from the original Euler simulation data on the 2048^3 mesh using filtering techniques, and then we can compare that spatial distribution with the result of a PPM computation with the turbulence model using a 256^3 mesh. The 256^3 mesh is sufficient to capture the energy containing scales in this flow, and therefore we might hope that using the turbulence model on this mesh we could arrive at a good approximation to the result for k_{SGS} from the 2048^3 Euler simulation. These two spatial distributions are compared in Figure 3 in a thin slice through the simulation volume at a time of 1.1, which is about half way to the time when the turbulence is fully developed and its spectrum is not changing in shape. The comparison seems to be quite good.

The turbulence model described here in initial experiments appears to damp the smallest scales of motion that the PPM Euler scheme tends to produce, and then to replace them with a relatively smooth distribution of modeled subfilter-scale turbulent energy. It remains to be established that this combination of resolved and modeled motions is more dynamically faithful to the very high Reynolds number limit of viscous flows in any particular situation. In future work we will investigate this issue. It should also be interesting to compare this model with results of different modeling approaches, such as the dynamical models of Germano et al. [10], Moin et al. [11] or the subgrid-scale model of Misra and Pullin [12], all of which begin with difference schemes that, unlike PPM, have vanishing formal dissipation.

References
1. Woodward, P. R., and P. Colella, "The Numerical Simulation of Two-Dimensional Fluid Flow with Strong Shocks," *J. Comput. Phys.* **54**, 115-173 (1984).
2. Colella, P., and P. R. Woodward, "The Piecewise-Parabolic Method (PPM) for Gas Dynamical Simulations," *J. Comput. Phys.* **54**, 174-201 (1984).
3. Woodward, P. R., "Numerical Methods for Astrophysicists," in *Astrophysical Radiation Hydrodynamics*, eds. K.-H. Winkler and M. L. Norman, Reidel, 1986, pp. 245-326.
4. Woodward, P. R., "The PPM Compressible Gas Dynamics Scheme," up-to-date algorithm description available on the LCSE Web site at www.lcse.umn.edu (2005).
5. Sytine, I. V., D. H. Porter, P. R. Woodward, S. W. Hodson, and K.-H. Winkler 2000, "Convergence Tests for Piecewise Parabolic Method and Navier-Stokes Solutions for Homogeneous Compressible Turbulence," *J. Comput. Phys.*, **158**, 225-238 (2000).
6. Woodward, P. R., D. H. Porter, I. Sytine, S. E. Anderson, A. A. Mirin, B. C. Curtis, R. H. Cohen, W. P. Dannevik, A. M. Dimits, D. E. Eliason, K.-H. Winkler, and S. W. Hodson, "Very High Resolution Simulations of Compressible, Turbulent Flows," in *Computational Fluid Dynamics*, Proc. of the 4th UNAM Supercomputing Conference, Mexico City, June, 2000, edited by E. Ramos, G. Cisneros, R. Fernández-Flores, A. Santillan-González, World Scientific (2001); available at www.lcse.umn.edu/mexico.
7. Cohen, R. H., W. P. Dannevik, A. M. Dimits, D. E. Eliason, A. A. Mirin, Y. Zhou, D. H. Porter, and P. R. Woodward, "Three-Dimensional Simulation of a Richtmyer-Meshkov Instability with a Two-Scale Initial Perturbation," *Physics of Fluids*, **14**, 3692-3709 (2002).
8. Woodward, P. R., Porter, D. H., and Jacobs, M., "3-D Simulations of Turbulent, Compressible Stellar Convection," Proc. 3-D Stellar Evolution Workshop, Univ. of Calif. Davis I.G.P.P., July, 2002; also available at www.lcse.umn.edu/3Dstars.
9. Woodward, P. R., D. H. Porter, and A. Iyer, "Initial experiences with grid-based volume visualization of fluid flow simulations on PC clusters," accepted for publ. in Proc. Visualization and Data Analysis 2005 (VDA2005), San Jose, CA, Jan., 2005.
10. Germano, M., Piomelli, U., Moin, P., and Cabot, W. H. "A dynamic subgrid-scale eddy viscosity model," *Physics of Fluids A* **3**, 1760-1765 (1991).
11. Moin, P., Squires, K., Cabot, W., and Lee, S. "A dynamic subgrid-scale model for compressible turbulence and scalar transport," *Physics of Fluids A*, **3**, 2746 (1991).
12. Misra, A., and Pullin, D. I. "A vortex-based subgrid model for large-eddy simulation," *Physics of Fluids* **9**, 2443-2454 (1997).

Parallel large-eddy simulations of turbulent flows with complex moving boundaries on fixed Cartesian grids

Jianming Yang [a] and Elias Balaras [a] *

[a]Department of Mechanical Engineering, University of Maryland,
College Park, MD 20742, USA

A parallel embedded-boundary approach for large-eddy simulations of turbulent flows with complex geometries and dynamically moving boundaries on fixed orthogonal grids is presented. The underlying solver is based on a second-order fractional step method on a staggered grid. The boundary conditions on an arbitrary immersed interface are satisfied via second-order local reconstructions. The parallelization is implemented via a slab decomposition. Several examples of laminar and turbulent flows are presented to establish the accuracy and efficiency of the method.

1. INTRODUCTION

Today with the advent of inexpensive parallel computers state-of-the-art modeling strategies for simulating turbulent and transitional flows -i.e. direct numerical simulations (DNS) and large-eddy simulations (LES)- are continuously expanding into new areas of applications. Particularly attractive examples, from the perspective of the required spatio-temporal resolution, are from the field of medicine and biology, where in most cases transitional and low Reynolds numbers flows are of interest. The challenge in this case, however, is to properly account for the complex fluid-structure interactions that dominate the dynamics of these flows. The use of numerical methods based on unstructured grids is a possibility, but usually the transition to parallel platforms is fairly complex. Mesh partition, load balance, and the inversion of large sparce matrixes are the most commonly encountered bottlenecks in efficient parallelization. Furthermore, in most cases the intrinsic dissipative discretizations make them problematic in DNS and LES.

The use of embedded boundary formulations, where a complex moving boundary can be modeled on a structured Cartesian grid is an attractive strategy especially for problems with boundaries undergoing large motions/deformations. A basic advantage of this strategy is that highly efficient and energy conserving numerical methods can be used. In addition, their implementation on parallel platforms is usually fairly straightforward. In the this paper we will present an efficient, parallel, embedded-boundary formulation applicable LES of complex external flows. In the next section the numerical method and parallelization strategy will be briefly described. Then, some results demonstrating the efficiency and range of applicability of the approach will be given.

---

*Financial support from the National Institutes of Health (Grant R01-HL-07262) and the National Science Foundation (Grant CTS-0347011) is gratefully acknowledged.

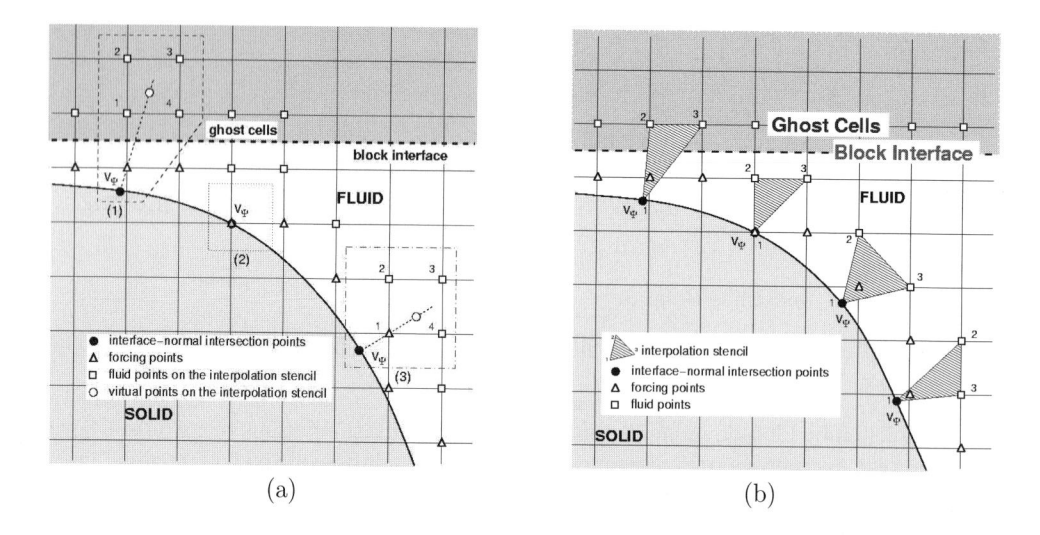

Figure 1. (a) Interpolation stencil in [1], where locations (1) to (3) illustrate possible reconstruction stencils depending on the interface topology and local grid size. (b) The modified interpolation stencil used in the parallel code that utilizes information form one layer of cells near the sub-domain boundary, keeping communication cost to at minimal level.

2. METHODOLOGY

In the LES approach the resolved, large-scale, field is obtained from the solution of the filtered Navier-Stokes equations, where the effect of the unresolved scales is modeled. In the present method, a top-hat filter in physical space is implicitly applied by the finite-difference operators. The resulting subgrid scale (SGS) stresses are modeled using the Lagrangian dynamic SGS model [3]. The filtered equations are discretized using second-order central difference scheme on a staggered grid. Time advancement is done using a fractional-step method, with a second-order Crank-Nicolson scheme for the diffusive terms and a low-storage third-order Runge-Kutta scheme for the convective terms. Both Cartesian and cylindrical coordinate grids can be used.

To compute the flow around complex objects, which are not aligned with the underlying grid we have developed a methodology that practically reconstructs the solution in the vicinity of the body according to the target boundary values [1,7]. The approach allows for a precise imposition of the boundary conditions without compromising the accuracy and efficiency of the Cartesian solver. In particular, an immersed boundary of arbitrary shape is identified by a series of material-fixed interfacial markers whose location is defined in the reference configuration of the solid (see Fig. 1). This information is then used to identify the Eulerian grid nodes involved in the reconstruction of the solution near the boundary in a way that the desired boundary conditions for the fluid are satisfied to the desired order of accuracy. The reconstruction is performed around the points in the fluid phase closest to the solid boundary (all points that have least one neighbor in the solid phase). An advantage of this choice is that it simplifies the treatment of the points that emerge from the solid as the boundary moves through the fixed grid. An example is shown in In Fig. 1.

3. PARALLELIZATION

The parallelization is implemented via slab decomposition and the parallel communication between among blocks is done using the MPI library. To minimize the communications at the block (slab) interface, the decomposition is carried out in the direction with the largest number of points (usually the streamwise direction). Each block is equally sized, which greatly simplifies the mesh partition and provides optimal load balance. Minor modifications are introduced to the original serial solver, as ghost cells at the block interfaces are utilized to provide information from neighboring blocks. In the current implementation, only one layer of ghost cells from the neighboring block is required as a standard second-order central difference scheme is used for the spatial discretization.

Usually the solution of the pressure Poisson equation is the most expensive part of a Navier-Stokes solver, and the parallelization via domain decomposition adds further complications. In the present study the computational grid is always uniform in spanwise direction and, therefore, fast Fourier transforms (FFT) can be used to decompose the Poisson equation in a series of decoupled two-dimensional problems which are then solved using a direct solver [5]. The FFT is performed in each block and no information from neighboring blocks is required. However, global communication is required to swap the slabs in streamwise direction into slabs in wavenumber space, and swap back after the direct solution of the two dimensional problem for each wavenumber. Nevertheless, the overhead for the global communication is small and, as shown in results part, the speedup for the problem under consideration is linear.

The solution reconstruction on immersed boundary points in the neighborhood of the slab interface introduces additional complications in the parallelization strategy. In particular, the interpolation stencil proposed in [1], as shown in Fig. 1a, may contain grid points that do not belong to the neighboring grid lines of the forcing point considered. When a domain decomposition strategy is employed additional communication is required to perform the interpolations. For this reason we have developed an interpolation strategy that utilizes a more compact stencil (see Fig. 1b) that can be implemented in the basic parallel solver without further modifications.

4. RESULTS

4.1. Simulations of flows with stationary boundaries

To validate the accuracy of proposed method, the flow past a sphere for Reynolds numbers ranging from $Re = 50$ to $Re = 1000$ is simulated using cylindrical coordinates ($Re = UD/\nu$, where U is the incoming freestream velocity, D is the cylinder diameter and ν is the kinematic viscosity of the fluid). The flow is steady and axisymmetric for Reynolds number up to 200. For higher Reynolds numbers the symmetry brakes down and the wake is dominated by periodic vortex shedding. On a grid of $384 \times 64 \times 96$

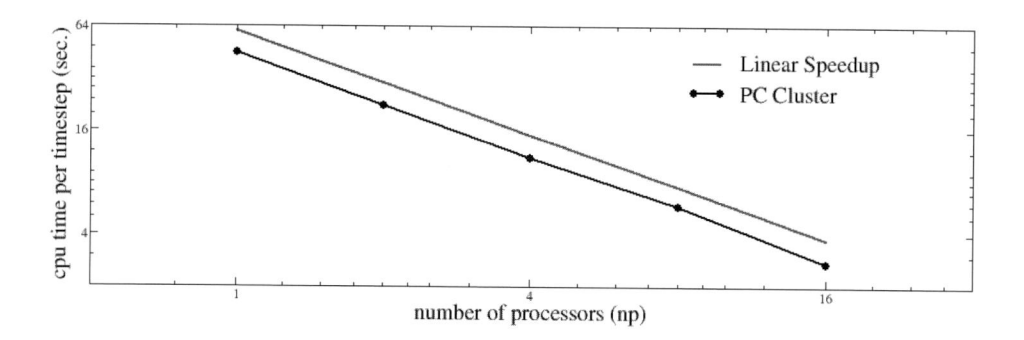

Figure 2. Flow around a sphere at $Re = 300$. Parallel performance for a $384 \times 64 \times 96$ grid.

(streamwise × azimuthal × radial direction) points, the predicted mean drag and lift coefficients for $Re = 300$ are $C_D = 0.655$ and $C_L = 0.064$, respectively, which are within 1% to the values reported in [2], in which a body-fitted method was adopted. The parallel speedup for this case is shown in Fig. 2. Linear speedup is obtained up to 16 processors, which was the limit of the cluster where the preliminary tests were carried out.

A DNS of the transitional flow past a sphere at $Re = 1000$ was also performed on a grid with nearly 15 million points. At this Reynolds number small scales are present in the wake and the flow becomes chaotic further downstream. In Fig. 3a the instantaneous vortical structures visualized by iso-surfaces of the second invariant of the velocity gradient tensor, Q, are shown together with azimuthal vorticity contours. The roll-up of the shear layer, the development of large hairpin structures, and the breakdown of larger structures into smaller scales can be clearly observed. The flow becomes turbulent in the far wake. The results are in good agreement with the DNS using a high-order spectral method reported in [6].

A more challenging test-case is the computation of transitional flow around an Eppler E387 airfoil. The angle of attack is 10° and the Reynolds number based on the cord length and the incoming velocity is 10,000. LES have been performed in this case on a grid with approximately 18×10^6 nodes ($672 \times 48 \times 560$ in the streamwise, spanwise and cross-stream directions respectively). An instantaneous snapshot of the flow field which is characterized by a massive separation on the top surface is shown in Fig. 4.

4.2. Simulations of flows with moving boundaries

In this section LES of turbulent flow over a flexible wavy wall undergoing transverse motion in a form of streamwise traveling wave is presented. The immersed boundary in this case has a non-uniform prescribed velocity varying with time. Also, the availability of accurate DNS data in the literature [4], allows for a quantitative analysis of the accuracy the proposed algorithm in turbulent flows. The parametric space and computational box in our LES were selected to match the reference DNS by Shen et. al. [4], where the

Figure 3. Transitional flow past a sphere at $Re = 1000$. (a) Instantaneous vortical structures visualized using iso-surfaces of $Q = 0.1$ colored by the streamwise vorticity; (b) Instantaneous azimuthal vorticity contours.

location of the wall boundary as a function of time is given by $y_w(t) = a \sin k(x - ct)$ (a is the magnitude of the oscillation, $k = 2\pi/\lambda$ is the wavenumber with λ the wavelength, and c is the phase-speed of the traveling wave). The wave steepness is, $ka = 0.25$, and Reynolds number, $Re = U\lambda/\nu = 10,170$ (U is the mean freestream velocity). A grid of $288 \times 88 \times 64$ in the streamwise, wall normal, and spanwise direction, respectively is used for all simulations. The grid is uniform in the streamwise and spanwise homogeneous directions, and is stretched in the vertical direction. In Fig. 5 the instantaneous vortical structures visualized using iso-surfaces , Q, are shown for $c/U = 0.0$ and $c/U = 0.4$ (c/U is the ratio of the wave speed to the freestream velocity). It can be seen that the strong streamwise vortices that are characteristic of stationary wavy walls ($c/U = 0$) are suppressed as c/U increases from 0.0 to 0.4. Similar behavior has been observed in the reference DNS [4]. Quantitative comparisons are shown in Fig. 6, where the variation of F_f and F_p as a function of the phase speed of the traveling wavy wall, c/U, are shown (F_f is the total friction force, F_p is total pressure force on the wall in the streamwise direction, respectively). Our data are in good agreement with the reference simulation [4].

To further demonstrate the robustness of the present method in handling realistic turbulent and transitional flow problems that involve complex three-dimensional boundaries consisting of multiple moving parts, we have computed the flow past a mechanical bileaflet, heart valve in the aortic position. The shape and size of the leaflets roughly mimics the St. Jude Medical (SJM) standard bileaflet, which is commonly used in clinical practice. The movement of the leaflets is prescribed according to a simplified law that resembles

106

(a)

(b)

Figure 4. Flow around an Eppler E387 airfoil at angle of attack 10° and $Re = 10,000$. (a) Instantaneous vortical structures visualized using iso-surfaces of Q colored by the streamwise vorticity; (b) Instantaneous spanwise vorticity contours.

the real movement of the leaflets as determined by their interaction with the blood flow. The configuration which involves the interaction of the flow with a stationary (the aortic chamber) and two moving (the leaflets) boundaries is a great challenge for any structured or unstructured boundary-conforming method. In the present computation the overall geometry is immersed into a Cartesian grid and the boundary conditions on the stationary and moving boundaries are imposed using the method described in the previous sections. The Reynolds number is 4000 based on the peak systole velocity and the artery diameter, which is in the physiologic regime. The computational grid involves $420 \times 200 \times 200$ nodes in streamwise, spanwise (direction parallel to the rotation axis of the leaflets), and transverse directions, respectively. The total number of nodes in this computation is 16.8 million.

The flow just downstream of the leaflets is very complex, and it is dominated by intricate vortex-leaflet and vortex-vortex interactions. The present method can accurately capture the thin shear layers that form on both moving and stationary immersed boundaries. To better illuminated the highly three-dimensional nature of the complex vortex interactions, iso-surfaces of Q are shown in Fig 7. Two consecutive instantaneous realizations have been selected ($t/T = 0.7$ and $t/T = 0.8$) which correspond to instances in time near the end of the opening of the valve. In Fig 7a the early stage of the formation of a ring vortex

(a) $c/U = 0.0$ (b) $c/U = 0.4$

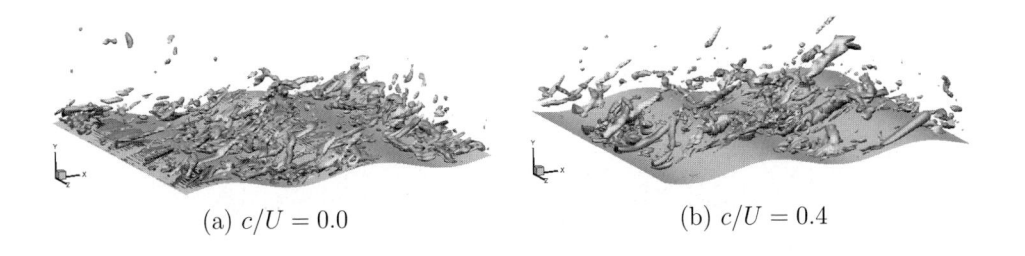

Figure 5. Instantaneous vortical structures visualized using iso-surfaces of $Q = 8.0$ colored by the streamwise vorticity.

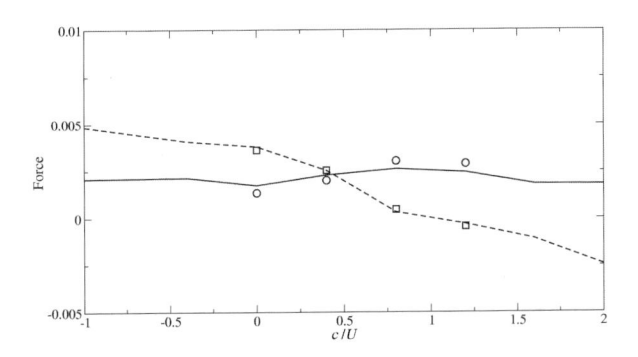

Figure 6. Mean force acting on the traveling wavy boundary as a function of c/U. solid line and circle is total friction force, F_f; dash line and square is the total pressure force, F_p. Lines are the DNS results in [4], and symbols are the present results.

at the expansion plane (structure A) can be seen. At the same time the motion of the leaflets also generate another strong vortex (structure B). At a later time (Fig 7b) both structures evolve and interact with each other and the surrounding flow giving rise to the complex vortex dipoles.

5. CONCLUSIONS

A parallel embedded-boundary formulation for LES of turbulent flows with complex geometries and dynamically moving boundaries is presented. The overall strategy is to take advantage of the accuracy and scalability of structured Cartesian solves to generate a robust tool applicable to problems that would be a great challenge to any boundary conforming formulation. The computation of the flow around the heart valve for example, has been performed on a Linux desktop workstation equipped with four AMD Opteron

(a) $t/T = 0.7$ (b) $t/T = 0.8$

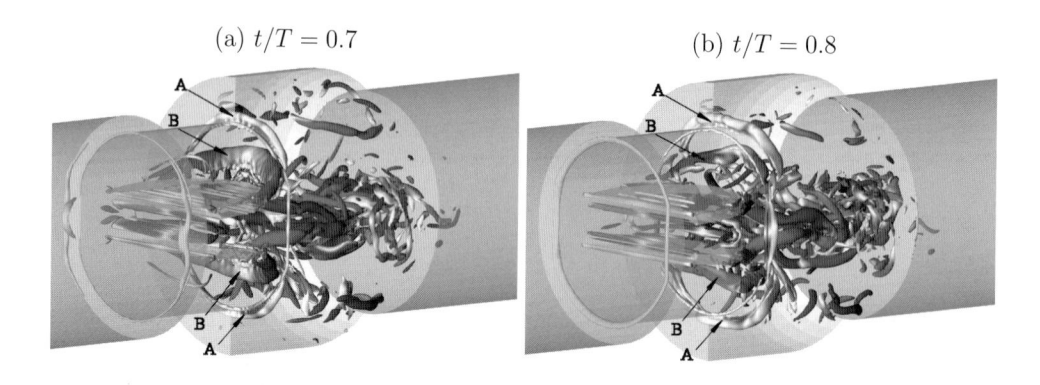

Figure 7. Instantaneous vortical structures visualized using iso-surfaces of $Q = 80$ colored by the streamwise vorticity.

846 2GHz processors (each with 1M cache, and 8GB memory in total) and takes about 6 CPU hours for one pulsatile cycle. A 10 cycle simulation which with 17 million points can be completed in less than 3 days. For a variety of problems, especially from biology where low/moderate Reynolds numbers are encountered, methods such the present one have a lot of advantages.

REFERENCES

1. E. Balaras, Modeling complex boundaries using an external force field on fixed Cartesian grids in large-eddy simulations. Computers & Fluids, 33 (2004):375–404.
2. T. A. Johnson and V. C. Patel, Flow past a sphere up to a Reynolds number of 300. J. Fluid Mech. 378 (1999):353–385.
3. C. Meneveau, C. S. Lund, and W. H. Cabot, A Lagrangian dynamic subgrid-scale model of turbulence. J. Fluid Mech. 319 (1996):353–385.
4. L. Shen, X. Zhang, D. K. P. Yue, and M. S. Triantafyllou, Turbulent flow over a flexible wall undergoing a streamwise traveling wave motion. J. Fluid Mech. 484 (2003):197–221.
5. P.N. Swartzrauber, Direct method for the discrete solution of separable elliptic equations. SIAM J. Num. Anal., 11 (1974):1136–.
6. A. G. Tomboulides and S. A.Orszag, Numerical investigation of transitional and weak turbulent flow past a sphere. J. Fluid Mech. 416 (2000):45–73.
7. J. Yang and E. Balaras, An embedded-boundary formulation for large-eddy simulation of turbulent flows interacting with moving boundaries. J. Comput. Phys., (2005) in review.

Direct Numerical Simulation of Turbulent Flows on a low cost PC Cluster

F.X. Trias[a], M. Soria[a], A. Oliva[a] and C.D. Pérez-Segarra[a]

[a]Centre Tecnològic de Transferència de Calor (CTTC)
ETSEIT, c/ Colom 11, 08222 Terrassa, Spain
e-mail: cttc@cttc.upc.edu, web page: http://www.cttc.upc.edu

A code for the direct numerical simulation (DNS) of incompressible turbulent flows, that provides fairly good scalability even on a low cost PC clusters has been developed. The spatial discretization of the incompressible Navier-Stokes equations is carried out with the fourth-order symmetry-preserving scheme by Verstappen and Veldman [1]. The main bottleneck, the Poisson equation, is solved with a Direct Schur-Fourier Decomposition (DSFD) [2,3] algorithm. This method allows to solve each Poisson equation to almost machine accuracy using *only one all-to-all communication*. Here, an overview of the extension of the DSFD algorithm to high-order Poisson equations is given; its advantages and limitations are discussed focusing on the main differences from its second-order counterpart [2,3]. Benchmark results illustrating the robustness and scalability of method on a PC cluster with 100 Mbits/s network are also presented and discussed. Finally, illustrative direct numerical simulations of wall-bounded turbulent flows are presented.

Keywords: Direct Numerical Simulation; PC clusters; parallel Poisson solver; Schur complement method;

1. INTRODUCTION

Direct numerical simulation (DNS) of transition and turbulence flows is one of the typical examples of an application with huge computing demands, that needs parallel computers to be feasible. Efficient and scalable algorithms for the solution of Poisson equation that arises from the incompressibility constraint are of high interest in this context. The most efficient sequential methods for the Poisson equation, such as Multigrid, are difficult to parallelize efficiently because they need a large number of communications between the processors. Due to the network latency, the time needed for each of them is not negligible, even if the messages are small, such as for the halo update operation at the coarsest levels. Krylov-subspace methods also need several communication episodes (for matrix-vector products, scalar products and preconditioning) in order to solve each Poisson equation. In low cost PC clusters, with a convectional 100 Mbits/s network these communications have a very high cost. Therefore, parallel algorithms tolerant to low network performance are of high interest for DNS of incompressible flows.

The main purpose of this work is to describe the extention of the Direct Schur-Fourier Decomposition (DSFD) [2,3] algorithm for the solution of higher order Poisson equations.

In previous versions of our DNS code [4,5], the DSFD algorithm was used to solve the well-known standard second-order Poisson equation. This general extension have allowed to use the fourth-order symmetry-preserving schemes by Veldman and Verstappen [1] for the spatial discretization. Finally, illustrative direct numerical simulations of wall-bounded turbulent flows are presented.

2. OVERVIEW OF THE NAVIER-STOKES SOLVER

2.1. Governing Equations

We consider a cavity height L_z, width L_y and depth L_x (height aspect ratio $A_z = L_z/L_y$ and depth aspect ratio $A_x = L_x/L_y$) filled with an incompressible Newtonian viscous fluid of thermal diffusivity α and kinematic viscosity ν. To account for the density variations, the Boussinesq approximation is used. Thermal radiation is neglected. Under these assumptions, the dimensionless governing equations in primitive variables are

$$\nabla \cdot \mathbf{u} = 0 \tag{1a}$$

$$\frac{\partial \mathbf{u}}{\partial t} + (\mathbf{u} \cdot \nabla)\mathbf{u} = \frac{Pr}{Ra^{0.5}}\Delta \mathbf{u} - \nabla p + \mathbf{f} \tag{1b}$$

$$\frac{\partial T}{\partial t} + (\mathbf{u} \cdot \nabla)T = \frac{1}{Ra^{0.5}}\Delta T \tag{1c}$$

where Ra is the Rayleigh number $(g\beta\Delta T L_z^3)/(\nu\alpha)$, $Pr = \nu/\alpha$ and the body force vector is $\mathbf{f} = (0,0,PrT)$. The cavity is subjected to a temperature difference across the vertical isothermal walls ($T(x,0,z) = 1$, $T(x,1/A_z,z) = 0$) while the top and bottom walls are adiabatic. At the four planes $y = 0, y = \frac{1}{4}, z = 0, z = 1$, null velocity is imposed. Periodic boundary conditions are imposed in the x direction.

2.2. Spatial and temporal discretizations

Equations (1a-1c) are discretized on a staggered grid in space by second- or fourth-order symmetry-preserving schemes [1]. Such family of discretizations preserves the underlying differential operators symmetry properties. These global discrete operator properties ensure both stability and that the global kinetic-energy balance is exactly satisfied even for coarse meshes if incompressibility constraint is accomplished. More details about these schemes can be found [1].

For the temporal discretization, a central difference scheme is used for the time derivative term, a fully explicit second-order Adams-Bashforth scheme for both convective and diffusive terms, and a first-order backward Euler scheme for the pressure-gradient term and mass-conservation equation. To solve the velocity-pressure coupling we use a classical fractional step projection method.

3. DIRECT SCHUR-FOURIER DECOMPOSITION

Spatial decomposition is the typical approach to parallelize CFD codes. As the momentum and energy equations (1b,1c) are discretized in a fully explicit method, their parallelization is straightforward. Thus, the main problem is the Poisson equation arising from the incompressibility constraint (1a).

In this work, we propose a parallel Direct Schur-Fourier Decomposition (DSFD) method for the direct solution of the arbitrary order Poisson equation. It is based on a combination of a Direct-Schur [6,7] method and a Fourier decomposition. Fourier decomposition allows to decouple the unknowns in the periodic direction. As FFT parallelization is not efficient on loosely coupled parallel computers, DSFD uses a domain decomposition only in the two directions orthogonal to the Fourier decomposition. Each bidimensional decoupled problem is solved with a parallel Direct Schur decomposition (DSD) [6], that is based on the fact that the matrix of coefficients remains constant during all the fluid flow simulation. This allows us to evaluate and store the inverse of the interface matrix of each equation in a pre-processing stage. Then, in the solution stage, all the systems are solved together. Only *one all-to-all communication episode* is needed to solve each three-dimensional Poisson equation to machine accuracy.

3.1. Problem Definition

Once the Poisson equation is discretized the problem can be written in the form[1]

$$A^{3d} x^{3d} = b^{3d} \qquad (2)$$

where $A^{3d} \in \mathbb{R}^{N \times N}$ is a singular symmetric matrix, and vectors $x^{3d} \in \mathbb{R}^N$ and $b^{3d} \in \mathbb{R}^N$ can be divided into $n_{p2} n_{p3}$ sub-vectors with n_{p1} components each. With this partition and considering a column- or row-wise ordering of grid points (lexicographical ordering) the resulting matrix A^{3d} is a block $(6M+1)$-diagonal matrix with a regular sparsity pattern

$$A^{3d} = \begin{pmatrix} A^p_{1,1} & A^{n_1}_{1,1} & A^{n_2}_{1,1} & \cdots & A^{t_1}_{1,1} & A^{t_2}_{1,1} & \cdots & & & & \\ A^{s_1}_{2,1} & A^p_{2,1} & A^{n_1}_{2,1} & A^{n_2}_{2,1} & \cdots & A^{t_1}_{2,1} & A^{t_2}_{2,1} & \cdots & & & \\ & & \ddots & & & & & & & & \\ \cdots & A^{b_2}_{j,k} & A^{b_1}_{j,k} & \cdots & A^{s_2}_{j,k} & A^{s_1}_{j,k} & A^p_{j,k} & A^{n_1}_{j,k} & A^{n_2}_{j,k} & \cdots & A^{t_1}_{j,k} & A^{t_2}_{j,k} & \cdots \\ & & & & & & \ddots & & & & \\ & & & & \cdots & A^{b_2}_{n_{p2},n_{p3}} & A^{b_1}_{n_{p2},n_{p3}} & \cdots & A^{s_2}_{n_{p2},n_{p3}} & A^{s_1}_{n_{p2},n_{p3}} & A^p_{n_{p2},n_{p3}} \end{pmatrix} \in \mathbb{R}^{N \times N} \quad (3)$$

where $A^{nd}_{j,k}$, $A^{sd}_{j,k}$, $A^{td}_{j,k}$ and $A^{bd}_{j,k}$ are $n_{p1} \times n_{p1}$ diagonal matrices $A^{nd}_{j,k} = a^{nd}_{j,k} I \in \mathbb{R}^{n_{p1} \times n_{p1}}$, where $a^{nd}_{j,k}$ is a constant and $A^p_{j,k}$ are $n_{p1} \times n_{p1}$ symmetric circulant [8] matrices of the form

$$A^p_{j,k} \equiv \begin{pmatrix} a^p_{j,k} & a^{e_1}_{j,k} & a^{e_2}_{j,k} & \cdots & a^{e_1}_{j,k} \\ a^{e_1}_{j,k} & a^p_{j,k} & a^{e_1}_{j,k} & \cdots & a^{e_2}_{j,k} \\ a^{e_2}_{j,k} & a^{e_1}_{j,k} & a^p_{j,k} & \cdots & a^{e_3}_{j,k} \\ \vdots & \vdots & \vdots & \ddots & \vdots \\ a^{e_1}_{j,k} & a^{e_2}_{j,k} & a^{e_3}_{j,k} & \cdots & a^p_{j,k} \end{pmatrix} \in \mathbb{R}^{n_{p1} \times n_{p1}} \qquad (4)$$

---

[1] The superindex $3d$ is used here to express that each unknown is coupled with neighbouring unknowns in the three directions.

Using the previous block matrices, equation (2) can be expressed in the form

$$A^p_{j,k}\, x_{j,k} + \sum_{d=1}^{M} \left(A^{b_d}_{j,k}\, x_{j,k-d} + A^{s_d}_{j,k}\, x_{j-d,k} + A^{n_d}_{j,k}\, x_{j+d,k} + A^{t_d}_{j,k}\, x_{j,k+d} \right) = b_{j,k} \qquad (5)$$

where $1 \le j \le n_{p2}$, $1 \le k \le n_{p3}$, $M = o - 1$ is the scheme stencil size and o the order of accuracy of the scheme, and the terms that correspond to non-existent block vectors (outside the domain) should be eliminated.

3.2. Fourier Decomposition

As we showed in the previous section, all the $A^p_{j,k}$ matrices are symmetric real-valued circulant. Thus, their diagonalization is straightforward

$$Q_{\mathbb{R}}^{-1} A^p_{j,k} Q_{\mathbb{R}} = \Lambda_{j,k} \qquad (6)$$

where $\Lambda_{j,k} \in \mathbb{R}^{n_{p1} \times n_{p1}}$ is a diagonal matrix with the eigenvalues of $A^p_{j,k}$ down the diagonal.

The previous relation allows to decompose the original block diagonal system (2) into a set of n_{p1} independent block diagonal equations. To do so, Eq.(5) is premultiplied by $Q_{\mathbb{R}}^{-1}$ and the sub-vectors $x_{j,k}$ are expressed as $Q_{\mathbb{R}} \hat{x}_{j,k}$. After these operations, the $A^p_{j,k}$ matrices become diagonal while the $A^{nb}_{j,k}$ matrices, that are the identity matrix multiplied by a scalar, are not affected (e.g., $Q_{\mathbb{R}}^{-1} a^{n_d}_{j,k} I Q_{\mathbb{R}} = a^{n_d}_{j,k} I$). The resulting equations are

$$\lambda_{j,k}\, \hat{x}_{j,k} + \sum_{d=1}^{M} \left(A^{b_d}_{j,k}\, \hat{x}_{j,k-d} + A^{s_d}_{j,k}\, \hat{x}_{j-d,k} + + A^{n_d}_{j,k}\, \hat{x}_{j+d,k} + A^{t_d}_{j,k}\, \hat{x}_{j,k+d} \right) = \hat{b}_{j,k} \qquad (7)$$

This expression, like Eq.(5), is a block diagonal equation with $n_{p2} n_{p3}$ unknowns, each of them with n_{p1} components. But in Eq.(7), as the non-diagonal entries of matrices $A^p_{j,k}$ have been eliminated, unknown $x_{i,j,k}$ is only coupled with unknowns in the same plane i. Therefore, selecting the i component of the $n_{p2} n_{p3}$ block equations, we obtain a block diagonal scalar equation system that can be expressed as

$$\hat{a}^p_{i,j,k}\, \hat{x}_{i,j,k} + \sum_{d=1}^{M} \left(a^{b_d}_{j,k}\, \hat{x}_{i,j,k-d} + a^{s_d}_{j,k}\, \hat{x}_{i,j-d,k} + a^{n_d}_{j,k}\, \hat{x}_{i,j+d,k} + a^{t_d}_{j,k}\, \hat{x}_{i,j,k+d} \right) = \hat{b}_{i,j,k} \qquad (8)$$

where the term of the main diagonal is the i entry of diagonal matrix $\Lambda_{j,k}$, $\hat{a}^p_{i,j,k} = \lambda_{i,j,k}$. Therefore, Eq.(8) can be expressed more compactly as

$$\hat{A}_i \hat{x}_i = \hat{b}_i \qquad i = 1 \cdots n_{p1} \qquad (9)$$

where $\hat{A}_i \in \mathbb{R}^{n_{p2} n_{p3} \times n_{p2} n_{p3}}$ is a block diagonal matrix associated with the transformed equation for plane i. The operations to be performed to solve the equation system are

1. Calculate the n_{p1} transformed right-hand-side sub-vectors, $\hat{b}_{j,k} = Q_{\mathbb{R}}^{-1} b$.
2. Solve the n_{p1} decoupled block diagonal equation systems $\hat{A}_i \hat{x}_i = \hat{b}_i$.
3. Carry out the anti-transformation of the n_{p1} solution sub-vectors $x_{j,k} = Q_{\mathbb{R}} \hat{x}_{j,k}$.

3.3. Direct Schur Decomposition (DSD)

Each of the n_{p1} system of linear equations (9) is solved with a Direct Schur Decomposition (DSD) method [6,2]. To simplify the notation, the hats and sub-indices are dropped and each of the n_{p1} block diagonal equations (9) to be solved is denoted as

$$Ax = b \qquad (10)$$

where $A \in \mathbb{R}^{n_{p2}n_{p3} \times n_{p2}n_{p3}}$ and $x, b \in \mathbb{R}^{n_{p2}n_{p3}}$. The unknowns in vector x are partitioned into a family of P subsets, called inner domains, that are labelled from 0 to $P-1$, plus one interface, labelled s, defined so that the unknowns of each subset are not directly coupled with the other subsets. With this partition, the system (10) can be expressed using block matrices as

$$\begin{pmatrix} A_{0,0} & 0 & \cdots & 0 & A_{0,s} \\ 0 & A_{1,1} & \cdots & 0 & A_{1,s} \\ \vdots & & & \vdots & \vdots \\ 0 & 0 & \cdots & A_{P-1,P-1} & A_{P-1,s} \\ A_{s,0} & A_{s,1} & \cdots & A_{s,P-1} & A_{s,s} \end{pmatrix} \begin{pmatrix} x_0 \\ x_1 \\ \vdots \\ x_{P-1} \\ x_s \end{pmatrix} = \begin{pmatrix} b_0 \\ b_1 \\ \vdots \\ b_{P-1} \\ b_s \end{pmatrix} \qquad (11)$$

For the solution of the reordered system, the interface unknowns are isolated using block Gaussian elimination. System (11) is transformed to

$$\begin{pmatrix} A_{0,0} & 0 & \cdots & 0 & A_{0,s} \\ 0 & A_{1,1} & \cdots & 0 & A_{1,s} \\ \vdots & & & \vdots & \vdots \\ 0 & 0 & \cdots & A_{P-1,P-1} & A_{P-1,s} \\ 0 & 0 & \cdots & 0 & \tilde{A}_{s,s} \end{pmatrix} \begin{pmatrix} x_0 \\ x_1 \\ \vdots \\ x_{P-1} \\ x_s \end{pmatrix} = \begin{pmatrix} b_0 \\ b_1 \\ \vdots \\ b_{P-1} \\ \tilde{b}_s \end{pmatrix} \qquad (12)$$

The last block equation, involving only unknowns in $x_s \in \mathbb{R}^{n_s}$, is the interface equation

$$\tilde{A}_{s,s} x_s = \tilde{b}_s \qquad (13)$$

with the modified right-hand side $\tilde{b}_s = b_s - \sum_{p=0}^{P-1} A_{s,p} A_{p,p}^{-1} b_p$. And the Schur complement matrix $\tilde{A}_{s,s} = A_{s,s} - \sum_{p=0}^{P-1} A_{s,p} A_{p,p}^{-1} A_{p,s}$. Therefore, the interface equation (13) can be used to solve exactly x_s before the inner domains. Once x_s is known, each of the x_p can be obtained independently by its owner p, solving its original equation

$$A_{p,p} x_p = b_p - A_{p,s} x_s \qquad (14)$$

Finally, note that $\tilde{A}_{s,s} \in \mathbb{R}^{n_s \times n_s}$ is a dense matrix. Both $\tilde{A}_{s,s}$ and $\tilde{b}_s$ can be evaluated without the explicit calculation of $A_{p,p}^{-1}$, as described in [6].

3.4. Solution of $A^{3d}x^{3d} = b^{3d}$ system

After the completion of the pre-processing stage, where the n_{p1} block diagonal transformed matrices $\hat{A}_i$ are evaluated and their Schur decomposition computed, the solution of each three-dimensional equation is carried out in three steps:

1. Evaluate the transformed n_{p1} right-hand side vectors $\hat{b}_{j,k} = Q_{\mathbb{R}} b_{j,k}$.
2. Solve the n_{p1} block diagonal equations $\hat{A}_i \hat{x}_i = \hat{b}_i$ using Schur decomposition to obtain $\hat{x}$.
3. Evaluate the final solution undoing the transformation, $x = Q_{\mathbb{R}}^{-1} \hat{x}_{j,k}$.

The second step could be performed invoking n_{p1} times DSD algorithm outlined in previous sections. In a straightforward implementation n_{p1} global summation operations would be needed. However, all the messages can be grouped into just one to save latency time proceeding as follows (the notation $[X]_i$ refers to the repetition of operation X for i-index values between 1 and n_{p1}):

Step 2. Solution of $\left[\hat{A}\hat{x} = \hat{b}\right]_i$ {

    2.1-Solve $[A_{p,p} t = b_p]_i$

    2.2-Evaluate the local contribution to the r.h.s. of the interface equations, $\left[\tilde{b}_s^p = A_{s,p} t\right]_i$

    2.3-Carry out the global summation $\left[t = \sum_{p=0}^{P-1} \tilde{b}_s^p\right]_i$

    2.4-Evaluate the r.h.s of the interface equations, $\left[\tilde{b}_s = b_s - t\right]_i$

    2.5-Evaluate the interface nodes $\left[x_s = \tilde{A}_{s,s}^{-1} \tilde{b}_s\right]_i$ where needed

    2.6-Solve locally the inner nodes from $[A_{p,p} x_p = b_p - A_{p,s} x_s]_i$

}

Vector t is used to denote a temporary storage area, different for each i plane. To perform the communication operation (step 2.3), the n_{p1} local vectors are concatenated into a single vector with $n_{p1} n_s$ components, that is added in a *single all-to-all communication* operation. So, a significant reduction of latency time is achieved grouping all messages into just one.

4. PARALLEL PERFORMANCE AND ILLUSTRATIVE RESULTS

4.1. The DSFD Benchmark Results

The computing times have been measured using a PC cluster with 24 standard PCs (AMD K7 CPU 2600 MHz and 1024 Mbytes of RAM and a conventional network (100 Mbits/s 3COM cards and a CISCO switch), running Debian Linux 3.0 kernel version 2.4.23. MPI has been used as message-passing protocol (LAM 6.5.8).

Speed-up results. The second and fourth-order speed-ups corresponding to different meshes have been represented in Fig. 1(left). The times for sequential executions have been estimated because they can not be directly measured for lack of RAM memory. Also, under the previous considerations respect to the computing time, for the same n_{p2} and n_{p3}, S can be estimated to be at least as high as in the case of DSD algorithm [6].

Comparison with other algorithms. The behaviour of DSFD can be analyzed comparing it with a well-known sequential alternative method (allowing it to take advantage of the periodicity in one direction too), such as a combination of a Fourier decomposition and a conventional algebraic multigrid (ACM [9]) solver for each penta-diagonal equation. In [6], DSD was estimated to be about 30 times faster than ACM for $P = 24$, a mesh of 235×470 control volumes and stopping the iterations when reduction of the initial residual by a factor 10^{-3} was achieved. The same estimation holds for problems of size $n_{p1} \times 235 \times 470$ solved with the present method, with an accuracy several orders of magnitude higher. However, the Fourier/ACM approach does not involve a pre-processing stage like DSFD.

4.2. Illustrative Results

The direct numerical simulation (DNS) of a buoyancy-driven turbulent natural convection flow in an enclosed cavity has been used as a problem model. Several illustrative DNS results of instantaneous temperature maps and mean values of second-order statistics obtained with $Ra_z = 6.4 \times 10^8$, 2×10^9 and 10^{10}, with $Pr = 0.71$ (air), height and width aspect ratios 4 and 1 respectively are shown in Fig. 1. A detailed description of this cases can be found in [3].

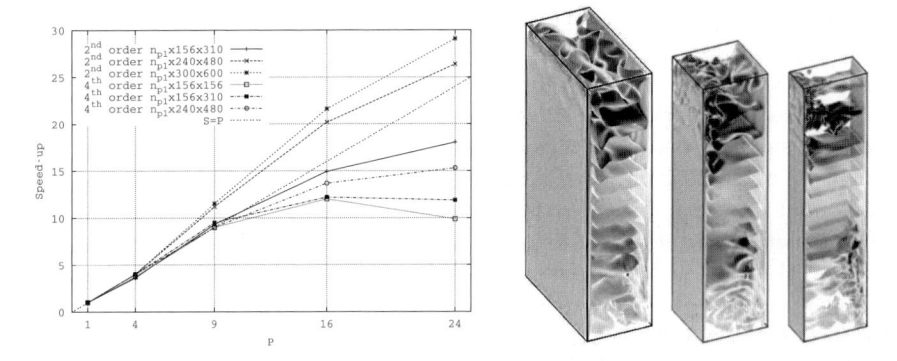

Figure 1. Left: Estimated speed-ups for different meshes and orders. Right: (from left to right), three instantaneous temperature fields at $Ra_z = 6.4 \times 10^8$, 2×10^9 and 10^{10} respectively, with $Pr = 0.71$ (air).

Despite the geometric simplicity of these configurations, it has to be noted that turbulent natural convection is a very complex phenomenon, that is not still well understood [10]. These configurations are of high interest because the majority of the previous works concerning direct numerical simulations of turbulent natural convection flows assumed a bidimensional behaviour.

5. CONCLUSIONS

A robust and fairly scalable direct parallel algorithm for the solution of the three-dimensional high-order Poisson equations arising in incompressible flow problems with a periodic direction has been described. After a pre-processing stage, it provides the solution of each right-hand-side using only one all-to-all communication episode per time-step, so it is specially intended for loosely-coupled parallel computers, such as PC clusters with a low-cost network. It is based on a combination of a direct Schur Complement method and a Fourier decomposition. The Schur Complement part is based on the method presented in [6,2] extended to higher order problems.

The good efficiency and accuracy of the proposed method have been shown by performing several numerical experiments on a PC cluster of 24 nodes with 1024 Mbytes of RAM each and a conventional 100 Mbits/s network. To benchmark the DSFD algorithm a direct numerical simulation (DNS) of turbulent natural convection problem used as a problem model. According to the measured computing times and numerical accuracies, the algorithm proposed is a suitable method for the use on low cost PC clusters for DNS/LES simulations. Speed-up results up to 24 processors have proved that the algorithm scalability for large enough meshes is good.

REFERENCES

1. R. W. C. P. Verstappen and A. E. P. Veldman. Symmetry-Preserving Discretization of Turbulent Flow. *Journal of Computational Physics*, 187:343–368, May 2003.
2. M. Soria, C. D. Pérez-Segarra, and A.Oliva. A Direct Schur-Fourier Decomposition for the Solution of the Three-Dimensional Poisson Equation of Incompressible Flow Problems Using Loosely Parallel Computers. *Numerical Heat Transfer, Part B*, 43:467–488, 2003.
3. F. X. Trias, M. Soria, C. D. Pérez-Segarra, and A. Oliva. A Direct Schur-Fourier Decomposition for the Efficient Solution of High-Order Poisson Equations on Loosely Coupled Parallel Computers. *Numerical Linear Algebra with Applications*, (in press).
4. M. Soria, F. X. Trias, C. D. Pérez-Segarra, and A. Oliva. Direct Numerical Simulation of Turbulent Natural Convection Flows Using PC Clusters. In B. Chetverushkin, A. Ecer, J. Periaux, N. Satofuka, and P. Fox, editors, *Parallel Computational Fluid Dynamics, Advanced Numerical Methods, Software and Applications*, pages 481–488. Elsevier, May 2003.
5. M. Soria, F. X. Trias, C. D. Pérez-Segarra, and A. Oliva. Direct numerical simulation of a three-dimensional natural-convection flow in a differentially heated cavity of aspect ratio 4. *Numerical Heat Transfer, part A*, 45:649–673, April 2004.
6. M. Soria, C. D. Pérez-Segarra, and A. Oliva. A Direct Parallel Algorithm for the Efficient Solution of the Pressure-Correction Equation of Incompressible Flow Problems Using Loosely Coupled Computers. *Numerical Heat Transfer, Part B*, 41:117–138, 2002.
7. Natalja Rakowsky. The Schur Complement Method as a Fast Parallel Solver for Elliptic Partial Differential Equations in Oceanography. *Numerical Linear Algebra with Applications*, 6:497–510, 1999.
8. P.J.Davis. *Circulant Matrices*. Chelsea Publishing, New York, 1994.
9. B.R.Hutchinson and G.D.Raithby. A Multigrid Method Based on the Additive Correction Strategy. *Numerical Heat Transfer*, 9:511–537, 1986.
10. J. Salat, S. Xin, P. Joubert, A. Sergent, F. Penot, and P. Le Quéré. Experimental and numerical investigation of turbulent natural convection in large air-filled cavity. *International Journal of Heat and Fluid Flow*, 25:824–832, 2004.

Large Eddy Simulation of turbulent Couette-Poiseuille flows in a square duct

Wei Lo[a], and Chao-An Lin[a*]

[a]Department of Power Mechanical Engineering, National Tsing Hua University, Hsinchu 300, TAIWAN

The fully developed turbulent Couette-Poiseuille flow in a square duct is simulated using Large Eddy Simulation(LES). Focus has been taken on the secondary flow near the moving wall. Results of present simulation show one dominate vortex and a relatively small vortex presented near the moving wall. They block the high momentum fluid near the center of the duct into the corner regions and have different effects on both the wall shear stress distribution and turbulence statistics compared with that near the stationary wall. The generation of the secondary flow is further investigated by studying the streamwise vorticity transport equation. It is found that the major generation mechanism of the dominate vortex is the normal Reynolds stress anisotropy and the viscous diffusion and Reynolds shear stress anisotropy act as transport mechanisms for the mean vorticity.

1. Introduction

The turbulent flow within a square duct is rather complex, due to the presence of two inhomogeneous directions. Near the corner, a transverse circulatory flow arises resulting from the an-isotropy of the turbulent stresses. The secondary motion is identified as secondary flow of the second kind by Prandtl [2], which is directed towards the wall in the vicinity of the corner bisector. As a consequence, it convects momentum and scalar quantities from the central region to the walls. The secondary velocities are usually only about 1 to 3 percent of the streamwise bulk velocity in magnitude but have been observed to appreciably alter the rate of momentum and heat transfer near the walls.

Gavrilakis [6]performed a direct numerical simulation (DNS) of Poiseuille flow within a square duct at a Reynolds number 4410 (based on bulk values), where 16.1×10^6 grid points with second order finite-difference techniques were employed. Gavrilakis's simulation were in reasonable agreement with the mean flow and turbulence statistics obtained from experiment. DNS study of square duct flow at higher Reynolds number, 10320 based on bulk values, was investigated by Huser and Biringen [1] using a spectral/high-order finite-difference scheme and a time-splitting integration method. Mechanisms responsible for the generation of the secondary flow are studied by quadrant analysis. By connecting the ejection and burst structures to the generation of the anisotropic Reynolds stresses, further physical understanding was provided about the generation of secondary flow.

---

*Email address: calin@pme.nthu.edu.tw

To reduce the computational cost, Madabhushi and Vanka [4] performed large eddy simulation of a square duct flow with Reynolds number of 5810 (based on bulk values). Comparison between the LES and DNS results clearly demonstrates that LES is capable of capturing most of the energy carried by turbulence eddies and can accurately predict the generally accepted pattern of the turbulence-driven secondary flow.

In the present study, the LES method is applied to simulate turbulent Couette-Poiseuille flows in a square duct. The moving wall would effectively reduce the mean shear near the wall, and hence the turbulent structures and the secondary flow patterns within the duct are modified. The formation of the secondary motion is further investigated by analyzing the streamwise mean vorticity transport equation.

2. Governing Equations and Modeling

The governing equations for the LES simulation are obtained by applying the filtering operation. The grid-filtered, incompressible continuity and Navier-Stokes equations assume the following forms:

$$\frac{\partial \rho \bar{u}_j}{\partial x_j} = 0 \qquad (1)$$

$$\frac{\partial \rho \bar{u}_i}{\partial t} + \frac{\partial (\rho \bar{u}_i \bar{u}_j)}{\partial x_j} = -\frac{\partial \bar{P}}{\partial x_i} + \frac{\partial}{\partial x_j}[\mu(\frac{\partial \bar{u}_i}{\partial x_j} + \frac{\partial \bar{u}_j}{\partial x_i})] - \frac{\partial \tau^s_{ij}}{\partial x_j} \qquad (2)$$

where, $\tau^s_{ij} = \rho(\overline{u_i u_j} - \bar{u}_i \bar{u}_j)$ is the sub-grid stress due to the effects of velocities being not resolved by the computational grids and has to be modeled.

In the present study, the Smagorinsky model[5] has been used for the sub-grid stress(SGS) such that,

$$\tau^s_{ij} = -(C_s \Delta)^2 \frac{1}{\sqrt{2}} \sqrt{(S_{kl} S_{kl})} S_{ij} + \frac{2}{3} \rho k_{sgs} \delta_{ij} \qquad (3)$$

where $C_s = 0.1$, $S_{ij} = \frac{\partial \bar{u}_i}{\partial x_j} + \frac{\partial \bar{u}_j}{\partial x_i}$, and $\Delta = (\Delta x \Delta y \Delta z)^{1/3}$ is the length scale. It can be seen that in the present study the mesh size is used as the filtering operator. A Van Driest damping function accounts for the effect of the wall on sub-grid scales is adopted here and takes the form as, $l_m = \kappa y[1 - exp(-\frac{y^+}{25})]$, where y is the distance to the wall and the length scale is redefined as, $\Delta = Min[l_m, (\Delta x \Delta y \Delta z)^{1/3}]$.

3. Numerical Algorithms

The framework of the present numerical procedure incorporates the finite volume method and the staggered grid arrangement. The integration method is based on a semi-implicit, fractional step method proposed by Kim and Moin [3]. The non-linear terms and wall parallel diffusion terms are advanced with the Adams-Bashfoth scheme in time, whereas the Crank-Nicholson scheme is adopted for the diffusion terms in the wall-normal direction. The explicit treatment of the non-linear terms eliminates the need of linearization and the implicit treatment of the diffusion terms removes the severe diffusion time step constrain in the near wall region.

The computer code based on above numerical algorithms was written in Fortran language and had been parallelized by the message passing interface (MPI) library. In the present parallel implementation, the single program multiple data (SPMD) environment is adopted. The domain decomposition is done on the last dimension of the three dimensional computation domain due to the explicit numerical treatment on that direction. The parallel performance on the IBM-P690 and Hp-Superdome is shown in Figure 2. The IBM-P690 has better performance than the Hp-Superdome which is primarily caused by the higher internal system bandwidth.

4. Results

Three cases were simulated: case P is a turbulent Poiseuille flow in a square duct, while CP's are turbulent Couette-Poiseuille flow. The Reynolds numbers, $U_w D/\nu$, of the Couette-Poiseuille flows are 9136 and 11420 for CP1 and CP2, respectively. The simulation technique is first validated by simulating the Poiseuille flow at a comparable Reynolds number. The LES results of the Poiseuille flow exhibit reasonable agreement with DNS results from Huser and Biringen [1]. Numerical mesh of present simulation is (128x128x96) in the spanwise, wall normal, and streamwise direction,respectively.

4.1. Mean flow field in turbulent Couette-Poiseuille flow

Figure 3 shows the mean cross-plane velocity vector and streamwise velocity contour of case CP1 and CP2 in half of the duct for the geometric symmetry. The streamwise velocity contour distribution near the top and bottom of the square duct have very different spatial shape which suggests that the secondary flow on these regions is different. The contour lines near the bottom corners have similar distribution as Poiseuille flow while contour lines near top of the duct show bulge towards the center of the duct from the corners. The vector plots in Figure 3 confirms that secondary flow is different between top and bottom section of the square duct.

From Figure 3(b), which demonstrates the secondary flow near the top corners, one dominate vortex and a relatively small vortex are present near the moving wall. The size of the smaller vortex increases in tandem with the increase of U_w and the maximum streamwise velocity is observed to move upward from the center of the duct because the additional input of streamwise momentum caused by the moving wall. The secondary motion of the Couette-Poiseuille flow on the top corners has very different effect on the streamwise momentum transport compared with the Poiseuille flow. Unlike the Poiseuille flow which the secondary motion convects the streamwise momentum into the corners, the secondary motion of Couette-Poiseuille flow actually "blocks" this momentum transport into the top corners. This "block" effect is caused by the smaller vortex which is centered at the corner bisector. With the smaller vortex occupied the corner region, the direction of the larger vortex near the corners now been redirected to the vertical wall of the relative corner. The streamwise momentum therefore, is convected to the vertical wall and can be identified by the distorted contour lines in Figure 3 near y=0.8. By virtue of continuity the larger vortex carries the low momentum fluid from the near wall region and transports them to the center of the duct along the upper moving wall. This transportation of low momentum by the larger vortex is more evident in Case CP2 which the lid speed (U_w) is faster and consist with the contour distribution in Figure 3(a) near y=0.9.

These flow features give rise to the average wall stress ($\tau_w/\overline{\tau_w}$) distribution shown in Figure 4. The distribution at the stationary wall follows the DNS results of a turbulent Poiseuille flow because the similar vortex structure near the stationary wall. However, distributions at the moving wall show a dramatic increase near the corner region at x=0.1 and x=0.3 of Case CP1 and Case CP2, respectively. This increase is caused by the transportation of the low momentum fluid onto the moving wall by the secondary flow. From Figure 3, the contour lines show a bulge from the corner towards the center region of the duct. The length of the bulged feature is evident about 0.1 of Case CP1 and 0.3 of Case CP2. It is consistent with the increase of the wall stress ratio at the same horizontal position. Therefore, the wall stress distribution at the stationary and moving wall is very different from each other and is expected to be effective to the turbulence production.

4.2. Turbulence statistics in turbulent Couette-Poiseuille flow

The rms values of fluctuating velocity components along the wall bisector are shown in Figure 5. Near the stationary wall the fluctuating components have similar profiles as the Poiseuille flow. Away from the stationary wall (y=0) ,however, the velocity fluctuations decrease monotonously towards the moving wall (y=1). Therefore, the turbulence intensities near the wall bisector is higher on the stationary wall than on the moving wall. Because the secondary motion is weakest near the corresponding region the turbulence intensities are affected considerably by the turbulence coming from the stationary wall. Figure 6 shows the turbulence intensities at x=0.2 where the secondary motion on both the moving and stationary wall is evident. Away from the top and bottom corners intensities show an increase compared to that at the wall bisector because this position near the high turbulence generation near-wall region. Near the stationary wall the intensities still follow the Poiseuille case P with the peak values of cross-plane components smaller and higher than that on the moving wall. Near the moving wall($y > 0.75$) the intensities of Couette-Poiseuille flow are larger than that at the wall bisector and this increase does not appear in the Poiseuille flow. Refer to Figure 3, the line of x=0.2 is near the center of the larger vortex on the upper half of the square duct. Therefore, this increase in the turbulent intensities may be related to the effect of the larger vortex near the top corners.

4.3. Origin of the secondary flow

The mechanism governing the formation of the secondary flow can be identified by noting the streamwise mean vorticity equation, $\Omega_z = \frac{\partial U}{\partial y} - \frac{\partial V}{\partial x}$:

$$U\frac{\partial \Omega_z}{\partial x} + V\frac{\partial \Omega_z}{\partial y} = \nu(\frac{\partial^2 \Omega_z}{\partial x^2} + \frac{\partial^2 \Omega_z}{\partial y^2}) + \underbrace{(\frac{\partial^2}{\partial x^2} - \frac{\partial^2}{\partial y^2})(\overline{uv})}_{\text{Shear stress contribution}} + \underbrace{\frac{\partial^2}{\partial y \partial x}(\overline{v^2} - \overline{u^2})}_{\text{normal stress contribution}} \quad (4)$$

Anisotropy of the turbulence stresses is the driving force that generates the secondary flow. It can be observed that both the shear and normal stresses contribute to the generation of the streamwise vorticity. Because the turbulence is highly anisotropic near the corners therefore the production events of the secondary flow is also originated near the corners. The results of LES simulation is used to calculate the generation terms in Eq. 4. Focus will be on the effect of the moving wall on the vorticity generation and therefore the distribution is plotted near the top-left corner of the square duct.

For the Poiseuille square duct flow the shear and normal stresses contributions are approximately the same values around the corner region. The difference between the absolute maximum of shear and normal stress contribution is about 4 percent.

However, for Couette-Poiseuille flow the maximum production from the normal stress is larger than the shear stress as much as 52 percent and 67 percent for Case CP1 and CP2, respectively.

A more direct comparison between all terms in Eq. 4 is made in Figure 7 and Figure 8 for case CP1 and CP2, respectively. Each term is plotted along the line x=0.2 near the horizontal position of the center of the larger vortex on top corners. In the region of $d < 0.2$ the normal stress production is the dominate production for both case CP1 and CP2. The viscous diffusion and shear stress production play a major part of the transport of mean vorticity. In case CP1 the shear stress is weak and this transport is most done by viscous diffusion and in CP2 is by the shear stress contribution. Away from the top wall $d > 0.5$, viscous diffusion and convection effect become relatively small and the mean vorticity is governed by the shear and normal stress production only.

5. Conclusions

Turbulent Couette-Poiseuille flows in a square duct are simulated by Large eddy simulation. The secondary flow pattern of the turbulent Couette-Poiseuille flow is different from the Poiseuille flow. Near the moving wall the vortex system is largely modified. One dominate vortex and a relatively small vortex are now present near the moving wall. The secondary motion on the upper corners has very different effect on the streamwise momentum transport compared with the Poiseuille flow. Unlike the Poiseuille flow which the secondary motion convects the streamwise momentum into the corners, the secondary motion of Couette-Poiseuille flow actually "blocks" this momentum transport into the upper corners. This has made the wall stress distribution on the moving and stationary wall very different.

The turbulence level near the moving wall is reduced compared to other bounding walls which is caused by the damped turbulence production. This is attributed the decreased mean shear rate on the top moving wall.

The evident secondary motion in the square duct is examined by the streamwise vorticity transport Eq 4. Focus has been taken on the top corners where the vortex system is very different from the Poiseuille flow. The maximum shear stress contribution is lessen as much to 67 percent to normal stress for case CP2. For the generation of the larger vortex near the top corners the normal stress contribution is the dominate production mechanism where the viscous diffusion and shear stress contribution are act as transport mechanisms.

6. Acknowledgments

The computational facilities are provided by the National Center for High-Performance Computing of Taiwan which the authors gratefully acknowledge.

REFERENCES

1. A. Huser and S. Biringen, Direct numerical simulation of turbulent flow in a square duct, J. Fluid Mech. **257** (1993) 65-95.
2. L. Prandtl, Uber die ausgebildcte turbulenz, Verh. 2nd Intl. Kong. Fur Tech. Mech., Zurich,[English translation NACA Tech. Memo. 435, 62]. (1926)
3. Kim, J. and Moin, P., Application of a fractional-step method to incompressible NavierVStokes equations. J. Comput. Phys. **177** (1987) 133V166.
4. R. K. Madabhushi and S. P. Vanka, Large eddy simulation of turbulencedriven secondary flow in a square duct, Phys. Fluids A **3**, (1991) 2734-2745.
5. Smagorinsky, J., General circulation experiments with the primitive equations. I. The basic experiment. Mon. Weather Rev. **91** (1963) 499-164.
6. S. Gavrilakis, Numerical simulation of low-Reynolds-number turbulent flow through a straight square duct, J. Fluid Mech. **244**, (1992) 101-129.

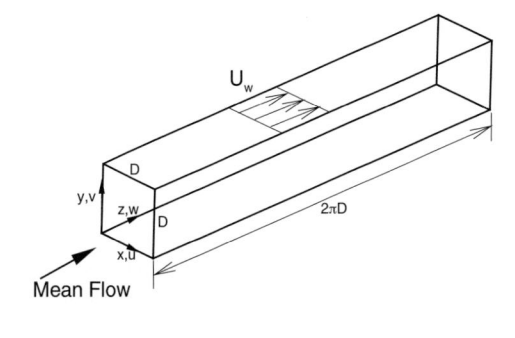

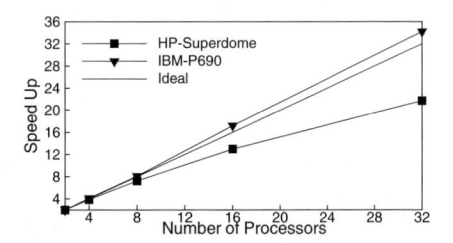

Figure 1. Schematic plot of the square duct with a moving wall.

Figure 2. Parallel performance of present simulation.

123

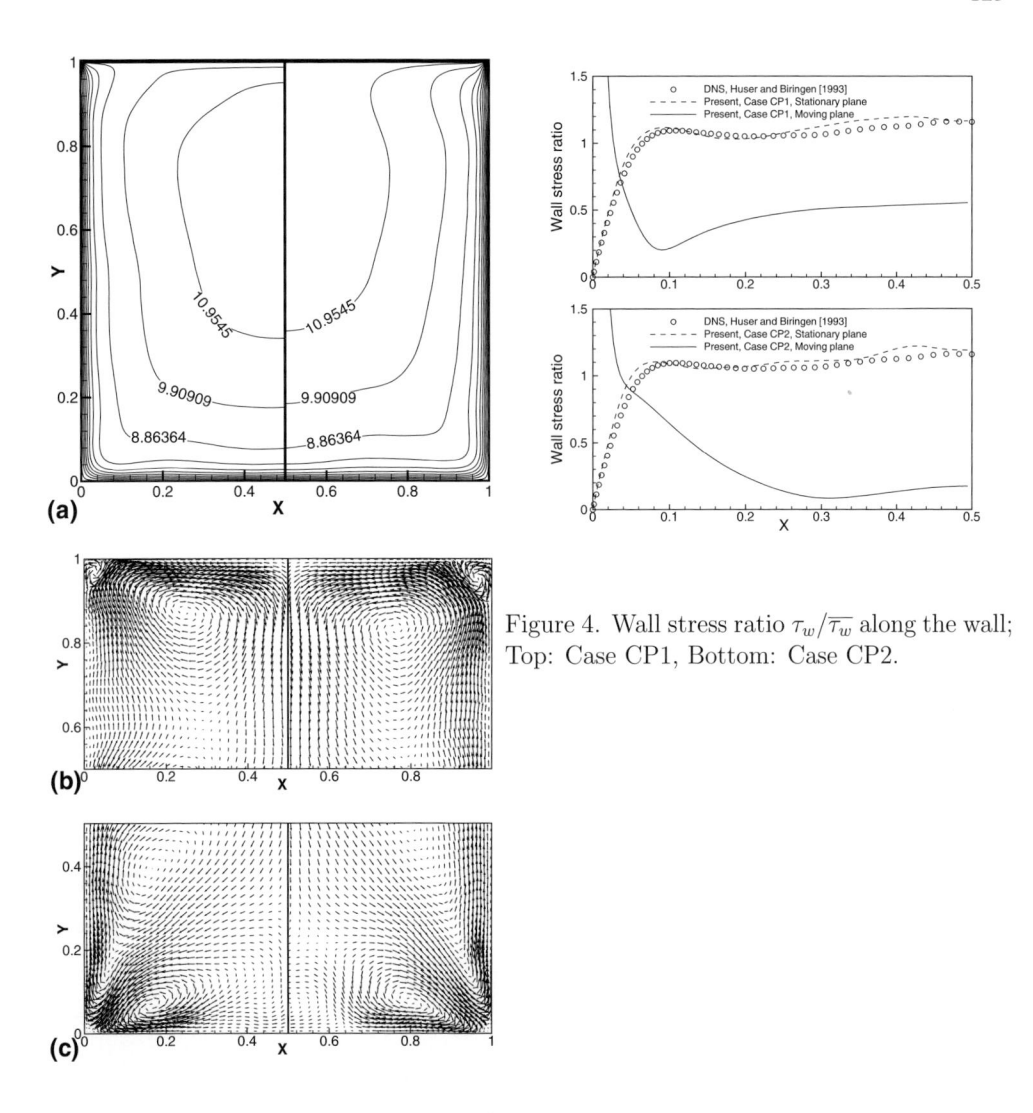

Figure 4. Wall stress ratio $\tau_w/\overline{\tau_w}$ along the wall; Top: Case CP1, Bottom: Case CP2.

Figure 3. (a)Mean streamwise velocity contour, (b)Mean secondary velocity vectors near top corners (c)Mean secondary velocity vectors near bottom corners, Only half of the duct is shown. Left:case CP1,Right:case CP2.

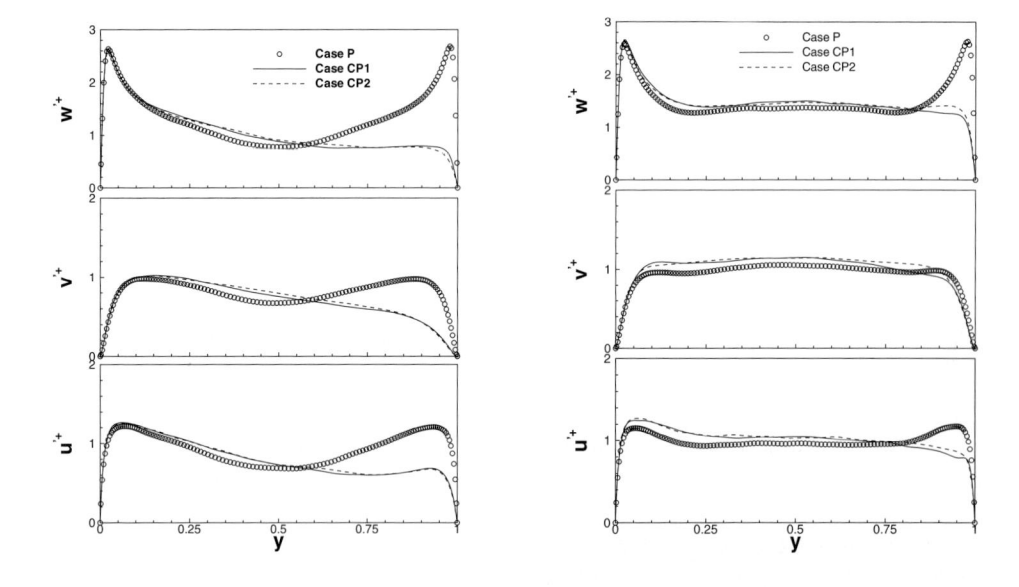

Figure 5. Turbulence intensities along the wall bisector.

Figure 6. Turbulence intensities along the line of x=0.2.

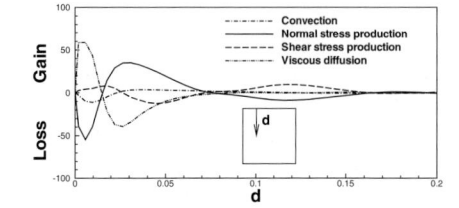

Figure 7. Budget of mean streamwise vorticity transport of case CP1 along the line x=0.2

Figure 8. Budget of mean streamwise vorticity transport of case CP2 along the line x=0.2

Development of a framework for a parallel incompressible LES solver based on Free and Open Source Software.

R. Giammanco and J. M. Buchlin [a]

[a]Department of Environmental and Applied Fluid Dynamics, Von Karman Institute, Chaussèe de Waterloo 72, B-1640 Rhode-St-Genèse, Belgium

A new natively parallel LES solver has been developed in VKI in order to relieve a legacy code [1] used until now for fundamental turbulence research; all the shortcomings of the legacy code have been addressed and modern software technologies have been adopted both for the solver and the surrounding infrastructure, delivering a complete framework based exclusively on Free and Open Source Software (*FOSS*) to maximize portability and avoid any dependency from commercial products.

The role and importance of the freely available software in the development of the new framework will be discussed in the present article; the philosophy and reasons behind the current approach, together with the current status of the solver and some samples of computations, will be presented as well.

Keywords: Software Development, FOSS, LES, Parallelism

1. Solver characteristics

The new solver wanted to provide as much continuity as possible with the legacy one in order to insure a smooth transition, but it was developed to include a series of potential improvements that could be tested and taken advantage of in the future whenever possible.

The basic characteristics are the same: finite differences, staggered grid cell arrangement, multi-domains for moderately complex geometries, grid conformity across the domains, fractional time step method for incompressible flows, explicit viscous terms, Adam-Bashforth time stepping, basic Smagorinsky model for SGS modeling and central convective schemes.

The previous code was *formally* 2nd order accurate in time and space, as long the grid was weakly stretched, and was sporting the Arakawa formulation for the convective terms; the new solver can replicate exactly the same behavior but it can provide some enhancements in the discretization of the velocity derivatives and in the way the incompressibility constrain is satisfied: the convective derivatives can be computed using the a FOSS CAS (see § 2.1 and § 1.1 respectively) to assure 2nd order accuracy, or a modified version of the high-resolution central schemes proposed in [2] can be used. In the first case Arakawa is still needed to stabilize the solution. The pressure derivatives can be computed with two points along one direction in the classical way [3], or with a not centered stencil involving three points. More details will be provided in oncoming publications.

An easy parallelization is achieved thanks to the multi-domain approach assigning each domain to a different processor, at the price of deteriorating the parallel efficiency of the code, always deeply influenced by the initial choice of the domains; the new solver, on the other hand, has been enriched with the capability of auto-partitioning a domain across different processors, as will be seen in § 2.2.

1.1. MAXIMA

All the discretized formulas used in the solver have been automatically generated from the FOSS CAS[1] MAXIMA [5]. Using this software for symbolic computations, is possible to perform a series of interesting tasks, like computing the discretized formula of a mathematical entity over a certain stencil. The basic principle used is the Taylor expansion of the given function around the desired point between the ones of the chosen stencil, and in the determination of all the possible solutions with their relative truncation errors. See below for an snipped of code for the computation of the Taylor expansion.

```
Maxima                                <conv:Taylor>
F:tgt_f;
f1:tgt_f;
for l:1 thru (order) do
(F:diff(F,t),
  F:subst(expand(xyz_new[0]-x_0),diff(x,t),F),
  F:subst(expand(xyz_new[1]-y_0),diff(y,t),F),
  F:subst(expand(xyz_new[2]-z_0),diff(z,t),F),
  f1:f1+F/(l!));
```

A series of programs have been developed and inserted in the CVS (§ 2.5) repository of the framework to allow the developer to derive all the formulas he could need automatically, and after to simply *copy paste* them from one window to another, avoiding all possible errors that could be make coding the coefficients manually.

This procedure achieves the purpose of reducing the possibility of error and allowing the developer to be able to test rapidly different schemes, evaluating their behavior in terms of accuracy, speed, stability etc. etc. Steps in this direction, of evaluating automatically the stability and efficiency for the schemes from within MAXIMA are underway.

This approach can be extended to the non-linear terms as well: this time there are two functions that are expanded in Taylor series in *different* discrete points, and the derivations are appropriately more complicated. Implicit formulas (compact schemes) have been successfully derived as well.

2. Framework characteristics

In the previous § the LES solver characteristics, as they are currently implemented, have been briefly reviewed. In the design phase of the new framework, most of the exigencies of the solver were known: parallel execution, simultaneous execution of pairs of master-slave simulations [2], multi-domain approach, time advancement scheme etc. etc. On the base of those needs and of the past experiences, the new framework was designed

---

[1] *C*omputer *A*lgebra *S*ystem
[2] Consider the case of a Backward Facing Step being feed its inlet profile from a simultaneously running turbulent plane channel simulation.

to satisfy a series of requirements: being written in a programming language more flexible than Fortran77/90, being natively parallel, while incorporating a basic load-balancing and auto-partitioning mechanism, being extensively documented *both* from the code and the algorithms point of view, establishing a well defined and reusable framework for parallel communications, maximizing code reuse and preferring code clarity to performance, including extensive error handling, logging *Q*uality *A*ssurance capabilities, supplying examples for functions optimization, providing a version managing and bug tracking system, providing data in well defined formats readable by a set of approved programs in serial and parallel context and finally using exclusively Free and Open Source Software.

For the first point, the *C* language has been chosen: the new framework needs to interact easily with the underlying OS, to perform a series of data and file manipulation tasks easily and efficiently, and to interface to a series of libraries for different purposes. Most of the other points requires sections of their owns.

2.1. Free Open Source Software

While even the title of present article boldly recalls the use of *F*ree and *O*pen *S*ource *S*oftware (commonly referred as FOSS), we will just briefly explain this concept at the present, suggesting a visit to the Free Software Foundation web site http://www.fsf.org for further informations.

A crude idea behind the FOSS movement is that software, like knowledge, should be free, as in free speech, not as in free of charge; the software should be seen as a mean, not as an end, of a production cycle, or, as taken from the FSF site:

> Free software is a matter of the users' freedom to run, copy, distribute, study, change and improve the software.

From a scientific and academic point of view, making use of FOSS for not commercial purposes is probably the most efficient way of using software and libraries, and to avoid being tied to any particular platform or extension. The FOSS, providing its source code, implies open standards : all the data stored will always be accessible from both commercial and free software, and a code based on FOSS has no portability limits, since all the supporting software can be replicated at a new site at the cost of a download from a ftp repository. While it is true that commercial software is sometimes more polished and more efficient than FOSS, the freedom of being able to replicate the software installation in the office, in a massively parallel cluster or in a private PC thousands of kilometers away, is without equal. Insisting on the educative aspect of FOSS, there is no better way to learn to use a given program or library, being it PETSc, OpenDX or else, than being able to see its internals and how they actually work, and eventually modifying them.

Each software comes with a specific license, and while the license number proliferation required the appointment of a specific body for the license examinations [3] all the software used for the current solver does not pose any limitation for the purposes of developing, running and publishing results on scientific publications and journals.

2.2. Load Balancing and domain auto-partitioning

As seen in § 1, relatively complex geometries can be achieved via the multi-domain approach; generally speaking, this technique does not allow for a fine control of the com-

---

[3] *OSI certification mark and program*, http://www.opensource.org.

putational expenses for each single domain. Furthermore, the conformity of the grid across domains can induce an excessive presence of cells in some regions of the flow that slows down the computations. As said before, there was no way to alleviate this inconvenience intrinsic to the multi-domain approach, if not indirectly, with a judicious management of the computational resources at disposal.

The task of distributing a certain number of domains to a set of processors is equivalent to the weighted partitioning of a graph whose nodes are not anymore single cells as in the case of unstructured grids, but whole domains. If this procedure is of course more coarse, it allows to use all those algorithms and relative libraries already developed in literature. In the current case, the METIS [6] library is used to perform the graph partitioning of the computation at run-time. In Fig. 1, there is a graphic representation of the way a five domains computation is partitioned across two processors. The graph is a file actually produced by the solver at run-time via the GRAPHVIZ suite for monitoring purposes.

In the norm there are obviously more processors than domains: in this case, *at the price of not trivial programming*, the code has been enriched with the possibility of splitting one or more domains, once or recursively, and to distribute the sub-domains around. The extreme example is a single domain subdivided across multiple processors, like the case of a three-dimensional lid driven cavity (Fig. 2).

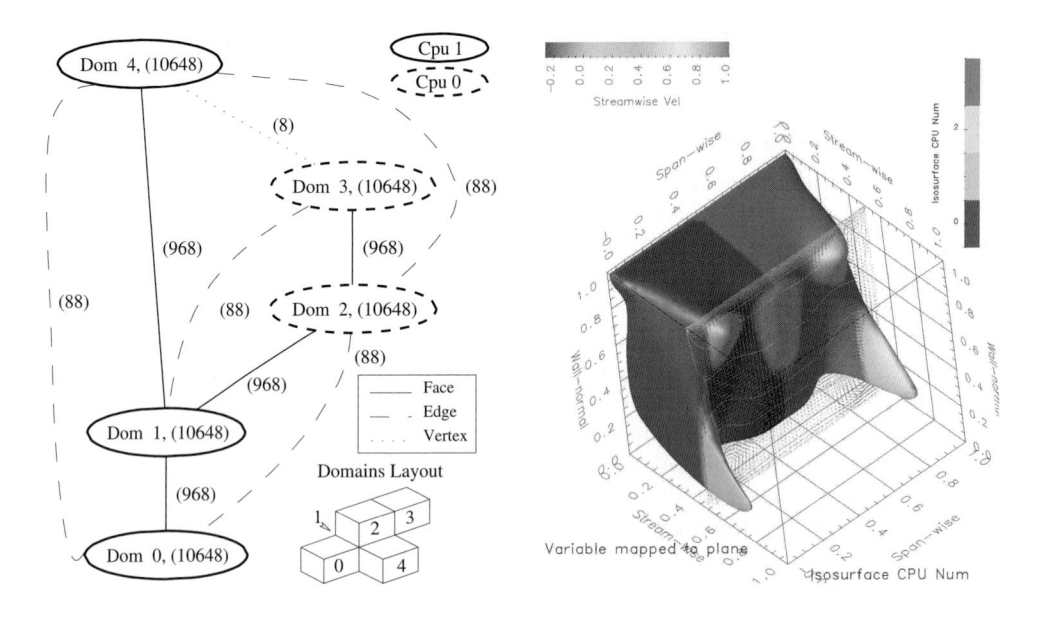

Figure 1. Load balancing performed using METIS: five domains are divided between two processors based on the number of cells they have to process and exchange between them (the weights in the graph).

Figure 2. OpenDX used to visualize the iso-surface of wall-normal velocity (colored based on the cpu owning the volume) in a recirculation cavity, with velocity vectors along a plane cut.

2.3. Coding standards, support software and data formats

The new framework utilizes general guidelines [4] regarding programming styles, indentation, variable naming, error and warnings handling, command line parsers, etc. etc. in order to increase the consistency across the code-base.

As stated in § 2.1, the reliance on commercial software for data manipulation and processing seriously limits the portability of the code, and a limited number of licenses could be aggravating. To overcome these problems, OpenDX [7], an ex commercial application released under a free and open source license is selected as primary tool for data visualization, with the added benefit of being compatible with the reference data format chosen.

The main advantage of this software is the professional quality of the results (Fig. 2 and 7), although its use could be problematic for inexperienced users: OpenDX is a visual program editor/executor, it is not intended as a general purpose data analysis tool. It is, for all intend and purposes, a language interpreter: in order to function it needs a visual program with all the instructions, and writing such a visual program is a task in its own right.

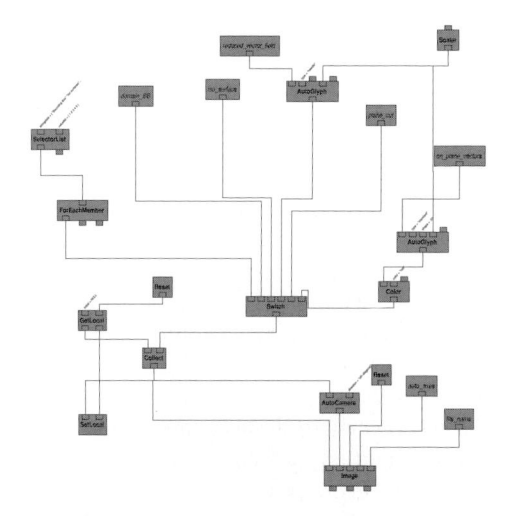

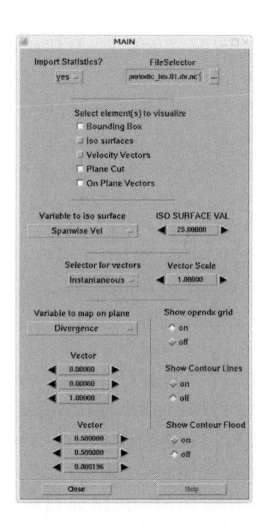

Figure 3. A simple module of a visual program of OpenDX dealing with image composition: a visual program is a collection of modules.

Figure 4. Example of user interaction window in OpenDX: based on the Motif© toolkit, offers the available actions.

Building these programs could be daunting for new users, so a set of pre-made programs for the most generic geometries is provided as part of the framework, see for example Fig. 3 for a small portion of a visual program, and Fig. 4 for the current interface the users utilize to explore the data, beside some basic built-in capabilities.

An important step is the choice of a single data format for *all* the files involved in the solver: simple input files with user input parameters or full blown data files with time series of flow solutions with multiple variables alike.

Different formats have been evaluated, and the one that has been chosen is the Unidata NetCDF [8] Version 3: it is platform independent, self described, with ancillary data [4], mature and well suited for the class of data in consideration, and compatible with a large array of programs, both NetCDF specific or generalist, like Grace or OpenDX. Its parallel extension [5] will be implemented in the near future, for the moment data is stored domain wise and reassembled in a master node in case of domain auto-partitioning: only recently the parallel extension has reached a freeze in the APIs and released the first stable version.

2.4. Core Libraries

The parallel context does not requires only the capability of resolving linear systems across multiple processors, it mainly requires exchanging data for multiple purposes: boundary conditions updates, data scattering, data gathering and so on. While generic MPI implementations like mpich can be used for such operations, defining all the parallel communications via a single library helps creating homogeneity across the solver and encouraging code reuse. The parallel library used is PETSc [10]: it provides, beside a linear solver, a series of functions that create a solid and tested framework for parallel communication, easy to reuse for future developers, well documented and ante posing clarity and simplicity to performance whenever necessary. The structured nature of the grid allows the abundant use of centralized macros that are reused and minimize coding errors. The library is open source, uses documentation embedded into the code, and it is easy to read and eventually modify.

2.5. Version Management, Bug tracking and Documentation

In order to have a development process able to keep track of all the code evolution, with possibility of code roll-back and multiple branches, the use of a version system has been introduced, the stable CVS [9]. This system allows to store all the changes performed on a file by the different developers together with the information they enter explaining the reasons of their changes. Different developers can work on different branches of the source being aware of the work that is being done by others.

If the CVS repository contains the history of the code evolution, bugzilla [6] keeps track of the bugs evolution, from their discovery to their elimination, as well of feature requests.

The documentation is a crucial point in every software project, but it is evermore so for developers that have a limited amount of time for a given task. It is a despicable but unfortunately common practice that the code documentation is very often left at the end of the whole process, if ever produced: the final result is often a rushed out document, separate and independent from the code. In the latter snips of documentation are sparse and far between, with the difficulties associated to expressing concept via text while a graph or an equation would be best suited.

In this context, documenting the code is important as much, if not more for certain aspects, than the code in itself: a clear documentation means ease of use, and in the long

---

[4]Or data about the data, or meta-data, as time stamp, protocols, file version number, etc. etc.
[5]http://www-unix.mcs.anl.gov/parallel-netcdf/
[6]http://www.bugzilla.org/

term helps retaining the original code philosophy.

Therefore, the strategy adopted in the current context is to document early, document often, document not only with text, and, most importantly, embed documentation inside the code. This practice is successfully employed by large and small projects alike: while big projects use their own way to produce documentation [10], others use some standards FOSS like Doxygen [11] to generate contemporaneously man pages, HTML and LaTeX documentation. The possibility of creating hyper-textual documentation, embedding images and LaTeX, with auto-generated dependency graphs (see Fig. 5), is an invaluable help to the current and future developers.

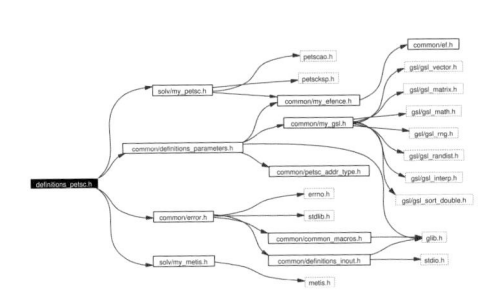

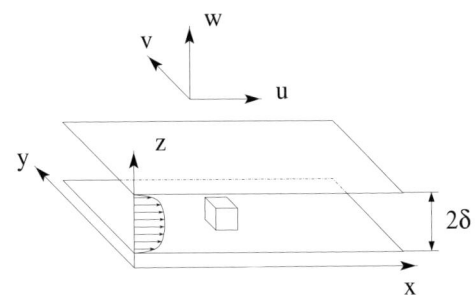

Figure 5. Doxygen generates a series of useful informations, like dependencies between includes and source files.

Figure 6. Wall mounted cube in a channel. First step towards wind engineering and severe test for the LES solver.

3. Solver status and Perspectives

The whole framework and the solver has recently reached the first feature freeze (namely release 1.0 Beta), having reached all the design goals, the last being the concurrent execution of a pair of master-slave simulations. For the time being therefore only fixes to the bugs found by the developer or reported to him through bugzilla will be introduced in the existing code-base, and all the efforts will be devoted to the validation of the solver for different test cases. In the same time the framework will be opened for common fruition in VKI to extend the user base.

Flows currently under investigation include turbulent plane channel, lid driven three dimensional recirculation cavity, square duct and flow around a wall mounted cube in a plane channel [12] (Fig. 6). The turbulent plane channel test case has already been validated successfully [13], and the flow around the wall mounted cube is now under exam; both cases of laminar inlet as found in literature [12] and proper turbulent inlet via slave simulation are under exam, for different configurations (see Fig 7 for an example of ongoing simulations).

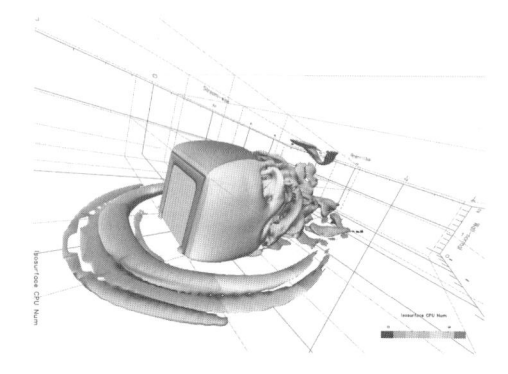

Figure 7. Q (coherent structures indicator) iso-surface colored by the different processors in the simulation for a wall mounted cube inside a channel.

Figure 8. Diagram with all the FOSS related to the LES solver framework, directly used in the solver or used as a support tool. Current solver code name: MiOma.

4. Conclusions

In the present paper a new framework for a new parallel LES solver built from the ground up has been illustrated; based on FOSS software (Fig. 8) it has been built in order to be developer-friendly and built according to modern software development paradigms. The framework has reached a mature stage and the code-base starts to be validated successfully against reference data and being opened for fruition to the intended target user base in the von Karman Institute.

REFERENCES

1. E. Simons, von Kármán Institute for Fluid Dynamics Doctoral Thesis, 2000.
2. A. Kurganov and E. Tadmor, J. Computational Physics 160 (2000)
3. Harlow, F.H. and Welsh, J.E., The Physics of Fluids 8 (1965)
4. R.M. Stallman et al., http://www.gnu.org/prep/standards/
5. W. Schelter, http://maxima.sourceforge.net/
6. G. Karypis and V. Kumar, http://www-users.cs.umn.edu/~karypis/metis
7. G. Abram et al, http://www.opendx.org/
8. Unidata, http://www.unidata.ucar.edu/packages/netcdf/
9. P. Cederqvist, http://savannah.nongnu.org/projects/cvs/
10. S. Balay and K. Buschelman and W. D. Gropp and D. Kaushik and M. G. Knepley and L. C. McInnes and B. F. Smith and H. Zhang, http://www.mcs.anl.gov/petsc
11. D. van Heesch, http://www.stack.nl/~dimitri/doxygen/
12. W. Rodi and J. Ferziger and M. Breuer and M. Pourquié, J. Fluids Eng., 1192 (1997)
13. R. Giammanco, Proceedings of Third International Conference on Advanced computational Methods in Engineering (2005)

On a Fault Tolerant Algorithm for a Parallel CFD Application

M. Garbey[a] and H. Ltaief[a]

[a]Dept of Computer Science, University of Houston, Houston, TX 77204, USA

1. Introduction

The objective of this paper is to present one new component of **G - Solver** [1],which is a general framework to efficiently solve a broad variety of PDE problems in grid environments.

A grid can be seen as a large and complex system of heterogeneous computers [2–4], where individual nodes and network links can fail. A Grid Solver must be efficient and robust in order to solve the large problems that justify grid environments.

This implies that a Grid Solver should maintain a high level of numerical efficiency in a heterogeneous environment while being tolerant to high latency and low bandwidth communication [5–8] as well as system and numerical failures. We will focus on fault tolerance in this paper. Our model problem is the two dimensional heat equation:

$$\frac{\partial u}{\partial t} = \Delta u + F(x,y,t), \ (x,y,t) \in \Omega \times (0,T), \ u_{|\partial\Omega} = g(x,y), \ u(x,y,0) = u_o(x,y). \tag{1}$$

We suppose that the time integration is done by a first order implicit Euler scheme,

$$\frac{U^{n+1} - U^n}{dt} = \Delta U^{n+1} + F(x,y,t^{n+1}), \tag{2}$$

and that Ω is partitioned into N subdomains Ω_j, $j = 1..N$.

The G-Solvers must be tested on a testbed that represents a realistic grid computing environment. We are using computing resources at the University of Houston (Itanium2 cluster and Appro-AMD beowulf cluster), the University of Florida (SGI-Altix and Appro-dual Xeon beowulf cluster), the High Performance Computing Center in Stuttgart (HLRS, Cray T3E and Cray-Operon cluster), RRZE (GBit-Xeon Cluster, Erlangen) in Germany and the Academy of Sciences (Cluster 128 nodes, Beijing) in China. All these computers are combined into a friendly grid environment that we use routinely.

2. Fault Tolerant Algorithms

Several interfaces for communication softwares exist which allow the application to continue running more or less, even though a hardware failure has occurred [9]. However, at the end of the application, we do not know if we will get the correct solution because data can be lost during the failure. We have developed and combined numerical algorithms with existing fault tolerant libraries to recover efficiently from catastrophic failures during execution.

The state of the art in fault tolerance for long running applications on a grid of computers is to checkpoint the state of the full application and then rollback when a node fails. However, this approach does not scale. Indeed, as the number of nodes and the problem size increases, the cost of checkpointing and recovery increases, while the mean time between failures decreases.

The approach we have taken is as follows: redundant processors provided by the grid are used as spare processors to efficiently store the subdomain data of the application during execution in *asynchronous mode*. We cannot guarantee then that we get a copy of the solution u_j^n for all subdomains Ω_j that corresponds to the same time step n.

In other words, we assume that spare processors have stored copies of subdomain data $u_j^{n(j)}, j = 1..N$, with no missing block on disks but a priori $n(j) \neq n(k)$ for $j \neq k$.

In case of failure, a spare processor can then take over for the failed processor without the entire system having to roll back to a globally consistent checkpoint. So, the first step of the algorithm is to allocate to the spare processor(s), which will become the application processor(s), the subdomain's data corresponding to the processor(s) that would fail(s).

Now, the mathematical difficulty is to reconstruct in parallel u at a time step $M \in (min_j\{n(j)\}, max_j\{n(j)\})$, from subdomains data at different but near time steps.

3. Reconstruction of the solution

In this section, we discuss several algorithmic ideas to solve the reconstruction problem in one dimension and then, we will present some results got with our two dimensional model problem (1).

3.1. Interpolation method

Let $M = \frac{1}{N}\Sigma_{j=1..N} n(j)$. We look for an approximation of U_j^M in Ω. We assume that we have at our disposal $U^{n(j)}$ and $U^{m(j)}$ at two time steps $n(j) < m(j)$, in each subdomain Ω_j.

Then, if $||U^{n(j)} - U^{m(j)}||_{\Omega_j}$ is below some tolerance number, we may use a second order interpolation/extrapolation in time to get an approximation of U^M. The numerical error should be of order $((m(j) - n(j))dt)^2$.

This simple procedure reduces the accuracy of the scheme and introduces small jump at the interfaces between subdomains. This method is perfectly acceptable when one is not interested in accurately computing transient phenomena. However, this method is not numerically efficient in the general situation.

3.2. Forward Time Integration

Let us assume that for each subdomain, we have access to $U^{n(j)}$. For simplicity we suppose that $n(j)$ is a monotonically increasing sequence. We further suppose that we have stored in the memory of spare processors the time history of the artificial boundary conditions $I_j^m = \Omega_j \cap \Omega_{j+1}$ for all previous time steps $n(j) \leq m \leq n(j+1)$.

We can then reconstruct with the forward time integration of the original code (2), the solution $U^{n(N)}$, as follows:

- Processor one advances in time u_1^n from time step $n(1)$ to time step $n(2)$ using boundary conditions I_1^m, $n(1) \leq m \leq n(2)$.

- Then, Processors one and two advance in parallel u_1^n and u_2^n from time step $n(2)$ to $n(3)$ using neighbor's interface conditions, or the original interface solver of the numerical scheme.

- This process is repeated until the global solution $u^{n(N)}$ is obtained.

This procedure can be easily generalized in the situation of Figure 1 where we do not assume any monotonicity on the sequence $n(j)$. The thick line represents the data that need to be stored in spare processors, and the intervals with circles are the unknowns of the reconstruction process.

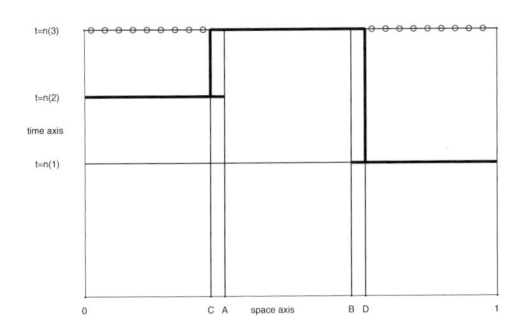

Figure 1. Reconstruction procedure in one dimension using forward time integration.

The advantage of this method is that we can easily reuse the same algorithm as in the standard domain decomposition method, but restricted to some specific subsets of the domain decomposition. We reproduce then the exact same information U^M as the process had no failures.

The main drawback of the method is that saving at each time step the artificial boundary conditions may slow down the code execution significantly. While interface conditions are still one dimension lower than the solution itself, message passing might be time consuming and may slow down the application.

For example, Figure 2 gives a rough idea of checkpointing time cost on the 3D code resolving the Heat Equation with an implicit first order Euler scheme using a Krylov method. As a matter of fact, for a large problem, accumulating the boundary conditions slows dramatically on AMD Beowulf Cluster Stokes with Gigabit Ethernet and adds a high penalty on Ithanium2 Cluster Atlantis with Myrinet Network.

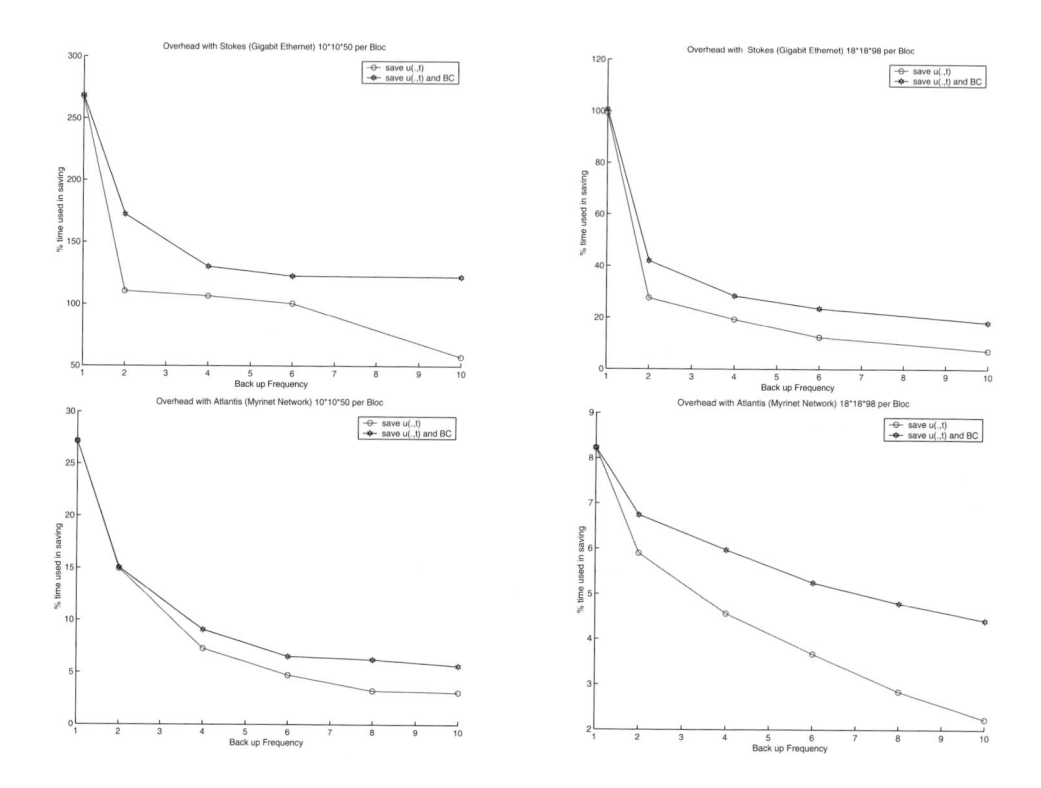

Figure 2. Percentage of the elapsed time used in saving the solution and then the solution with the Boundary Conditions at different back up frequency. Tests done on AMD Beowulf Cluster Stokes (GiGabit Ethernet Network) and Ithanium2 Cluster Atlantis (Myrinet Network) with a size per bloc 10*10*50 and 18*18*98.

3.3. Backward Integration and Space Marching

Let us suppose now that we asynchronously store only the subdomain data, and not the chronology of the artificial interface condition. In fact, this efficient method does not require storing the boundary conditions of each subdomain at each time step. To be more specific, we suppose that we have access to the solution for each subdomain at two different time steps $n(i), m(i)$ with $m(i) - n(i) = K \gg 1$.

The Forward Implicit scheme (2) in one dimension provides an explicit formula when we go backward in time:

$$U_j^n = U_j^{n+1} - dt \frac{U_{j+1}^{n+1} - 2U_j^{n+1} + U_{j-1}^{n+1}}{h^2} - F_j^{n+1}. \tag{3}$$

The existence of the solution is granted by the forward integration in time. Two difficulties are first the instability of the numerical procedure and second the fact that one is restricted to the cone of dependence as shown in Figure 3.

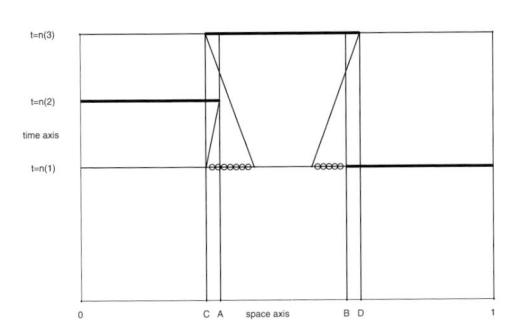

Figure 3. Reconstruction procedure in one dimension using explicit backward time stepping.

We have in Fourier modes
$$\hat{U}_k^n = \delta_k \hat{U}_k^{n+1},$$
with
$$\delta_k \sim -\frac{2}{h}(\cos(k\,2\,\pi\,h) - 1), \ |k| \leq \frac{N}{2}.$$

The expected error is at most in the order $\frac{\nu}{h^K}$ where ν is the machine precision and K the time step. Therefore, the backward time integration is still accurate up to time step K with
$$\frac{\nu}{h^K} \sim h^2.$$

To stabilize the scheme, one can use the Telegraph equation that is a perturbation of the heat equation:
$$\epsilon\frac{\partial^2 u}{\partial t^2} - \frac{\partial^2 u}{\partial x^2} + \frac{\partial u}{\partial t} = F(x,t), x \in (0,1), t \in (0,T). \tag{4}$$

The asymptotic convergence can be derived from [10] after time rescaling. The general idea is then to use the previous scheme (3) for few time steps and pursue the time integration with the following one
$$\epsilon\frac{U_j^{n+1} - 2\,U_j^n + U_j^{n-1}}{\tilde{dt}^2} - \frac{U_{j+1}^n - 2\,U_j^n + u_{j-1}^n}{h^2} + \frac{U_j^{n+1} - U_j^n}{\tilde{dt}} = F_j^n. \tag{5}$$

Let us notice that the time step $\tilde{dt}$ should satisfy the stability condition $\tilde{dt} < \epsilon^{1/2}h$. We take in practice $\tilde{dt} = dt/p$ where p is an integer. The smaller is ϵ, the more unstable is

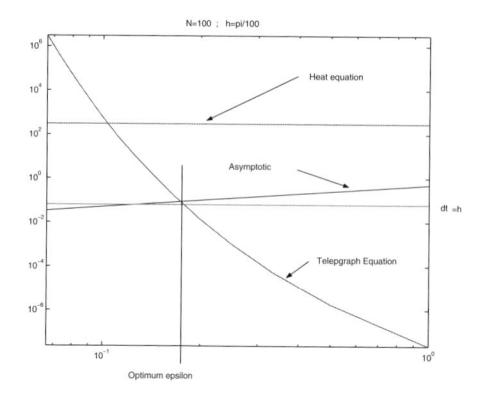

Figure 4. Stability and error analysis with Fourier

the scheme (5) and the flatter is the cone of dependence. The smaller is ϵ, the better is the asymptotic approximation.

We have done a Fourier analysis of the scheme and Figure 4 shows that there is a best compromise for ϵ to balance the error that comes from the instability of the scheme and the error that comes from the perturbation term in the telegraph equation. We have obtained a similar result in our numerical experiments.

To construct the solution outside the cone of dependencies we have used a standard procedure in inverse heat problem that is the so called space marching method [11]. This method may require a regularization procedure of the solution obtained inside the cone using the product of convolution

$$\rho_\delta * u(x, t),$$

where

$$\rho_\delta = \frac{1}{\delta\sqrt{\pi}} \exp(-\frac{t^2}{\delta^2}).$$

The following space marching scheme:

$$\frac{U_{j+1}^n - 2U_j^n + U_{j-1}^n}{h^2} = \frac{U_j^{n+1} - U_j^{n-1}}{2\,dt} + F_j^n, \tag{6}$$

is unconditionally stable, provided $\delta \geq \sqrt{\frac{2\,dt}{\pi}}$.

The last time step $U^{n(i)+1}$ to be reconstructed uses the average

$$U^{n(i)+1} = \frac{U^{n(i)} + U^{n(i)+2}}{2}.$$

We have observed that filtering as suggested in [11] is not necessary in our reconstruction process.

We have extended to two dimensions the explicit backward scheme (3). This method to retrieve the solution inside the cone is quite obvious to set up. Then, we have applied the space marching scheme (6) in each direction, i.e. X, Y and oblique directions, to get the solution outside the cone. The scheme remains the same except the space step, which changes depending on the direction: hx for the X direction, hy for the Y direction and $\sqrt{hx^2 + hy^2}$ for the oblique direction (Figure 5).

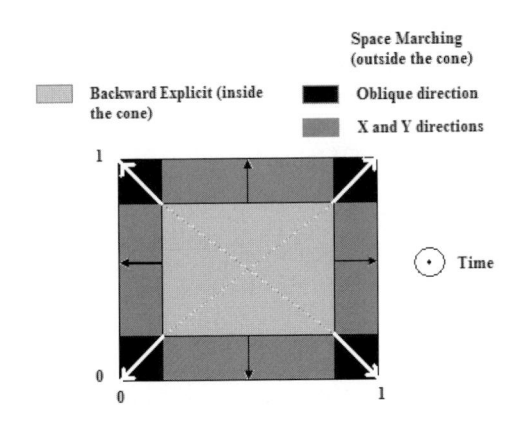

Figure 5. Reconstruction procedure in two dimensions.

Figure 6 illustrates the numerical accuracy of the overall reconstruction scheme for $\Omega = [0,1] \times [0,1]$, $dt = 0.25 \times hx$, $K = 7$ and F such that the exact analytical solution is $cos(q_1 (x+y))(sin(q_2 t) + \frac{1}{2}cos(q_2 t))$, $q_1 = 2.35, q_2 = 1.37$.

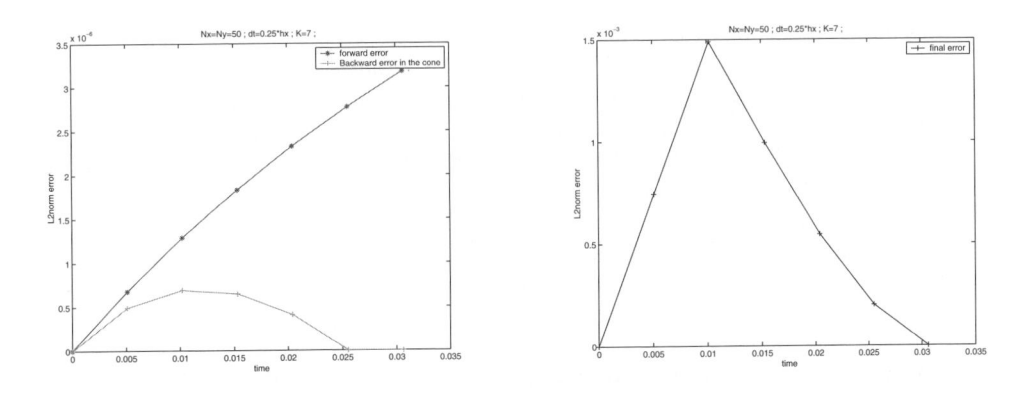

Figure 6. Numerical accuracy of the overall reconstruction scheme in two dimensions.

In this specific example our method gives better results than the interpolation scheme provided that $K \leq 7$. For larger K we can use the scheme (5) for time steps below $m(i) - 7$. However the precision may deteriorate rapidly in time.

4. Conclusion

We have presented the Fault Tolerant specificity of the G-solver framework. It combines numerical analysis and computer science. We have reviewed several procedures to reconstruct the solution in each subdomain from a set of subdomain solutions given at disparate time steps. This problem is quite challenging because it is very ill posed. We found a satisfactory solution by combining explicit reconstruction techniques that are a backward integration with some stabilization terms and space marching. Moreover, in addition to checkpoint the application, one can think about another job for the spare processors which could be solution verification [12].

REFERENCES

1. M.Garbey, V.Hamon, R.Keller and H.Ltaief, *Fast parallel solver for the metacomputing of reaction-convection-diffusion problems*, to appear in parallel CFD04.
2. I. Foster, N. Karonis, *A Grid-Enabled MPI: Message Passing in Heterogeneous Distributed Computing Systems*, Proc. 1998 SC Conference, 1998.
3. I. Foster, C. Kesselman, J. Nick, S. Tuecke, *Grid Services for Distributed System Integration*, Computer, 35(6), 2002.
4. I.Foster, *The Grid: A New Infrastructure for 21st Century Science*, Physics Today, 55(2):42-47, 2002.
5. F.Dupros, W. E.Fitzgibbon and M. Garbey, *A Filtering technique for System of Reaction Diffusion equations*, preprint COSC, University of Houston, submitted.
6. M. Garbey, H.G.Kaper and N.Romanyukha, *A Some Fast Solver for System of Reaction-Diffusion Equations*, 13th Int. Conf. on Domain Decomposition DD13, Domain Decomposition Methods in Science and Engineering, CIMNE, Bracelona, N.Debit et Al edt, pp. 387–394, 2002.
7. M. Garbey, R. Keller and M. Resch, *Toward a Scalable Algorithm for Distributed Computing of Air-Quality Problems*, EuroPVM/MPI03 Venise, 2003.
8. J. G. Verwer, W. H. Hundsdorfer and J. G. Blom, *Numerical Time Integration for Air Pollution Models*, MAS-R9825, http://www.cwi.nl, Int. Conf. on Air Pollution Modelling and Simulation APMS'98.
9. Gropp and Lusk, *Fault Tolerance in Message Passing Interface Programs*, International Journal of High Performance Computing Applications.2004; 18: 363-372.
10. W.Eckhaus and M.Garbey, *Asymptotic analysis on large time scales for singular perturbation problems of hyperbolic type*, SIAM J. Math. Anal. Vol 21, No 4, pp867-883, 1990.
11. D.A.Murio, *The Mollification Method and the Numerical Solution of Ill-posed Problems*, Wiley, New York, 1993.
12. M.Garbey and W.Shyy, A Least Square Extrapolation Method for improving solution accuracy of PDE computations, J. of Comput. Physic, 186, pp1-23, 2003.

Computational fluid dynamics applications on TeraGrid

R.U. Payli[*], H.U. Akay[*], A.S. Baddi[*], A. Ecer[*], E. Yilmaz[*], and E. Oktay[†]

[*]Computational Fluid Dynamics Laboratory
Department of Mechanical Engineering
Indiana University-Purdue University Indianapolis (IUPUI)
Indianapolis, Indiana 46202 USA
http://www.engr.iupui.edu/cfdlab

[†]EDA Engineering Design and Analysis Ltd. Co.
Technopolis, METU Campus
06531, Ankara, Turkey
http://www.eda-ltd.com.tr

Keywords: TeraGrid, Grid Computing, Parallel CFD.

In this paper, we document our experiences with TeraGrid (http://www.teragrid.org), which is an NSF-supported national grid-computing environment in USA, and show results of large-scale parallel CFD applications using our parallel codes. We explore benefits of using the TeraGrid environment in distributed and parallel fashion, including code-coupling applications. We also explore the data management and visualization aspects of the results on the Grid, which are essential for CFD applications. The capabilities available and future needs for large-scale parallel computing applications will be reviewed.

1. INTRODUCTION

Solving the large scale CFD applications are challenging task. They require great deal of processing power. Recent advancement on hardware and software made solving large-scale applications possible. NSF-supported TeraGrid is a grid computing environment to build an infrastructure for open scientific research.

TeraGrid integrates a distributed set of the highest capability computational, data management, and visualization resources through high performance network connection, grid computing software, and coordinated services. Currently eight U.S supercomputing centers provide resources to the TeraGrid. These centers are the University of Chicago/Argonne National Laboratory (UC/ANL), Indiana University(IU), the National Center for Supercomputing Applications (NCSA), Oak Ridge National Laboratory (ORNL), Pittsburgh Supercomputing Center (PSC), Purdue University (PU), the San Diego Supercomputing Center (SDSC), and the Texas Advanced Computing Center (TACC).

These institutions link to one another over a high-speed network. The backbone of this network is a 40-gigabit-per-second optical fiber channel. TeraGrid resources have 20 teraflops of computing power and capable of managing and storing nearly one petabyte of data. In this research, we used NCSA and UC/ANL TeraGrid resources for running our parallel codes and processing the results for visualization.

1.1. NCSA IA-64 Linux Cluster

NCSA's IA-64 TeraGrid Linux cluster consists of 887 IBM cluster nodes: 256 nodes with dual 1.3 G Hz Intel® Itanium® 2 processors (Phase 1) and 631 nodes with dual 1.5 G Hz Intel® Itanium® 2 processors (Phase 2). Half of the 1.3 G Hz nodes and all the 1.5 G Hz nodes are equipped with 4 gigabytes of memory per node; the other half of the 1.3 G Hz nodes are large-memory processors with 12 gigabytes of memory per node, making them ideal for running memory-intensive applications. The cluster is running SuSE Linux and is using Myricom's Myrinet cluster interconnect network.

1.2. NCSA SGI Altix

NCSA's SGI Altix consists of two systems each with 512 Intel® Itanium® 2 processors running the Linux® operating system. The systems have 1 and 2 terabytes of memory respectively. 370 terabytes of SGI InfiniteStorage serve as the shared file system.

1.3. UC/ANL IA-32 Linux Cluster

UC/ANL's IA-32 TeraGrid Linux cluster consists of 96 nodes with dual 2.4 Ghz Intel® Xeon® processors. Each node equipped with 4 gigabyte of memory. The cluster is running SuSE Linux and is using Myricom's Myrinet cluster interconnect network.

1.4. SDSC IA-64 Linux Cluster

SDSC's IA-64 TeraGrid Linux cluster consists of 256 cluster nodes with dual 1.5 G Hz Intel® Itanium® 2 processors. Each node equipped with 2 gigabytes of memory. The cluster is running SuSE Linux and is using Myricom's Myrinet cluster interconnect network.

2. PARALLEL COMPUTING ON TERAGRID

We have used two different codes in our applications with TeraGrid. The first application is an unstructured grid CFD code. The second application is a coupled fluid-structured analysis code, which couples CFD, and CSD codes.

2.5. PACER3D: An Unstructured Mesh CFD Solver

The first application is with a finite volume Euler flow solver on the unstructured tetrahedral meshes [1]. The solver, PACER3D, uses cell-vertex based numerical discretization and has capability for adaptive remeshing. It employs artificial dissipation terms in flux calculation. Local time stepping is used for steady state flow solutions. To accelerate solutions, local time stepping, residual averaging technique, and enthalpy damping are used. MPI is used for parallel implementation.

2.6. FSIeda: Fluid-Solid Interaction Solver

A loosely-coupled approach is presented for solution of aerolelastic problems. A Computational Fluid Dynamics (CFD) code developed for unsteady solution of Euler equations on unstructured moving meshes is coupled with a Computational Structural Dynamics (CSD) code for solution of structure-fluid interaction (SFI) problems to predict aeroelastic flutter [2–5]. The loose coupling is employed for transfer of fluid pressures from CFD code to CSD code and the transfer of structural displacements from CSD code to CFD code. A parallel code-coupling library, SINeda, is developed to obtain the coupling across non-similar meshes of these two codes using search and interpolation schemes. An implicit cell-centered finite volume Euler solver with Arbitrary Lagrangian Eulerian (ALE) formulation is used to model flow equations.

The dynamic response of the structure is obtained by mode superpositioning. The fluid mesh is dynamically deformed every time step based on a linear spring analogy. The CFD solver uses three-dimensional tetrahedral elements, whereas the CSD solver uses quadrilateral shell elements with mid-surface representation of the wing-like geometries. SINeda employs an efficient Alternating Digital Tree (ADT) [6] geometric search and interpolation algorithm. The fluid domain is partitioned into sub domains using domain-decomposition approach for parallel computing to speedup the coupled solution process.

3. VISUALIZATION ON TERAGRID

TeraGrid offers visualization services for users and provides tools and libraries for researchers in visualization and graphics. One of the visualization tools available on TeraGrid useful for CFD applications is ParaView (*http://www.paraview.org*). ParaView is an extensible, open source multi-platform application. On top of the conventional visualization features ParaView is suitable for distributed computing for processing large data sets. ParaView runs parallel on distributed and shared memory systems using MPI. It uses the data parallel model in which the data is divided into pieces to be process by different processes. It supports both distributed rendering and local rendering, or combination of both. This provides scalable rendering for large data.

4. RESULTS

4.7. PACER3D Application

To measure parallel efficiency of different TeraGrid sites, PACER3D is used with up to 256 numbers of processors. Test case chosen is Onera M6 wing with 3.5 Millions of tetrahedral mesh elements. Partitioning is done up to 256 sub-blocks. Parallel solutions are obtained for 8, 16, 32, 64, 128, 256 processors. Figure 1 shows speedup results for two different sites. Speedup for computation only is also measured to see the effect of the communication on overall results. Up to 128 processors, PACER3D shows super-linear speedup on the SGI cluster. This can be noticed in the efficiency plots given in Figure 2 as well. However, IA64 shows almost linear speed up to 64 processors.

Speedup for computation only shows super-linear trend for both computers. This is assumed to be due to paging size effect, as individual blocks still seem to be big to fit into the memory hence it pages. However, communication overhead increases as the number of partitions increases since number of partitions to communicate at interface increases as well. It makes overall gain of more partitioning not desirable since total timing of the solution increases more than ideal one.

Figure 3 shows the effect of allocating one or two partitions (solution blocks) per node of IA64 machines, which has dual processors at one node. Surprisingly, two-partition allocation has lower speedup than that of one partition allocation. This might be associated with sharing same bus for the data communications and computations. All bandwidth is used by one processor while in two partition case bandwidth is shared by two partitions.

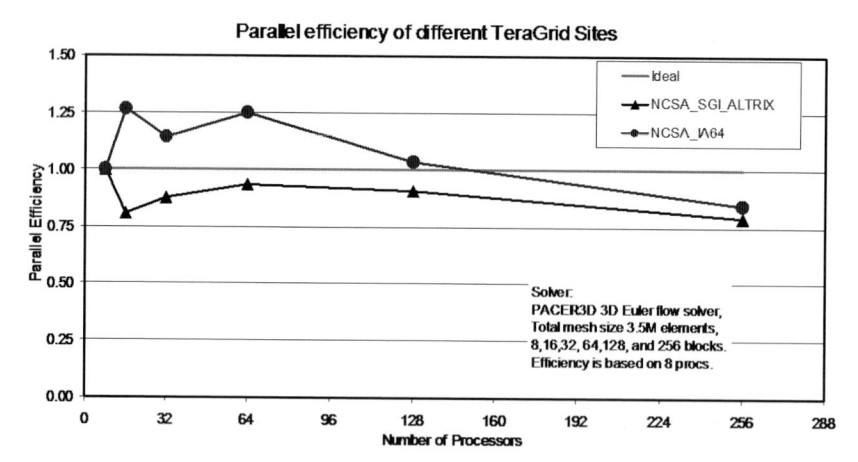

Figure 1: Efficiency comparison for TeraGrid sites used.

4.8. Multi-Site Execution of SINeda

SFIeda is used on TeraGrid to experiment with multi-site job execution. Shown in Figure 4 is the structure of FSIeda for coupling of meshes and parallel computing. Execution of FSIeda across the multiple sites of grid environments is depicted in Figure 5. For this purpose, MPICH-G2 that is the grid-enabled implementation of popular Message Passing Interface (MPI) is used. MPICH-G2 uses grid capabilities for starting processes on remote systems. It selects appropriate communication methods such as, high performance local interconnect (Myrinet, InfiniBand, etc.) for local communication and TCP/IP for remote communication.

In this experiment (Figure 5), SDSC and NCSA Linux Clusters are used. The flow of multi-site job execution as follows: First, job request is submitted to SDSC and NCSA Globus Gatekeepers from NCSA login shell using Globus Resource Specification Language (RSL). Then, Globus Gatekeepers contact with Globus Job Managers and submit jobs via local schedulers (PBS). Finally, flow solver code runs on NCSA, structural solver and search and interpolation codes run on SDSC cluster simultaneously.

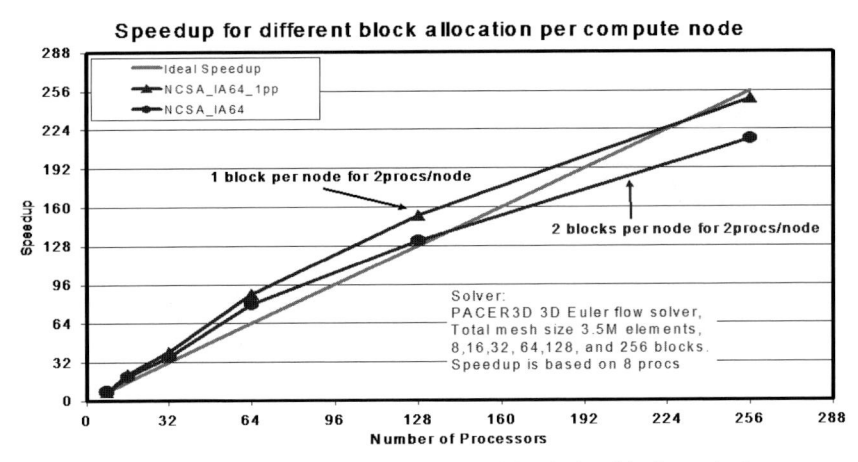

Figure 2: Speedup comparison for running 1 and 2 solution blocks at dual processor.

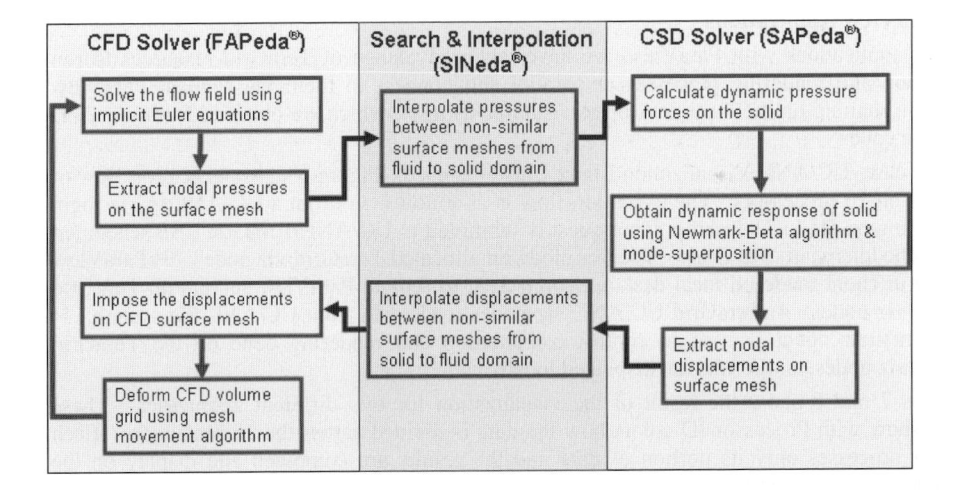

Figure 3: FSIeda structure for parallel computing.

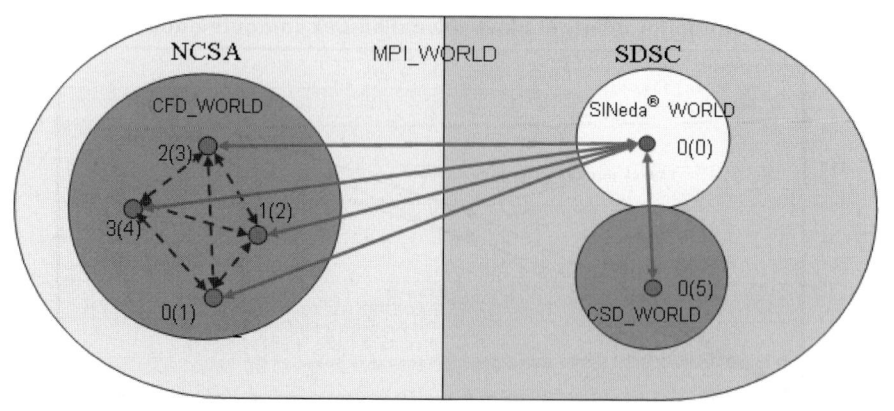

Figure 4: Execution of FSIeda® across multiple sites.

4.9. ParaView Applications

In our applications with ParaView, we have taken advantage of TeraGrid resources to run computation and rendering processes in parallel and connect to them via a client ParaView application running on the local desktop to visualize data set, which we obtained from our CFD simulation code.

In this case, UC/ANL Visualization Linux Cluster is used. Figure 6 shows execution flow of ParaView on 16 processors. The execution flow is as follows: 1) From a UC/ANL tg-viz-login shell running on the local desktop, job request is submitted to UC/ANL TeraGrid PBS scheduler. 2) PBS scheduler starts ParaView in server mode on allocated TeraGrid viz nodes. 3) ParaView is started in client mode on local desktop. connecting to remote ParaView servers on TeraGrid UC/ANL viz nodes. 4) TeraGrid UC/ANL viz nodes reads data from UC/ANL file system. 5) Local ParaView client is used to control computation and rendering done on the TeraGrid UC/ANL viz nodes, and results are displayed locally.

Figures 7 and 8 show the result of the visualization for two different geometries. These pictures show with Processor ID colors how the data is divided across the 16 processors. Each processor processes only its portion of data and the results are combined and display on the client side of the ParaView.

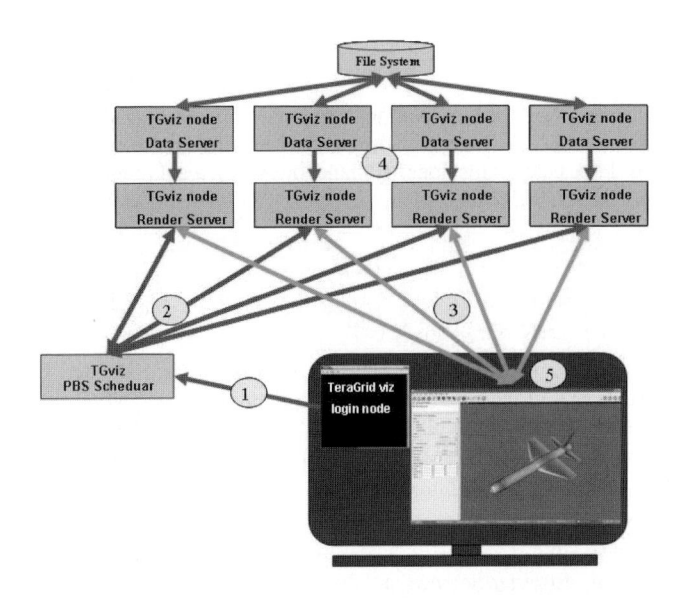

Figure 5: Execution flow of ParaView on UC/ANL Linux Cluster.

Figure 6: 16 Processor ID colors for post-processing of 50-partitions with 625 tetrahedral mesh elements.

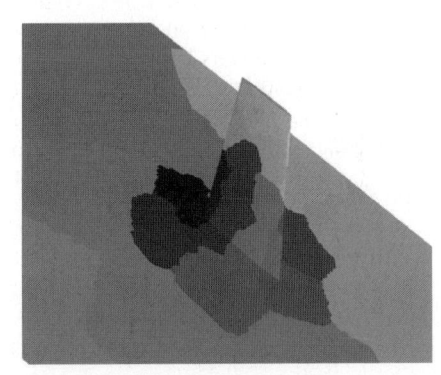

Figure 7: 16 Processor ID colors for post-processing of 64-partitions 3.5M tetrahedral mesh elements.

5. CONCLUSIONS

We explored the potential applications of TeraGrid. Our short experience with the environment has been positive. TeraGrid offers a good platform for researchers to run and visualize the large data sets which results of the solution of the large-scale problems. Even though some of the features are still in the testing stage, the stability of the system has been observed to improve. More performance test has to be conducted for large-scale applications.

6. ACKNOWLEDGEMENT

TeraGrid access was supported by the National Science Foundation (NSF) under the following programs: Partnerships for Advanced Computational Infrastructure, Distributed Terascale Facility (DTF) and Terascale Extensions: Enhancements to the Extensible Terascale Facility, with Grant Number: TG-CTS050003T.

7. REFERENCES

1. E. Yilmaz, M.S. Kavsaoglu, H.U. Akay, and I.S. Akmandor, "Cell-vertex Based Parallel and Adaptive Explicit 3D Flow Solution on Unstructured Grids," *International Journal of Computational Fluid Dynamics*, Vol. 14, pp. 271-286, 2001.
2. E. Oktay, H.U. Akay, and A. Uzun, "A Parallelized 3D Unstructured Euler Solver for Unsteady Aerodynamics," *AIAA Journal of Aircraft*, Vol. 40, No. 2, pp. 348-354, 2003.
3. H.U. Akay, E. Oktay, X. He, and R.U. Payli, "A Code-Coupling Application for Solid-Fluid Interactions and Parallel Computing," *Proceedings of Parallel CFD'03*, Edited by B. Chetverushkin, et al., Elsevier Science, pp. 393-400, 2004.
4. H.U. Akay, E. Oktay, Z. Li, and X. He, "Parallel Computing for Aeroelasticity Problems," *21st AIAA Applied Aerodynamics Conference*, Orlando, FL, June 23-26, 2003.
5. H.U. Akay and E. Oktay, "Parallel Adaptivity for Solution of Euler Equations Using Unstructured Solvers," *Proceedings of Parallel CFD'02*, Edited by K. Matsuno, et al., Elsevier Science, pp. 371-378, 2003.
6. J. Bonet and J. Peraire, "An Alternating Digital Tree (ADT) Algorithm for Geometric Searching and Intersection Problems", *Int. J. Num. Meth. Eng.*, Vol. 31, 1991.

Mapping LSE method on a grid : Software architecture and Performance gains

Christophe Picard [a], Marc Garbey [a] and Venkat Subramaniam [a]

[a]Department of Computer Science, University of Houston, Houston, TX 77204, USA

In CFD community, a popular technique for a Posteriori error estimators consists of combining mesh refinements with Richardson Extrapolation (RE). Despite the fact this method is intuitive and simple, the validity and robustness remain questionable. In a series of papers ([1],[2]), the Least Square Extrapolation method (LSE), that provides a convenient framework to improve RE has been presented. Rather than an a priori Taylor expansion like model of the error, LSE makes use of the PDE discretization information, and thus on the research of the best linear combinations of coarse meshes solution. The computation of these solutions presents a strong asynchronous parallelism. On the other hand, there is an increasing interest in grid computation involving heterogeneous distribution of computing resources. In this paper, we will focus on the application of LSE method with a commercial CFD software on a network of heterogeneous computers. We will present out approach to software architecture to map the LSE on to the computer grid, and present measurement of performance gains compared to sequential execution.

1. Introduction and Motivations

In Computational Fluid Dynamic (CFD), *a Posteriori* error estimators are widely produced using Richardson extrapolation (RE) and variations of it ([6], [5], [7]). All these methods rely on the a priori existence of an asymptotic expansion of the error such as a Taylor formula, and make no direct use of the PDE formulation. As a consequence RE methods are extremely simple to implement.

But in practice, meshes might not be fine enough to satisfy accurately the a priori convergence estimates that are only asymptotic in nature. RE is then unreliable ([8]) or fairly unstable and sensitive to noisy data ([2]).

On the other hand, a Posteriori estimates in the framework of finite element analysis have been rigorously constructed. While most of the work has been limited to linear elliptic problems in order to drive adaptive mesh refinement, more recently a general framework for finite element a Posteriori error control that can be applied to linear and non-linear elliptic problem has been introduced in ([4]).

RE method can be embedded into an optimization framework to cope with RE's limitations while retaining its simplicity. Least Square Extrapolation (LSE) ([1], [2], [3]) uses grid meshes solutions that can be produced by any discretization. LSE sets the weights of the extrapolation formula as the solution of a minimization problem. This approach might be combined to existing a Posteriori estimate when they are available, but is still applicable as a better alternative to straightforward RE when no such stability estimate is available.

The extrapolation procedure is simple to implement and should be incorporated into any computer code without requiring detailed knowledge of the source code. Its arithmetic cost should be modest compare to a direct computation of a very fine grid mesh solution. Finally, the procedure should overall enhance the accuracy and trust of a CFD application in the context of solution verification.

In this paper, we show how LSE can perform on a commercial code, without having any knowledge on the source code. We emphasize this approach by using a basic network of workstation to compute the weighted solutions and then solve the minimization problem over the produce results.

In Section 2, the general idea of LSE method for steady problems is recalled. In Section 3, we discuss how the network grid can be set to match LSE requirements. In Section 4, we present some general numerical results about LSE when using a commercial code and what are the advantages of using a grid of computers instead of performing a sequential resolution.

2. LSE Method

Let G_i, $i = 1..3$, be three embedded grid meshes that do not necessarily match, and their corresponding solutions U_i. Let M^0 be a regular mesh that is finer than the meshes G_i. Let $\tilde{U}_i$ be the coarse grid meshes solutions interpolated on the fine grid M^0.

The main idea of the LSE method is to look for a consistent extrapolation formula based on the interpolated coarse grid meshes solutions $\tilde{U}_i$ that minimizes the residual, resulting from $\tilde{U}_i$ on a grid M^0 that is fine enough to capture a good approximation of the continuous solution.

Defining the Navier-Stokes operator by $N(u) = 0$ for example, we can rewrite the problem as follow :

$P_{\alpha,\beta}$: Find α, β such that $N(\alpha\, U^1 + \beta\, U^2 + (1 - \alpha - \beta)\, U^3)$ is minimum in L_2 in some Banach space.

Since minimizing the consistency error lead to the minimization of the solution error.

In practice, since we have access to solutions of discretized PDE problem, we work with *grid meshes functions*. The idea is now to use the PDE in the RE process to find an improved solution on a given fine grid mesh M^0, by post-processing the interpolated functions $\tilde{U}^i$, by few steps of the relaxation scheme

$$\frac{V^{k+1} - V^k}{\delta t} = L_h[V^k] - f_h, \quad V^0 = \tilde{U}^i, \tag{1}$$

with appropriate artificial time step δt. For elliptic problems, this may readily smooth out the interpolant. For a more complete description of the method and mathematical aspects, one should refers to the work in [1], [2], [3], [8].

3. Mapping LSE method onto a grid of computer

The LSE method can be split in 4 main computational steps:

Step 1 - Consists on defining the problem and solve it on 2 (or 3) different meshes. While this step may be parallelized, in reality we may realize reduce error by solving it sequentially since the result of one computation can be used as a guess for the next one.

Step 2 - Once the solutions are obtained on coarse grids meshes, we interpolate them on a fine grid mesh G_0, and form linear combinations with 1 (or 2 parameters) as initial guess for step 3.

Step 3 - This is the most time and resource consuming step. The problems may be solved independently on different nodes, and the error and residual of solutions are the only output of interest.

Step 4 - In this part we find the minimum of residual and error with respect to parameters. We generate curve (or surface) responses. This should be kept open since it will be subject to modification and optimization.

We would like to be able to re-launch calculation based on the same fine grid mesh, but with different weights around the potential minimum, or plot the map of the residual/error for optimal parameters in order to perform manual mesh refinements.

To solve the problem in Step 3 we are employing a grid of computers. Our grid consists of a master node and several heterogeneous slave nodes. The steps 1 and 4 run on the master. Step 3 is distributed on several nodes, depending on the number of nodes available and the problem size.

The distribution of the computation across nodes involves some network challenges. The result of step 2 is a set of files that can be pretty large (any where from 7 MB to 21 MB or more). Moving such large files across the network will be a major bottleneck and will limit our capability to realizing any performance gains from parallelizing the computations across grids of computers. By reordering some computations from step 2 to step 3, we expect that the files size to be transmitted will be much smaller. Even though each node may have to perform some extra computation, we think that the overall gain from parallelization offsets that extra time spent.

In order to get a better utilization of nodes and to adjust for variations in performance and availability of nodes, we employ dynamic load balancing of the computational task across the nodes.

Figure 1 is a symbolic representation of the computational procedure describe in the previous paragraph.

4. Realizing LSE method

In order to apply LSE on a grid of computers, we are looking for some applications that can be used on different types of architectures and operating systems, and allowing us to use a scripting shell language to launch and manipulate the data. A first step in this process was to implement a 2D non-homogeneous diffusion equation.

The idea here is to have an implementation that is independent from the computational code. Fig .2 shows what are the languages used to perform the different tasks.

4.1. Application to Laplace equation

We have tried different test cases. Here we show some results for non homogeneous Laplace equations with a strong gradient in the diffusion coefficient (Fig. 4.1). The grid meshes are regular Cartesian meshes. We first compute the solutions with ten different

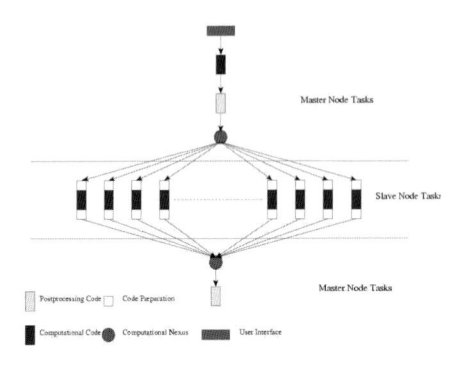

Figure 1. Software architecture realizing LSE

Figure 2. Language architecture realizing LSE

value for α. The grid meshes were chosen as follow : $G1 = 30 \times 30$, $G2 = 40 \times 40$, $G2 = 50 \times 50$ and $G0 = 100 \times 100$, where $G0$ is the fine grid.

To verify our solution, we also compute the solution on the finest grid mesh G_0. This solution allows us to compute the error we obtain using LSE versus the solution on the fine grid mesh. In figure 4.1 we show residuals obtained after few relaxations steps, and the corresponding error with the fine grid mesh solution. As we can see, there is an alpha that minimizes the residual, but also the error.

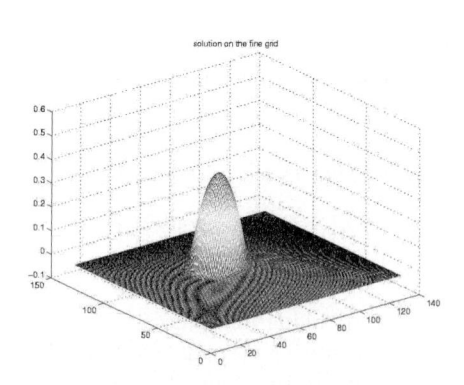

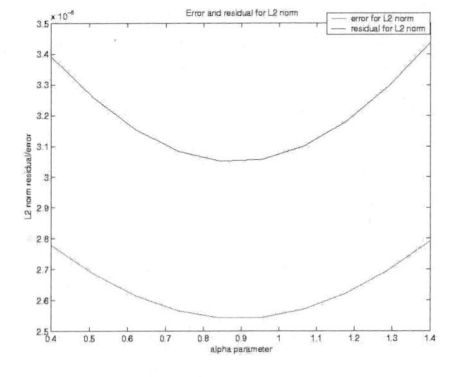

Figure 3. Solution of Laplace equation

Figure 4. Residual and error curve for Laplace equation

For a two dimensions optimum design space for LSE optimization parameters, the results obtain are shown on figure. As we can see, to obtain a good resolution on the

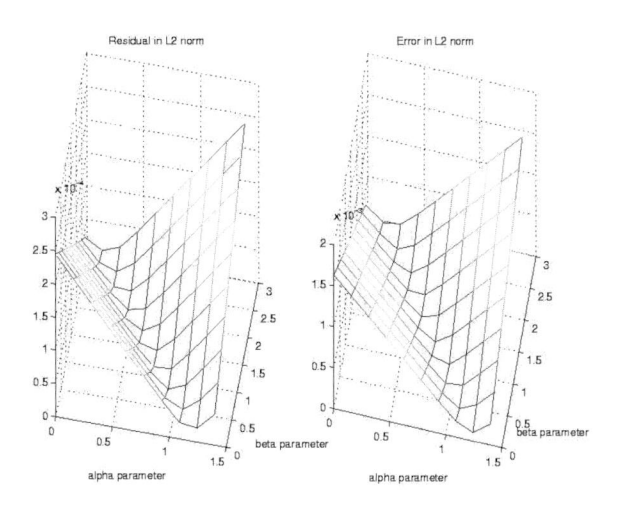

Figure 5. Residual and error curve for Laplace equation for a 2D optimum design space

value of the optimal parameters, the number of of initial step parameters can be important.

When solving LSE in a sequential way, with 100 parameters, the time required for this test case was 11.26 minutes , with roughly $80s$ for the Step 1, $5s$ for the Step 2, $600s$ for the Step 3, and $10s$ for the Step 4. It appears that Step 3 is a bottleneck in the method. Overall the solution verification procedure takes 615 s.

When mapping LSE onto a grid of computer, we manage to improve the performances of the method. In order to check the efficiency of this mapping, we run our code with three different scheme :

- on a single computer running Windows OS, using sequential LSE
- on a cluster of AMD Athlon running Linux OS, using a parallel version of LSE,
- on a heterogeneous grid running either Linux OS or Windows OS.

The speedup and scalability we obtain for the two parallel schemes are shown on fig. 4.1 and fig. 4.1. The speedup and scalability have a comportment close to the ideal one for the grid application. The main issue in that case is the quality of the network that is not homogeneous. For the cluster scheme, the comportment is becoming worst as we increase the number of nodes. Indeed, LSE required a lot of disk access since all the data have to be stored. Unfortunately, we didn't use a cluster with a parallel I/O system. Therefore, the scalability and the speedup reflect this characteristic.

The speed-up and scalability presented here are also relevant to the size of the grid meshes: given a number of weight, if we increase the size of the mesh the quality of the speedup should improve.

Figure 6. Speedup for LSE mapped onto a cluster and on a grid

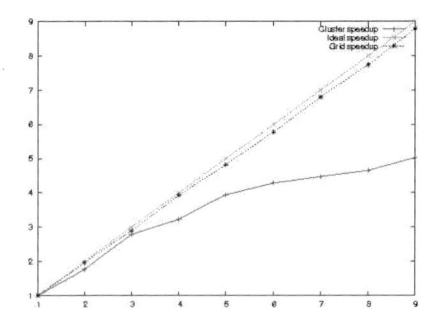

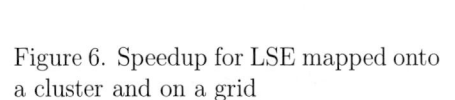

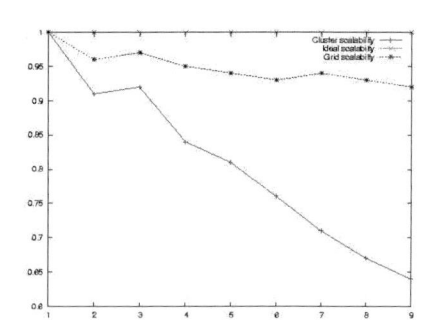

Figure 7. Scalability for LSE mapped onto a cluster and on a grid

4.2. Application to Navier-Stokes equations

The next step was to apply the same method to velocity equations in velocity-pressure formulation of Navier-Stokes equations. The software used to solve Navier-Stokes equation is ADINA R&D, a modeling software oriented to fluid-structure interaction. The test case we tried in the driven cavity flow in square box with a Reynold number of $Re = 100$ and a wall velocity of $v_0 = 1.0 m.s^{-1}$. The grids meshes are three nodes finite elements meshes. We first compute the solutions with ten different value for α. The grid meshes were chosen as follow : $G1 = 30 \times 30$, $G2 = 40 \times 40$, $G2 = 50 \times 50$ and $G0 = 100 \times 100$, where $G0$ is the fine grid.

The figures 8 and 9 show the coarse grid mesh G_1 and the solution in velocity norm we obtained before any post-processing on step 1 of the presented algorithm. We obtained similar results for grids meshes G_2 and G_3.

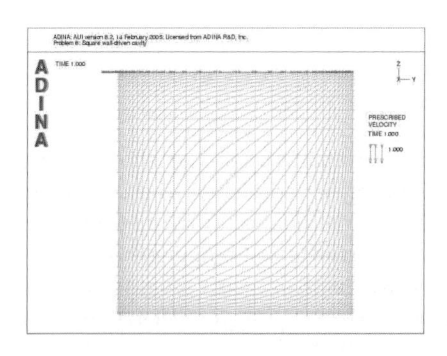

Figure 8. Coarse grid mesh G_1

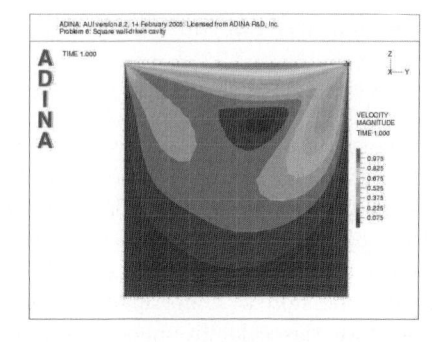

Figure 9. Solution on Grid G_1

To verify our solution, we also compute the solution on the finest grid mesh G_0. This solution allows us to compute the error we obtain using LSE versus the solution on the fine grid mesh. In figure 10 we show residuals obtained after few relaxations steps, and the corresponding error with the fine grid mesh solution. As we can see, there is an alpha that minimizes the residual, but also the error. On figure 11 the solution for the optimal parameter $\alpha = 1.5$ is presented.

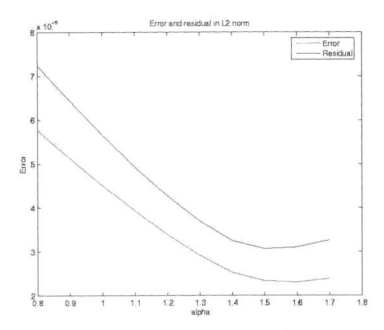

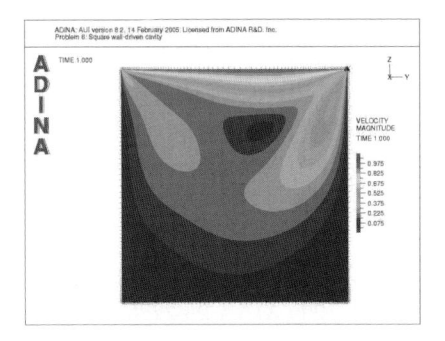

Figure 10. Error and residual in L_2 norm for $\alpha \in [0.8; 1.7]$

Figure 11. Solution on Grid G_0 for the optimal parameter

When solving LSE in the sequential time we obtained is 22 minutes, with roughly $130s$ for Step 1, $26s$ for step 2, $1200s$ for step 3 and $10s$ for step 4. For a parallelism on 10 nodes, Step 1 and Step 3 will have the same time as before, but the time for Step 2+3 will be $150s$, since we will compute the interpolation only once per node. We expect the speedup to be then 8.4 for the solution verification.

5. Conclusion

LSE is new extrapolation method for PDE that can be a solution verification method with hands off coding. It has be shown to be more efficient than Richardson extrapolation when the order of convergence of the CFD code is space dependent. In this paper, our objectives were to map the LSE method on a grid of computer and report the performance gains. We successfully managed to achieve those goal for a simple application.

This first step toward a grid application with Navier-Stokes equations is encouraging. It is obviously an interesting advantage to be able to perform a solution verification on 3D grid meshes using a grid of computer: most of the application time will then be spend into computation and not in communication. Some other aspects of LSE that are currently under consideration and that can influence the mapping onto a grid of computer are the optimum design space objective functions that can be higher order.

REFERENCES

1. M. Garbey. Some remarks on multilevel method, extrapolation and code verification. In N. Debit, M. Garbey, R. Hoppe, D. Keyes, Y. Kuznetsov, and J. Périaux, editors, *13th Int. Conf. on Domain Decomposition DD13, Domain Decomposition Methods in Science and Engineering*, pages 379–386. CIMNE, Barcelona, 2002.
2. M. Garbey and W. Shyy. A least square extrapolation method for improving solution accuracy of pde computations. *Journal of Computational Physics*, 186(1):1–23, 2003.
3. M. Garbey and W. Shyy. A least square extrapolation method for the a posteriori error estimate of the incompressible navier stokes problem. *Int. Journal of Fluid Dynamic*, to appear.
4. L. Machiels, J. Peraire, and A.T. Patera. A posteriori finite element output bounds for incompressible Navier-Stokes equations; application to a natural convection problem. *Journal of Computational Physics*, 2000.
5. W. L. Oberkampf and T. G. Trucano. Verification and validation in computational fluid dynamics. Technical report, Sandia National Laboratory, 2002.
6. W.L. Oberkampf, F.G. Blottner, and D.P. Aeshliman. Methodology for computational fluid dynamics code verification/validation. In *26th AIAA Fluid Dynamics Conference*, 1995.AIAA 95-2226.
7. P.J. Roache. *Verification and Validation in Computational Science and Engineering*. Hermosa Publishers, Albuquerque, New Mexico, 1998.
8. W. Shyy, M. Garbey, A. Appukuttan, and J. Wu. Evaluation of richardson extrapolation in computational fluid dynamics. *Numerical Heat Transfer: Part B: Fundamentals*, 41(2):139 – 164, 2002.

Acceleration of fully implicit Navier-Stokes solvers with Proper Orthogonal Decomposition on GRID architecture

D. Tromeur-Dervout [a] [*] and Y. Vassilevski [†] [b]

[a]CDCSP/UMR5208, University Lyon 1, 15 Bd Latarjet, 69622 Villeurbanne, France

[b]Institute of Numerical Mathematics, 8, Gubkina str., 119991 Moscow, Russia

Keywords: Parallel computation, Acceleration of convergence, Algorithms for specific class of architecture, distributed system, Nonlinear PDE of parabolic type, fluid mechanics. **MSC:** 65Y05, 65Y10, 65B99, 68H14, 35K55, 76.

1. Motivations

In the field of high performance computing the architecture characteristics always impact the development of computational methods. Novel numerical methodology may be attributed to the computer architecture and the associated software constraints/capabilities. The development of the communication network performances and infrastructures as the low cost of computing resources, authorizes to define computing architectures that gather several distant located distributed computers to solve one application. Metacomputing and GRID computing refer to such computing architectures. Metacomputing uses few stable high performance computers with a secured environment through dedicated or not dedicated communication network. GRID computing uses computing resources that are shared by several users, may be subject to hardware failures, through a non dedicated communication network. From the engineering point of view, coupling huge high performance computers seems not to be realistic in terms of day-to-day practice and infrastructure costs. Nevertheless, due to the material replacement staggering, industrial companies often have several medium computing resources with different performance characteristics. These computing resources can constitute a GRID architecture with high latencies and slow communication networks with fluctuating bandwidth shared by several users. The main drawback of this kind of computing architecture that processor persistent availability is not guaranteed during the computing. This difficulty becomes very painful, especially in computational Fluid Dynamics where time integration scheme on huge data problems have to run on long periods of time. On an other side, the large number of computing resources as well as the huge amount of associated memory, allow to dedicate spare processors to enhance the solution, to give estimates on the quality of the obtained

---
[*]This author was partially supported thru the GDR MOMAS project:"simulation et solveur multidomaine"
[†]This work is backward to the Région Rhône-Alpes thru the project: "Développement de méthodologies mathématiques pour le calcul scientifique sur grille".

solutions, and to speed the computing.

We propose in this paper a technique to accelerate the convergence of the Newton method to solve non-linear unsteady CFD problems. The principle relies on an effective way to obtain a better solution for the current time step than the solution of the previous time step. This is performed on a reduced model computed with a Proper Orthogonal Decomposition of some solution computed at some previous time step. The developed methodology fits the grid computing design with a client server approach to compute the POD components necessary for the acceleration. We will shows on several flows configurations the effective computational saving obtained with the proposed technique.

Let us recall briefly the fully implicit discretizations of unsteady nonlinear problems. Let $L(u)$ be a nonlinear discrete operator representing a spatial approximation of a parabolic boundary value problem. The simplest robust technique for time approximation of unsteady problems is the backward Euler time stepping:

$$\frac{u^i - u^{i-1}}{\Delta t} + L(u^i) = g^i. \qquad (1)$$

The main advantage of the method is its unconditional stability. Being the first order scheme (in time), it may be generalized to the higher order approximations (e.g., BDF time stepping). In general, the ith time step of a fully implicit scheme may be represented by the nonlinear system $F^i(u^i) = 0$, where F^i contains all the problem data and previous solutions. For instance, in the case of scheme (1),

$$F^i(u^i) = u^i + L(u^i)\Delta t - u^{i-1} - g^i \Delta t. \qquad (2)$$

The price to be paid for the robustness of the method is its arithmetical complexity: at each time step, a nonlinear system has to be solved. In last decade, several robust nonlinear solvers have been proposed, analyzed, and implemented ([5] p. 303). The inexact Newton backtracking [5] method offers global convergence properties combined with potentially fast local convergence. Algorithm INB presumes the choice of initial guesses u_k for nonlinear iterations. We recall that the arithmetical complexity of the method is expressed in the total number of function evaluations n_{evF} and total number of preconditioner evaluations n_{evP} (if any), the remaining overheads are negligible.

2. Initial guess for the Newton solution with POD

Proper orthogonal decomposition (POD) produces an orthonormal basis for representing the data series in a certain least squares optimal sense [4]. Combined with the Galerkin projection, POD is a tool for generation of reduced models of lower dimension. The reduced models may give a better initial guess for the Newton solution at the next time step. POD provides the most close (in the terms of the least squares) m-dimensional subspace $S \subset R^N$ to the given set of vectors $\{u^i\}_{i=1}^n$,

$$S = \arg \min_{S \in R^{N \times m}} \sum_{i=1}^{n} \|u^i - P_S u^i\|^2?$$

Here P_S is the orthogonal projection onto S. Define the correlation matrix $R = XX^T$, $X = \{u^1 \ldots u^n\}$, and find m eigenvectors of the problem

$$Rw_j = \lambda_j w_j, \quad \lambda_1 \geq \cdots \geq \lambda_N \geq 0$$

corresponding to m largest eigenvalues $\lambda_1 \geq \cdots \geq \lambda_m$. Then $S = \text{span}\{w_j\}_{j=1}^m$ and

$$\sum_{i=1}^n \|u^i - P_S u^i\|^2 = \sum_{j=m+1}^N \lambda_j. \qquad (3)$$

Computational cost of finding m-largest eigenvalues of symmetric matrix R is not high. Indeed, our experience shows that for $m = O(10)$ the application of the Arnoldi process requires a few tens of R-matrix-vector multiplications in order to retrieve the desirable vectors with very high accuracy [2]. In spite of large dimension N and density of R, the matrix-vector multiplication is easy to evaluate, due to the factored representation $R = XX^T$. The arithmetical cost of the R-matrix-vector multiplication costs at most $4Nn$ flops.

Each time step of the scheme (1) generates the equation (2) which we call the original model. A reduced model is generated on the basis of POD for a sequence of solutions at time steps $\{u^i\}_{i=i_b}^{i_e}$, $i_e - i_b + 1 = n$. The eigenvectors $\{w_j\}_{j=1}^m$ may be considered as the basis of m-dimensional subspace $V_m \in R^N$. The reduced model is the Galerkin projection onto this subspace: $V_m^T F^i(V_m \hat{u}^i) = 0$, or, equivalently with $\hat{u}^i \in R^m$ and $\hat{F}^i : R^m \to R^m$:

$$\hat{F}^i(\hat{u}^i) = 0, \qquad (4)$$

The reduced model is the nonlinear equation of very low dimension m. For its solution, we adopt the same INB algorithm with finite difference approximation of Jacobian-vector multiplication. Being the formal Galerkin projection, each evaluation of function $\hat{F}^i(\hat{u}_k^i)$ is the sequence of the following operations: $u_k^i = V_m \hat{u}_k^i$, $f_k^i = F^i(u_k^i)$, $\hat{f}_k^i = V_m^T f_k^i$. Therefore, the overhead is matrix-vector multiplications for V_m and V_m^T, i.e., $4Nm$ flops. We notice that usually $m = O(10)$ and the evaluation of function $F(u)$ is much more expensive than $40Nm$ which implies a negligible weight of the overheads. Another important consequence of low dimensionality of (4) is that INB algorithm may be applied without any preconditioner.

2.1. Fully implicit solver with POD-reduced model acceleration

Coupling POD and Galerkin projection for the generation of the reduced model gives a powerful tool for acceleration of the fully implicit schemes. Let n, the length of data series be defined, as well as the desirable accuracy ϵ, for $F^i()$: $\|F^i(u^i)\| \leq \epsilon$. For any time step $i = 1, \ldots,$ perform:

ALGORITHM
    IF $i \leq n$, SOLVE $F^i(u^i) = 0$ BY PRECONDITIONED INB
          WITH THE INITIAL GUESS $u_0^i = u^{i-1}$ AND ACCURACY ϵ
    ELSE

1. IF$(mod(i, n) = 1)$:

    (A) FORM $X = \{u^{i-n} \ldots u^{i-1}\}$;

    (B) FIND SO MANY LARGEST EIGENVECTORS w_j OF $R = XX^T$ THAT $\sum_{j=m+1}^N \lambda_j \leq \epsilon$;

(c) FORM $V_m = \{w_1 \ldots w_m\}$

2. SET $\hat{u}_0^i = V_m^T u^{i-1}$

3. SOLVE $\hat{F}^i(\hat{u}^i) = 0$ BY NON-PRECONDITIONED INB WITH THE INITIAL GUESS $\hat{u}_0^i$ AND ACCURACY $\epsilon/10$

4. SET $u_0^i = V_m \hat{u}^i$

5. SOLVE $F^i(u^i) = 0$ BY PRECONDITIONED INB WITH THE INITIAL GUESS u_0^i AND ACCURACY ϵ

Several remarks are in order. The absence of the preconditioner for the reduced model is dictated by two reasons: a) it is not clear how to construct a preconditioner for the reduced model, b) it is hardly needed if m is small. The reduced model is slightly oversolved: this provides better initial guess u_0^i. The number of eigenvectors is chosen adaptively in the above algorithm: it allows to form such a reduced model that approximate the original model with the desirable accuracy ϵ. Actually, this condition may be replaced by a more rough $N\lambda_{m+1} < \epsilon$ or even a fixed number m, $m = 10 - 40$. Solution of the eigenvalue problem may be performed asynchronously with the implicit solution: as soon as V_m is formed, the reduced model becomes the active sub-step. The latter observation allows to use a client server approach for the grid implementation.

3. POD-reduced model acceleration and its GRID applications

The present acceleration is well designed for the GRID computing. The target architecture is represented by large amount of low cost computational resources (clusters) connected with standard Ethernet communication network. We consider the basic features the GRID architecture and appropriate modes of POD usage in this context:

- Slow communication network with high latency time between clusters of resources. It is usual to have one or two order of magnitude gap between the communication speeds inside and outside a cluster. The POD acceleration can be used wherever POD data are available. The asynchronous non blocking communications between the POD generator and the solver resource provide computation of the time step without idling. Therefore, a slow network with high latency time is affordable for the proposed technology.

- High probability of failure of some part of the computing resources. This is typical for GRIDs where the resources are not dedicated to single application and single user. The conventional way to cope with a hardware failure is on-fly backup of the solution which deteriorates the performance. Being activated on a separate computational resource, the POD generator gives a natural way to restart the computation on the solver resource. The MPI-2 process model allows to spawn a new set of MPI processes. The POD can be the external resource control that can start the spawn processes of the application in case of failure of the solver resources. Upon an interruption of the data supply, the POD generator can restart the solver processes

	without POD			with POD m=10			
i	10	20	30	40	50	60	80
	$u_0^i = u^{i-1}$			$u_0^i = V_m \hat{u}^i$			
$\|F(u_0^i)\|$	0.3	0.8	0.1	$8 \cdot 10^{-6}$	10^{-5}	10^{-6}	10^{-6}
n_{evF}	193	184	173	42+39	43+36	46+11	36+12
n_{precF}	187	178	167	0 +36	0 +32	0 + 9	0 +10
time	16	15	15	0.9+3.1	1+2.6	1+0.7	0.8+0.8
	full model			Reduced model + full model			

Table 1
Effect of POD acceleration for the lid driven cavity 2D problem modelled by the biharmonic streamfunction equation with a periodic drive

and deliver them the appropriate last solution recovered from the reduced model representation.

- The POD generator task is waiting data from the solver resource and computing the POD basis when sufficient data are gathered. For the sake of more efficient use of the POD generation resource, it may work with other tasks as well. For instance, the POD generator can be used by several solvers and perform the POD on different sets of solutions tagged by the generating solver. Besides, the reduced basis may be used for other postprocessing tasks such as data visualization or a posteriori error estimation.

4. Numerical experiments on homogeneous and grid architecture

In order to illustrate the basic features of the proposed methodology, we choose the backward Euler approximation of the unsteady 2D Navier-Stokes equations. We consider the classical driven cavity problem in the streamfunction formulation:

$$\frac{\partial}{\partial t}(\Delta \psi) - \frac{1}{Re}\Delta^2 \psi + (\psi_y(\Delta \psi)_x - \psi_x(\Delta \psi)_y) = 0 \text{ in } \Omega, \quad (5)$$

$$\psi|_{t=0} = 0 \text{ in } \Omega, \; \psi = 0 \text{ on } \partial\Omega, \; \frac{\partial \psi}{\partial n}\Big|_{\partial\Omega} = \begin{cases} v(t) & \text{if } y = 1 \\ 0 & \text{if } 0 \leq y < 1 \end{cases}$$

Here, $\Omega = (0,1)^2$, $Re = 1000$, and $v(t)$ will denote the unsteady boundary condition leading the flow. Two different types of flows are simulated: quasi-periodic in time, and quasi-periodic in time with variable periods (arrhythmic). These two cases are defined by the unsteady boundary velocity $v(t)$. We set $v(t) = 1 + 0.2\sin(t/10)$ for the quasi-periodic flow, and $v(t) = 1 + 0.2\sin([1+0.2*\sin(t/5)]*t/10)$. We motivate the chosen parameters as follows. In the case of $v(t) = 1$, the unsteady solution saturates within $t_s \sim 150$. Therefore, to get a quasi-periodic solution, we need the periodic forcing term with the period $T < t_s$ but comparable with t_s, $T \sim t_s$. Indeed, if $T \ll t_s$, the inertia of the dynamic system will smear out the amplitude of the oscillations; if $T > t_s$, the dynamic system will have enough time to adapt to the periodic forcing term and demonstrate periodic behavior. The function $\sin(t/10)$ has the period $T = 20\pi$ which perfectly fits the above restrictions for the quasi-periodicity. The arrhythmic flow is a simple modification

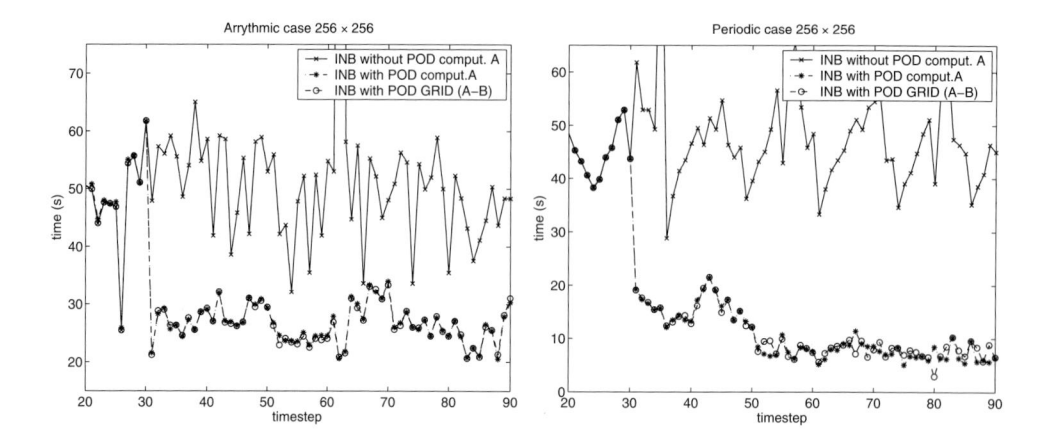

Figure 1. Arrhythmic case (left) Periodic case (right): comparison of elapsed time of INB on computer A, INB with POD initial guess computer A, INB with POD initial guess GRID computer A (POD computed on computer B)

of the quasi-periodic flow by arrhythmic scaling of the time $t \to [1 + 0.2 * \sin(t/5)] * t$. It is well known that the feasible time step Δt for approximation of periodic solutions satisfies $12\Delta t = T$, i.e., $\Delta t \sim 5$.

Table 1 shows that the initial function value drops from 10^{-1} without POD acceleration to 10^{-6} with. The elapsed time is then reduced by a factor 6 on a alpha ev67 processor. The mesh size is $h = 256^{-1}$ leading to 65025 dof, the stopping criterion for INB is $\|F^i(u_k^i)\| < 10^{-7}\|F^0(0)\|$. The time step is $\Delta t = 5$ leading to have 13 time steps per period and $m = 10$. V_m are produced starting from 20th step.

GRID experiments involve a computer A (SGI Altix350 with Ithanium 2 processors cadenced at 1.3Ghz/3Mo, 1.3Gb/s network bandwidth) and a computer B (6 nodes Linux cluster with AMD BiAthlon 1600+ MP processors, with 256KB L2-cache, 1GB of RAM per node, 100Mb/s Ethernet internal network). The mpich communication library is compiled with the ch_p4 interface. A latency of 140 μs and a maximum bandwidth of $71Mb/s$ have been measured for the communication network between the computers A and B. For compilation we use the g77 Fortran compiler.

This section compares the conventional solver INB with the initial guess from the previous time step, the INB-POD solver running on computer A, and the INB solver running on computer A and communicating with the POD generator running on computer B. The POD acceleration begins at the 30th time step.

Figure 1 give the elapsed time of the INB solver for the arrhythmic and the periodic cases on the GRID architecture context. The figures show that:

- the INB-POD gives quite similar results when it performs on the GRID context or on an homogeneous computer. Consequently the computer B can be used with no penalty on the performances;

- the elapsed time is greater than that from Table 1 due to the poor performance of the g77 compiler on IA64 processor. Nevertheless, the acceleration factor of the INB-POD remains high.

Figure 2 represents the GRID computation in detail: in addition to the total elapsed time of each time step, they exhibit the time t_1 spent for the initial guess computation by INB applied to the reduced model and the time t_2 of the INB solution of the original model. We observe that t_1 is almost constant with the mean value 3.7s for the periodic case and 3.9s for the arrhythmic case. In contrast, t_2 manifests strong (more than 100%) fluctuations with the mean value 6.6s for the periodic case and 22.5s for the arrhythmic case. The stability of t_1 is the consequence of the stable convergence of INB due to the low dimension ($m = 10$) of the reduced model.

The implicit solution may be advanced without the reduced model POD acceleration, or with the obsolete (not updated) reduced model acceleration. The technology of the asynchronous data exchanges provides the tools for launching the reduced model usage only when the respective reduced basis will be available (*i.e.*, computed somewhere else and received to the processor carrying out the implicit solution). On the other hand, upon each time step termination, the solution may be sent asynchronously to the processor resolving eigenproblems. Therefore, the number of time steps n_{delay} accelerated with obsolete data (or not accelerated) depends on the time for solving the partial eigenproblem for matrix $R = XX^T$ and the time of data (solution series and reduced model basis) exchange. We consider two configurations of resources. Configuration I uses one processor for NITSOL and two processors for POD on computer A. Configuration II uses one processor for NITSOL on computer A and four processors for POD on computer B. The mesh size is $h = 128^{-1}$. In both cases the number of delayed time steps n_{delay} is equal to 1. These results demonstrate the appealing features of the approach in day-to-day engineering computations. First, it is possible to run the POD and the INB on the same processor. However, there exists the risk of performance deterioration due to cache reuse and/or code swapping as the POD needs sufficient data to be relevant. Second, the cost of processors of computer B dedicated to the POD is much less than those of processors of computer A. Consequently, it is beneficial to load processors B with useful but not critical task of POD generation and dedicate the advanced processors to heavy computation of nonlinear systems. Third, the spare processors on computer B can postprocess the solution (a posteriori error estimates, visualization). Fourth, the spare processors can monitor the computation and restart it in case of failure of computer A.

We note that the reported GRID implementation suggests certain improvements. MPICH-Madelaine [1] should allow to use different mpich protocols of communications inside and outside a cluster. Besides, the employment of the Intel Fortran compiler on the IA64 processor architecture should improve the performance of the INB-POD solver.

Conclusions

The method for the acceleration of the fully implicit solution of nonlinear boundary value problems is presented. The use of the reduced model to compute much better initial guess reduces the computational time as well as the numbers of nonlinear and linear iterations. The asynchronous communications with the POD generator make the algo-

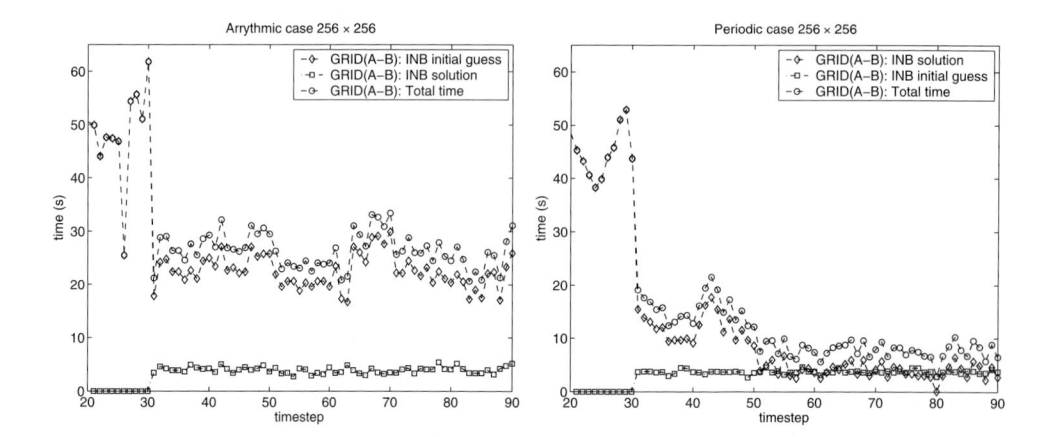

Figure 2. Elapsed times for the GRID computation of the arrhythmic case (left) and the periodic case (right): total time for the time step, time of computation of the initial guess by INB, time of the INB solution of the original model.

rithm appealing for the GRID computing. The parallel efficiency in the GRID context is understood as the computational speed-up on the solver resource due to the communications with the POD generator, and not the parallel speed-up with respect to the number of processors. Another appealing feature of the approach is its hardware failure protection: upon failure, the solver resource can recover its computation by spawning the MPI processes and using the POD data from the POD generator resource.

REFERENCES

1. O. Aumage, L. Bougé, L. Eyraud, G. Mercier, R. Namyst, L. Prylli, A. Denis, J.-F. Méhaut, *High Performance Computing on Heterogeneous Clusters with the Madeleine II Communication Library*, Cluster Computing, 5-1 (2002) 43–54.
2. R. Lehoucq, D. C. Sorensen and C. Yang, *ARPACK Users' Guide: Solution of Large-Scale Eigenvalue Problems with Implicitly Restarted Arnoldi Methods*, Software, Environments, and Tools 6, SIAM, 1998.
3. M. Pernice and H. Walker, *NITSOL: a Newton iterative solver for nonlinear systems*. SIAM J. Sci. Comput., 19 (1998) 302–318.
4. M. Rathinam and L. Petzold, *A new look at proper orthogonal decomposition*. SIAM J. Numer.Anal., 41 (2003) 1893–1925.
5. M. Pernice, H. F. Walker, *NITSOL: a Newton iterative solver for nonlinear systems*, Special issue on iterative methods (Copper Mountain, CO, 1996), SIAM J. Sci. Comput., 19(1), pp. 302–318, 1998.

A Gentle Migration Path to Component-Based Programming

Craig E Rasmussen[a]*, Matthew J. Sottile[a] Christopher D. Rickett[a], and Benjamin A. Allan[b]

[a]Advanced Computing Laboratory, Los Alamos National Laboratory,
P.O. Box 1663, Los Alamos, NM 87545 USA

[b]Sandia National Laboratories,
Livermore, CA 94551-0969 USA

Paradoxically, software components are widely recognized as providing real benefits to the scientific computing community, yet are generally ignored in practice. We seek to address this discrepancy by proposing a simplified component model (CCAIN) that departs only slightly from current programming practices. This component model can be gradually extended in a staged approach to obtain full-blown implementations of component-based applications using standard component models and frameworks. We present an example of how CCAIN components can be used as a migration path to Common Component Architecture (CCA) components.

1. Introduction

Software components hold the promise of increasing the productivity of scientific application programmers through code modularity and increased software reuse. Unfortunately, through the proliferation of component models, differing application frameworks, and a general lack of accepted standards, this promise is still largely unmet. The path to a component model is simply too formidable a task for many. This paper introduces a simplified component programming model that provides a gentle migration path to component-based application development.

Most existing component models require a relatively large leap from current software practices. The simple component model proposed here allows scientific software developers to take an evolutionary and staged approach for component usage, rather than a revolutionary plan that requires extensive software rewriting. The proposed model will still require a movement towards increased software modularization, but this, to a large extent is a good idea and is happening anyway.

---

*Los Alamos National Laboratory is operated by the University of California for the National Nuclear Security Administration of the United States Department of Energy under contract W-7405-ENG-36, LA-UR No. 05-1907.

2. CCAIN Component Model

The first task is to define what a component is. Out of the many existing (and sometimes conflicting) definitions, we choose one that is particularly simple and useful within the high-performance computing community. A CCAIN (CCA INtegration framework) component instance is a set of procedures that are compiled and linked into a static (or dynamic) library. Generic interfaces for the procedure set are declared in both a C header file and a Fortran BIND(C) interface (see below).

Component frameworks used for scientific computing must specify and provide for three general tasks:

- Language interoperability.

- Invocation of component procedures.

- Composition of a component set into a running application.

Furthermore, the framework *must not* prohibit or disrupt parallelism within or between components.

2.1. Language Interoperability

The existing practice for language interoperability is to develop (for the most part) within one language. For large scientific applications, this language is frequently Fortran. If there is a need to call functions in another language, it is done through implicit Fortran interfaces that allow calls to C by pretending that C functions are "old" style Fortran 77 procedures. For better language interoperability, intermediary bridging code can be generated to provide the glue between languages. These tools include, Babel [?] (for interoperability between a common suite of languages), Chasm (for interoperability between language pairs, such as C/C++ and Fortran 95) [?], or F2PY (for interoperability between Fortran and Python) [?].

Most existing Common Component Architecture (CCA) [?] frameworks have chosen one of the above mechanisms for language interoperability. The first CCA framework (CCAFFEINE [?]), was written in C++ (classic version) and assumed all components were also written in or wrapped by C++. This was followed by a new version allowing components using Babel for argument marshalling and function invocation between a suite of languages. Other language-specific frameworks are XCAT-JAVA [?] (a distributed framework supporting Java components) and Dune (used for the rapid prototyping of components using Python and employing Python for argument marshalling and function invocation) [?].

Problems arise if one pretends that C functions are F77 procedures, as Fortran and C compilers use different naming conventions for symbols and pass parameters differently. Bridging code may be used to get around this but this adds additional complexity. Fortunately, the Fortran standard now includes the ability to declare interfaces to C procedures within Fortran [?] itself. This is done by declaring procedure interfaces as BIND(C) and employing only interoperable arguments. BIND(C) interfaces are callable from either Fortran or C and may be implemented in either language.

Fortran BIND(C) interfaces are also available to C++ through an `extern "C"` declaration. Interoperable types include primitive types, derived types or C structures (if all fields are interoperable), and C pointers (which may be associated with Fortran arrays or Fortran procedures). Interoperability between C structures and Fortran derived types is important, because Fortran and C programmers often encapsulate data in this manner (note, currently fields in structures are only available with set/get functions using Babel).

As briefly touched on earlier, CCAIN components shall use Fortran BIND(C) interface declarations for language interoperability with Fortran. Components shall be implemented in either Fortran, C, or C++ (using an `extern "C"` declaration). Arguments to component procedures shall be Fortran BIND(C) interoperable types. Note that Ada95 [?] can also be used to create CCAIN components by using the Interfaces.C Ada package.

This standard greatly simplifies language interoperability because it places the burden on the Fortran compiler to marshal procedure arguments and to create interoperable symbols for the linker, rather than placing the burden on the programmer. If implementing in C, no additional work is needed other than to declare an interface and provide an implementation. For interfacing to C++ classes, a light-weight wrapper may need to be created to provide an `extern "C"` interface with interoperable BIND(C) types.

2.2. Component Procedure Invocation

A software component must have a clearly defined way to call a procedure on another component. This is not as simple as it at first seems, because the goal is to call a *generic component* procedure within a *concrete* component implementation. The problem is that differing implementations of a component class cannot all use the same generic interface, because of name conflicts that will appear during application linkage.

The standard way to avoid symbol name conflicts in languages like C (without a Fortran module or C++ namespace mechanism) is to provide name modifications through prepending or appending additional characters to a procedure name. CCAIN components adopt this standard. For example, if the component class is MHD and the procedure name is init, the generic procedure name would be `MHD_init`. An actual implementation of the generic procedure name must append an implementation name. For example, a specific `MHD_init` procedure might be named `MHD_init_NASA`. This choice, *i.e.* to not use C++ to implement namespace management, simplifies the development tool chain and eliminates additional interoperability issues between code built with different C++ compilers.

A problem immediately arises. If a component uses an MHD component, how does one call a specific MHD instance without using the specific procedure name? This is a fundamental component requirement, because one would like to create a component, test it, provide a version number, and file it away semi-permanently in a repository. This is not possible if one calls `MHD_init_NASA` explicitly, because one may want to try another MHD implementation and call `MHD_init_ESA`, for example. To do this, it is not optimal to dig out the component, replace every call to `MHD_init_NASA` with `MHD_init_ESA`, recompile and retest the component, provide a new version number, and finally replace the component version in the repository.

The solution chosen for CCAIN components is to store procedure pointers in a derived type. Each component class shall have a corresponding derived type (and C struct) defined for it containing function pointers to each procedure in the procedure set defining the component class. Each procedure shall declare this type as its first argument.

This simple requirement allows for a high degree of flexibility. For instance, the CCA getPort API can be used to: 1. Get the component type handle (in C, a pointer to the component class struct) using the string name of the component; and 2. Use the type handle to invoke the desired procedure. For example, in C,

```
#ifdef USE_COMPONENTS
    mhd = svc->getPort(svc, "MHD Model");
    mhd->init(mhd);
#else
    MHD_init_NASA(mhd);
#endif
```

where svc is the services object for the component (obtained during the component creation phase). The Fortran code is similar, but requires an additional step to convert from a C pointer to a Fortran procedure pointer. Note that now the truly generic name, init can be used, rather than the specific (and actual) name MHD_init_NASA. No change is required of the callee component; it requires only the slight modification shown above for the caller component.

2.3. Component Composition

It still remains to be determined how to create an application based on CCAIN components. The easiest way is entirely standard. Use specific procedure names (MHD_init_NASA) throughout and implement a "main" routine to create, initialize, and run all components. Then simply link the components with the main routine to create an application.

A more general way is to implement the CCA BuilderServices API for component composition (at the framework level) and setServices for setting function pointers (in each component). For example, the following code creates two components:

```
mhdID = bs->createInstance(bs, "MHD", "Nasa MHD Model",
                    nasa_mhd_comp);
heaterID = bs->createInstance(bs, "Heater", "Nasa Heating Model",
                    nasa_heating_comp);
```

where bs is a BuilderServices object and nasa_mhd_comp and nasa_heating_comp are pointers to component structures.

Suppose the MHD component calls (uses) a heating component for data. Then one must provide this information to the framework so that the correct heating model instance can be provided to the MHD component (as shown below):

```
connID = bs->connect(bs, mhdID, "MHD Model", heaterID, "Heating Model")
```

This call on the framework associates the uses (caller) interface of the MHD model with the provides (callee) interface in the heating model.

A component-based application is created by implementing a "main" routine that uses the CCA BuilderServices routines shown above and linking in all of the separate component libraries that are used. Complete examples of component usage may be found on the CCAIN website [?].

3. Summary

The proposed CCAIN component standard allows one to take an evolutionary and staged approach for the adoption of component-based applications by a scientific community. Usage of current component frameworks (like CCA) requires the immediate adoption of an unfamiliar programming model by users. It is suggested a staged and flexible approach (see below) can lead to broader acceptance within a scientific community.

The stages for the migration from a set of traditional scientific applications to a full-fledged CCA component application is as follows:

1. Create software modules. Begin by refactoring the set of applications into software modules and carefully declaring interfaces for each high-level class of component. This is an important task, because the creation of a set of well-defined interfaces will pay dividends later on [?]. Provide users with both Fortran BIND(C) and standard C declarations of the interfaces. If the component is implemented in C++, create `extern "C"` interfaces. Finally, provide an application "main" routine that creates, initializes and runs the set of components using standard compilation and link techniques.

2. Modify the components from stage one to allow usage of generic interfaces in a light-weight CCA/BIND(C) framework as described in the previous section. Implementation of this stage will allow multiple instances of a component class to be run at a given time using generic interfaces. Finally, create a new "main" application using the `BuilderServices` API to create, initialize, connect and run the components.

3. Migrate to a full-fledged CCA/SIDL (Scientific Interface Definition Language) environment using Babel for increased language interoperability beyond C and Fortran. Provide a SIDL interface for each generic component class. (Note that derived types as procedure arguments will not be interoperable between languages at this stage, so care must be taken in stage one in defining component interfaces, if migration to stage three is desired.) Run the Babel compiler to provide language binding code (stubs and skeletons) and edit the skeletons to call into the components created in stage one. Compile the stubs and skeletons and link with the standard CCAIN components implemented in stage two.

This staged development model allows for a gradual, evolutionary path to be taken from existing programming practices to a CCA/SIDL environment. These stages may be taken at a rate that is acceptable to a scientific community, as understanding and acceptance of the component programming model increases. Note that at no point are components from stage one greatly changed. In fact, if the stage two components are developed using preprocessing directives (as shown earlier), the same component code can by used in all three stages.

This approach greatly simplifies component development:

- Language interoperability is provided by compilers; only interfaces need be defined in both C and Fortran. Thus, there is no need for the generation, implementation, and management of extra code for language integration.

- The same component code can be run in either a traditional mode or a CCA mode. Only the application "main" need change.

- Components may reside in static libraries. The creation and loading of dynamic components is operating system dependent and may not even be available on a given operating system. It is also not clear that dynamic components are truly needed in production codes.

- The CCAIN component framework implementation is also simplified because it does not need to dynamically load components. It effectively only acts as a component factory and naming service. It need only implement five functions for most uses (`getPort`, `createInstance`, `connect`, `addProvidesPort`, and `registerUsesPort` (the latter two have not been discussed).

CCAIN components may not be written in java or python as this would require support for dynamic components; this simplifies the effort required to maintain a CCA framework (the CCA specification does not require the ability to dynamically load components). However, the CCAIN framework, BuilderService, and CCAIN components could be wrapped using SWIG [?] or F2PY to provide top-level application scripting, if desired.

4. Future Work

It is believed that the proposed simple CCAIN component model can be used within other scientific frameworks and execution modules. Certainly this can be done with slight code modifications (stage 2) or with wrappers (Stage 3). However, there may be ways in which CCAIN components can be recognized by other frameworks and used directly. For example, CCAIN function pointers can be called by the Babel IOR (Internal Object Representation) call mechanism, thus bypassing and eliminating the need for the Babel skeleton wrappers in Stage 3. It is left to future research to determine to what extent this is possible for other scientific frameworks such as Cactus [?] and the Earth System Modeling Framework (ESMF) [?].

Acknowledgements

We would like to thank Tamara Dahlgren for helpful conversations regarding the use of CCAIN components with Babel.

PyNSol: A Framework for Interactive Development of High Performance Continuum Models

Michael Tobis

Department of the Geophysical Sciences, University of Chicago, 5734 S Ellis Av, Chicago IL 60637, USA.

Keywords: Parallel computing, continuum modeling, finite differences, high performance, high productivity, code comprehension, domain specific language

1. THE PYNSOL WORKING ENVIRONMENT

"Our goal is to make it easier to make a piece of code simultaneously clear and fast in a high level language than to make it fast in a low-level language" [11].

This ambitious goal, stated over a decade ago by the CLOS group at MIT, also summarizes the objectives of PyNSol (a Python Numerical Solver)[14]. PyNSol is a framework architecture for the incremental development of a high productivity environment for developing high performance codes. Its target domain is computational fluid dynamics problems with simple geometries and schemes of modest order, but with multiple physical processes, multiple time scales, and complex forcing.

From the end-user's point of view, PyNSol is a competitor to Parallel Matlab. Numerous efforts to provide distributed Matlab are underway, e.g., [10]. These tools provide the scientist with an interactive development environment while offloading the complexities of targeting the parallel platform to an infrastructure layer that is not explicitly presented to the user. PyNSol intends to provide higher performance as well as comparable ease of use by approaching a less general set of problems.

The PyNSol end user, a physical scientist, investigates physical problems of various sorts, and is presented with extensions to Python to facilitate this process. Such a user chooses discretization, time differencing, and parallelization strategies from menus of existing strategies, and specifies a physical description of the system of interest in a terse and accessible format.

An interactive layer is presented to the scientist, built around the IPython working environment [13] and the matplotlib graphics library [9]. These tools, while in active development, are already widely deployed and used. A notebook presentation layer is also under development.

For the developer of PyNSol extensions, however, the picture is of a framework architecture. Here we use the word "framework" as defined by Gamma et al.: *"A framework is a set of cooperating classes that make up a reusable design for a specific class of software"* [6]. The PyNSol approach is also informed by the elaboration of this definition in the introduction to Carey and Carlson's practical guide to the subject [5].

Numerical analysts and software engineers, in a support role to the scientific process, extend the PyNSol framework by creating plugins to expand the set of geometries, time differencing schemes, and architectures supported. As PyNSol grows, the menus of schemes and strategies presented to the application scientist will become extensive, so that an increasingly large array of problems can be addressed using the existing framework components. A working prototype for the PyNSol environment and framework provides a promising beginning for a toolkit of considerable practical value.

2. MOTIVATION AND REQUIREMENTS

Computational science is faced with a peculiar sort of crisis, a crisis of opportunity. We increasingly have more computational power than we can use effectively. Even though the total utility of our computational infrastructure continues to increase, the fraction of its potential that is achieved continues to plummet. This quandary, one of missed opportunity, has been called the "software crisis" [7].

The continuing and remarkably rapid decrease in computational cost has made enormous computational power available to individual researchers. In general, the capacity of the human mind to manage complexity has not comparably increased and presumably will not comparably increase in the foreseeable future. As a result, all else being equal there is a tendency for the utility of our computational resources to lag dramatically behind their availability. A strategy to make the most of computational resources is needed, one which must necessarily include an increase in effectiveness in software tools. The work we propose here is intended to contribute to this effort.

In the commercial sector, a formal discipline called software engineering is emerging to cope with these matters, and a formal body of knowledge is being maintained [4]. Efforts to bring this progress into scientific computation have not been entirely successful. A primary reason for this is that the distinction between developer and user is not as clear-cut in the research environment as in the commercial one.

Arguably the most important type of end user of many types of scientific software is the application scientist. Such a user best understands the application domain, and is willing to put substantial talent and effort into manipulating the software, but has little formal training in software practice and rather low expectations of the usability of the available tools, especially when high-performance computing is involved.

In PyNSol's target discipline of climate science, software engineering principles are put into practice by a few teams of professional programmers at nationally funded laboratories such as the Goddard Institute, the Geophysical Fluid Dynamics Laboratory, and the National Center for Atmospheric Research. These models are used internally at these centers, and are also released to the scientific community at large.

Researchers, untrained in software engineering principles and unfamiliar with the code structure, typically make ad hoc modifications to these models, sometimes with considerable difficulty and in ways that are difficult to verify and validate. Individual investigators are required to develop expertise in their domain of interest and in adjacent sciences. In the case of environmental sciences, this can be a very broad terrain indeed. The demands of this work are significant and consuming, and as with any active science, ever-increasing. Expecting the scientific investigator to additionally have ever-increasing expertise in software engineering practice is unreasonable.

The result is that business as usual as practiced by the university scientist or student who makes heavy use of community models becomes increasingly difficult as the models themselves become more complex and inclusive. Already a great deal of effort has to be devoted to tasks related to system administration and to keeping up with the various parallel extensions to the standard programming languages provided by various vendors and targeting various architectures.

The software engineering methodology, which treats the scientist as an end user, does not capture the actual usage pattern of the scientific community. Many application scientists require the ability to modify the model being simulated, and yet do not want to learn sophisticated software abstraction techniques. Scientific modeling environments that support rather than increasingly thwart this type of user are needed.

Scientists with demanding computations consequently spend a great deal of time and effort on problems on which they are inexpert. This is an inappropriate division of labor. PyNSol refactors many problems of this sort so that these concerns are separated from the study of the physical system being modeled. It is the productivity of the scientist-as-programmer (rather than the productivity of the software engineering teams) that the PyNSol project addresses.

Some scientists limit themselves to problems of a size that is accessible using an interactive interpreted platform such as Matlab or Python. The resulting productivity gains can outweigh the sometimes severe performance limitations of the platform. We draw upon this observation and require a maximization of the interactivity of the development environment.

In typical climate simulations, spatial discretizations used are of low order, both because the uncertainty in forcings and boundary conditions does not warrant high accuracy and because the computational demands of the very long integrations of interest argue against it. On the other hand, the variety of important processes requires complex time differencing strategies. Traditional models of climate leave the control layer cognitively inaccessible because performance-efficient compiler codes cannot efficiently represent the separation of concerns between time and space differencing in critical loops. This is a persistent source of complexity that could be avoided by a refactoring of the problem.

These sorts of considerations lead to the following requirements for an interactive environment and domain specific language to support planetary scientists:

- Promote end-user productivity with appropriate domain specific language constructs
- Factor out and systematize an object model for extending the framework
- Support an exploratory programming style with an interactive environment
- Inherit pre- and post-processing tools from the Python world
- Maximize exposure of the control level to the scientific programmer
- Minimize end user attention to spatial discretizations
- Avoid end user attention to parallelization on simple grids

PyNSol approaches these requirements subject to the constraint that there is no significant impact on the performance of the resulting executable when a simulation is run in a production mode on a distributed memory cluster.

Intuitively it may appear that such a constraint is in contention with the objectives. In fact, there is no fundamental reason why this must be so; economy of expression is not necessarily traded off for performance. PyNSol instead proceeds by trading off generality. It is possible to achieve economy of expression and high performance with domain specific languages. PyNSol additionally seeks to provide a convenient interactive development environment.

3. ARCHITECTURE

The application scientist, upon hearing of the modularity of the object-oriented approach, and the software engineer, on first hearing of the scientist's problems, tend to choose as the objects of the system those components that would likely be swapped out in an inter-model comparison. The atmospheric scientist, for example, tries to design a class for radiation objects, another for convection, another for boundary layers, and so on. Unfortunately, in fact these phenomena are not loosely coupled at all, so that they are not ideal candidates for an object-oriented decomposition.

The ideal object takes responsibility for its own data, is passed a very brief message, modifies its data according to that message, and then returns brief messages to other objects. On the other hand, physical phenomena in complex environmental systems are very tightly coupled and unavoidably share a vast amount of state. The object abstraction doesn't fit very well.

Consider, instead, how the fluid dynamical cores of computational models are constructed. Construction of such models essentially begins from a terse mathematical definition of the system. This description passes through a set of intermediate descriptions. It is first elaborated by a spatial discretization and boundary conditions, next by time differencing, then by the bookkeeping of input and output operations, and finally by applying a parallelization.

Each of these steps is in principle formalizable. As such, each step is representable itself as the main method of an object whose responsibility is the transformation of one textual object to another. Each of these transformations is essentially independent of the other transformations. It is therefore possible to imagine a set of abstract classes that perform each of the transformations, and then to instantiate concrete objects fulfilling the responsibilities. These transformations are indeed loosely coupled.

PyNSol addresses a large and useful set of models where the decomposition described above is straightforward. Formal intermediate methods of model transformation from continuous to discretized representations are specified. Abstract classes are implemented to perform the transformations, and the resulting representation is then passed to a factory object, which creates a Fortran 77 representation and Python bindings to it. The framework is extended by providing classes that inherit from the abstract classes and provide implementations of the required methods.

The UML (Unified Modeling Language [3]) diagram in Figure 1 shows the salient features of PyNSol's class hierarchy. From the end user's point of view, using PyNSol consists of instantiating an instance of a Model class. That model's constructor takes a very high-level textual description of the model and maintains it as an attribute. The user is presented with two ways of using the Model instance, as a Python object that can perform simulations directly and as a code generator that emits code in a foreign language (initially, Fortran 77).

It is the responsibility of PyNSol extensions to ensure that the results of either operation will be mathematically identical, neglecting roundoff error due to different evaluation order. PyNSol's framework extension points are concrete classes inheriting from Scheme and Grid metaclasses, respectively responsible for time differencing and space differencing numerics.

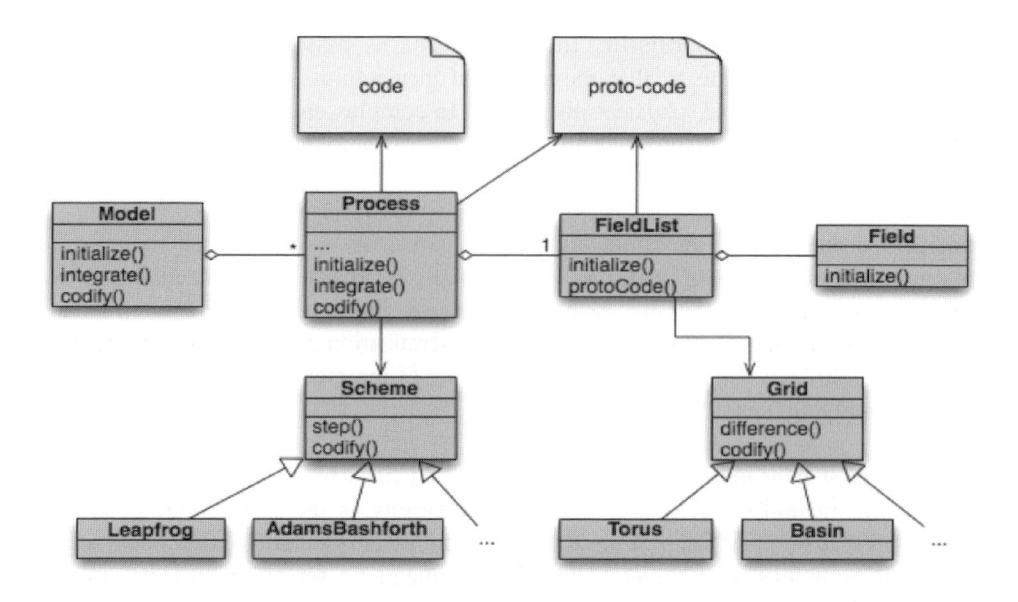

Figure 1: UML representation of the most important PyNSol classes

In instantiating a model, the typical end user selects a Scheme class and a Grid class from menus of existing Schemes and Grids, i.e., of implemented temporal and spatial discretization strategies. Of these, the larger responsibility falls to the Scheme class. The Grid class, when its codify method is invoked, transforms the model description, which is in a continuum form, into a spatially discrete representation in an intermediate representational language ("proto-code") which represents spatial differencing directly. This proto-code representation is further transformed by the selected Scheme class. The Scheme class must provide two separate ways of running the code. There is on-the-fly interpretation within Python simulation code (for experimentation and debugging) and code generation into text files that are suitable for compilation into compiled executables.

The choice of Fortran 77 as the language of execution has both intrinsic and opportunistic motivations. Intrinsically, the simplicity of the language provides a simple and straightforward target for the code generation step. Opportunistically, for sufficiently regular grids, an external transformation tool, FLIC already exists to automatically transform Fortran 77 codes into distributed memory multiprocessor codes calling using the MPI standard [12]. FLIC is in regular use in the WRF (Weather Research and Forecasting) model of the National Center for Atmospheric Research. The existence of FLIC relieves early instantiations of PyNSol from the obligation to instantiate a parallelization metaclass.

4. INNOVATIONS

PyNSol is not unique in its objectives. Indeed, there is considerable effort toward high productivity scientific programming environments currently underway, as documented recently by Zima et al. [15]. However, it combines several strategies in novel ways, providing newly convenient approaches to continuum modeling.

PyNSol gains several advantages from extending the existing Python scripting language rather than by developing a new language from a blank slate. This strategy not only avoids a great deal of rather subtle and complex effort in developing an entire language; it also allows immediate support of an interactive development style with an easy-to-use graphics environment.

Another innovation is the framework architecture underlying PyNSol. The intention of this approach is to allow extensibility of the domain specific tools. While much focus on the advantage of frameworks is directed at the value of being able to mix and match components into a full application, an equally important advantage of the framework is that it provides a specification for adding new components.

Using Fortran 77 as the generated code layer facilitates the use of a range of source analysis and modification tools, which are more problematic to implement in richer and more complex languages. In addition to automatic parallelization with FLIC and easily automated Python bindings, the availability of ADIFOR [2] and similar mathematically based source transformations allows a PyNSol model to obtain further benefit from Fortran 77's simplicity and its accessibility to automatic code transformation tools.

PyNSol has no ambitions to be an interoperability framework. It can, indeed, be designed to facilitate the use of such frameworks, but it is no competitor to them. Indeed, by interposing PyNSol as an interoperability layer with emerging component architectures such as the Earth System Modeling Framework (ESMF) [8] and the Common Component Architecture (CCA) [1] and delegating interoperability framework compliance to the code generation layer of PyNSol, we can relieve the application scientist of the burdens of participating in these structures while retaining the advantages.

Finally, there is a feature common to continuum modeling domain specific languages that we eschew. PyNSol refrains from hiding the timing loops from view in the model description language. Scientists who study environmental continuum problems often work with multiple phenomena that operate on multiple time scales. Typical representations of such codes are difficult to produce via conventional methods and often leave the simulation's underlying logical structure opaque and difficult to work with. By exposing these control structures in the domain specific language, while hiding discretization, PyNSol provides control with transparency.

PyNSol draws upon the flexibility of a modern scripting language and upon the application of a non-obvious object decomposition of the fluid dynamics modeling problem. As a consequence it promises to considerably enhance the productivity of professional scientists who may be competent but not expert programmers and who make extensive use of computations within their research.

5. ACKNOWLEDGEMENTS

The author is grateful for interesting conversations on this matter with Ian Bicking, Matt Knepley, Jay Larson, Ray Pierrehumbert, Mike Steder and George Thiruvathukal. Assistance from John Hunter, Konstantin Laufer and John Michalakes is also much appreciated. This work is supported under NSF grant ATM-0121028.

REFERENCES

1. Armstrong, T., et al., *Toward a Common Component Architecture for High Performance Scientific Computing.* Proceedings of Conference on High Performance Distributed Computing, 1999.
2. Bischof, C., et al. *ADIFOR – Generating Derivative Codes from FORTRAN Programs.* Scientific Computing vol 1 no 1 pp 11-29, 1992.
3. Booch, G., Rumbaugh, J. and Jacobson, I. *The Unified Modeling Language User Guide.* Addison-Wesley, Boston MA, 1999.
4. Bourque, P., and DuPuis, R., eds., *Guide to the Software Engineering Body of Knowledge.* IEEE Computer Society, Los Alamitos CA 2004.
5. Carey, J. and Carlson, B. *Framework Process Patterns.* Addison-Wesley, Boston MA, 2002.
6. Gamma, E., Helm, R., Johnson, R. and Vlissides, J. *Design Patterns: Elements of Reusable Object-Oriented Software.* Addison-Wesley, Reading MA, 1994.
7. Gibbs, W., *Trends in Computing: Software's Chronic Crisis.* Scientific American pp 68 ff, September 1994.
8. Hill, C., DeLuca, C., Balaji, Suarez, M., and da Silva, A. *The Architecture of the Earth System Modeling Framework.* Computing in Science and Engineering vol 6 no 1, pp 18 – 29, 2004.
9. Hunter, J. The Matplotlib User's Guide. http://matplotlib.sourceforge.net/
10. Kepner, J. and Ahalt, S. *MatlabMPI.* Journal of Parallel and Distributed Computing, vol 64 no 8, pp 997-1005, 2004.
11. Kiczales, G., Des Riveres, J., and Bobrow, D. *The Art of the Meta-Object Protocol.* MIT Press, Cambridge MA 1992.
12. Michalakes, J. *FLIC, A Translator for Same-Source Parallel Implementation of Regular Grid Applications.* Mathematics and Computer Science Division of Argonne National Laboratory Tech Report A-297. http://www-fp.mcs.anl.gov/division/tech-memos.htm
13. Perez, F. *Interactive Work in Python: IPython's Present and Future.* SciPy Meeting, Pasadena CA 2004. http://www.scipy.org/wikis/scipy04/ConferenceSchedule
14. Tobis, M., *PyNSol: Objects as Scaffolding.* Computing in Science and Engineering, Vol 7 No 4, pp 84-91, 2005.
15. Zima, H., ed., *Workshop on High Productivity Languages and Models,* Santa Monica CA, 2004. http://www.nitrd.gov/subcommittee/hec/hecrtfoutreach/bibliography

The Exchange Grid: a mechanism for data exchange between Earth System components on independent grids

V. Balaji[a*], Jeff Anderson[b], Isaac Held[c], Michael Winton[c], Jeff Durachta[c], Sergey Malyshev[a], Ronald J. Stouffer[c]

[a]Princeton University
PO Box 308, GFDL
Princeton NJ 08542 USA

[b]NCAR Institute for Math Applied to Geophysics
PO Box 3000
Boulder CO 80307 USA

[c]NOAA Geophysical Fluid Dynamics Laboratory
PO Box 308, GFDL
Princeton NJ 08542 USA

We present a mechanism for exchange of quantities between components of a coupled Earth system model, where each component is independently discretized. The exchange grid is formed by overlaying two grids, such that each exchange grid cell has a unique parent cell on each of its antecedent grids. In Earth System models in particular, processes occurring near component surfaces require special surface boundary layer physical processes to be represented on the exchange grid. The exchange grid is thus more than just a stage in a sequence of regridding between component grids.

We present the design and use of a 2-dimensional exchange grid on a horizontal planetary surface in the GFDL Flexible Modeling System (FMS), highlighting issues of parallelism and performance.

1. Introduction

In climate research, with the increased emphasis on detailed representation of individual physical processes governing the climate, the construction of a model has come to require large teams working in concert, with individual sub-groups each specializing in a different component of the climate system, such as the ocean circulation, the biosphere, land hydrology, radiative transfer and chemistry, and so on. The development of model code now requires teams to be able to contribute components to an overall coupled system, with no single kernel of researchers mastering the whole. This may be called the *distributed development process*, in contrast with the monolithic small-team process of earlier decades.

---

*Corresponding author: balaji@princeton.edu

These developments entail a change in the programming framework used in the construction of complex Earth system models. The approach is to build code out of independent modular components, which can be assembled by either choosing a configuration of components suitable to the scientific task at hand, or else easily extended to such a configuration.

A specific issue that arises is how different components of the Earth system, say atmosphere and ocean, are discretized. Earlier generations of climate models used the *same* discretization, or simple integer refinement, for all components: thus, data exchange between components was a relatively simple point-to-point exchange. But any limitations on resolution of one component necessarily imposed itself on the other as well. Now it is increasingly common for each model component to make independent discretization choices appropriate to the particular physical component being modeled. In this case, how is, say a sea surface temperature from an ocean model, made available to an atmosphere model that will use it as a boundary condition? This is the *regridding problem*[6], subject to the following constraints when specialized to Earth system models:

- Quantities must capable of being *globally conserved*: if there is a flux of a quantity across an interface, it must be passed conservatively from one component to the other. This consideration is less stringent when modeling weather or short-term (intraseasonal to interannual) climate variability, but very important in models of secular climate change, where integration times can be $\mathcal{O}(10^6) - \mathcal{O}(10^8)$ timesteps.

- The numerics of the flux exchange must be stable, so that no limitation on the individual component timestep is imposed by the boundary flux computation itself.

- There must be no restrictions on the discretization of a component model. In particular, resolution or alignment of coordinate lines cannot be externally imposed. This also implies a requirement for *higher-order interpolation* schemes, as low-order schemes work poorly between grids with a highly skewed resolution ratio. Higher-order schemes may require that not only fluxes, but their higher-order spatial derivatives as well, be made available to regridding algorithms.

  The independent discretization requirement extends to the time axis: component models may have independent timesteps. (We do have a current restriction that a coupling timestep be an integral multiple of any individual model timestep, and thus, timesteps of exchanging components may not be co-prime).

- The exchange must take place in a manner consistent with all physical processes occurring near the component surface. This requirement is highlighted because of the unique physical processes invoked near the planetary surface: in the atmospheric and oceanic boundary layers, as well as in sea ice and the land surface, both biosphere and hydrology.

- Finally, we require computational efficiency on parallel hardware: a solution that is not rate-limiting at the scalability limits of individual model components. Components may be scheduled serially or concurrently between coupling events.

The GFDL Flexible Modeling System (FMS)[2] [1] deploys software known as an *exchange grid* to solve this problem. We define an exchange grid in Section 2. We introduce the no-

---
[2]http://www.gfdl.noaa.gov/~fms

tion of implicit coupling in Section 3, and demonstrate the coupling sequence used in FMS. In Section 4 we describe the parallelization of the exchange grid. Finally, in Section 5, we show basic performance characteristics of the FMS exchange grid at typical climate model resolutions. FMS and its exchange grid have been used in production for the models run at GFDL for the 2007 round of IPCC simulations (e.g [3]).

2. Definition of an exchange grid

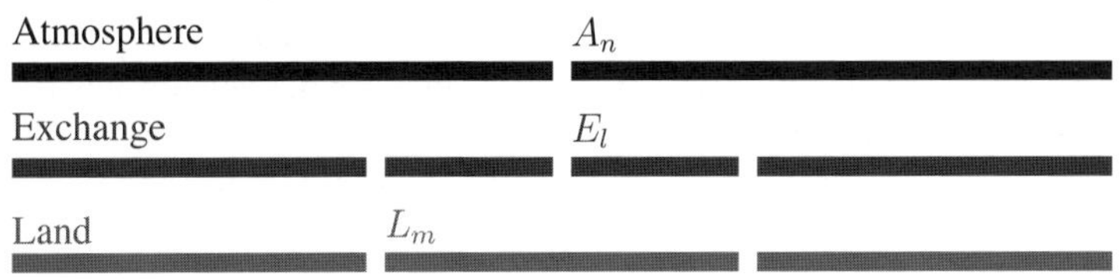

Figure 1. One-dimensional exchange grid.

A *grid* is defined as a set of *cells* created by *edges* joining pairs of *vertices* defined in a discretization. Given two grids, an *exchange grid* is the set of cells defined by the union of all the vertices of the two parent grids. This is illustrated in Figure 1 in 1D, with two parent grids ("atmosphere" and "land"). (Figure 2 shows an example of a 2D exchange grid, most often used in practice). As seen here. each exchange grid cell can be uniquely associated with exactly one cell on each parent grid, and *fractional areas* with respect to the parent grid cells. Quantities being transferred from one parent grid to the other are first interpolated onto the exchange grid using one set of fractional areas; and then averaged onto the receiving grid using the other set of fractional areas. If a particular moment of the exchanged quantity is required to be conserved, consistent moment-conserving interpolation and averaging functions of the fractional area may be employed. This may require not only the cell-average of the quantity (zeroth-order moment) but also higher-order moments to be transferred across the exchange grid.

Given N cells of one parent grid, and M cells of the other, the exchange grid is, in the limiting case in which every cell on one grid overlaps with every cell on the other, a matrix of size $N \times M$. In practice, however, very few cells overlap, and the exchange grid matrix is extremely sparse. In code, we typically treat the exchange grid cell array as a compact 1D array (thus shown in Figure 1 as E_l rather than E_{nm}) with indices pointing back to the parent cells. Table 1 shows the characteristics of exchange grids at typical climate model resolutions. The first is the current GFDL model CM2 [3], and the second for a projected next-generation model still under development. As seen here, the exchange grids are extremely sparse.

The computation of the exchange grid itself could be time consuming, for parent grids on completely non-conformant curvilinear coordinates. In practice, this issue is often sidestepped by precomputing and storing the exchange grid. The issue must be revisited if either of the parent grids is adaptive.

The FMS implementation of exchange grids restricts itself to 2-dimensional grids on the

Atmosphere	Ocean	Xgrid	Density	Scalability
144×90	360×200	79644	8.5×10^{-5}	0.29
288×180	1080×840	895390	1.9×10^{-5}	0.56

Table 1
Exchange grid sizes for typical climate model grids. The first column shows the horizontal discretization of an atmospheric model at "typical" climate resolutions of 2° and 1° respectively. The "ocean" column shows the same for an ocean model, at 1° and $\frac{1}{3}$°. The "Xgrid" column shows the number of points in the computed exchange grid, and the density relates that to the theoretical maximum number of exchange grid cells. The "scalability" column shows the load imbalance of the exchange grid relative to the overall model when it inherits its parallel decomposition from one of the parent grids.

planetary surface. However, there is nothing in the exchange grid concept that prevents its use in exchanges between grids varying in 3, or even 4 (including time) dimensions.

2.1. Masks

A complication arises when one of the surfaces is partitioned into *complementary components*: in Earth system models, a typical example is that of an ocean and land surface that together tile the area under the atmosphere. Conservative exchange between *three* components may then be required: crucial quantities like CO_2 have reservoirs in all three media, with the *total* carbon inventory being conserved.

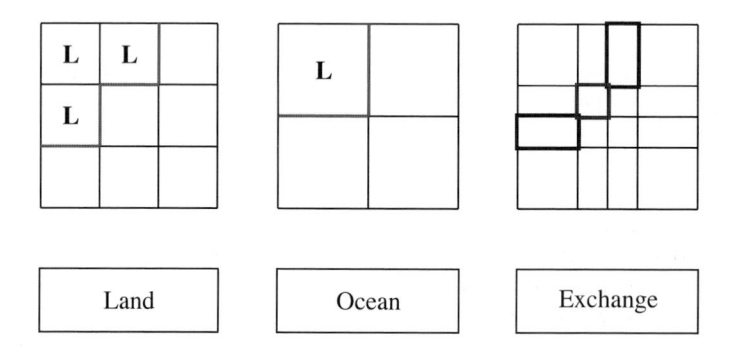

Figure 2. The mask problem. The land and atmosphere share the grid on the left, and their discretization of the land-sea mask is different from the ocean model, in the middle. The exchange grid, right, is where these may be reconciled: the red "orphan" cell is assigned (arbitrarily) to the land, and the land cell areas "clipped" to remove the doubly-owned blue cells.

Figure 2 shows such an instance, with an atmosphere-land grid and an ocean grid of different resolution. The green line in the first two frames shows the *land-sea mask* as discretized on the two grids, with the cells marked **L** belonging to the land. Due to the differing resolution, certain exchange grid cells have ambiguous status: the two blue cells are claimed by both land and ocean, while the orphan red cell is claimed by neither.

This implies that the mask defining the boundary between complementary grids can only be accurately defined on the exchange grid: only there can it be guaranteed that the cell areas exactly tile the global domain. Cells of ambiguous status are resolved here, by adopting some ownership convention. For example, in the FMS exchange grid, we generally modify the land model as needed: the land grid cells are quite independent of each other and amenable to such transformations. We add cells to the land grid until there are no orphan "red" cells left on the exchange grid, then get rid of the "blue" cells by *clipping* the fractional areas on the land side.

2.2. Tile dynamics

A further complication arises when we consider tiles within parent grid cells. *Tiles* are a refinement within physical grid cells, where a quantity is partitioned among "bins" each owning a fraction of it. Tiles within a grid cell do not have independent *physical* locations, only their associated fraction. Examples include different vegetation types within a single land grid cell, which may have different temperature or moisture retention properties, or partitions of different ice thickness representing fractional ice coverage within a grid cell.

As the fractional quantity associated with each tile is different, we could associate an array dimensioned by tile with each exchange grid cell. The issue is that while the tile dimension may be large, the number of actual tiles with non-zero fraction is generally small. For instance, vegetation models often count 20 or more vegetation types (see e.g [2]); yet a single grid cell generally contains no more than two or three. It would therefore be inefficient to assign a tile dimension to the exchange grid: instead we model each tile as an independent cell and use the compaction as described above to eliminate tiles of zero fraction.

This implies that the compaction used to collapse the exchange grid sparse matrix is dynamic. We generally update the exchange grid every time a coupling event occurs. Compaction costs are negligible.

3. Implicit coupling

Fluxes at the surface often need to be treated using an implicit timestep. Vertical diffusion in an atmospheric model is generally treated implicitly, and stability is enhanced by computing the flux at the surface implicitly along with the diffusive fluxes in the interior. Simultaneously we must allow for the possibility that the surface layers in the land or sea ice have vanishingly small heat capacity. This feature is key in the design of the FMS coupler. Consider simple vertical diffusion of temperature in a coupled atmosphere-land system:

$$\frac{\partial T}{\partial t} = -K\frac{\partial^2 T}{\partial z^2} \qquad (1)$$

$$\Rightarrow \quad \frac{T_k^{n+1} - T_k^n}{\Delta t} = -K\frac{T_{k+1}^{n+1} + T_{k-1}^{n+1} - 2T_k^{n+1}}{\Delta z^2} \qquad (2)$$

$$\Rightarrow \quad \mathbf{A}\mathbf{T}^{n+1} = \mathbf{T}^n \qquad (3)$$

This is a tridiagonal matrix inversion which can be solved relatively efficiently using an up-down sweep, as shown in Figure 3. The problem is that some of the layers are the atmosphere and others are in the land. Moreover, if the components are on independent grids, the key flux computation at the surface, to which the whole calculation is exquisitely sensitive, is a physical process (e.g [7]) that must be modeled on the finest possible grid without averaging. Thus, the exchange grid, on which this computation is performed, emerges as an independent model component for modeling the surface boundary layer.

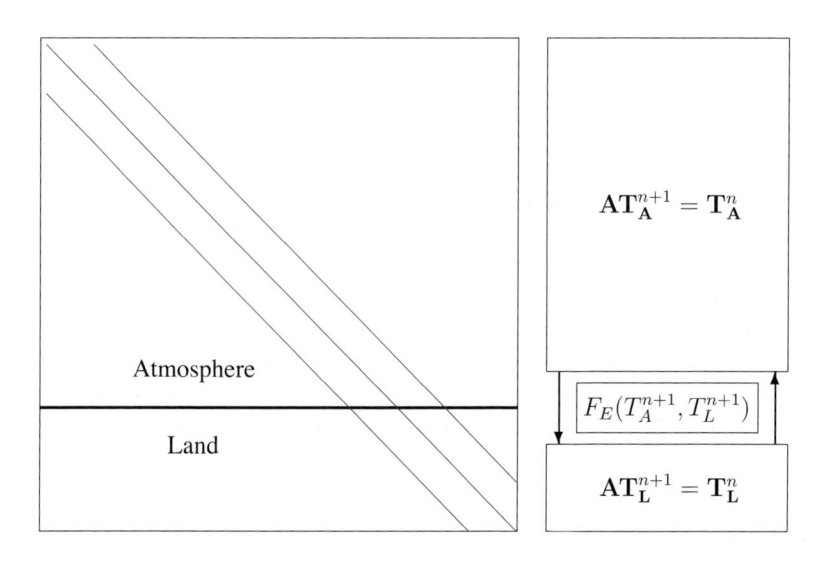

Figure 3. Tridiagonal inversion across multiple components and an exchange grid. The atmospheric and land temperatures T_A and T_L are part of an implicit diffusion equation, coupled by the implicit surface flux on the exchange grid, $F_E(T_A^{n+1}, T_L^{n+1})$.

The general procedure for solving vertical diffusion is thus split into separate up and down steps. Vertically diffused quantities are partially solved in the atmosphere (known in FMS as the "`atmosphere_down`" step) and then handed off to the exchange grid, where fluxes are computed. The land or ocean surface models recover the values from the exchange grid and continue the diffusion calculation and return values to the exchange grid. The computation is then completed in the up-sweep of the atmosphere. Note that though we are computing vertical diffusion, some spurious horizontal mixing can occur as the result of regridding. More details are available in [4].

4. Parallelization

Now we consider a further refinement, that of parallelization. In general, not only are the parent grids physically independent, they are also parallelized independently. Thus, for any exchange grid cell E_{nm}, the parent cells A_n and L_m (see Figure 1) may be on different processors. The question arises, to which processor do we assign E_{nm}? The choices are,

1. to inherit the parallel decomposition from one of the parent grids (thereby eliminating communication for one of the data exchanges); or

2. to assign an independent decomposition to the exchange grid, which may provide better load balance.

In the FMS exchange grid design, we have chosen to inherit the decomposition from one side. Performance data (not shown here) indicate that the additional communication and synchronization costs entailed by choosing (2) are quite substantial, and of the same order as the computational cost as the flux computations on the exchange grid itself. Should the computational cost of the exchange grow appreciably, we may revisit this issue.

In choosing (1) we have also chosen to inherit the exchange grid parallelism from the side that uses dynamic tiling (generally, land and ocean surfaces; the atmosphere uses static tiling). The calls to the exchange grid are written to be entirely local, and all the communication is internal and on the static side. Note that when communication is on the side extracting data from the exchange grid onto the parent, we have a choice of performing on-processor sums *before* communication, which reduces data transfer volume. This results in loss of bit-reproducibility when changing processor counts, because of changes in summation order. We provide an option to send all data prior to summation as an option, when bitwise-exact results are needed (e.g for regression testing). This option adds about 10% to the cost at typical climate model resolutions.

5. Conclusions

We have described the design of a parallel exchange grid, used for conservative exchange of quantities between components of a coupled climate model, with sensitive flux computations in an intervening surface boundary layer. The physically most meaningful design is to perform these computations on the finest grid defined on the basis of the two parent grids. The exchange grid thus emerges as an independent model component, not just an intermediary in a regridding computation.

The exchange grid is designed not as a sparse matrix between all possible pairs of parent grid cells, but as a compact array pointing back to the parents. When parent grids have dynamic tiles, each tile appears as an independent cell on the exchange grid, rather than an extra tile dimension.

The grid may be precomputed and stored. If such grids are to be shared across a wide range of users, a standard grid specification must be accepted across the community. Efforts to develop a standard grid specification are underway. Masks defining the boundary between complementary grids are also best defined on an exchange grid.

Parallelization of the exchange grid is done on the basis of inheriting the decomposition from one of the parents, thus eliminating communication on one side of the exchange. The alternative, that of computing an independent decomposition for the exchange grid, was seen to be introducing much code complexity with the dubious promise of reward in the form of better load balance.

Current coupled climate computations in FMS [3] are carried out with a wide disparity in resolution ($2° \times 2.5°$ in the atmosphere; $1° \times \frac{1}{3}°$ in the ocean near the equator). The cost of the exchange grid component, including the flux computation, is about 12% at high scalability (180 processors). Much of the cost can be attributed to load imbalance: of a total of ~ 80000

exchange grid cells, ~3000 are on one processor in an extreme instance: thus limiting the scalability of the exchange grid relative to the overall model to about 0.29, as shown in Table 1.

The Flexible Modeling System website[3] contains the public-domain software for FMS, including its exchange grid. The site also contains links to documentation and papers on software features and design. The FMS exchange grid is a design prototype for the Earth System Modeling Framework [5], an emerging community standard for the construction of coupled Earth system models.

Acknowledgments

V. Balaji is funded by the Cooperative Institute for Climate Science (CICS) under award number NA17RJ2612 from the National Oceanic and Atmospheric Administration, U.S. Department of Commerce. The statements, findings, conclusions, and recommendations are those of the author and do not necessarily reflect the views of the National Oceanic and Atmospheric Administration or the Department of Commerce.

REFERENCES

1. V. Balaji. Parallel numerical kernels for climate models. In ECMWF, editor, *ECMWF Teracomputing Workshop*, pages 184–200. World Scientific Press, 2001.
2. Wolfgang Cramer, Alberte Bondeau, F. Ian Woodward, I. Colin Prentice, Richard A. Betts, Victor Brovkin, Peter M. Cox, Veronica Fisher, Jonathan A. Foley, Andrew D. Friend, Chris Kucharik, Mark R. Lomas, Navin Ramankutty, Stephen Sitch, Benjamin Smith, Andrew White, and Christine Young-Molling. Global response of terrestrial ecosystem structure and function to CO2 and climate change: results from six dynamic global vegetation models. *Global Change Biology*, 7(4):357–373, April 2001.
3. T.L Delworth, A.J. Broccoli, A. Rosati, R.J. Stouffer, V. Balaji, J.T. Beesley, W.F. Cooke, K.W. Dixon, J. Dunne, K.A. Dunne, J.W. Durachta, K.L. Findell, P. Ginoux, A. Gnanadesikan, C.T. Gordon, S.M. Griffies, R. Gudgel, M.J. Harrison, I.M. Held, R.S. Hemler, L.W. Horowitz, S.A. Klein, T.R. Knutson, P.J. Kushner, A.R. Langenhorst, H.-C. Lee, S.-J. Lin, J. Lu, S.L. Malyshev, P.C. D. Milly, V. Ramaswamy, J. Russell, M.D. Schwarzkopf, E. Shevliakova, J.J. Sirutis, M.J. Spelman, W.F. Stern, M. Winton, A.T. Wittenberg, B. Wyman, F. Zeng, and R. Zhang. GFDL's CM2 Global Coupled Climate Models. Part 1: Formulation and Simulation Characteristics. *Journal of Climate*, accepted for publication, 2005.
4. Isaac Held. Surface fluxes, implicit vertical diffusion, and the exchange grid. http://www.gfdl.noaa.gov/fms/pdf/surf.pdf, NOAA/GFDL, 1998. http://www.gfdl.noaa.gov/fms/pdf/surf.pdf.
5. Chris Hill, Cecelia DeLuca, V. Balaji, Max Suarez, Arlindo da Silva, and the ESMF Joint Specification Team. The Architecture of the Earth System Modeling Framework. *Computing in Science and Engineering*, 6(1):1–6, January/February 2004.
6. P. Knupp and S. Steinberg. *Fundamentals of Grid Generation*. CRC Press, 1993.
7. A.S Monin and A.M Obukhov. Basic laws of turbulent mixing in the ground layer of the atmosphere. *Tr. Geofiz. Inst. Akad. Nauk. SSSR*, 151:163–187, 1954.

---

[3]http://www.gfdl.noaa.gov/~fms

A Generic Coupler for Earth System Models

Shujia Zhou,[a] Joseph Spahr[b]

[a]*Northrop Grumman Corporation, 4801 Stonecroft Blvd., Chantilly, VA 20151, USA*

[b]*Dept. of Atmospheric and Oceanic Sciences, UCLA, USA*

Keywords: Coupler; Earth System Modeling Framework; distributed data broker;

1. Introduction

One of the distinct features of Earth system models (e.g., atmosphere, ocean, land, sea-ice) is that the models need to exchange data, typically along the model interface and on different grids, to obtain an accurate prediction. In addition, these Earth system models typically run on parallel computers, at least in the production mode. Therefore, various parallel software tools have been developed to support coupling operations such as the UCLA Distributed Data Broker (DDB) [1], the NASA Goddard Earth Modeling System (GEMS) [2], the GFDL Flexible Model System (FMS) [3], and the NCAR Flux Coupler [4]. Although those tools have been used in production, they are typically used only within their own organizations. It still remains challenging to couple models across organizations.

The Earth System Modeling Framework (ESMF) is funded by NASA's Earth-Sun System Technology Office/Computational Technologies Project to facilitate coupling Earth system models across organizations [5, 6]. So far, several cross-organization, sequential model couplings have been successfully enabled by ESMF [7]. That is partly because ESMF defines a way of coupling models based on the consensus of the Earth system model community and provides utilities such as parallel regridding. However, a coupler has to be custom-built for each coupled system, which hinders the usage of ESMF.

In addition, we find that there are a lot of similarities among the couplers for ESMF-enabled coupled Earth system models. In particular, the couplers for the combinations of the UCLA atmosphere model [8] with the LANL POP ocean model [9] and the MITgcm ocean model [10] are the same, except for the name and the number of exchanged variables. Couplers for the NASA-NCAR fvCAM atmosphere-NCEP SSI analysis [11, 12] and the GFDL FMS B-grid atmosphere-MITgcm ocean combinations are nearly identical as well [3, 10].

In this paper we will discuss the generic coupling features of Earth system models and introduce a data registration service to enable a generic coupler for the cases where only regridding is involved.

2. Model Coupling issues

There are four major issues associated with the coupling of Earth system models:

1. The function (subroutine) names in model components are different.
2. The names of import variables of a "consumer" component are different from the names of export variables of a "producer" component.
3. The number of import variables of a "consumer" component is different from the number of export variables of a "producer" component.
4. Maintaining a consistent inter-component list of the data elements to be transferred between components.

To cope with the first issue, ESMF standardizes three basic operations in Earth system models with ESMF_GridCompInitialize(), ESMF_GridCompRun(), and ESMF_GridCompFinalize(). Each interface also has a stage option. In addition, ESMF provides a "producer" component with a function registration service, ESMF_GridCompSetEntryPoint(), to link its subroutines to those three standard interfaces, and provides the "consumer" component with an accessing service, ESMF_GridCompSetServices(), to use the functionality of the "producer" component through these three standard interfaces.

However, to resolve the second, third, and fourth issues, a data registration service has to be developed to match the variables of the "consumer" component with the variables of the "producer" component. This is a "consumer-producer" relationship, which has been coped with by the Common Component Architecture (CCA) in a generic way [13, 14, 15]. CCA uses "Use-Port" and "Provide-Port" to establish a generic interface between a consumer component and a producer component and hide the implementation details. CCA's approach will be used to guide the development of the generic part of our generic coupler.

With the data registration, the fourth issue can be resolved by building the required ESMF data structures: ESMF_State for the inter-component data exchange and ESMF_Field for regridding. The consistency of a user-defined variable name is enforced in the process of registration to coupling (regridding).

3. A generic coupler

As we discussed previously, a generic coupler for Earth system models needs to have a general and flexible data registration. In the following, we review portion of the data registration of the UCLA Distributed Data Broker (DDB).

The DDB is a software tool for coupling multiple, possibly heterogeneous, parallel models. One of the outstanding characteristics of the DDB is its data registration ability, at initialization time, to reconcile all requests ("consume") and offers ("produce") for data exchange between components.

There are two phases of registration. Each producer model component sends the "registration broker" an "offer" message, and each consumer model component sends a "request" message. (Of course, any process may simultaneously produce some quantities and consume others in a concurrent-run mode.) Once the "registration broker" has received all offers and requests, it starts the second phase in which it creates the lists of the matched offer-request pairs, forwards them to relevant producer model components, and notifies each consumer model component that each request will be satisfied by a certain number of data tiles.

Since our generic coupler is aimed at coupling models from different groups or organizations, a standard or well-known name convention should be used to facilitate matching names of exchanged variables from different models. The NetCDF Climate and Forecast Metadata Convention, the so-called CF Convention, is the most popular in the Earth system model community [16]. Thus, we will use it for illustration of our generic coupler. Of course, other name conventions can be used in our generic coupler. One implementation is to read the name convention in through a file IO.

3.1. Approach

Our generic coupler provides two services: 1) data registration and 2) is data transfer and regridding. The data registration service will be derived from the DDB since its consumer-producer registration scheme is flexible and dynamic (real-time), and the service of data transfer and regridding will be based on ESMF.

Our generic coupler is designed to enable coupling of ESMF-compatible model components in an easy and general way. (An ESMF-compatible model has three standard interfaces, ESMF_GridCompInitialize(), ESMF_GridCompRun(), and ESMF_GridCompFinalize() and has a list of import and export variables for coupling with other models. But an ESMF-compliant model needs to use ESMF import/export states to exchange data in addition to having three standard interfaces.) The easy-to-use feature is implemented through relieving a user from manually building ESMF_Field and ESMF_State and coding the coupler. A user still deals with his/her familiar data type, namely Fortran arrays. Moreover, the general (abstraction) feature is implemented by adopting a well-known or standard name convention such as CF convention. The usage of a standard data variable name is similar to the usage of three standard interfaces mentioned above for functions.

3.2. Data registration service

Essentially each model component provides two lists of variable names: one for import and another for export. In each list, the model component needs to provide its user-defined name along with the corresponding CF name for import variables as well as export variables. In addition to the CF name, the pointer to the grid metadata (i.e., starting and ending value as well as interval) is also needed in the list. That grid metadata is needed for building ESMF_field and then performing regridding operation.

The data registration service of our generic coupler takes these two kinds of lists, maps the import variable name with the corresponding export name based on their corresponding CF name, and provides the matched name list to the relevant consumer and producer model components.

Specifically, the following steps are taken consecutively:

1. Provide the variable name of exchanged data. Each component notifies the generic coupler of its variable requests, offers to supply, and any other required information (e.g., component logical name). For a concurrent-run mode, it is efficient for the generic coupler to use one computer node, while for a serial-run mode the generic coupler uses all the nodes.
2. Match the name of an exchanged variable pair. The generic coupler (a) collects the messages of request and offer from all the components; (b) matches the request with the corresponding offer message; (c) builds the lists of "produce" and "consume" for each pair of the coupled components; (d) aggregates all the lists of "produce" and "consume" for each component. (Here we choose one import/export state for each component, that is, this import/export state is the aggregate of all the import/export variables from/to components. We believe that this aggregated approach will lead to a simple user interface); (e) distributes these lists and data structures to all the nodes in a serial-run mode and relevant components in a concurrent-run mode.
3. Process the grid metadata and prepare regridding. The generic coupler (a) queries the grid meta data of each variable to be exchanged; (b) creates ESMF fields, import, and export states; (c) creates the data structures to hold information needed for regridding (e.g., pointers to import/export state, consumer/producer gridded component, ESMF VM). In particular, an ESMF regridding handle is created.

3.3. Data transfer and regridding

There are a few steps to be performed by our generic coupler during the run time. The three major steps are as follows:

1. Select the appropriate list to perform the requested data exchange. Specifically, the generic coupler determines which direction the data is to flow and which component pair will be performing the data transfer.
2. Call the ESMF_StateReconcile() subroutine to synchronize the data among the coupled component in a concurrent-run mode.
3. Call a component of our generic coupler to perform the inter-component data exchange as well as the regridding operations. A component of our generic coupler is developed to enable the functionality of our generic coupler to fit into the regular ESMF running flow and shift the flow from a gridded component to the generic coupler.

3.4. Run sequence of a generic coupler within a user driver

For a coupled system, a user will not call any subroutines or utilities of an ESMF coupler directly while calling the subroutines or utilities of ESMF gridded components as usual. Instead, a simple set of API's of the generic coupler are called directly. Specifically,

1. In the phase of component registration (e.g., call ESMF SetService subroutines), no call is needed.
2. In the phase of initialization (e.g., call ESMF Initialize subroutines), call the data registration subroutines of the generic coupler. Inside that registration subroutines, the subroutine of ESMF setService for the coupler component of the generic coupler is called and the component data structure is set up (e.g., call ESMF_CplCompCreate()). At the end of the registration function, the subroutine of ESMF_CplCompIntialize() is called to complete the data registration and component initialization.
3. In the phase of Run (e.g., call ESMF Run subroutines), call the "doCouple" subroutines of the generic coupler to perform data transfer and regridding. Inside "doCouple," the subroutine of ESMF_CplCompRun() is called.
4. In the phase of Finalization (e.g., call ESMF Finalize subroutines), call the "finalizeCouple" subroutines of the generic coupler's to clean up. Inside "finalizeCouple," the subroutine of ESMF_CplCompFinalize() is called.

4. Discussion

So far, we have described the coupling issues and our solutions mostly for a sequential-run mode, although we also discussd the concurrent-run mode in some places. This is partly because ESMF has not completed its development for a concurrent-run mode. We believe that the data registration service of our generic coupler is applicable for the sequential - as well as the concurrent-run mode since it is derived from the DDB, which has been proven to be applicable for both modes. In the following, we clarify a few points for these two modes:

1. In the sequential-run mode, the generic coupler is operated in the combined set of computer nodes associated with all the model components. That is to ensure all the nodes are available to the generic coupler. This is the mode that the current ESMF is supporting.
2. In the concurrent-run mode, the generic coupler is operated in the joint set of computer nodes, that is, the combined nodes of coupling model component pairs. For example, the joint set of nodes is 18 for the case in which model A uses 10 nodes and model B uses 8. The point is that the generic coupler is only operated for the component pair, not necessarily involving all the other components. In this case, the generic coupler between model A and model B will use one node to collect the name list from components, match names, create a name-matched list, and distribute the list to those needed models, which is similar to the DDB. When a component receives such a list, it can perform data transfer and regridding after calling ESMF_StateReconcile() subroutine to synchronize the data for regridding.

A software utility layer between ESMF and model components, called "Geo-generic", is being developed [17]. Its goal is also to simplify the usage of ESMF. However, our generic coupler focuses on the generic part of coupling through a data registration service. In addition, our generic coupler is applicable to sequential as well as concurrent-run modes.

5. Summary

We find that it is possible to build a generic coupler to simplify the usage of ESMF in coupling Earth system models. We have described our approach and discussed the implementation details based on CCA, DDB, and ESMF.

6. Acknowledgement

This is project is supported by the NASA Earth-Sun System Technology Office Computational Technologies Project. We would like to thank C. Roberto Mechoso for helpful discussions and Tom Clune and Jarrett Cohen for reading the manuscript.

References

1. UCLA Distributed Data Broker, www.atmos.ucla.edu/~mechoso/esm/ddb_pp.html.
2. Goddard Earth Modeling System, unpublished
3. Flexible Modeling System, http://www.gfdl.noaa.gov/~fms
4. NCAR Flux Coupler, http://www.ccsm.ucar.edu/models/cpl/
5. ESMF, http://www.esmf.ucar.edu
6. C. Hill, C. DeLuca, V. Balaji, M. Suarez, A. da Silva, and the ESMF Joint Specification Team, "The Architecture of the Earth System Modeling Framework," Computing in Science and Engineering, Volume 6, Number 1, 2004.

7. S. Zhou et al., "Coupling Weather and Climate Models with the Earth System Modeling Framework," Concurrency Computation: Practice and Experience, to be submitted.
8. UCLA atmosphere model, http://www.atmos.ucla.edu/~mechoso/esm/
9. LANL POP ocean model, http://climate.lanl.gov/Models/POP/
10. MITgcm ocean model, http://mitgcm.org
11. fvCAM atmosphere model, http://www.ccsm.ucar.edu/models/atm-cam/
12. M. Kanamitsu, "Description of the NMC global data assimilation and forecast system," Wea. and Forecasting, Volume 4, 1989, pp. 335-342.
13. Common Component Architecture, http://www.cca-forum.org/
14. S. Zhou et al., "Prototyping of the ESMF Using DOE's CCA," NASA Earth Science Technology Conference 2003.
15. S. Zhou, "Coupling Climate Models with Earth System Modeling Framework and Common Component Architecture," Concurrency Computation: Practice and Experience, in press.
16. CF convections, http://www.cgd.ucar.edu/cms/eaton/cf-metadata/CF-20010808.html
17. Geo-generic, Max Suarez and Atanas Trayanov, private communication.

CFD Analyses on Cactus PSE

Kum Won Cho,[a] Soon-Heum Ko,[b] Young Gyun Kim,[c] Jeong-su Na,[a] Young Duk Song,[a] and Chongam Kim[b]

[a]*Dept. of Supercomputing Application Technology, KISTI Supercomputing Center, Dae-jeon 305-333, Korea*

[b]*School of Mechanical and Aerospace Eng., Seoul National University, Seoul 151-742, Korea*

[c]*School of Computer Eng., Kumoh National Institute of Technology, Gu-mi 730-701, Korea*

Keywords: Cactus ; PSE(Problem Solving Environment) ; The Grid

1. INTRODUCTION

Many researches have been conducted on developing user-friendly interfaces for application scientists to easily utilize advanced computing environment, i.e., the Grid[1]. A specialized computer system which offers computational conveniences required for a specific problem is called PSE(Problem Solving Envi-ronment) and, nowadays Nimrod/G[2], Triana[3], and Cactus[4] are actively applied. Of the PSEs, Cactus was firstly developed for collaboration among astrophysicists, but it has been expanded to be applicable to many scientific researches like CFD(Computational Fluid Dynamics) analysis, MHD(Magneto Hydrodynamics), etc. Utilizing this framework for the numerical analysis gives lots of advantages to users. By just porting their application solver to Cactus framework, user can utilize the toolkit for checkpointing, automatically parallelize, easily visualize the solution, and so on, by the support of Cactus framework. Additionally, Cactus operates in various computer platforms and various operating systems. And, application researchers can utilize many softwares and libraries on the Cactus frame, like Globus toolkit, HDF5 I/O, PETSc library, and visualization tools.

Thus, present research focuses on improving Cactus framework for CFD and utilize it on the Grid computing environment. So, a CFD flow solver is applied to Cactus framework and modules for general coordinate I/O, specific boundary conditions for CFD analysis are newly developed. Additionally, researches on developing computational toolkits including advanced visualization, job migration and load balancing, and portal service, are conducted by computer scientists. Using improved Cactus framework, a lot of validations are carried out through a portal service. From these studies, CFD flow solver is successfully implemented to the existing Cactus framework and the collaboration between computer scientists and CFD researchers is accomplished.

2. CACTUS-BASED CFD ANALYSIS

2.1. Cactus and CFD

To apply Cactus frame to CFD analysis, CFD researchers should implement their flow solver into Cactus frame. For application scientists, implementing their own flow solver into a new frame may seem absurd at a glance. Thus, the advantages and disadvantages of applying Cactus to CFD analysis are to be investigated in the present paragraph.

The advantages of Cactus-based CFD analysis are as follows. Firstly, application scientists can perform the Grid computing without profound knowledge of the Grid. Additionally, the collaborative research among multiple disciplines can be accomplished easily as all the researchers modularize their application solver to the Cactus format. And, the most advanced computer science technologies can be utilized without associated knowledge by using Cactus toolkits provided by computer scientists. However, when researchers analyze small-scaled problems or develop numerical techniques, Cactus is not necessary as analysis can be conducted inside a local organization. Rather, it can have a bad influence as learning how to use Cactus and implementing their flow solver into Cactus frame requires additional time. And, implementation process is somewhat complex as flow solver should be generalized while each flow solver is developed for the analyses of specific problems. Finally, as CFD researches by Cactus are not activated until now, supports for CFD analysis, like mesh generation using body-fitted coordinate and visualization of CFD result, are not sufficient.

2.2. 3-D Inviscid Flow Analysis Using Cactus

The three-dimensional compressible Euler Equations are adopted as governing equations. As a spatial discretization, AUSMPW+(modified Advection Upstream Splitting Method Press-based Weight function)[5] has been applied and LU-SGS (Lower-Upper Symmetric Gauss-Seidel) scheme[6] is used as the implicit time integration method. Time scale of each cell is determined by using a local time stepping method, where the eigenvalues of linearized matrix at each cell satisfies CFD stability condition. A 2-D RAE-2822 airfoil and a 3-D Onera-M6 wing mesh are generated using body-fitted coordinate system.

The procedure of 3-D inviscid flow analysis is as follows. At the pre-processing level, mesh generation is accomplished and at initial level, initialization of flow conditions, inputting mesh points and transformation to curvilinear coordinate are performed. And then, at iteration step, determination of time scale by local time stepping method, flux calculation by spatial discretization, time integration and application of boundary conditions are accomplished. Then, at the post-processing level, resultant data are visualized and analyzed. In present research, each component comprehended in the analysis process is made as a separate thorn of Cactus frame. Figure 1 shows the flow chart of 3-D inviscid flow analysis after converting each subroutine to thorn and scheduling thorns to the time bins of Cactus.

Each thorn is configured by CCL(Cactus Configuration Language) scripts and, analysis subroutine of flow solver is stored inside the src/ directory with the modification of parameters and functions to Cactus format.

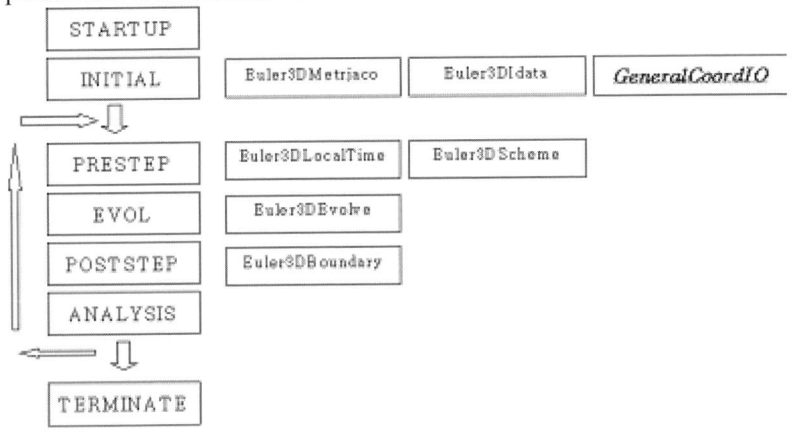

Fig. 1. Time Bins and Thorns for 3-D Inviscid Analysis

2.3. Numerical Results

The flowfield around a wing is analyzed by using Cactus-based CFD analyzer. Firstly, to confirm the possibility of CFD analysis using body-fitted coordinates, flow analysis around RAE-2822 airfoil is conducted on single processor. As shown in Fig. 2, the present configuration is basically 2-dimensional, but the mesh is extended to z-direction, making a 3-D mesh with symmetric condition along the z-direction. Total mesh points are 161×41×3. Mach number is set to be 2.0 and 0.78, depending on flow conditions. Angle of attack is set to be 0°. The result of flow analyses are presented in Fig. 3 and they show qualitatively valid results.

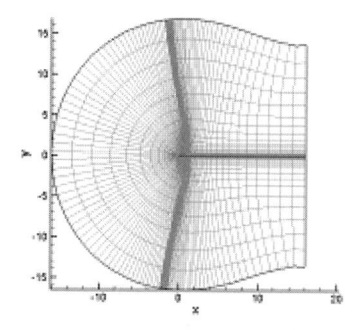

Fig. 2. RAE-2822 Mesh

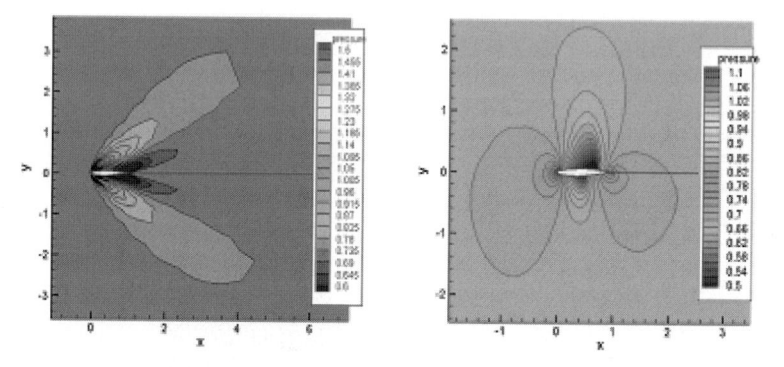

Fig. 3. Supersonic(L) and Transonic(R) Analysis

As Cactus CFD solver is proved to be valid from the analysis of RAE-2822 airfoil, the flowfield around a 3-D Onera-M6 wing is analyzed by using K* Grid, a Korean Grid testbed. The mesh system is shown in figure 4 and total mesh points are 141×33×65. 6 processors are co-worked and, Cactus frame automatically partitions the whole domain by 1×2×3 zones. Mach number is 0.78 and angle of attack is 0 degrees. And, the resultant pressure contour is shown in figure 5. Pressure increase near the leading edge of the wing and gradual decrease of pressure shows the correctness of the flow analysis. However, when more than 8 processors are utilized, periodic boundary condition should be applied along the tangential direction of wing surface and the result is not correct when periodic boundary condition is applied. Thus, the improvement of periodic condition for existing Cactus frame is required.

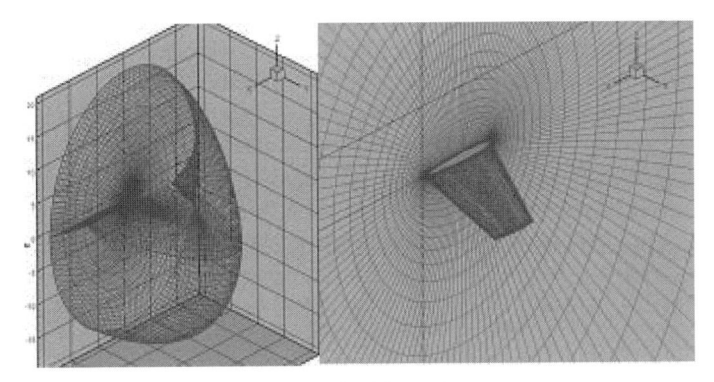

Fig. 4. 3-D Mesh around Onera-M6 Wing(O-type)

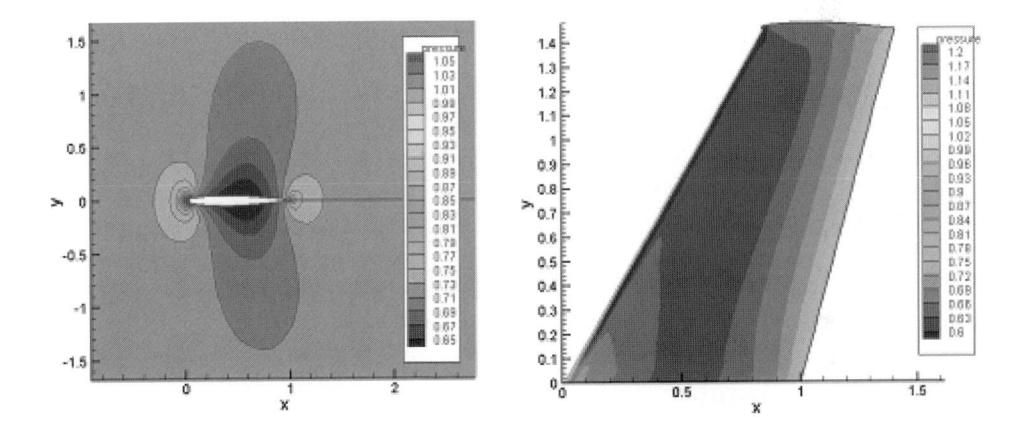

Fig. 5. Pressure Contours along Root of the Wing and Wing Surface

3. Computational Supports for CFD Analyses

3.1. GridOne Web Portal

The GridOne portal is developed to support the end-users with the Grid service. Cactus-based CFD analysis can be processed through the portal service. The frame of a portal service is built on the basis of GridSphere.[7] The GridOne Web portal is implemented as shown in Fig. 6 and 7. The work procedure via GridOne portal is as follows. At first, users register to the GridOne portal and login. After logging-in, user will configure their own information like e-mail, password, and application field. Users will upload their own Cactus-based application modules and then, compile and link uploaded Cactus program. After compiling and linking process, users start their CFD analysis program through the portal.

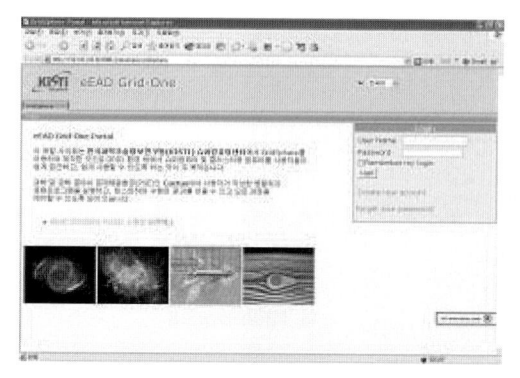

Fig. 6. The GridOne Web Portal

Fig. 7. Job Submission and Execution on the GridOne Portal

3.2. Remote Visualization

Recently, so many visualization programs are developed which supports multiple advanced functions. However, most of those softwares require high-performance computing resources and most of the functions are of no use to common users except for some specific application areas. Thus, a remote visualization software is developed which is relatively simple and runs fast on the web. The developed program does not make images at a server. It works in a way that the executing program and data are transferred and visualization program operates on the client machine. This kind of behavior will increase calculation of the client computer, but it avoids a load on server computer where many users access to and jobs are conducted on. The present program is developed using a Java Applet of JDK 1.3 and GL4Java library. And the remote visualization result on the GridOne portal is shown in Fig. 8

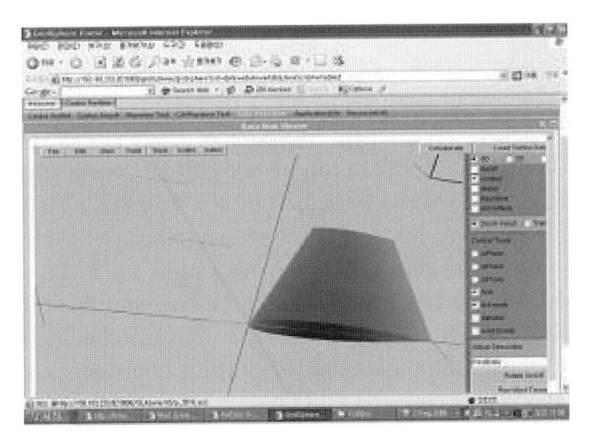

Fig. 8. The Remote Visualization through Web Portal

3.3. Job Migration

Job migration with Cactus CFD simulation is accomplished through GridOne portal as shown in Fig. 9. Firstly, Cactus CFD solver is submitted to the portal server, located in the K*Grid gateway. After compiling, CFD analysis starts in the gateway server. While job is running, migration manager continually searches for the better resources where the current CFD job will migrate to and when the manager finds the better resource in the Grid, job migration process is conducted. Nowadays, a migration test on the K* Grid has been conducted and, as a result, migration procedure from the gateway to the resource in KISTI Supercomputing center and, finally to the resource in Konkuk University has been observed with the correct simulation result.

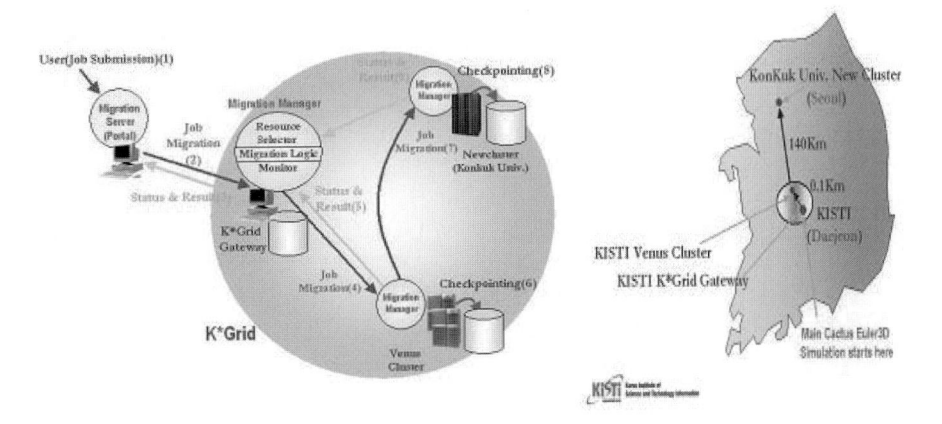

Fig. 9. Job Migration Procedure and Map

4. CONCLUSION

The present research applied Cactus PSE to the analysis of CFD problems. From the understandings of the structure of Cactus, researchers developed CactusEuler3D arrangement with General Coordinate I/O routine that can utilize body-fitted coordinate in the Cactus frame. And Cactus-based CFD solver is validated by the analysis of the flowfield around the wing. From these researches, Cactus PSE is shown to be applicable to CFD analysis and the possibility of multi-disciplinary collaboration by Cactus framework is presented. Simultaneously, the construction of computational environment including the efficient visualization, job migration and portal service are researched and successfully implemented. Eventually, those computational supporting environments with CFD solver on the Cactus framework will ease the CFD researchers in using the advanced computing technologies.

REFERENCES

1. M. M. Resch, "Metacomputing in High Performance Computing Center," IEEE 0-7659-0771-9/00, pp. 165-172 (2000)
2. D. Abramson, K. Power, L. Kolter, "High performance parametric modelling with Nimrod/G: A killer application for the global Grid," Proceedings of the International Parallel and Distributed Processing Symposium, Cancun, Mexico, pp. 520–528 (2000)
3. http://www.triana.co.uk
4. http://www.cactuscode.org
5. K. H. Kim, C. Kim and O. H. Rho, "Accurate Computations of Hypersonic Flows Using AUSMPW+ Scheme and Shock-aligned-grid Technique," AIAA Paper 98-2442 (1998)
6. S. Yoon and A. Jameson, "Lower-Upper SymmetricGauss-Seidel Method for the Euler and Navier-Stokes Equations," AIAA Journal, Vol. 26, No. 9, pp. 1025-1026 (1988)
7. http://www.gridsphere.org

The Scalability Impact of a Component-based Software Engineering Framework on a Growing SAMR Toolkit: A Case Study

Benjamin A. Allan,[*] S. Lefantzi and Jaideep Ray

We examine the evolution and structure of a growing toolkit for scientific simulations, implemented via a Component Based Software Engineering (CBSE) approach. We find that CBSE, in particular the use of the Common Component Architecture (CCA), enhances the scalability of the development process while detracting nothing from the performance characteristics. The CBSE approach enables a wide range of contributions by team members with varying (software engineering) sophistication to the toolkit in a short time. We find that the root cause of these efficiencies is a simplification of software complexity. We also examine the question of publications: Does writing good software help productivity?

1. Introduction

The Common Component Architecture [2] is a component-based software model for scientific simulation codes. It is not restricted to any particular field of science or type of simulation, does not impose a parallel computing model and enables high performance computing. The Ccaffeine framework [1] used in this work provides for low single-cpu overheads and parallel computing using MPI, PVM, OpenMP and other communication techniques. The CCA has been employed in the design of a few scientific simulation toolkits and the first scientific results are beginning to appear in print [6,10]. In this paper, we examine the impact of CBSE on the evolution and structure of the CFRFS [11] project, one of the early adopters of the CCA.

The CFRFS project develops a toolkit for the high fidelity simulation of laboratory-sized (10cm^3) flames. The components of the toolkit solve the low Mach number formulation of the Navier-Stokes equation at moderate Reynolds numbers. Each of the chemical species has its own evolution equation. A complete description of the equations is in [5]. The basic numerical strategy adopted by the toolkit is in [10] while the design is in [9]. Though still a work in progress, it contains 61 components spanning the range from simple numerical algorithms to involved physical models and large frameworks for meshing and data management on block-structured adaptively refined meshes (SAMR). Thus the current toolkit, though incomplete, is sufficiently large and diverse for one to compute metrics which will be approximately representative of the final product. We use these metrics to examine the core precepts of the CCA concept, particularly with respect to productivity and performance. We also present the data as a benchmark, there being few if any presentations in the literature of how component-oriented software engineering affects the design of a parallel scientific code. Most of the costs and issues involved in building scientific research codes are rarely discussed in the science literature that

---

[*]Sandia National Laboratories, MS9158, Livermore, CA 94550-3425
baallan,jairay@ca.sandia.gov

presents the results of the simulations. As the complexities of studied phenomena are exploding, however, we must start examining the underlying problems in building codes to simulate these phenomena.

1.1. The research software problems

Creating a CFD toolkit as part of a research project presents many practical issues. While scientific questions regarding consistency and correctness of the solutions are well known, large multiphysics simulation codes bring to the fore may software issues which are poorly understood (and appreciated) in the computational science community. Some of them are:

- The limited time available requires the use of externally controlled legacy libraries, of mixed language programming (since rewriting all the legacy code in a more favored language is not an option), and of a team of developers of mixed abilities many of whom may be only transiently involved in the project.

- The lead developer (usually also the software architect and a hero programmer) must be able to rigorously enforce subsystem boundaries and to control boundary changes.

- The team is usually required to continue publishing while coding.

- In a high-performance computing (HPC) code, coping with shifting platforms (either hardware or software) is required.

- For better scientific validation, swapping out developed package elements for alternative code (possibly developed by competing research groups) is often required.

- Over the course of any project, requirements change and various package elements will have to be recoded. Less frequently, but usually more expensively, an interface between elements may need to be extended or redefined.

The cause of these problems is that there is no simple way of solving multiphysics problems. When complicated solution strategies are realized in software, the complexity is fully reflected in its structure. The problem may be exacerbated because few computational scientists are trained software architects.

1.2. The CCA approach to HPC CBSE

Thus the solution to the problem of software complexity (in a scientific simulation context) lies in adopting a software development pattern and tools that support a modular design and automated enforcement of the design. Component models seek to do precisely this and CCA is no exception. Our development process and software products follow the Common Component Architecture (CCA) [7] methods. The CCA defines, among other things, a set of generic rules for composing an application from many software components. In our work, the components all support single-program-multiple data (SPMD) parallel algorithms. Each component is a black-box object that provides public interfaces (called *ports*) and uses ports provided by other components. In object-oriented terms, a port is merely the collection of methods into an interface class. The CCA leaves the definition of SAMR-specific ports to us.

A component is never directly linked against any other. Rather, at run-time each component instance informs the framework of which ports it provides and which it wishes to use. At the user's direction, the framework exchanges the ports among components. A component sees other components only through the ports it uses. Implementation details are always fully hidden.

2. Software development scalability

In this section we present summary information about the software produced to date. The task flow of the project, broken down by products and developers, shows that the CCA approach enables diverse contributors to produce all the required products as a team. The raw data, in the form of a Gantt chart, is too large to present here, but appears in [8]. Over the four year course of the project nine developers have participated, with the average head-count at any moment being four.

2.1. How the CCA model manifests as a SAMR framework

The component design of the CFRFS project [11] allows encapsulation and reuse of large legacy libraries written in C and FORTRAN. Of the 61 components produced in the project so far, 13 require some aspect of parallel computing, while the rest handle tasks local to a single compute node. As with most research projects, the lead developer is critical: in the work-flow chart we see that 35 of the components were created by developer A, 13 by developer B, 8 by developer C, and the remaining components by others.

Port interfaces get created to handle new functionality; the present count is 31. Four of the eight developers contributed port definitions. In defining ports, the lead developer often plays a consulting role, but the details of a port encapsulating new functionality are usually dictated by an expert on the underlying legacy libraries rather than by the lead developer.

As we see in figure 2, the typical component source code is less than 1000 lines of C++ (excluding any wrapped legacy code); the exceptions being the GrACE [12] library wrapping component at almost 1500 lines and the adapter for Chombo [4] inter-operation at almost 2700 lines. The average component is therefore reasonably sized to be maintainable. The best measure of the legacy codes encapsulated by the components may be obtained by examining the size of the resulting component libraries in figure 1. The Chombo, HDF, and thermochemical properties libraries are the three giants at the end of the histogram; most components compile to under 100 kilobytes. An often proposed metric of quality for an object-oriented code is the number of lines per function. Figure 3 shows that the distribution of lines per *port* function varies widely across components, being particularly high in those components which are complicated application drivers or wrappers which hide complexity in an underlying legacy library.

A question often asked by people considering adopting the component approach is "how complex should my ports be?", by which they mean how many functions are in one port. The correct answer is "it depends on what the port does", but in figure 4 we see that most ports are small with fewer than 10 functions. The largest port, with 39 functions is the AMRPort which encapsulates the bulk of the GrACE functionality.

CCA component technology is often described as providing plug-and-play functionality such that alternative implementations can be plugged in for testing and optimization. We test this claim in the CFRFS context by examining how many different components implement each kind of port, in figure 5. In this combustion work, the most often implemented port turns out to be the diffusion functionality, with 10 components providing alternate models.

In CFD modeling, there is often a central mesh concept that is reused in nearly all code modules. This is apparent in our case from figure 6, where the most frequently used port is the AMRPort with 29 clients. The next most frequently used port is the properties port which allows a driver component to tune any component that exports the properties port.

The component model supported by CCA encourages incremental development and orderly expansion of functionality in our toolkit. This progress over time is illustrated in figure 7, where we see bursts in the number of port interfaces defined by the software lead followed by steadier growth as component implementations and occasionaly new ports are developed by the many other contributors. Further, there are fewer ports than components, indicating that a few well-designed interfaces can accomodate a large complexity in implementation details.

2.2. Productivity

We have seen the impact of the CCA on the CFRFS software development, but what of the project intellectual productivity? One of the principal goals of the project is to produce a reusable software toolkit, of course, but what of other metrics of career success? Software developers in the course of the work have produced: six scientifically interesting in-house test applications, 11 conference papers, a book chapter, and 3 journal papers all related directly to this work. A SAMR framework interoperability standard has also been drafted and tested with the Chombo framework. Rather than distracting the development team from the intellectual issues the software is designed to address, the component structure of the SAMR framework enables faster testing of new ideas because any needed changes are usually isolated to one or two components.

3. Performance

In the previous section, we have shown that the adoption of a component-based software paradigm enhances productivity within the context of simulation science. This is in line with the widespread use of such a paradigm in the business world. In this section, we examine if this productivity comes at the cost of performance - none of the business oriented component models support high performance or complex parallelism. We now examine if (a) CCA codes are scalable - i.e. the CCA framework imposes no restrictions on the parallel nature of the code, and (b) single processor performance is not affected significantly by CCA-compliance. For this purpose, we will use two codes, assembled from our CRFRS toolkit.

RDiff is a code which solves a reaction-diffusion system of a non-uniform mixture of H_2 and O_2 reacting in a non-uniform temperature field. The gases (and the intermediate radicals) react and diffuse through the domain. The simulation is memory intensive (each grid point has 10 variables) and involves time-integration of a stiff chemical system at each grid point, making it compute intensive. This code will be used for scalability tests. The reaction-diffusion system solved can be expressed as

$$\frac{\partial \Phi}{\partial t} = \nabla^2 \Phi + W(\Phi) \tag{1}$$

subject to zero-gradient boundary condition on a unit square. The vector $\Phi = T, Y_i \quad i = 1 \ldots 8$, each i corresponding to the reactants, products and radicals in a 9-species 5-reaction $H_2 - O_2$ mechanism. W is a source term affected only by chemistry (production/consumption of species via chemical reactions) while the Laplacian term models Fickian diffusion. The system was integrated using an operator-split method where a chemistry-only time-integration over Δt is sandwiched between 2 diffusion-only integrations of $\Delta t/2$. The diffusion advance is done via

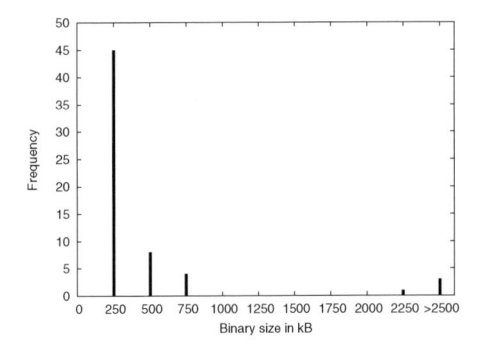

Figure 1. Histogram of component binary library sizes.

Figure 2. Histogram of component source code sizes.

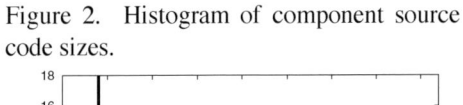

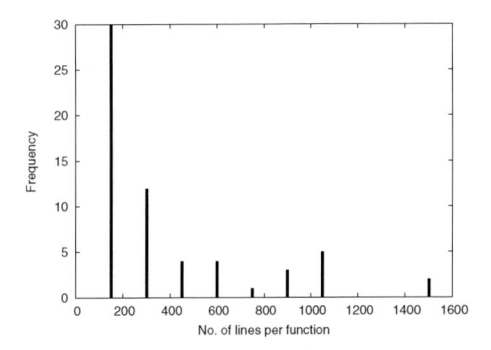

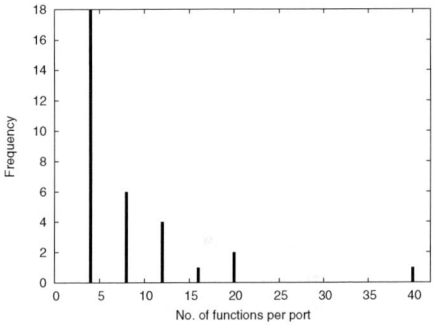

Figure 3. Histogram of function size.

Figure 4. Histogram of port size in number of functions.

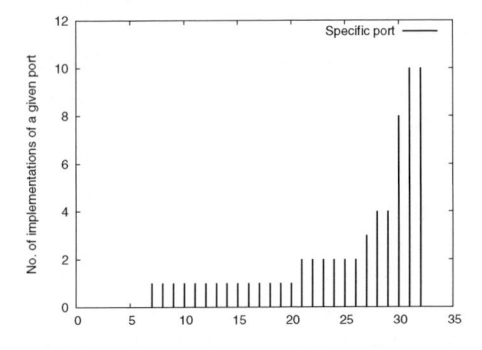

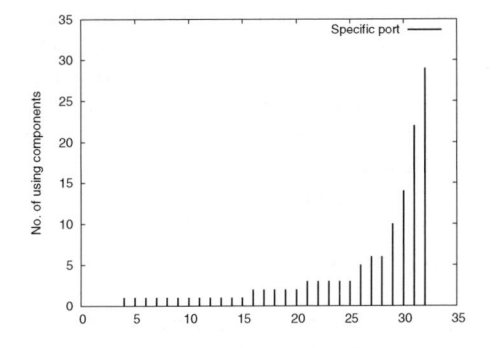

Figure 5. Number of components implementing each port type, sorted by increasing frequency.

Figure 6. Number of components using each port type, sorted by increasing frequency.

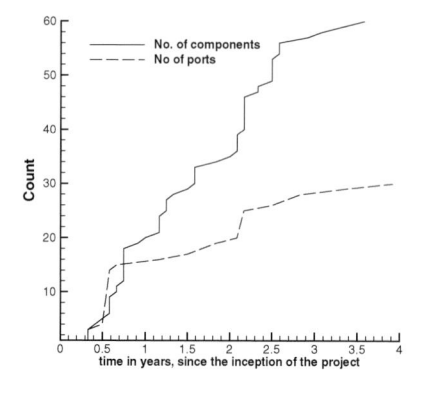

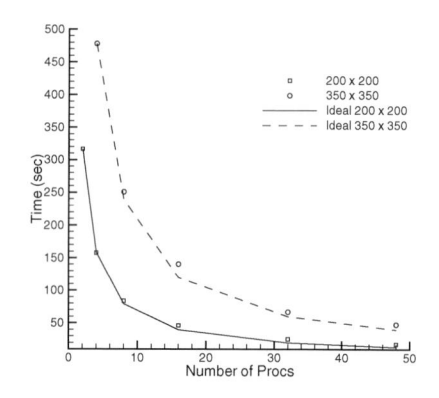

Figure 7. The growth in the number of components and ports with time (in years).

Figure 8. Scaling tests for 200×200 and 350×350 problems sizes over 2 to 48 CPUs

an explicit Runge-Kutta algorithm; the chemistry integration is done using Cvode [3], suitably wrapped into a component. No adaptation was done since this code is being used for scaling studies and deterministic loads were needed.

Cvode is a code we test to address the question of whether a CCA-compliant component incurs a performance penalty vis-a-vis a C code. This model is actually a truncated version of RDiff. Both adaptation and diffusion were turned off and we simulated the chemical evolution of the $H_2 - O_2$ and temperature fields. The initial condition was spatially *uniform* i. e. the computational load per grid point was identical. The CvodeComponent was used to do the time-implicit integration of the unknowns. The right-hand-side (RHS) was supplied by a ChemicalRates component (also used in RDiff). Briefly, the ODE system is transformed into an **A x = b** problem and solved iteratively. Each iteration involves the calling the ChemicalRates component to evaluate the RHS. Test were done for various grid sizes (size of the integration vector) as well as Δt (number of times the RHS was evaluated) in a effort to exploit the known virtual pointer lookup overhead that CCA components incur. Results were compared with a plain C code which used the non-wrapped Cvode library.

3.1. Results

In Fig. 8 we present performance results from the **RDiff** code. The problem was integrated for 5 timesteps of 1×10^{-7}. We conducted a weak scaling study where the problem was scaled up along with the number of processors were increased to keep the per-processor work load constant. The mean, median and standard deviations of the run-times showed that the average run-time as the number of processors increase did not change much. Also, as the problem size on a given processor increased, the run times increase in proportion. Thus the machine can be considered to be a "homogeneous machine" – the communication time for nearest-neighbor communications is not very different from a distant-processor communication, at least for the problem sizes considered. In Fig 8 we present a strong scaling study where the global problem size remains constant and scaling is studied by using a progressively finer domain decomposition. The comparison with ideal scaling shows that the scaling efficiency, at worst (200×200 mesh on 48 processors) is 73 %. No adaptation was done - the problem remained unchanged on each processor for all timesteps.

In Table 1 we present a comparison of **Cvode** code run as a component on a single processor and compared to a plain C library version. Results are presented for 2 Δts and for a problem with 100, 500 and a 2000 cells. The problem was run for 100 timesteps. We see that the percentage difference in time taken is well below 5 %, and usually below 1 %. The results improve as the problem size becomes bigger.

$\Delta t = 0.1$				
Ncells	NFE	Component	C-code	% diff
100	66	0.19	0.2	-3.5
500	66	1.18	1.23	-3.9
2000	66	5.51	5.66	-2.8

$\Delta t = 0.01$				
100	150	0.37	0.39	-4.46
500	150	2.34	2.38	-1.85
2000	150	10.53	10.6	-0.63

Table 1
Timings for the **Cvode** code. Ncells refers to the number of cells/points on the mesh. NFE = number of RHS evaluation i.e the number of times the ChemicalRates component was called. Timings are in seconds. % diff is the percentage difference in execution time.

The machine used for **Cvode** had Intel 550 MHz processors with 256 MB of RAM, connected via a 100 Mbps Fast Ethernet switch. **RDiff** was run on a cluster of 433 MHz Compaq Alphas with 192 MB of RAM and a 1GB/s messaging fabric based on 32-bit Myrinet PCI32c cards. The codes were compiled with `g++ -O2`, using egcs version 2.91 compilers, both on Alpha and the Intel machines.

4. Conclusions

In this paper we have characterized the CFRFS toolkit. The individual components tend to be small, with 1-2 ports, each of which have 5-10 methods. Components mostly tend to call on 2-3 ports, making it possible to compose simulation codes which are sparsely connected. This is characteristic of systems which are built hierarchically from smaller subsystems, each approximately self contained and modular. The modularization (and simplicity!) enforced by componentization pervades throughtout the toolkit and is reflected in assembled codes.

From the performance perspective, these preliminary runs show that the CCA architecture presents no constraints to developing high-performance scientific codes. Single processor performance is barely affected (less than 5 % performance degradation) while scalability is a property of the assembled components and hardware rather than the Ccaffeine implementation.

5. Acknowledgments

This work was supported by the United States Department of Energy, in part by the Office of Advanced Scientific Computing Research: Mathematical, Information and Computational Sciences through the Center for Component Technology for Terascale Simulation Software and in part by the Office of Basic Energy Sciences: SciDAC Computational Chemistry Program. This is an account of work performed under the auspices of Sandia Corporation and its Contract No. DE-AC04-94AL85000 with the United States Department of Energy.

REFERENCES

1. Benjamin A. Allan, Robert C. Armstrong, Alicia P. Wolfe, Jaideep Ray, David E. Bernholdt, and James A. Kohl. The CCA Core Specification in a Distributed Memory SPMD Framework. *Concurrency and Computation: Practice and Experience*, 14(5):323–345, 2002.
2. CCA Forum homepage. http://www.cca-forum.org/, 2004.
3. S. D. Cohen and A. C. Hindmarsh. Cvode, a stiff/nonstiff ode solver in c. *Computers in Physics*, 10(2):138–143, 1996.
4. P. Colella et al. Chombo – Infrastructure for Adaptive Mesh Refinement. http://seesar.lbl.gov/anag/chombo.
5. H. N. Najm et al. Numerical and experimental investigation of vortical flow-flame interaction. SAND Report SAND98-8232, UC-1409, Sandia National Laboratories, Livermore, CA 94551-0969, February 1998. Unclassified and unlimited release.
6. Joseph P. Kenny et al. Component-based integration of chemistry and optimization software. *J. of Computational Chemistry*, 24(14):1717–1725, 15 November 2004.
7. Lois Curfman McInnes et al. Parallel PDE-based simulations using the Common Component Architecture. In Are Magnus Bruaset, Petter Bjørstad, and Aslak Tveito, editors, *Numerical Solution of PDEs on Parallel Computers*, volume 51 of *Lecture Notes in Computational Science and Engineering*. Springer-Verlag, 2005.
8. PCFD 2005 Gantt chart of CFRFS software development processes. http://cca-forum.org/~baallan/pcfd05-snl/workflow, 2005.
9. S. Lefantzi, J. Ray, and H. N. Najm. Using the Common Component Architecture to Design High Performance Scientific Simulation Codes. In *Proceedings of the 17th International Parallel and Distributed Processing Symposium*, Los Alamitos, California, USA, April 2003. IEEE Computer Society.
10. Sophia Lefantzi, Jaideep Ray, Christopher A. Kennedy, and Habib N. Najm. A Component-based Toolkit for Reacting Flows with High Order Spatial Discretizations on Structured Adaptively Refined Meshes. *Progress in Computational Fluid Dynamics*, 5(6):298–315, 2005.
11. Habib N. Najm et al. CFRFS homepage. http://cfrfs.ca.sandia.gov/, 2003.
12. M. Parashar et al. GrACE homepage. http://www.caip.rutgers.edu/~parashar/TASSL/Projects/GrACE/, 2004.

Dynamics of biological and synthetic polymers through large-scale parallel computations

I. D. Dissanayake[a] and P. Dimitrakopoulos[a] * [†]

[a]Department of Chemical and Biomolecular Engineering,
University of Maryland, College Park, Maryland 20742, USA

We present the efforts of our research group to derive an optimized parallel algorithm so that we are able to efficiently perform large-scale numerical studies on the dynamics of biological and synthetic polymers. To describe the polymer macromolecules we employ Brownian dynamics simulations based on a semiflexible bead-rod chain model. Our algorithm has been parallelized by employing Message Passing Interface (MPI) on both shared- and distributed-memory multiprocessor computers with excellent parallel efficiency. By utilizing this methodology, we routinely employ up to 320 of the fastest processors to study polymer chains with more than 40,000 beads. Our optimized parallel algorithm facilitates the study of polymer dynamics over a broad range of time scales and polymer lengths.

1. INTRODUCTION

The dynamics of polymer solutions is a problem of great technological and scientific interest since these systems are encountered in a broad range of industrial, natural and physiological processes. Common examples include biopolymers such as DNA, actin filaments, microtubules and rod-like viruses, as well as synthetic polymers such as polyacrylamides, Kevlar and polyesters. These polymers show a wide range of stiffness which results in some unique properties of their solutions and networks. For example, networks of actin filaments can provide biological cells with mechanical stability while occupying a significantly smaller volume fraction of the cytosol than would be required for a flexible network. In the case of the stiff synthetic polymers, stiffness is responsible for the macroscopic alignment of the chains in the system which imparts unique mechanical properties for these materials. Thus, the study of these systems is motivated by both the biological relevance and engineering applications.

The numerical study of polymer dynamics is computationally expensive due to the large molecule lengths and the requirement of monitoring the polymer properties over extended time periods. To overcome this obstacle, we have developed an optimized parallel algorithm and thus we are able to efficiently perform large-scale numerical studies on

---

*Corresponding author, email: dimitrak@eng.umd.edu
[†]This work was supported in part by the M. Martin Research Fund, the National Center for Supercomputing Applications (NCSA) in Illinois, and by an Academic Equipment Grant from Sun Microsystems Inc.

polymer dynamics. Our interest lies on understanding the physical properties of important biopolymers such as DNA, actin filaments and microtubules as well as of synthetic polymers. Our studies are further motivated by the recent development of experimental techniques which study individual biological molecules as well as by the recent development of micro-devices involving stretched tethered biopolymers (e.g. [1–3]).

2. MATHEMATICAL FORMULATION

To describe a semiflexible polymer chain, we employ the Kratky-Porod wormlike chain model [4,5] based on a Brownian dynamics method developed in Ref.[6]. This method considers a (flexible) bead-rod model with fixed bond lengths and ignores hydrodynamic interactions among beads as well as excluded-volume effects. The polymer chain is modeled as $N_B = (N+1)$ identical beads connected by N massless links of fixed length b (which is used as the length unit). The position of bead i is denoted as $\boldsymbol{X}_i$, while the link vectors are given by $\boldsymbol{d}_i = \boldsymbol{X}_{i+1} - \boldsymbol{X}_i$.

To account for polymer stiffness, we add a bending energy proportional to the square of the local curvature. For a continuous chain the bending energy is given by

$$\phi^{bend} = \frac{\mathcal{E}b}{2} \int_0^L (\frac{\partial \hat{\boldsymbol{d}}}{\partial s})^2 ds = \frac{\mathcal{E}b}{2} \int_0^L (\frac{\partial^2 \boldsymbol{X}}{\partial s^2})^2 ds, \qquad (1)$$

where L is the (constant) contour length of the chain and $\hat{\boldsymbol{d}}$ the local unit tangent. The bending energy $\mathcal{E}$ is related to the persistence length L_p via $\mathcal{E}/k_B T \equiv L_p/b$, where k_B is the Boltzmann constant. The bending energy of the discrete model is given by

$$\phi^{bend} = \frac{\mathcal{E}b}{2} \sum_{i=2}^{N} (\frac{\boldsymbol{X}_{i+1} - 2\boldsymbol{X}_i + \boldsymbol{X}_{i-1}}{b^2})^2 b = \mathcal{E} \sum_{i=1}^{N-1} (1 - \frac{\boldsymbol{d}_i \cdot \boldsymbol{d}_{i+1}}{b^2}) \qquad (2)$$

and thus it depends on the angle θ_i between two successive links since $\boldsymbol{d}_i \cdot \boldsymbol{d}_{i+1} = b^2 \cos \theta_i$. For a fixed b, the properties of the polymer chain are specified by the number of links N and the dimensionless bending energy $E = \mathcal{E}/k_B T$.

Assuming that the bead inertia is negligible, the sum of all forces acting on each bead i must vanish, which leads to the following Langevin equation

$$\zeta \frac{d\boldsymbol{X}_i}{dt} = \boldsymbol{F}_i^{bend} + \boldsymbol{F}_i^{rand} + \boldsymbol{F}_i^{ten} + \boldsymbol{F}_i^{cor}, \qquad (3)$$

where ζ is the friction coefficient and $\boldsymbol{F}_i^{rand}$ the Brownian force due to the constant bombardments of the solvent molecules. The force $\boldsymbol{F}_i^{ten} = T_i \boldsymbol{d}_i - T_{i-1} \boldsymbol{d}_{i-1}$, where T_i is a constraining tension along the direction of each link $\boldsymbol{d}_i$, ensures the link inextensibility. $\boldsymbol{F}_i^{cor}$ is a corrective potential force added so that the equilibrium probability distribution of the chain configurations is Boltzmann [6]. The bending force $\boldsymbol{F}_i^{bend}$ is derived from the discrete form of the bending energy, Eq.(2),

$$\boldsymbol{F}_i^{bend} = -\frac{\partial \phi^{bend}}{\partial \boldsymbol{X}_i} = \frac{\mathcal{E}}{b^2} \sum_{j=i-2}^{i} [\delta_{j,i-2} \, \boldsymbol{d}_{i-2} + \delta_{j,i-1} (\boldsymbol{d}_i - \boldsymbol{d}_{i-1}) - \delta_{j,i} \, \boldsymbol{d}_{i+1}]. \qquad (4)$$

(In the equation above as well as in all the equations in this paper, a term exists only if its index can be defined within its permitted bounds.) The resulting system based on Eq.(3) may be solved in $O(N)$ operations facilitating the study of long and/or stiff chains. The polymer properties are determined as the ensemble averages of the corresponding instantaneous values by employing 10^4 to 10^5 independent initial configurations.

Note that although we model the polymer molecule as a bead-rod semiflexible chain, the mathematical formulation presented above is more general since it can also represent other polymer models including bead-spring models with Hooke, Fraenkel or FENE-type springs [7]. Our preference for the bead-rod semiflexible model results from the fact that this model allows a continuous crossover from a freely-jointed flexible chain to a rigid rod as the bending energy increases while it preserves the link lengths and thus the contour length of the entire chain. We emphasize that the tensions required to keep the link lengths constant play a significant role in the polymer dynamics both near and far from equilibrium as recent studies have revealed [6,8–10].

3. RELEVANT TIME SCALES

The Brownian forces give rise to a series of time scales associated with the diffusive motion of the chain's increasing length scales, from that of one bead $\tau_{rand} = \zeta b^2/k_B T$, up to the time scale for the entire chain length which is $\tau_f = N^2 \tau_{rand}$ for flexible chains and $\tau_s = N^3 \tau_{rand}$ for stiff ones. Similarly, the bending forces give rise to a series of time scales associated with the bending vibrations of portions of the polymer chain with increasing length. The smallest bending time scale is associated with the relaxation of the angle between two successive links given by $\tau_{bend} = \zeta b^2/\mathcal{E} = \tau_{rand}/E \ll \tau_{rand}$, while the largest time scale is associated with the entire polymer chain and given by $\tau_\perp = \zeta L^4/\mathcal{E}b^2 = (N^4/E)\tau_{rand}$. If we consider that the length N is large for polymer molecules, it is clear that all polymers, and especially the stiff chains, are associated with very small time scales while their behaviors cover extended time periods.

Therefore the numerical study of polymer chains is computationally expensive; even with the present powerful single-processor computers it may take months or even years to simulate a long polymer chain over an extended time period. To overcome this obstacle, we have developed an efficient parallel algorithm which utilizes the great computational power of current multiprocessor supercomputers, especially the relatively inexpensive PC clusters, as discussed in the next section.

4. PARALLEL ALGORITHM FOR POLYMER DYNAMICS

Due to the computationally expensive nature of Brownian dynamics simulations of polymer molecules, it is imperative that the corresponding algorithm is fully optimized. To optimize our code, we employ highly-tuned LAPACK routines for the solution of the tridiagonal systems resulting from the equation of motion, Eq.(3). For the calculation of the chain's normal modes we use highly-tuned FFT algorithms provided by system libraries. In addition, our FORTRAN code has been optimized using cache optimization techniques.

We note that further optimization efforts have been made. Many physical problems involve polymers at equilibrium. For example, a DNA molecule, before undergoing a shearing flow, may be at equilibrium. To determine the equilibrium configuration, one has to allow the chain to equilibrate first over many time-consuming Brownian dynamics steps. In this case it is more efficient to start directly from an equilibrium chain configuration. To produce these equilibrium configurations we apply the following procedure based on the chain bending energy. In particular, for completely flexible chains (i.e. $E = 0$), an equilibrium configuration is derived as a "random-walk" polymer configuration generated by choosing successive bead positions from random vectors distributed over the surface of a sphere. For chains with non-zero bending energy, we employ a Monte Carlo-Metropolis algorithm to calculate an equilibrium chain configuration based on the chain bending energy. The Monte-Carlo algorithm for calculating equilibrium configurations is much faster than the Brownian dynamics algorithm; thus, significant savings are achieved. We emphasize that for any chain's bending energy, our ability to efficiently produce equilibrium configurations is a consequence of the corrective potential force $\boldsymbol{F}^{cor}$ we described in section 2. (Thus, bead-rod models which do not include this correction need to be equilibrated first by employing Brownian dynamics time steps.)

To study the dynamic evolution of polymer macromolecules, we numerically determine the ensemble average value of properties of interest including the chain configuration, the polymer stress and the solution birefringence. This is achieved by starting from (different) initial polymer configurations and following a different sequence of Brownian forces to simulate different chain realizations. Our optimized algorithm can successfully be employed on single-processor computers for relatively short chains and time periods; for larger polymer molecules we employ multiprocessor supercomputers and suitable parallelization.

To describe this, observe that the ensemble average calculation is by nature parallel while we can use as many processors as the number of initial configurations which is of order 10^4–10^5. In practice, a moderate number of processors (of order 10–300) is adequate for most cases. For the parallelization we use Message Passing Interface (MPI) [11,12]. The great advantage of MPI is its portability to different platforms. With this parallelization, we have the ability to employ both shared-memory multiprocessor computers (e.g. IBM pSeries 690) and distributed-memory supercomputers (e.g. Linux clusters). On all machines, the efficiency is almost 100% even for a high number of processors as shown in Table 1. This is due to the limited time of communication among the processors compared with the CPU time consumed for calculations on each processor. Therefore, our numerical code is ideal for all types of the state-of-the-art supercomputers.

By utilizing this optimized parallel algorithm, we routinely employ up to 320 of the fastest processors to study polymer chains with more than 40,000 beads and thus we are able to identify the polymer behavior of very long chains at very small times and for extended time periods. For example as we discussed in Refs.[13,10] for the problem of the relaxation of initially straight flexible and stiff chains, we were able to determine the polymer evolution over 17 time decades for flexible chains and over 28 time decades for stiff ones. In addition, the length scales we study (up to at least $N = 40,000$ beads) correspond to DNA molecules up to 2.6 mm long and to synthetic molecules of polystyrene with molecular weight of $O(10^7)$!

Therefore, by employing this optimized parallel algorithm we are able to identify all physical behaviors even if they appear at very small times and span extended time periods. For example, we have studied the dynamics of chain stiffening due to temperature change

Table 1
Efficiency versus the number of processors N_p for a typical problem by employing our parallel Brownian Dynamics algorithm on the shared-memory IBM pSeries 690 and the distributed-memory Linux Xeon Supercluster provided by the National Center for Supercomputing Applications (NCSA) in Illinois. Note that efficiency denotes the ratio of the wall time for the serial execution T_s to that for the parallel execution T_p multiplied by the number of processors, i.e. efficiency $\equiv T_s/(T_p N_p)$.

NCSA IBM pSeries 690		NCSA Xeon Supercluster	
N_p	Efficiency (%)	N_p	Efficiency (%)
1	100.00	1	100.00
2	99.81	10	99.39
5	99.75	20	99.09
10	99.64	25	98.65
16	99.24	50	97.16
		160	97.02

or salt addition in the polymer solution [14–17]. An industrial application of this process involves the manufacturing of better trapping gel networks, micro-electronic devices from plastics and novel drug delivery systems [18]. In Figure 1 we present the shape evolution of an initially coil-like chain during stiffening towards its final rod-like shape. The associated transition from coil to helix and from helix to rod is obvious from this figure. These transitions are also shown in Figure 2 where we plot the time evolution of the polymer length and its widths. Observe that the time of the maximum chain's width corresponds to the helix creation while the time of the maximum chain's length corresponds to the rod-like final configuration.

5. CONCLUSIONS

The dynamics of polymer solutions is a problem of great technological and scientific interest but the numerical study of these systems is computationally expensive due to the large molecule lengths and the requirement of monitoring the polymer properties over extended time periods. To overcome this obstacle, we have developed an optimized parallel Brownian dynamics algorithm for the dynamics of biological and synthetic polymers via Message Passing Interface (MPI) based on ensemble average calculations. By employing both shared- and distributed-memory multiprocessor computers, we have the ability to determine and understand the polymer properties over a wide range of time scales and polymer lengths which has never been achieved before.

REFERENCES

1. T. T. Perkins, S. R. Quake, D. E. Smith and S. Chu, Science 264 (1994) 822.
2. T. T. Perkins, D. E. Smith and S. Chu, in Flexible Polymer Chains in Elongational Flow, T. Q. Nguyen and H.-H. Kausch (eds.), Springer, 1999.

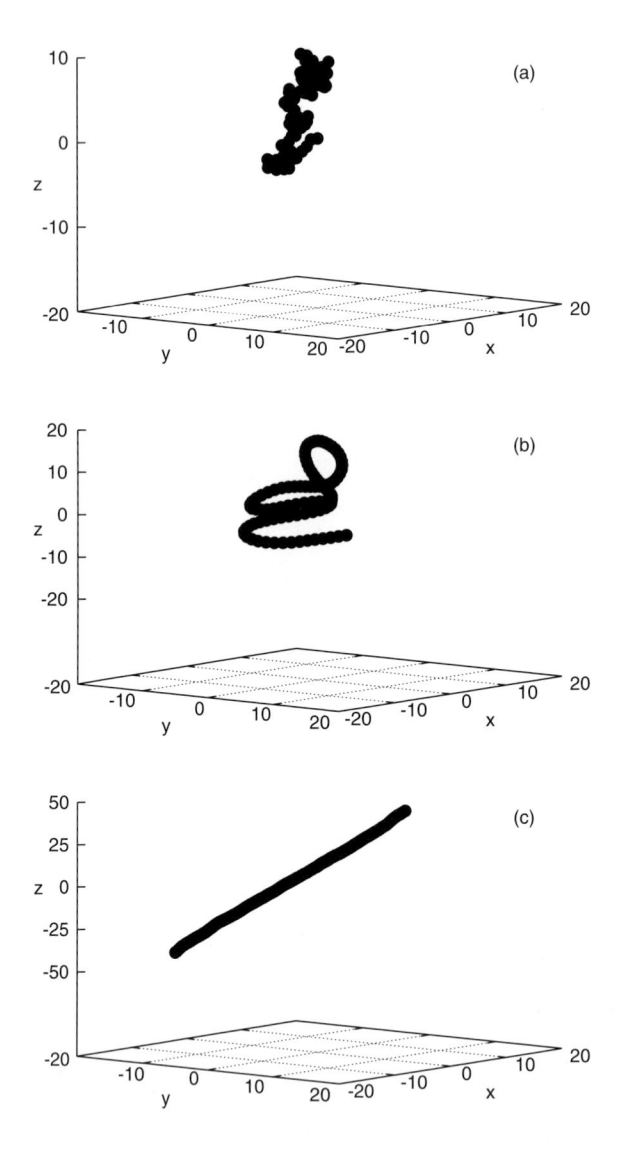

Figure 1. Evolution of a polymer molecule during backbone stiffening. The polymer length is $N = 100$ while the chain's final stiffness is $E/N = 10$. Three distinct configurations are presented: (a) the initial coil-like shape, (b) the intermediate helical shape, and (c) the final rod-like configuration.

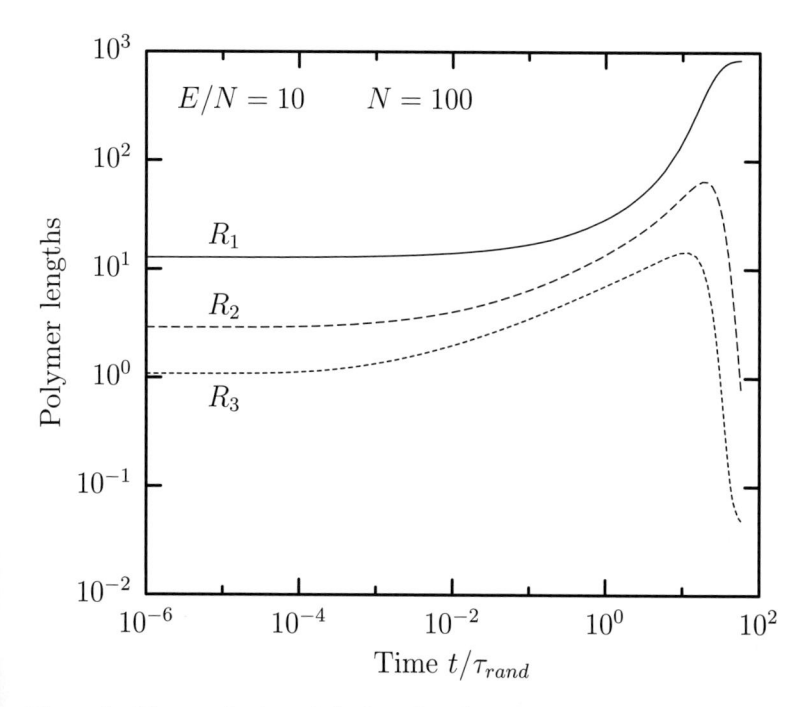

Figure 2. Time evolution of the lengths of a polymer molecule during backbone stiffening. The polymer's contour length is $N = 100$ while the chain's final stiffness is $E/N = 10$. Note that R_1 is the polymer length while R_2 and R_3 are the chain's widths.

3. G. J. L. Wuite, S. B. Smith, M. Young, D. Keller and C. Bustamante, Nature 404 (2000) 103.
4. H. Yamakawa, Helical Wormlike Chains in Polymer Solutions, Springer, 1997.
5. M. Doi and S. F. Edwards, The Theory of Polymer Dynamics, Clarendon, 1986.
6. P. S. Grassia and E. J. Hinch, J. Fluid Mech. 308 (1996) 255.
7. R. B. Bird, C. F. Curtiss, R. C. Armstrong and O. Hassager, Dynamics of Polymeric Liquids: Volume 2, Kinetic Theory, Wiley, 1987.
8. R. Everaers, F. Jülicher, A. Ajdari and A. C. Maggs, Phys. Rev. Lett. **82** (1999) 3717
9. V. Shankar, M. Pasquali and D. C. Morse, J. Rheol. **46** (2002) 1111
10. P. Dimitrakopoulos, Phys. Rev. Lett. 93 (2004) 217801.
11. Message Passing Interface Forum (eds.), Message Passing Interface, 2003 (available at `http://www.mpi-forum.org` and `http://www-unix.mcs.anl.gov/mpi`).
12. I. Foster, Designing and Building Parallel Programs, Addison-Wesley, 1995.
13. P. Dimitrakopoulos, J. Chem. Phys. 119 (2003) 8189.
14. Y. Y. Gotlib and L. I. Klushin, Polymer 32 (1991) 3408
15. C. G. Baumann, S. B. Smith, V. A. Bloomfield and C. Bustamante, Proc. Natl. Acad. Sci. 94 (1997) 6185
16. K. Terao, Y. Terao, A. Teramoto, N. Nakamura, M. Fujiki and T. Sato, Macromolecules 34 (2001) 4519.
17. J. R. Wenner, M. C. Williams, I. Rouzina and V. A. Bloomfield, Biophysical J. 82 (2002) 3160.
18. C. K. Ober, Science 288 (2000) 448.

Parallel Simulation of High Reynolds Number Vascular Flows

Paul Fischer,[a] Francis Loth, Sang-Wook Lee, David Smith, Henry Tufo, and Hisham Bassiouny

[a]Mathematics and Computer Science Division, Argonne National Laboratory
Argonne, IL 60439, U.S.A.

1. Introduction

The simulation of turbulent vascular flows presents significant numerical challenges. Because such flows are only weakly turbulent (i.e., transitional), they lack an inertial subrange that is amenable to subgrid-scale (SGS) modeling required for large-eddy or Reynolds-averaged Navier-Stokes simulations. The only reliable approach at present is to directly resolve all scales of motion. While the Reynolds number is not high (Re=1000–2000, typ.), the physical dissipation is small, and high-order methods are essential for efficiency. Weakly turbulent blood flow, such as occurs in post-stenotic regions or subsequent to graft implantation, exhibits a much broader range of scales than does its laminar (healthy) counterpart and thus requires an order of magnitude increase in spatial and temporal resolution, making fast iterative solvers and parallel computing necessities.

The paper is organized as follows. Section 2 provides a brief overview of the governing equations, time advancement scheme, and spectral element method. Section 3 describes boundary condition treatment for simulating transition in bifurcation geometries. Section 4 presents parallel considerations and performance results, and Section 5 gives results for transitional flow in an arteriovenous graft model.

2. Navier-Stokes Discretization

We consider the solution of incompressible Navier-Stokes equations in Ω,

$$\frac{\partial \mathbf{u}}{\partial t} + \mathbf{u} \cdot \nabla \mathbf{u} = -\nabla p + \frac{1}{Re}\nabla^2 \mathbf{u}, \qquad \nabla \cdot \mathbf{u} = 0, \tag{1}$$

subject to appropriate initial and boundary conditions. Here, $\mathbf{u}$ is the velocity field, p is the pressure normalized by the density, and $Re = UD/\nu$ is the Reynolds number based on the characteristic velocity U, length scale D, and kinematic viscosity ν.

Our temporal discretization is based on a semi-implicit formulation in which the nonlinear terms are treated explicitly and the remaining linear Stokes problem is treated implicitly. We approximate the time derivative in (1) using a kth-order backwards difference formula (BDFk, k=2 or 3), which for k=2 reads

$$\frac{3\mathbf{u}^n - 4\mathbf{u}^{n-1} + \mathbf{u}^{n-2}}{2\Delta t} = S(\mathbf{u}^n) + NL^n. \tag{2}$$

Here, $\mathbf{u}^{n-q}$ represents the velocity at time t^{n-q}, $q = 0, \ldots, 2$, and $S(\mathbf{u}^n)$ is the linear symmetric Stokes operator that implicitly incorporates the divergence-free constraint. The term NL^n approximates the nonlinear terms at time level t^n and is given by the extrapolant $NL^n := -\sum_j \alpha_j \mathbf{u}^{n-j} \cdot \nabla \mathbf{u}^{n-j}$. For $k = 2$, the standard extrapolation would use $\alpha_1 = 2$ and $\alpha_2 = -1$. Typically, however, we use the three-term second-order formulation with $\alpha_1 = 8/3$, $\alpha_2 = -7/3$, and $\alpha_3 = 2/3$, which has a stability region that encompasses part of the imaginary axis. As an alternative to (2), we frequently use the operator-integration-factor scheme of Maday et al. [10] that circumvents the CFL stability constraints by setting $NL^n = 0$ and replacing the left-hand side of (2) with an approximation to the material derivative of $\mathbf{u}$. Both formulations yield unsteady Stokes problems of the form

$$\begin{aligned} \mathcal{H}\mathbf{u}^n - \nabla p^n &= \mathbf{f}^n \\ \nabla \cdot \mathbf{u}^n &= 0, \end{aligned} \quad (3)$$

to be solved implicitly. Here, $\mathcal{H}$ is the Helmholtz operator $\mathcal{H} := \left(\frac{3}{2\Delta t} - \frac{1}{Re}\nabla^2\right)$. In Section 3, we will formally refer to (3) in operator form $S_{us}(\mathbf{u}^n) = \mathbf{f}^n$. In concluding our temporal discretization overview, we note that we often stabilize high-Re cases by filtering the velocity at each step ($\mathbf{u}^n = F(\mathbf{u}^n)$), using the high-order filter described in [3,5].

Our spatial discretization of (3) is based on the $\mathbb{P}_N - \mathbb{P}_{N-2}$ spectral element method (SEM) of Maday and Patera [9]. The SEM is a high-order weighted residual approach similar to the finite element method (FEM). The primary distinction between the two is that typical polynomial orders for the SEM bases are in the range $N=4$ to 16—much higher than for the FEM. These high orders lead to excellent transport (minimal numerical diffusion and dispersion) for a much larger fraction of the resolved modes than is possible with the FEM. The relatively high polynomial degree of the SEM is enabled by the use of tensor-product bases having the form (in 2D)

$$\mathbf{u}(\mathbf{x}^e(r,s))|_{\Omega^e} = \sum_{i=0}^{N}\sum_{j=0}^{N} \mathbf{u}^e_{ij} h_i^N(r) h_j^N(s), \quad (4)$$

which implies the use of (curvilinear) quadrilateral (2D) or hexahedral (3D) elements. Here, $\mathbf{u}^e_{ij}$ is the nodal basis coefficient on element Ω^e; $h_i^N \in \mathbb{P}_N$ is the Lagrange polynomial based on the Gauss-Lobatto quadrature points, $\{\xi_j^N\}_{j=0}^N$ (the zeros of $(1-\xi^2)L'_N(\xi)$, where L_N is the Legendre polynomial of degree N); and $\mathbf{x}^e(r,s)$ is the coordinate mapping from $\hat{\Omega} = [-1,1]^d$ to Ω^e, for $d=2$ or 3. Unstructured data accesses are required at the global level (i.e., $e = 1, \ldots, E$), but the data is accessed in i-j-k form within each element. In particular, differentiation—a central kernel in operator evaluation—can be implemented as a cache-efficient matrix-matrix product. For example, $\mathbf{u}_{r,ij} = \sum_p \widehat{D}_{ip}\mathbf{u}_{pj}$, with $\widehat{D}_{ip} := h'_p(\xi_i)$ would return the derivative of (4) with respect to the computational coordinate r at the points (ξ_i, ξ_j). Differentiation with respect to $\mathbf{x}$ is obtained by the chain rule [1].

Insertion of the SEM basis (4) into the weak form of (3) and applying numerical quadrature yields the discrete unsteady Stokes system

$$H\underline{\mathbf{u}}^n - D^T \underline{p}^n = B\underline{\mathbf{f}}^n, \quad D\underline{\mathbf{u}}^n = 0. \quad (5)$$

Here, $H = \frac{1}{Re}A + \frac{3}{2\Delta t}B$ is the discrete equivalent of $\mathcal{H}$; $-A$ is the discrete Laplacian, B is the (diagonal) mass matrix associated with the velocity mesh, D is the discrete divergence operator, and $\underline{\mathbf{f}}^n$ accounts for the explicit treatment of the nonlinear terms.

The Stokes system (5) is solved approximately, using the kth-order operator splitting analyzed in [10]. The splitting is applied to the *discretized* system so that ad hoc boundary conditions are avoided. For $k=2$, one first solves

$$H\,\hat{\underline{\mathbf{u}}} = B\,\underline{\mathbf{f}}^n + D^T\,\underline{p}^{n-1}, \qquad (6)$$

which is followed by a pressure correction step

$$E\delta\underline{p} = -D\hat{\underline{\mathbf{u}}}, \qquad \underline{\mathbf{u}}^n = \hat{\underline{\mathbf{u}}} + \Delta t B^{-1} D^T \delta\underline{p}, \qquad \underline{p}^n = \underline{p}^{n-1} + \delta\underline{p}, \qquad (7)$$

where $E := \tfrac{2}{3}\Delta t D B^{-1} D^T$ is the Stokes Schur complement governing the pressure in the absence of the viscous term. Substeps (6) and (7) are solved with preconditioned conjugate gradient (PCG) iteration. Jacobi preconditioning is sufficient for (6) because H is strongly diagonally dominant. E is less well-conditioned and is solved either by the multilevel overlapping Schwarz method developed in [2,4] or more recent Schwarz-multigrid methods [8].

3. Boundary Conditions

Boundary conditions for the simulation of transition in vascular flow models present several challenges not found in classical turbulence simulations. As velocity profiles are rarely available, our usual approach at the vessel inflow is to specify a time-dependent Womersely flow that matches the first 20 Fourier harmonics of measured flow waveform. In some cases, it may be necessary to augment such clean profiles with noise in order to trigger transition at the Reynolds numbers observed in vivo. At the outflow, our standard approach is to use the natural boundary conditions (effectively, $p=0$ and $\frac{\partial \mathbf{u}}{\partial n}=0$) associated with the variational formulation of (3). This outflow boundary treatment is augmented in two ways for transitional vascular flows, as we now describe.

Fast Implicit Enforcement of Flow Division. Imposition of proper flow division (or flow split) is central to accurate simulation of vascular flows through bifurcations (sites prone to atherogenesis). The distribution of volumetric flow rate through multiple daughter branches is usually available through measured volume flow rates. A typical distribution in a carotid artery bifurcation, for example, is a 60:40 split between the internal and external carotid arteries. The distribution can be time-dependent, and the method we outline below is applicable to such cases. A common approach to imposing a prescribed flow split is to apply Dirichlet velocity conditions at one outlet and standard outflow (Neumann) conditions at the other. The Dirichlet branch is typically artificially extended to diminish the influence of spurious boundary effects on the upstream region of interest. Here, we present an approach to imposing arbitrary flow divisions among multiple branches that allows one to use Neumann conditions at each of the branches, thus reducing the need for extraordinary extensions of the daughter branches.

Our flow-split scheme exploits the semi-implicit approach outlined in the preceding section. The key observation is that the unsteady Stokes operator, which is treated implicitly and which controls the boundary conditions, is *linear* and that superposition therefore applies. Thus, if $\tilde{\mathbf{u}}^n$ satisfies $S_{us}(\tilde{\mathbf{u}}^n) = \mathbf{f}^n$ and $\tilde{\mathbf{u}}_0$ satisfies $S_{us}(\tilde{\mathbf{u}}_0) = 0$ but with different boundary conditions, then $\mathbf{u}^n := \tilde{\mathbf{u}}^n + \tilde{\mathbf{u}}_0$ will satisfy $S_{us}(\mathbf{u}^n + \tilde{\mathbf{u}}_0) = \mathbf{f}^n$

with boundary conditions $\mathbf{u}^n|_{\partial\Omega} = \tilde{\mathbf{u}}^n|_{\partial\Omega} + \tilde{\mathbf{u}}_0|_{\partial\Omega}$. With this principle, the flow split for a simple bifurcation (one common inflow, two daughter outflow branches) is imposed as follows. In a preprocessing step:

(*i*) Solve $S_{us}(\tilde{\mathbf{u}}_0) = 0$ with a prescribed inlet profile having flux $\tilde{Q} := \int_{\text{inlet}} \tilde{\mathbf{u}}_0 \cdot \mathbf{n}\, dA$, and no flow (i.e., homogeneous Dirichlet conditions) at the exit of one of the daughter branches. Use Neumann (natural) boundary conditions at the the other branch. Save the resultant velocity-pressure pair $(\tilde{\mathbf{u}}_0, \tilde{p}_0)$.

(*ii*) Repeat the above procedure with the role of the daughter branches reversed, and call the solution $(\tilde{\mathbf{u}}_1, \tilde{p}_1)$.

Then, at each timestep:

(*iii*) Compute $(\tilde{\mathbf{u}}^n, \tilde{p}^n)$ satisfying (3) with homogeneous Neumann conditions on each daughter branch and compute the associated fluxes $\tilde{Q}_i^n := \int_{\partial\Omega_i} \tilde{\mathbf{u}}^n \cdot \mathbf{n}\, dA$, $i = 0, 1$, where $\partial\Omega_0$ and $\partial\Omega_1$ are the respective active exits in (*i*) and (*ii*) above.

(*iv*) Solve the following for (α_0, α_1) to obtain the desired flow split $Q_0^n : Q_1^n$:

$$Q_0^n = \tilde{Q}_0^n + \alpha_0 \tilde{Q} \qquad \text{(desired flux on branch 0)} \qquad (8)$$
$$Q_1^n = \tilde{Q}_1^n + \alpha_1 \tilde{Q} \qquad \text{(desired flux on branch 1)} \qquad (9)$$
$$0 = \alpha_0 + \alpha_1 \qquad \text{(change in flux at inlet)} \qquad (10)$$

(*v*) Correct the solution by setting $\mathbf{u}^n := \tilde{\mathbf{u}}^n + \sum_i \alpha_i \tilde{\mathbf{u}}_i$ and $p^n := \tilde{p}^n + \sum_i \alpha_i \tilde{p}_i$.

Remarks. The above procedure provides a fully implicit iteration-free approach to applying the flow split that readily extends to a larger number of branches by expanding the system (8)–(10). Condition (10) ensures that the net flux at the inlet is unchanged and, for a simple bifurcation, one only needs to store the difference between the auxiliary solutions. We note that S_{us} is dependent on the timestep size Δt and that the auxiliary solutions $(\tilde{\mathbf{u}}_i, \tilde{p}_i)$ must be recomputed if Δt (or ν) changes. The amount of viscous diffusion that can take place in a single application of the unsteady Stokes operator is governed by Δt, and one finds that the auxiliary solutions have relatively thin boundary layers with a broad flat core. The intermediate solutions obtained in (*iii*) have inertia and so nearly retain the proper flow split, once established, such that the magnitude of α_i will be relatively small after just a few timesteps. It is usually a good idea to gradually ramp up application of the correction if the initial condition is not near the desired flow split. Otherwise, one runs the risk of having reversed flow on portions of the outflow boundary and subsequent instability, as discussed in the next section. Moreover, to accommodate the exit "nozzle" ($\nabla \cdot \mathbf{u} > 0$) condition introduced below, which changes the net flux out of the exit, we compute $\tilde{Q}_i^n$ at an upstream cross-section where $\nabla \cdot \mathbf{u} = 0$.

Turbulent Outflow Boundary Conditions. In turbulent flows, it is possible to have vortices of sufficient strength to yield a (locally) negative flux at the outflow boundary. Because the Neumann boundary condition does not specify flow characteristics at the exit, a negative flux condition can rapidly lead to instability, with catastrophic results. One way to eliminate incoming characteristics is to force the exit flow through a nozzle, effectively adding a mean axial component to the velocity field. The advantage of using a nozzle is that one can ensure that the characteristics at the exit point outward under a wide range of flow conditions. By constrast, schemes based on viscous buffer zones require knowledge of the anticipated space and time scales to ensure that vortical structures are adequately damped as they pass through the buffer zone.

Numerically, a nozzle can be imposed without change to the mesh geometry by imparting a positive divergence to the flow field near the exit (in the spirit of a supersonic nozzle). In our simulations, we identify the layer of elements adjacent to the outflow and there impose a divergence $D(\mathbf{x})$ that ramps from zero at the upstream end of the layer to a fixed positive value at the exit. Specifically, we set $D(\mathbf{x}) = C[1 - (x_\perp/L_\perp)^2]$, where $x_\perp$ is the distance normal to the boundary and $L_\perp$ is maximum thickness of the last layer of elements. By integrating the expression for D from $x_\perp/L_\perp=1$ to 0, one obtains the net gain in mean velocity over the extent of the layer. We typically choose the constant C such that the gain is equal to the mean velocity prior to the correction.

Results for the nozzle-based outflow condition are illustrated in Fig. 1. The left panel shows the velocity field for the standard (uncorrected Neumann) condition near the outflow boundary of an internal carotid artery at $Re \approx 1400$ (based on the peak flow rate and stenosis diameter). Inward-pointing velocity vectors can be seen at the exit boundary, and the simulation becomes catastrophically unstable within 100 timesteps beyond this point. The center panel shows the flow field computed with the outflow correction. The flow is leaving the domain at all points along the outflow boundary and the simulation is stable for all time. The difference between the two cases (right) shows that the outflow treatment does not pollute the solution upstream of the boundary.

4. Parallel Performance

Our approach to parallelization is based on standard SPMD domain decomposition approaches, as discussed in [1,13]. Elements are distributed across processors in contiguous subgroups determined by recursive spectral bisection [11] and nearest neighbor data is exchanged with each matrix-vector product required for the iterative solvers. (See [1].)

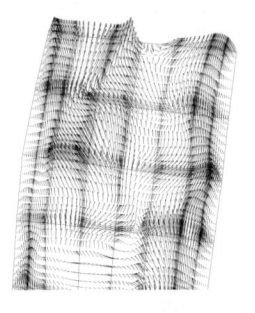

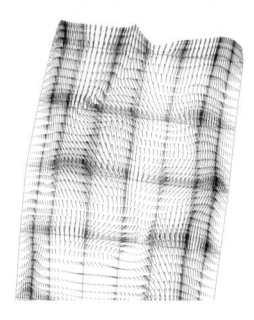

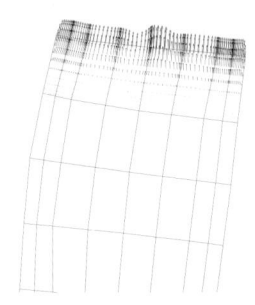

Figure 1. Velocity vectors near the outflow of an internal carotid artery: (left) uncorrected, (center) corrected, and (right) corrected-uncorrected.

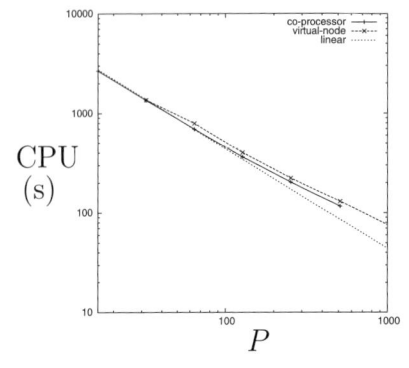

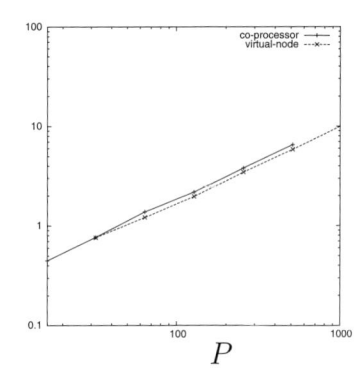

Figure 2. (left) CPU time for E=2640, N=10 for P=16–1024 using coprocessor and virtual-node modes on BGL; (right) percentage of time spent in the coarse-grid solve.

The only other significant communication arises from inner products in the PCG iteration and from the coarse-grid problem associated with the pressure solve.

The development of a fast coarse-grid solver for the multilevel pressure solver was central to efficient scaling to $P > 256$ processors. The use of a coarse-grid problem in multigrid and Schwarz-based preconditioners ensures that the iteration count is bounded independent of the number of subdomains (elements) [12]. The coarse-grid problem, however, is communication intensive and generally not scalable. We employ the projection-based coarse-grid solver developed in [14]. This approach has a communication complexity that is sublinear in P, which is a significant improvement over alternative approaches, which typically have communication complexities of $O(P \log P)$.

Figure 2 (left) shows the CPU time vs. P on the IBM BGL machine at Argonne for 50 timesteps for the configuration of Fig. 3. For the P=1024 case, the parallel efficiency is $\eta = .56$, which is respectable for this relatively small problem ($\approx$ 2600 points/processor). The percentage of time spent in the coarse-grid solver is seen in Fig. 2 (right) to be less than 10 percent for $P \leq 1024$. BGL has two processors per node that can be used in coprocessor (CO) mode (the second processor handles communication) or in virtual-node (VN) mode (the second processor is used for computation). Typically about a 10% overhead is associated with VN-mode. For example, for $P = 512$, we attain $\eta = .72$ in CO-mode and only $\eta = .64$ in VN-mode. Obviously, for a given P, VN-mode uses half as many resources and is to be preferred.

5. Transition in an Arteriovenous Graft

Arteriovenous (AV) grafts consist of a $\sim$15 cm section of 6 mm i.d. synthetic tubing that is surgically implanted to provide an arterial-to-vein round-the-clock short circuit. Because they connect a high-pressure vessel to a low-pressure one, high flow rates are established that make AV-grafts efficient dialysis ports for patients suffering from poor kidney function. The high speed flow is normally accompanied by transition to a weakly turbulent state, manifested as a 200–300 Hz vibration at the vein wall [6,7]. This high-

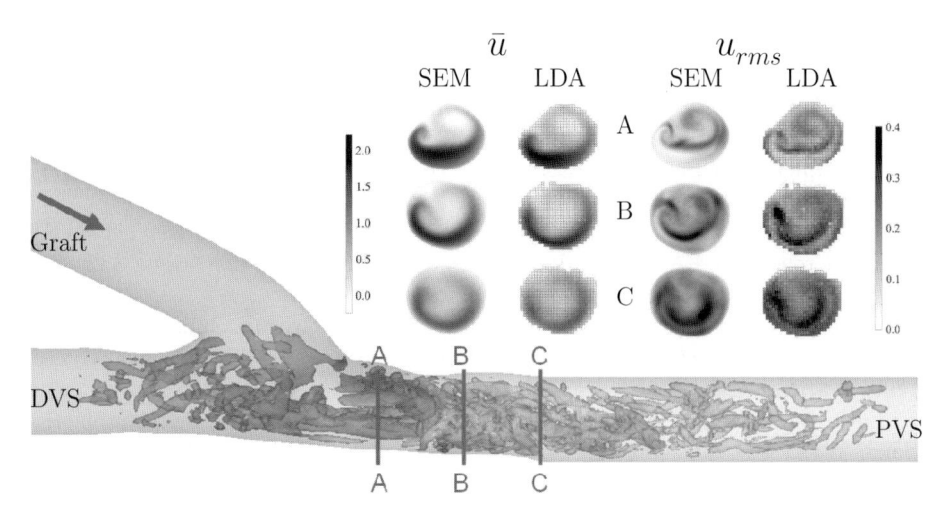

Figure 3. Transitional flow in an AV graft at $Re = 1200$: (bottom) coherent structures; (inset) mean and rms velocity distributions (m/s) at A–C for CFD and LDA measurements. The mesh comprized 2640 elements of order $N = 12$ with $\Delta t = 5 \times 10^6 s$.

frequency excitation is thought to lead to intimal hyperplasia, which can lead to complete occlusion of the vein and graft failure within six months of implant. We are currently investigating the mechanisms leading to transition in subject-specific AV-graft models with the aim of reducing turbulence through improved geometries. Figure 3 shows a typical turbulent case when there is a 70:30 split between the proximal venous segment (PVS) and distal venous segment (DVS). The SEM results, computed with 2640 elements of order 12 (4.5 million gridpoints), are in excellent agreement with concurrent laser Doppler anemometry results. The statistics are based on 0.5 seconds of (in vivo) flow time, which takes about 100,000 steps and 20 hours of CPU time on 2048 processors of BGL.

Acknowledgments

This work was supported by the National Institutes of Health, RO1 Research Project Grant (2RO1HL55296-04A2), by Whitaker Foundation Grant (RG-01-0198), and by the Mathematical, Information, and Computational Sciences Division subprogram of the Office of Advanced Scientific Computing Research, U.S. Department of Energy, under Contract W-31-109-Eng-38.

REFERENCES

1. M.O. Deville, P.F. Fischer, and E.H. Mund, *High-order methods for incompressible fluid flow*, Cambridge University Press, Cambridge, 2002.
2. P.F. Fischer, *An overlapping Schwarz method for spectral element solution of the incompressible Navier-Stokes equations*, J. Comput. Phys. **133** (1997), 84–101.

3. P.F. Fischer, G.W. Kruse, and F. Loth, *Spectral element methods for transitional flows in complex geometries*, J. Sci. Comput. **17** (2002), 81–98.
4. P.F. Fischer, N.I. Miller, and H.M. Tufo, *An overlapping Schwarz method for spectral element simulation of three-dimensional incompressible flows*, Parallel Solution of Partial Differential Equations (Berlin) (P. Bjørstad and M. Luskin, eds.), Springer, 2000, pp. 158–180.
5. P.F. Fischer and J.S. Mullen, *Filter-based stabilization of spectral element methods*, Comptes rendus de l'Académie des sciences, Série I- Analyse numérique **332** (2001), 265–270.
6. S.W. Lee, P.F. Fischer, F. Loth, T.J. Royston, J.K. Grogan, and H.S. Bassiouny, *Flow-induced vein-wall vibration in an arteriovenous graft*, J. of Fluids and Structures (2005).
7. F. Loth, N. Arslan, P. F. Fischer, C. D. Bertram, S. E. Lee, T. J. Royston, R. H. Song, W. E. Shaalan, and H. S. Bassiouny, *Transitional flow at the venous anastomosis of an arteriovenous graft: Potential relationship with activation of the ERK1/2 mechanotransduction pathway*, ASME J. Biomech. Engr. **125** (2003), 49–61.
8. J. W. Lottes and P. F. Fischer, *Hybrid multigrid/Schwarz algorithms for the spectral element method*, J. Sci. Comput. **24** (2005).
9. Y. Maday and A.T. Patera, *Spectral element methods for the Navier-Stokes equations*, State-of-the-Art Surveys in Computational Mechanics (A.K. Noor and J.T. Oden, eds.), ASME, New York, 1989, pp. 71–143.
10. Y. Maday, A.T. Patera, and E.M. Rønquist, *An operator-integration-factor splitting method for time-dependent problems: Application to incompressible fluid flow*, J. Sci. Comput. **5** (1990), 263–292.
11. A. Pothen, H.D. Simon, and K.P. Liou, *Partitioning sparse matrices with eigenvectors of graphs*, SIAM J. Matrix Anal. Appl. **11** (1990), 430–452.
12. B. Smith, P. Bjørstad, and W. Gropp, *Domain decomposition: Parallel multilevel methods for elliptic PDEs*, Cambridge University Press, Cambridge, 1996.
13. H.M. Tufo and P.F. Fischer, *Terascale spectral element algorithms and implementations*, Proc. of the ACM/IEEE SC99 Conf. on High Performance Networking and Computing (IEEE Computer Soc.), 1999, p. CDROM.
14. _____, *Fast parallel direct solvers for coarse-grid problems*, J. Parallel Distrib. Comput. **61** (2001), 151–177.

Large-Scale Multidisciplinary Computational Physics Simulations Using Parallel Multi-Zone Methods

Ding Li, Guoping Xia, and Charles L. Merkle

Mechanical Engineering, Purdue University, Chaffee Hall, 500 Allison Road, West Lafayette, IN 47907-2014

Keywords: Parallel Computing; Multiple physics computations; Linux Cluster; Multi-Physics Zone Method

Abstract

A parallel multi-zone method for the simulation of large-scale multidisciplinary applications involving field equations from multiple branches of physics is outlined. The equations of mathematical physics are expressed in a unified form that enables a single algorithm and computational code to describe problems involving diverse, but closely coupled, physics. Efficient parallel implementation of these coupled physics must take into account the different number of governing field equations in the various physical zones and the close coupling inside and between regions. This is accomplished by implementing the unified computational algorithm in terms of an arbitrary grid and a flexible data structure that allows load balancing by sub-clusters.

1. Introduction

High fidelity computational simulations of complex physical behavior are common in many fields of physics such as structures, plasma dynamics, fluid dynamics, electromagnetics, radiative energy transfer and neutron transport. Detailed three-dimensional simulations in any one of these fields can tax the capabilities of present-day parallel processing, but the looming challenge for parallel computation is to provide detailed simulations of systems that couple several or all of these basic physics disciplines into a single application.

In the simplest multidisciplinary problems, the physics are loosely coupled and individual codes from the several sub-disciplines can be combined to provide practical solutions. Many applications, however, arise in which the multidisciplinary physics are so intensely coupled that the equations from the various sub-branches of physics must likewise be closely coupled and solved simultaneously. This implies that the computational algorithm, data structure, message passing and load balancing steps must all be addressed simultaneously in conjunction with the physical aspects of the problem. In the present paper we outline a method for dealing with such multi-physics problems with emphasis on the computational formulation and parallel implementation. The focus is on applications that involve closely coupled physics from several or all of the sub-domains listed above. Because conservation laws expressed as field variable solutions to partial differential equations are central in all of these sub-domains of physics, the formulation is based upon a generalized implementation of the partial differential equations that unifies the various physical phenomena.

2. Computer Framework

2.1. Conservation Laws for Field Variables

The fundamental phenomena in nearly all fields of mathematical physics are described in terms of a set of coupled partial differential equations (pde's) augmented by a series of constitutive algebraic relations that are used to close the system. The pde's typically describe basic conservation relations for quantities such as mass, momentum, energy and electrical charge. In general, these pde's involve three types of vector operators, the curl, the divergence and the gradient that appear individually or in combination. When they appear alone, they typically represent wave phenomena, while when they appear in combination (as the div-grad operator) they typically represent the effects of diffusion. An important feature in simulating multi-physical phenomena is that the structure of the conservation relations is essentially parallel for all branches of mathematical physics.

In contrast to the conservation equations, the constitutive relations are most often algebraic in nature. Constitutive relations are used to relate thermodynamic variables through appropriate thermal and caloric equations of state and pertinent property relations; to relate electric and/or magnetic fields to currents, and to relate stresses and strains to velocity and displacement. The partial differential character of the conservation relations imply that these pde's set the global structure of the computational algorithm and the code, while the algebraic nature of the constitutive relations implies that this auxiliary data can be provided in subroutine fashion as needed, but does not impact the global structure of the code. These fundamental concepts provide much insight into parallel implementations as well.

Mathematically, the conservation relations for a general branch of mathematical physics may be written as a generic set of partial differential equations of the form:

$$\frac{\partial Q}{\partial t} + \nabla \bullet \vec{F}_D + \nabla \times \vec{F}_C + \nabla \Phi = 0 \qquad (1)$$

where Q and Φ are column vectors in the conservation system and $\vec{F}_D$ and $\vec{F}_C$ are tensors with similar column length whose rows correspond to the number of dimensions. (The subscripts, D and C, refer to 'Divergence' and 'Curl' tensors respectively.) Various components of these variables may be null. In writing these expressions, we have defined a generalized curl operator that applies to vectors of length larger than three. An important issue for computational purposes is that the length of the primitive variables vector can vary dramatically among different fields of physics. For example, in solids where mechanical and thermal stresses are to be determined, Q_p, will generally include three displacements, three velocities and one temperature for a total of seven equations. In simpler solid applications where only heat conduction is included, only one partial differential equation need be solved and Q_p contains only one component. In simple fluid dynamics problems, the variables will include three velocity components, the pressure and the temperature for a total of five components, but in more complex applications there may be as many as six additional partial differential equations for turbulence modeling and an essentially unlimited number of species continuity equations and phasic equations to describe finite rate chemical reactions and phase change. In electromagnetics, there are typically two vectors of length three for a total of six, while when the MHD approximation is used it is possible to get by with three. Finally, for radiation, the number of equations can vary from one three-dimensional equation to coupled six-dimensional equations.

2.2. Numerical Discretization and Solution Procedure

Partial differential equations must be discretized before they can be solved numerically. Because the conservation laws from nearly all branches of physics form matrices are wide-banded and, in addition, are nonlinear, iterative methods must be used to solve the resulting discretized systems. To accomplish the discretization and to define an appropriate iterative procedure, we add a pseudo-time derivative to the space-time conservation relations. The discretization procedure is then performed by integrating over a series of control volumes of finite size to obtain an integral relation of the form:

$$\Gamma \frac{\partial Q_p}{\partial \tau} \Omega + \int_\Omega \frac{\partial Q}{\partial t} d\Omega + \int_\Omega \nabla \bullet \overline{F}_D d\Omega + \int_\Omega \nabla \times \overline{F}_C d\Omega + \int_\Omega \nabla \Phi d\Omega = 0 \qquad (2)$$

By invoking the theorems of Green, Stokes and Gauss, the volume integrals can be written as surface integrals,

$$\Gamma \frac{\partial Q_p}{\partial \tau} \Omega + \frac{\partial}{\partial t} \left(\int_\Omega Q d\Omega \right) + \int_{\partial \Omega} \overline{n} \bullet \overline{F}_D d\Sigma + \int_{\partial \Omega} \overline{n} \times \overline{F}_C d\Sigma + \int_{\partial \Omega} \overline{n} \Phi d\Sigma = 0 \qquad (3)$$

The surface integrals require the specification of a 'numerical' flux across each face of the selected control volume and indicate that the rate of change of the primitive variables in pseudo time is determined by the sum of fluxes across the several faces of each control volume. An upwind scheme is employed to evaluate the numerical flux across the faces. The curl, divergence and gradient operators each generate unique flux functions at the faces. A key issue in the discretization and in the application to a multi-disciplinary procedure is that discretized expressions must be defined for each of the three vector operators. As a part of the discretization step, the pseudo-time can also be used to define the numerical flux across each face of the control volume.

In addition to defining the discretized equations, the coefficient matrix, Γ, in the pseudo time term also introduces an artificial property procedure that allows the eigenvalues of the convergence process to be properly conditioned thereby providing an efficient convergence algorithm to handle different time scale problems.

The introduction of the integral formulation for the discretized equation system allows the use of an arbitrary, structured/unstructured grid capability to enable applications to complex geometry. Specific data and code structures are implemented in a fashion that mimics the conventional mathematical notations given above and the corresponding operations for tensors, vectors and scalar functions. To allow for different numbers of conservation equations in different problems, the number of equations is chosen at input. In addition, individual problems which contain multiple zones in which different conservation equations must be solved are often encountered. The computational code that incorporates these general equations, the arbitrary mesh and the multiple physical zones is referred to as the General Equation and Mesh Solver (GEMS) code.

2.3. GEMS: General Equation and Mesh Solver

The GEMS code uses contemporary numerical methods to solve coupled systems of partial differential equations and auxiliary constitutive relations for pertinent engineering field variables (Fig. 1) on the basis of a generalized unstructured grid format. After converting the pertinent conservation equations from differential to integral form as noted above, the spatial discretization is accomplished by a generalized Riemann approach for convective terms and a Galerkin approach for diffusion terms. The numerical solution of these equations is then obtained by employing a multi-level pseudo-time marching algorithm that controls artificial dissipation and anti-diffusion for maximum accuracy, non-linear convergence effects at the outset of a computation and convergence efficiency in linear regimes. The multi-time formulation has been adapted to handle convection or diffusion dominated problems with similar effectiveness so that radically different field equations can be handled efficiently by a single algorithm. The solution algorithm complements the conservation equations by means of generalized constitutive relations such as arbitrary thermal and caloric equations of state for fluids and solution-dependent electrical and thermal conductivity for fluids and solids. For multi-disciplinary problems, GEMS divides the computational domain into distinct 'zones' to provide flexibility, promote load balancing in parallel implementations, and to ensure efficiency. The details of these techniques are given in the next section.

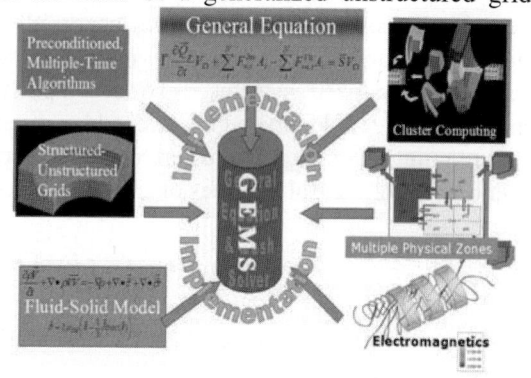

Fig. 1. Components of GEMS code

2.4. Multi-Physics Zone Method

In practical applications, we often face problems involving multiple media in which the pertinent phenomena are governed by different conservation laws. For such applications, we define a physics zone as a domain that is governed by a particular set (or sets) of conservation equations. For example, in a conjugate heat transfer problem, there are two physics zone. The continuity, momentum and energy equations are solved in the 'fluid' zone, while only the energy equation is solved in the 'solid' zone. Similarly, MHD problems can be divided into four physical zones, the fluid, electric conductor, dielectric rings and surrounding vacuum zones (see figure 2). The continuity, momentum, energy, species and magnetic diffusion equations are solved in the fluid zone; the energy and magnetic diffusion equations are solved in the electric conductor and dielectric solid zones, and only the magnetic diffusion equation is solved in the vacuum zone. In an arc-heater problem, the inner plasma region in which fluids, radiation and electromagnetics co-exist would be one zone; the heater walls in which electromagnetics and conjugate heat transfer are desired would be a second zone, and the external environment where only the EM equations are solved would be a third zone. To accomplish this effect, the number (and type) of equations to be solved in each zone is an input quantity. This zonal approach provides economies in terms of machine storage and CPU requirements while also simplifying load balancing. Regions with larger numbers of conservation equations are distributed to more processors and allocated more storage elements per cell, and etc.

This division into zones coupled with the unstructured grid makes load-balancing on parallel computers quite straightforward. The complete computational domain can be subdivided into several sub zones each of which is computed on a separate processor. In order to optimize the parallel computing time, each processor loading has to be balanced. Because each physics zone has different grid numbers and also different numbers of equations, the combination of the number of equations and the number of grids has to be balanced and optimized. The interface between multiple physics zones is treated as internal and external boundary conditions. Sharing information between two physics zones satisfies the internal boundary conditions while the external boundary conditions are treated as normal boundary conditions. The interface between processors in each sub cluster for a given physics zone has the same unknown variables and is treated as a normal inter-processor boundary. (Note that this physical zone definition is different from a multiple block grid of the type obtained from structured grid generators. Normally, all grid zones in a multiple block computation have the same conservation equations. There could be multiple grid blocks in each physical zone shown in Figure 3).

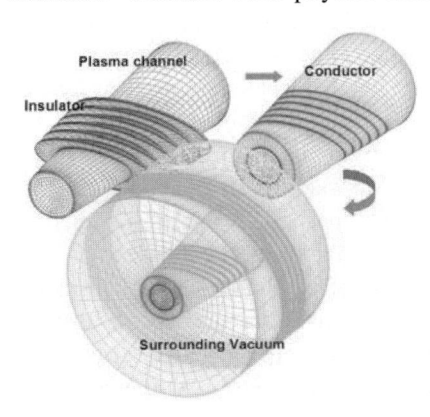

Fig. 2. Four physical zones in MHD power generator

3. Parallel Approach

As a parallel program GEMS must communicate between different processors. Our multi-physics zone method uses a fully implicit algorithm that is highly coupled inside each processor while loosely coupled between processors so that only small parts of data adjacent to the interface between two partitions need to be communicated between processors (see Fig. 3). A parallel point-to-point technique is used to reduce host control time and traffic between the nodes and also to handle any number of nodes in the cluster architecture. This technique should not introduce significant overhead to the computation until the number of processors in the cluster is increased to over 100. With the emergence of massively parallel computing architectures with potential for teraflop performance, any code development activity must effectively utilize the computer architecture in achieving the proper load balance with minimum inter-nodal data communication. The massively parallel processing has been implemented in GEMS for cross disciplines such as computational fluid dynamics, computational structural dynamics and computational electromagnetic simulations for both structured grid and unstructured grid arrangements. The hybrid unstructured grid-based finite-volume GEMS code was developed and optimized for distributed memory parallel architectures. The code handles inter-processor communication and other functions unique to the parallel implementation using MPI libraries. Very few MPI calls are used in GEMS code due to well defined data structure of shared data that is detailed later. Only the *mpi_sendrecv* subroutine is used for sending and receiving data between processors to update interface information after each iteration. GEMS loads mesh and other data into a master processor and then distributes this data to appropriate processors thereby making the code portable and flexible.

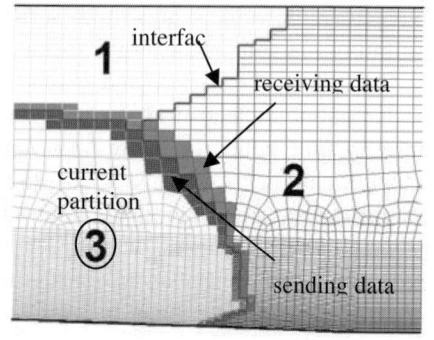

Fig. 3. Diagram of interface between

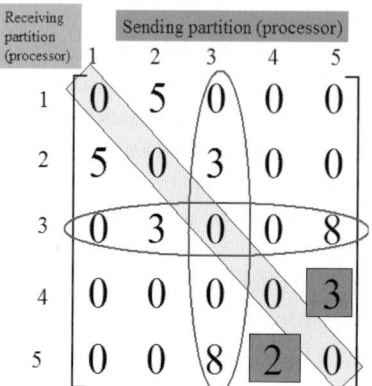

Fig. 4. The exchanging prototype matrix for sending and receiving data

The efficiency of the parallel methods considered in GEMS is rests upon the storage scheme used for the data shared between cluster nodes. Any single processor needs to send data to several other nodes while receiving data from an alternative array of nodes. The magnitude of the data sent from any processor is not necessarily equal to the amount it receives. To manage these inter-processor communications, we designed an exchanging prototype matrix (see figure 4.). In the figure 4 the index of the row represents the index of the sending processor while the index of the column represents the index of the receiving processor. A zero element in the matrix implies there is no communication between the two subject processors represented by the row and column indexes while nonzero elements of the matrix represent the number of data packets sent to the row index processor from the column index processor. The sum of row-wise elements is the total number of data received by the row index processor while the sum of columns-wise elements is the total number of data sent to the column index processor. The diagonal elements of the matrix are always equal to zero as there is no data sent or received inside the processor.

Having set up this communications matrix, we can now use effective compressed row storage (CRS) format to collect and pack data into a contiguous pointer array sorted in the order of the processors. The pointer array physically links to the storage in memory for the physical locations of those data. This exchange data structure is necessary to achieve the point-to-point message passing operation. In the GEMS code, two data pointer stacks, sending and receiving, are allocated to collect the exchange data between processors (Fig. 5). Each stack includes two arrays in CRS format: one is an index and the other is data. The index array stores the amount of data that is sent to individual processors that also indicate the current node has data to send to the node with non-zero value. Otherwise, there are no passing operation tokens.

Figure 6 shows the total wall clock time and the wall clock time per cell and per iteration vs. the number of processors for a two-dimensional hypersonic fluid flow calculation with about a half million cells. The computational domain is partitioned to 5, 10, 20 and 40 partitions respectively and tested in our SIMBA and MacBeth cluster. The SIMBA linux cluster built in early 2001 has a head node with 50 slave nodes which has a single Intel Pentium 4 1.8 Ghz CPU with 1 Gb memory and 10/100 BaseT network adapter. The MacBeth cluster is an Opteron cluster with 100 nodes, each of which has a dual AMD Opteron 24 1.6 GHz CPU, with 4 Gb of memory which we are just bringing on line. These processors are connected by high performance non-blocking infiniband network switches for low latency and high speed. We know the wall clock time is decreased when the number of processors increases while the average value of wall clock time of each cell and iteration (*wtime*) should be constant in the ideal case (in the absence of communication costs or other system operations). In the SIMBA cluster when the number of processors is less than 10, the *wtime* is almost constant while as the number of processors are increased above 20 the *wtime* is increased by about (*wtime*_30-*wtime*_10)/*wtime*_10=1.5% while there are almost no changes in MacBeth. The MacBeth cluster is also three time faster than SIMBA.

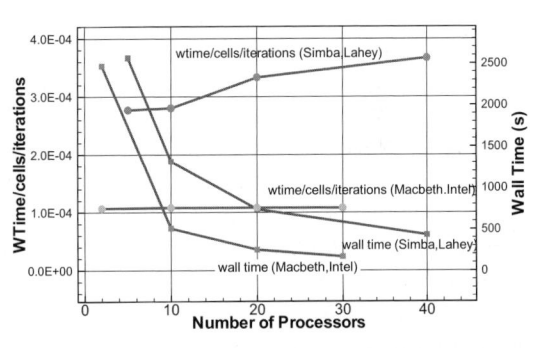

Fig. 6. Wall clock time and wall clock time per cell per iteration vs. number of processor

4. Representative Applications

The multi-physics zone GEMS code has been successfully applied to a variety of applications including a trapped vortex combustor (TVC) with liquid fuel spray, an MHD power generator with a plasma channel enclosed by dielectric/conductor walls of sandwich construction and the surrounding atmosphere in which the magnetic field decays to zero, the conjugate heat transfer in a rocket engine combustor and a combined pulsed detonation combustion with unsteady ejectors in operating in cyclic fashion. Results from a pulsed combustion turbine system are outlined below.

Constant volume combustion has the potential to provide substantial performance improvements in air-breathing propulsion systems although these improvements are difficult to realize in practical systems. In the present example, we look at a series of pulsed detonation tubes as a possible means for implementing constant volume combustion in a gas turbine system. The analysis involves both reacting and non-reacting flow solutions to the Navier-Stokes equations using the GEMS code that enables generalized fluids and generalized grids. The pressure contours of an unsteady, three-dimensional constant volume combustor are shown in Fig. 7 along with detailed diagnostics of the flow through a single PDE tube that is combined with a straight-tube ejector. The latter solution is compared with experiment to validate the PDE simulations.

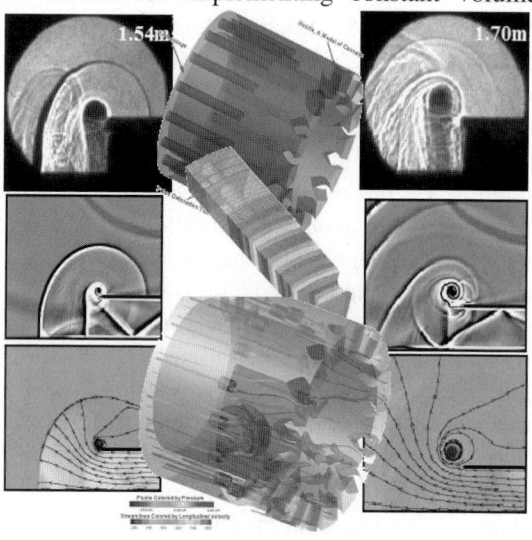

Fig. 11. Constant volume hybrid combustion system for turbine engine application

References

D. Li, "A User's Guide to GEMS," Internal Report, 2002

D. Li, S. Venkateswaran, J. Lindau and C. L. Merkle, "A Unified Computational Formulation for Multi-Component and Multi-Phase Flows," 43rd AIAA Aerospace Sciences Meeting and Exhibit, AIAA 2005-1391, Reno, NV, January 10-13, 2005.

G, Xia, D. Li, and C. L., Merkle, "Modeling of Pulsed Detonation Tubes in Turbine Systems," 43rd AIAA Aerospace Sciences Meeting and Exhibit, AIAA 2005-0225, Jan. 10-13, 2005.

Li, D., Keefer, D., Rhodes, R. Kolokolnikov, K., Merkle, C.L., Thibodeaux, R., "Analysis of Magnetohydrodynamic Generator Power Generation," Journal of Propulsion and Power, Vol. 21, No. 3, March, 2005, pp 424-432.

D. Li and C. L. Merkle, "Analysis of Real Fluid Flows in Converging Diverging Nozzles," AIAA-2003-4132, July 2003.

P.S. Pacheco, "A User's Guide to MPI," Dept. of Mathematics, University of San Francisco, San Francisco, CA 94117

Parallelization of Phase-Field Model for Phase Transformation Problems in a Flow Field

Ying Xu[a], J. M. McDonough[b] and K. A. Tagavi[a]

[a]Department of Mechanical Engineering
University of Kentucky, Lexington, KY 40506-0503

[b]Departments of Mechanical Engineering and Mathematics
University of Kentucky, Lexington, KY 40506-0503

We implement parallelization of a phase-field model for solidification in a flow field using OpenMP and MPI to compare their parallel performance. The 2-D phase-field and Navier–Stokes equations are presented with prescribed boundary and initial conditions corresponding to lid-driven-cavity flow. Douglas & Gunn time-splitting is applied to the discrete governing equations, and a projection method is employed to solve the momentum equations. Freezing from a supercooled melt, nickel, is initiated in the center of the square domain. The approach taken to parallelize the algorithm is described, and results are presented for both OpenMP and MPI with the latter being decidedly superior.

1. INTRODUCTION

Phase-field models have been applied to simulation of phase transformation problems for decades. Initially most researchers focused on pure substances in the 2-D case and did not consider convection induced by a velocity field or small-scale fluctuations. But phase transitions in binary alloys and solidification in the presence of convection has attracted increasing interest in recent studies. Anderson et al. [1] derived a phase-field model with convection and gave some simple examples of equilibrium of a planar interface, density-change flow and shear flow based on the model they obtained; they also applied the sharp-interface asymptotics analysis to the phase-field model with convection [2]. Beckermann et al. [3] provided a phase-field model with convection using the same kind of volume or ensemble averaging methods. They also presented more complex examples such as simulation of convection and coarsening in an isothermal mush of a binary alloy and dendritic growth in the presence of convection; phase-field and energy equations were solved using an explicit method in this research. Al-Rawahi and Tryggvason [4] simulated 2-D dendritic solidification with convection using a somewhat different method based on front tracking.

We remark that all previous research on solidification in a flow field involved length and time scales in microns and nanoseconds, respectively, which is not appropriate for studies of freezing in typical flow fields such as rivers and lakes, or industrial molds and castings. Therefore, the concepts of multiscale methods should be introduced to phase-field models with convection. Even such formulations still have the drawback of being very CPU intensive. Hence, it is necessary to apply parallelization to phase-field model algorithms in order to decrease wall-clock time to within practical limits.

We introduce the phase-field model with convection in the first part of this paper, followed by numerical solutions corresponding to growth of dendrites in a supercooled melt of nickel. Finally we discuss the approach to parallelization and the speedups obtained. Parallelization via both OpenMP and MPI has been implemented on a symmetric multiprocessor; the latter is found to be significantly more effective but required considerably more programming effort.

2. GOVERNING EQUATIONS OF PHASE-FIELD MODEL WITH CONVECTION

In this section we introduce the equations of the phase-field model including effects of convective transport on macroscopic scales. The difference in length scales between dendrites and flow field make it unreasonable to implement a dimensionless form of the governing equations since there is no single appropriate length scale. Therefore, dimensional equations are employed. Boundary and initial conditions required to formulate a well-posed mathematical problem are also prescribed.

The coupled 2-D Navier–Stokes equations and phase-field model are

$$u_x + v_y = 0, \tag{1a}$$

$$u_t + (u^2)_x + (uv)_y = -\frac{1}{\rho_0}p_x + \frac{\mu(\phi)}{\rho_0}\Delta u + X_1(\phi), \tag{1b}$$

$$v_t + (uv)_x + (v^2)_y = -\frac{1}{\rho_0}p_y + \frac{\mu(\phi)}{\rho_0}\Delta v + X_2(\phi) - \frac{[\rho(\phi,T) - \rho_L]}{\rho_0}g, \tag{1c}$$

$$\phi_t + (u\phi)_x + (v\phi)_y = \frac{\epsilon^2}{M}\nabla \cdot ((\boldsymbol{\xi} \cdot \nabla\phi)\boldsymbol{\xi}) - \frac{30\rho_0 L_0}{T_m M}\psi(\phi)(T_m - T)$$
$$- \frac{\rho_0}{aM}\psi'(\phi)T + Y(\phi, T), \tag{1d}$$

$$T_t + (uT)_x + (vT)_y = \frac{k}{\rho_0 c_p(\phi)}\Delta T + \left[\frac{\epsilon^2}{2}\nabla \cdot ((\boldsymbol{\xi} \cdot \nabla\phi)\boldsymbol{\xi}) - \frac{30 L_0 \psi(\phi)}{c_p(\phi)}\right]\frac{D\phi}{Dt}$$
$$+ W(u,v,\phi,T), \tag{1e}$$

to be solved on a bounded domain $\Omega \subseteq \mathbb{R}^2$. In these equations coordinate subscripts x, y, t denote partial differentiation, and ∇ and Δ are gradient and Laplace operators in the coordinate system imposed on Ω; D/Dt is the usual material derivative. Here u and v are velocities in x and y directions; p, T, ϕ are gauge pressure, temperature and phase field variable, respectively; L_0 is the latent heat per unit mass at the melting temperature T_m, and k is thermal conductivity; ρ_0 is the reference density. In our computations, density ρ, dynamic viscosity μ and specific heat c_p are no longer constants; they are constant in each bulk phase, but are functions of the phase-field variable over the thin interface separating the bulk phases. In addition, since the Boussinesq approximation is used to represent the buoyancy force term in Eq. (1c), density is also a function of temperature. We express the density, dynamic viscosity and specific heat in the forms:

$$\rho(\phi, T) = \rho_S + P(\phi)[\rho_L - \rho_S + \beta\rho_L(T - T_m)],$$
$$\mu(\phi) = \mu_S + P(\phi)(\mu_L - \mu_S),$$
$$c_p(\phi) = c_{pS} + P(\phi)(c_{pL} - c_{pS}),$$

where subscripts S and L denote solid and liquid respectively, and β is the coefficient of thermal volumetric expansion. $\psi(\phi)$ is a double-well potential, and $P(\phi)$ is a polynomial introduced to denote the interface; these functions are commonly given as polynomials in ϕ:

$$\psi(\phi) = \phi^2(1-\phi)^2,$$
$$P(\phi) = 6\phi^5 - 15\phi^4 + 10\phi^3.$$

Other parameters in Eqs. (1) are

$$\epsilon^2 = 6\sqrt{2}\sigma\delta, \qquad a = \frac{\rho_0 T_m \delta}{6\sqrt{2}\sigma}, \qquad M = \frac{\rho_0 L_0 \delta}{T_m \mu_k},$$

where ϵ^2 is the positive gradient coefficient related to the interfacial thickness δ in such a way that as $\delta \to 0$, the phase-field model approaches the modified Stefan model; μ_k is the kinetic coefficient, and M is the mobility which is related to the inverse of kinetic coefficient; a is a positive parameter occuring in the double-well potential which is related to the surface tension σ. Kinetic coefficient μ_k is a microscopic physical parameter reflecting kink density at steps in solid thickness and the atom exchange rate at each kink, as explained in Chernov [5] and Ookawa [6]. It is a function of temperature and orientation, and also depends on the material; the kinetic coefficient of metal is the highest among all materials. Linear variation of kinetic coefficient might be reasonable for a molecularly rough interface; however, it strongly depends on orientation of the interface for facetted interfaces, as shown by Langer [7]. For simplicity, the kinetic coefficient is usually assumed to be a constant as we do herein, but it is difficult to determine the value of this constant either theoretically or experimentally.

The forcing terms in Eqs. (1) take the forms

$$X_1(\phi) = -\frac{\epsilon^2}{\rho_0}\phi_x \left(\xi_1^2 \phi_{xx} + 2\xi_1\xi_2 \phi_{xy} + \xi_2^2 \phi_{yy}\right), \tag{5a}$$

$$X_2(\phi) = -\frac{\epsilon^2}{\rho_0}\phi_y \left(\xi_1^2 \phi_{xx} + 2\xi_1\xi_2 \phi_{xy} + \xi_2^2 \phi_{yy}\right), \tag{5b}$$

$$Y(\phi, T) = \frac{30\,p}{\rho_0 M}\psi(\phi)[\rho_L - \rho_S + \beta\rho_L(T - T_m)], \tag{5c}$$

$$W(u, v, \phi, T) = \frac{\mu(\phi)}{\rho_0 c_p(\phi)}\left[2u_x^2 + v_y^2 + (u_y + v_x)^2\right] + \frac{\epsilon^2}{4\rho_0 c_p(\phi)}\left(\xi_1^2 \phi_x^2 - \xi_2^2 \phi_y^2\right)(v_y - u_x)$$
$$- \frac{\epsilon^2}{2\rho_0 c_p(\phi)}\left[v_x\left(\xi_1\xi_2\phi_x^2 + \xi_2^2\phi_x\phi_y\right) + u_y\left(\xi_1^2 \phi_x\phi_y + \xi_1\xi_2\phi_y^2\right)\right]. \tag{5d}$$

In the above equations, we have introduced a vector

$$\boldsymbol{\xi} = (\xi_1, \xi_2) = |\boldsymbol{\xi}| \left(\frac{\phi_x}{|\nabla\phi|}, \frac{\phi_y}{|\nabla\phi|} \right)$$

$$= [1 + \epsilon_m \cos m(\theta + \alpha)] \left(\frac{\phi_x}{|\nabla\phi|}, \frac{\phi_y}{|\nabla\phi|} \right), \tag{6a}$$

to represent anisotropy in the interfacial energy and kinetics for a crystal of cubic symmetry with anisotropy strength ϵ_m; m determines the mode of symmetry of the crystal; $\theta = \arctan(\phi_y/\phi_x)$ is the angle between the interface normal and the crystal axis, and α denotes the angle of the symmetry axis with respect to the x-axis. The forcing terms X_1 and X_2 represent the effects introduced by the solidification process on the flow field; they are effective only over the interface region. $Y(\phi, T)$ is a modification to the Stefan condition caused by the buoyancy force term; $W(u, v, \phi, T)$ is viscous dissipation in the energy equation, and it can usually be neglected since it is small compared with other dissipative terms in that equation.

On the domain $\Omega \equiv [0, l] \times [0, l]$, prescribed boundary conditions are

$$u = 0 \quad \text{on } \partial\Omega \setminus \{(x,y)|y=l\},$$
$$u = U \quad \text{on } \{(x,y)|y=l\},$$
$$v \equiv 0 \quad \text{on } \partial\Omega,$$
$$\frac{\partial p}{\partial n} = 0 \quad \text{on } \partial\Omega \cup \partial\Omega_0,$$
$$\frac{\partial \phi}{\partial n} = 0 \quad \text{on } \partial\Omega,$$
$$\frac{\partial T}{\partial n} = 0 \quad \text{on } \partial\Omega.$$

Initial conditions are

$$u_0 = v_0 = 0 \quad \text{on } \Omega,$$
$$\phi_0 = 0 \text{ and } T_0 < T_m \quad \text{in } \Omega_0 \equiv \{(x,y) \mid |x|+|y| \leq l_c, (x,y) \in [-l_c, l_c]^2\},$$

where l_c is one half the length of the diagonal of a 45°-rotated square in the center of the domain.

3. NUMERICAL METHODS AND RESULTS

The governing equations (1b–e) are four coupled nonlinear parabolic equations in conserved form. We apply a projection method due to Gresho [8] to solve the momentum equations while preserving the divergence-free constraint (1a). The Shumann filter [9] is applied to solutions of the momentum equations (1b), (1c) prior to projection to remove aliasing due to under resolution. Since time-splitting methods are efficient for solving multi-dimensional problems by decomposing them into sequences of 1-D problems, a δ-form Douglas & Gunn [10] procedure is applied to the current model. Quasilinearization of Eqs. (1b–d) is constructed by Fréchet–Taylor expansion in "δ-form" as described by Ames [11] and Bellman and Kalaba [12].

The computations are performed on a square domain $\Omega \equiv [0, 63\,cm] \times [0, 63\,cm]$. Initially, the currently-used material, nickel, is supercooled by an amount $\Delta T = T_m - T = 224K$, and freezing begins from a small, rotated square with half-diagonal length $l_c = 1.89\,cm$ in the center of the domain. The numerical spatial and time step sizes are $\Delta x = \Delta y = 0.315\,cm$, $\Delta t = 10^{-6}s$, respectively, and the length scale for the interfacial thickness is $\delta = 0.105\,cm$. The kinetic coeffient μ_k is chosen to be $2.85\,m/s \cdot K$.

Figure 1 displays the velocity field at $t = 100.1s$. Lid-driven-cavity flow is introduced at $t = 100\,s$ with $U = 1cm/s$. Dendrite shape evolves from the initial rotated square to an approximate circle as shown in Fig. 1. We observe that velocity near the solid-liquid interface is greater than that nearby, a direct result of the forcing terms X_1 and X_2 in momentum equations. We also have found that the flow field has a significant effect on the growth rate of dendrites although this is not evident from Fig. 1; in particular, the growth rate is decreased by the flow field.

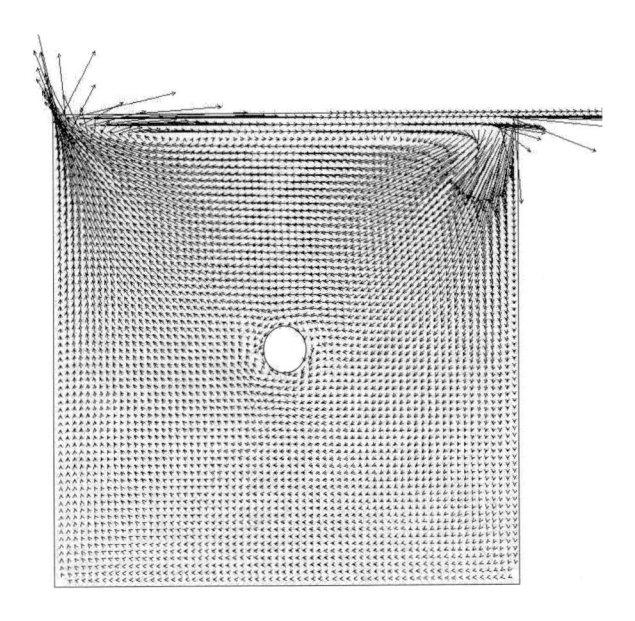

Figure 1. Lid-Driven-Cavity Flow with Freezing from Center at $t = 100.1s$

4. APPROACH TO PARALLELIZATION AND RESULTS

Parallelization of the numerical solution procedure is based on the shared-memory programming paradigm using the HP Fortran 90 HP-UX compiler. The program is parallelized using OpenMP and MPI running on the HP SuperDome at the University of Kentucky Computing Center to compare parallel performance of these two approaches. The maximum number of processors available on a single hypernode of the HP Super-

Dome is 64, and in the current study each processor is used to compute one part of the whole domain. For parallelization studies the grid is set at 201 × 201 points corresponding to the domain size $63\,cm \times 63\,cm$. The procedure for parallelizing two-step Douglas & Gunn time-splitting with MPI is to compute different parts of the domain on different processors, i.e., simply a crude form of domain decomposition. In particular, we divide the domain into n equal pieces along the separate directions corresponding to each split step, where n is the number of processors being used. That is, we first divide the domain in the x direction during the first time-splitting step, and then in the y direction during the second step. Therefore, transformations of data between each processor are required during the two steps of time-splitting. The sketch of this is shown in Fig. 2. Moreover, data transformations are also needed between adjacent boundaries of each processor. Since communication between processors increases with increasing number of processors for a fixed number of grid points, parallel performance is expected to decrease in such a case, resulting in only a sub-linear increase of speed-up for MPI.

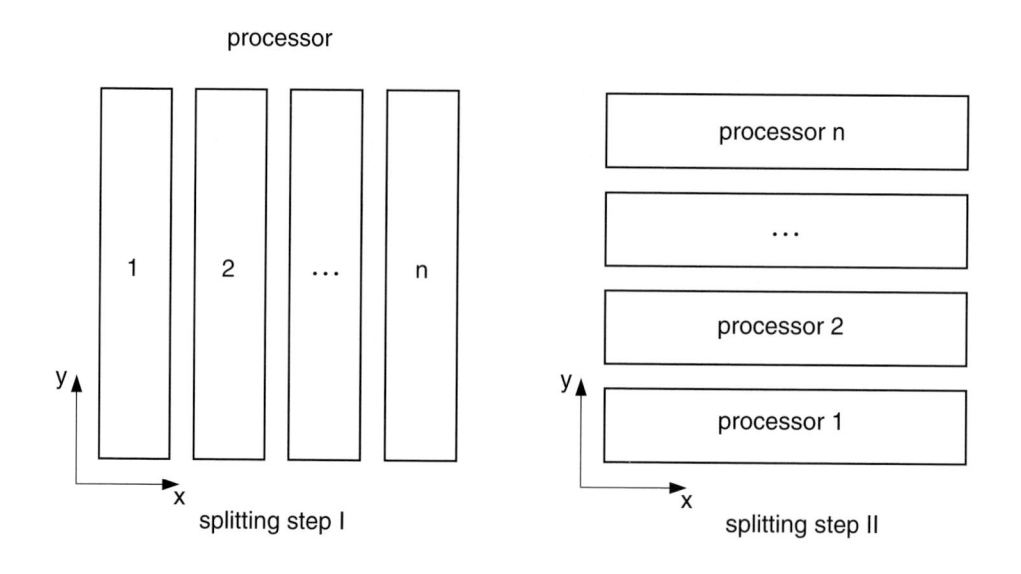

Figure 2. Distrubution of Processors for Two-Level Douglas & Gunn Time-Splitting

The implementation of OpenMP, on the other hand, is quite straightforward. It can be done by automatic parallelization of DO loops. All that is necessary is to share the information required by the parallelization within the DO loop, and this is easily handled with the OpenMP syntax.

To study the speed-up achieved by parallelization, different numbers n of processors ($n = 1, 2, 4, 8, 16, 32$) are used to execute the algorithm until $t = 5 \times 10^{-3}s$ for both OpenMP and MPI. Figure 3 displays the speed-up factor versus number of processors. It shows that, as the number of processors increases, the speed-up factors increase only sub-linearly for both OpenMP and MPI. Moreover, the speed-up performance of MPI is better than that of OpenMP. The curve for OpenMP in Fig. 3 also suggests that the speed-up factor attains its maximum at a number of processors only slightly beyond 32 for the present problem. Moreover, it is clear that parallel efficiency is quite low for OpenMP already by 16 processors, so this is possibly the maximum number that should be used. It should also be mentioned that even though the MPI implementation has not yet been completely optimized, the CPU time of MPI runs is somewhat less than that for OpenMP. Better performance could be achieved if further optimization of MPI is applied within the context of the current algorithm. Such optimization might include use of nonblocking communication, sending noncontiguous data using pack/unpack functions, decreasing unnecessary blocking, and optimizing the number of Douglas & Gunn split line solves simultaneously sent to each processor. The last of these can significantly alter the communication time to compute time tradeoff.

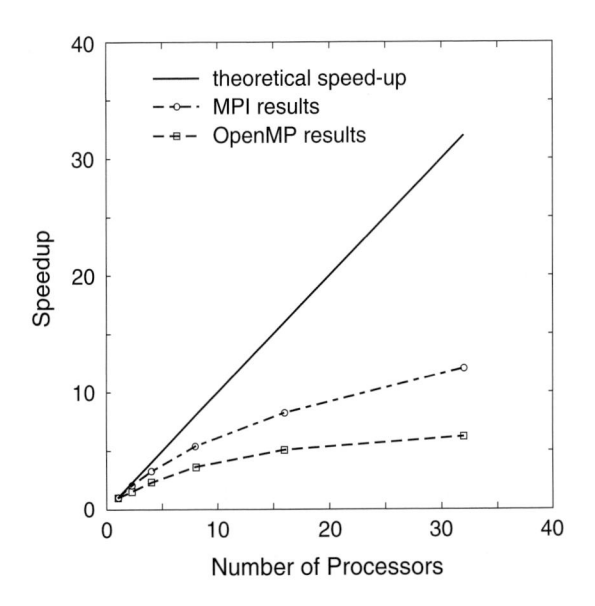

Figure 3. Speed-up Performance of Parallelized Phase-Field Model with Convection

5. SUMMARY AND CONCLUSIONS

In this paper we have compared parallel performance of OpenMP and MPI implemented for the 2-D phase-field model in a flow field. We found that MPI is both more efficient and also exhibited higher absolute performance than OpenMP. Moreover, it requires less memory than does OpenMP since MPI supports distributed memory while OpenMP supports shared-memory programming. Therefore, since memory requirements for our current problem are high, MPI is recommended for such problems. However, the implementation of MPI is more difficult than that of OpenMP. For example, programming effort for the current problem using MPI was approximately 100 times greater than that using OpenMP.

6. ACKNOWLEDGEMENTS

This work is supported by Center for Computational Science of University of Kentucky. We are also grateful to the University of Kentucky Computing Center for use of their HP SuperDome for all the computations.

REFERENCES

1. D. M. Anderson, G. B. McFadden, and A. A. Wheeler. A phase-field model of solidification with convection. *Physica D*, 135:175–194, 2000.
2. D. M. Anderson, G. B. McFadden, and A. A. Wheeler. A phase-field model with convection: sharp-interface asymptotics. *Physica D*, 151:305–331, 2001.
3. C. Beckermann, H.-J. Diepers, I. Steinbach, A. Karma, and X. Tong. Modeling melt convection in phase-field simulations of solidification. *J. Comput. Phys.*, 154:468–496, 1999.
4. Nabeel Al-Rawahi and Gretar Tryggvason. Numerical simulation of dendritic solidification with convection: Two-dimensional geometry. *J. Comput. Phys.*, 180:471–496, 2002.
5. A. A. Chernov. Surface morphology and growth kinetics. In R. Ueda and J. B. Mullin, editors, *Crystal Growth and Characterization*, pages 33–52. North-Holland Publishing Co., Amsterdam, 1975.
6. A. Ookawa. Physical interpretation of nucleation and growth theories. In R. Ueda and J. B. Mullin, editors, *Crystal Growth and Characterization*, pages 5–19. North-Holland Publishing Co., Amsterdam, 1975.
7. J. S. Langer. Models of pattern formation in first-order phase transitions. In G. Grinstein and G. Mazenko, editors, *Directions in Condensed Matter Physics*, pages 164–186. World Science, Singapore, 1986.
8. P. M. Gresho. On the theory of semi-implicit projection methods for viscous incompressible flow and its implementation via a finite element method that also introduces a nearly consistent mass matrix. part 1: Theory. *Int. J. Numer. Meth. Fluids*, 11:587–620, 1990.
9. F. G. Shuman. Numerical method in weather prediction: smoothing and filtering. *Mon. Weath. Rev.*, 85:357–361, 1957.
10. J. Douglas Jr. and J. E. Gunn. A general formulation of alternating direction methods, part 1. parabolic and hyperbolic problems. *Numer. Math.*, 6:428–453, 1964.
11. W. F. Ames. *Numerical Methods for Partial Differential Equations*. Academic Press, New York, NY, 1977.
12. R. E. Bellman and R. E. Kalaba. *Quasilinearization and Nonlinear Boundary-Value Problems*. American Elsevier Publishing Company, Inc., New York, NY, 1965.

A Parallel Unsplit Staggered Mesh Algorithm for Magnetohydrodynamics

Dongwook Lee[a] [*] and Anil E. Deane[a] [†]

[a]Institute for Physical Science and Technology,
University of Maryland, College Park, MD 20742

The parallel simulation of multidimensional, ideal MHD scheme is presented using an unsplit staggered mesh (USM-MHD) algorithm.[?] This new algorithm is based on the unsplit MHD algorithm recently studied by Crockett et al.[?] In their study a projection method was used for ensuring the divergence-free magnetic fields, whereas our approach takes a staggered mesh algorithm[?,?] for such numerical purpose. This algorithm utilizes the duality relation between the high-order Gudunov fluxes and electric fields and provides a robust numerical treatment to maintaining the divergence-free constraint of magnetic fields by keeping it up to computer's round-off error. We describe results of several test problems and discuss related performances for parallel computation.

1. Introduction

A dimensional splitting scheme has been used widely in most of multidimensional Godunov type finite-volume schemes due to its simplicity and relatively less expensive computational efforts for implementing multidimensional problems. The dimensional splitting scheme basically generalizes one dimensional algorithms to higher dimensional cases by taking one dimensional sweep in each direction. During each sweep one generally introduces splitting errors because of the non-commutativity of the linearized Jacobian flux matrices in most of the nonlinear multidimensional problems. To avoid this unphysical errors in computation it is desirable to use an unsplit approach. Compared to a dimensional splitting algorithm an unsplit scheme usually requires more storage for such as cell face and cell edge values and has limited the implementation of solving MHD problems. Crockett et al.[?] have recently studied an unsplit scheme using a projection scheme, generalized from the unsplit algorithm for hydrodynamics by Colella[?] that is encouraging.

In this paper we use an unsplit method which basically follows the formulation of Crockett et al. but we use a staggered mesh algorithm suggested by Balsara and Spicer[?] instead of using a projection method to maintain the $\nabla \cdot \mathbf{B} = 0$ constraint. The staggered mesh algorithm has a major advantage over the projection scheme in that it is computationally inexpensive. It also has many attractive features such as that it can be applied with different boundary conditions or different types of zoning,[?] while most of the FFT based projection schemes are restricted to comparatively simple boundary

---

[*]Also, Applied Mathematics and Scientific Computation (AMSC) Program, *dwlee@ipst.umd.edu*. This work has been partially supported by the NSF under grant DMS-0219282 to UMCP.
[†]Authors to whom correspondence should be directed, *deane@ipst.umd.edu*

conditions. The scheme can also be easily incoporated into parallel algorithms. Based on the duality relation the accuracy of round-off errors can be achieved in discretized form of $\nabla \cdot \mathbf{B} = 0$, provided the solenoidal initial and boundary conditions are satisfied at the initial problem setup. Alternatively, the 8-wave[?] (or divergence-wave) formulation can be used for the divergence-free constraint of magnetic fields. This formulation is found to be robust for many practical purposes, yet it is non-conservative and gives incorrect jump conditions across discontinuities. Comparative studies for such different MHD schemes can be found in Tóth.[?]

2. Numerical MHD

The study of MHD consists of the subject of plasma interaction with a magnetic field. Plasma, the ordinary state of matter in the Universe, is a completely ionized gas in which positively charged ions (or nuclei) and negatively chared electrons are freely moving. Among couple of different theoretical models, magnetohydrodynamics takes the fluid theory to describe plasma via macroscopic approach. Owing to the fact that charged particles stick to the magnetic field lines, the magnetic fields play an important roll to determine the geometry of the dynamics of the plasma and becomes a primary quantity of interest. Numerical formulations in MHD, therefore, are expected to predict correct physical behaviours of magnetic fields: keeping $\nabla \cdot \mathbf{B} = 0$ constraint as best as possible. To this end we use a staggered mesh algorithm, along with an unsplit temporal update.

2.1. Governing Equations of Ideal MHD

The ideal MHD equations can be formulated as a hyperbolic system of conservation laws. In general the resistive MHD equations in conservative form can be written as the following:

$$\frac{\partial \rho}{\partial t} + \nabla \cdot (\rho \mathbf{u}) = 0 \tag{1}$$

$$\frac{\partial \rho \mathbf{u}}{\partial t} + \nabla \cdot (\rho \mathbf{u}\mathbf{u} - \mathbf{B}\mathbf{B}) + \nabla p_{tot} = 0 \tag{2}$$

$$\frac{\partial \mathbf{B}}{\partial t} + \nabla \cdot (\mathbf{u}\mathbf{B} - \mathbf{B}\mathbf{u}) + \nabla \times (\eta \mathbf{J}) = 0 \tag{3}$$

$$\frac{\partial E}{\partial t} + \nabla \cdot (\mathbf{u}e + \mathbf{u}p_{tot} - \mathbf{B}\mathbf{B} \cdot \mathbf{u} - \mathbf{B} \times \eta \mathbf{J}) = 0 \tag{4}$$

The above equations represent the continuity equation, the momentum equation, the induction equation, and the energy equation, respectively. $\mathbf{J} = \nabla \times \mathbf{B}$ represents the current density, $p_{tot} = p + \mathbf{B}^2/2$ the total pressure and the thermal pressure $p = (\gamma - 1)(E - \frac{1}{2}\rho \mathbf{u}^2 - \frac{1}{2}\mathbf{B}^2)$. The parameters are the ratio of specific heats γ and the resistivity η. The resistivity $\eta = 0$ is for perfectly conducting MHD (ideal MHD), whereas $\eta > 0$ is valid non-ideal cases. We focus a numerical study in ideal MHD in this paper. The initial condition on Faraday's law should satisfy $\nabla \cdot \mathbf{B} = 0$ and this equation should remain as a restriction in qualifying the correct numerical magnetic fields at all times. For zero resistivity the ideal MHD equations become hyperbolic and admit wave-like solutions that propagate without dissipation. The wave structure of ideal MHD consists of slow, Alfvén, and fast waves, which results more complicated structures than in the pure hydrodynamics case. The Alfvén wave is caused by tension of the magnetic fields and provides a restoring mechanism of the distorted geometry of the fields to the initial shape. One of the significant property of Alfvén wave is that it involves the information of the overall magnetic geometry in the macroscopic MHD. It is linearly degenerate and propagates with a speed $|B_x|/\sqrt{\rho}$, whereas other two fast and slow waves are associated with compression of plasma and ordinary sound waves propagating with the sound speed.

2.2. Overview of USM-MHD

The formulation of an unsplit staggered mesh algorithm (USM-MHD) can be broken up into several steps. The first step is a *quasilinearization* of the nonlinear system of ideal MHD governing equations (??)-(??). The approximated, quasilinearized equations are then solved to compute the boundary extrapolated evolutions in the normal direction by a half time step (*predictor step*) using a characteristing tracing method. A second-order TVD slope limiter is used in constructing a MUSCL-Hancock type of evolution. Another set of quasilinearized equations are solved to account for the contributions from the transversal fluxes, again via a characteristic tracing and we call it a *corrector step*. The transversal flux updates together with the evolution in normal direction provide two Riemann state values which are of second-order. The next step is a *Riemann solver*, where we obtain approximated intermediate fluxes according to the left and right Riemann state values from the predictor-corrector steps. Such fluxes are second-order accurate located at the cell interfaces. The second-order fluxes are then used to construct the electric fields at each cell corner using the duality relation between the fluxes and electric fields. Simultaneously, unsplit time integrations are performed in a next *solution update* step and advance the cell-centered variables to the next time step. In general, we are left with non divergence-free magnetic fields at cell centers after the solution update and they are to be corrected. To obtain divergence-free magnetic fields we solve a discrete form of the induction equation (??) using second-order accurate electric fields which are collocated at cell corners. These divergence-free magnetic fields are then used for the next time step. More detailed description of USM-MHD algorithm will appear in Lee and Deane.[?]

2.3. $\nabla \cdot \mathbf{B} = 0$ treatment using Staggered Mesh Algorithm

The simplest form of our staggered mesh algorithm, proposed by Balsara *et al.*,[?] is to use a staggered mesh system coupled to a high-order Godunov fluxes. That is,

$$\begin{aligned} E_{z,i+1/2,j+1/2} &= \frac{1}{4}\Big\{-F^{*,n+1/2}_{6,i+1/2,j} - F^{*,n+1/2}_{6,i+1/2,j+1} + G^{*,n+1/2}_{5,i,j+1/2} + G^{*,n+1/2}_{5,i+1,j+1/2}\Big\} \\ &= \frac{1}{4}\Big\{E^{n+1/2}_{z,i+1/2,j} + E^{n+1/2}_{z,i+1/2,j+1} + E^{n+1/2}_{z,i,j+1/2} + E^{n+1/2}_{z,i+1,j+1/2}\Big\} \end{aligned} \quad (5)$$

where $F^{*,n+1/2}_{6,i+1/2,j}$ and $G^{*,n+1/2}_{5,i,j+1/2}$ are the sixth and fifth components in the corresponding flux functions. A discrete form of the induction equation (??) now allows the update of the cell interface centered magnetic fields $b^{n+1}_{x,i+1/2,j}$ and $b^{n+1}_{y,i,j+1/2}$ using

$$b^{n+1}_{x,i+1/2,j} = b^n_{x,i+1/2,j} - \frac{\Delta t}{\Delta y}\Big\{E_{z,i+1/2,j+1/2} - E_{z,i+1/2,j-1/2}\Big\}, \quad (6)$$

$$b_{y,i,j+1/2}^{n+1} = b_{y,i,j+1/2}^{n} - \frac{\Delta t}{\Delta x}\left\{-E_{z,i+1/2,j+1/2} + E_{z,i-1/2,j+1/2}\right\}. \tag{7}$$

Finally we update the components of the cell-centered fields $\mathbf{B}^{n+1}$ by interpolating $\mathbf{b}^{n+1}$,

$$B_{x,i,j}^{n+1} = \frac{1}{2}\left\{b_{x,i+1/2,j}^{n+1} + b_{x,i-1/2,j}^{n+1}\right\}, \quad B_{y,i,j}^{n+1} = \frac{1}{2}\left\{b_{y,i,j+1/2}^{n+1} + b_{y,i,j-1/2}^{n+1}\right\}. \tag{8}$$

We can see that the discretized numerical divergence of $\mathbf{B}$,

$$(\nabla \cdot \mathbf{B})_{i,j}^{n+1} = \frac{b_{x,i+1/2,j}^{n+1} - b_{x,i-1/2,j}^{n+1}}{\Delta x} + \frac{b_{y,i,j+1/2}^{n+1} - b_{y,i,j-1/2}^{n+1}}{\Delta y} \tag{9}$$

remains zero to the accuracy of machine round-off errors, provided $(\nabla \cdot \mathbf{B})_{i,j}^{n} = 0$.

3. Numerical Results

We present parallel computational work performed on SGI 1200 linux cluster. It is equipped with 26 Intel Xeon dual Pentium III 700 Mhz processors (13 nodes) communicating via myrinet architecture, 1GB memory at each node. Our test problem includes 2D Orszag-Tang's MHD vortex problem. The square computational domain is subdivided into the available number of processors keeping the same number of processors on each dimension. This subdivision yields that each processor takes smaller *square* subregion at each parallel computation task, with which, in general, better scalability is obtained than other decomposition.

3.1. Parallel Implementation and Performance

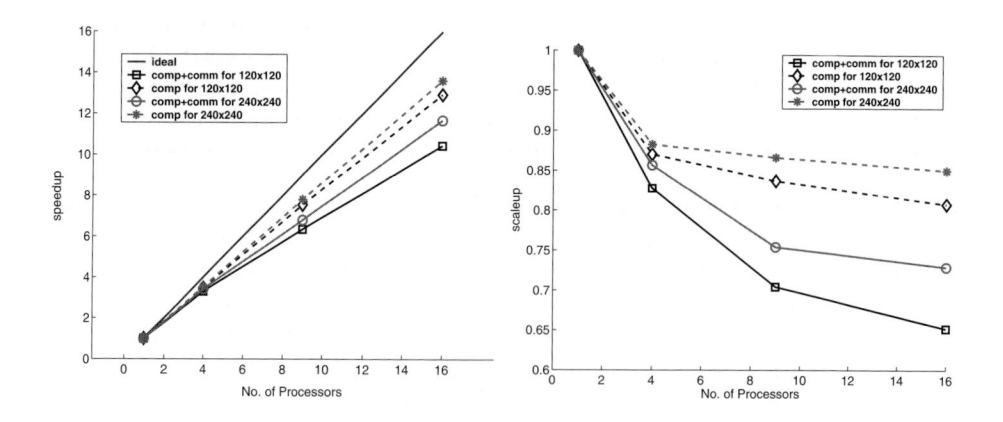

Figure 1. Speedup and scaleup performances of two different problem sizes in Orszag-Tang's MHD problem.

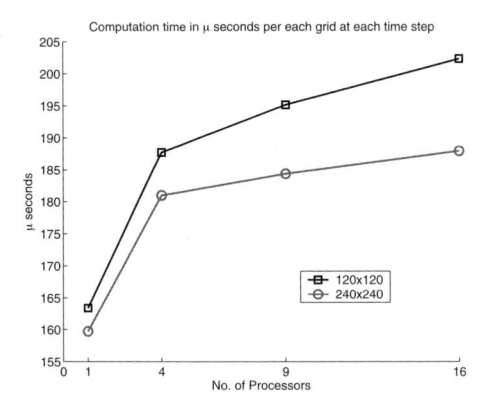

Figure 2. Computation time for two different problem sizes of Orszag-Tang MHD vortex problem.

To explore the scalability of our USM-MHD algorithm we consider two quantities of relative speedup and relative scaleup (or efficiency). Two plots in figure ?? include such performances, considering effects of including and excluding parallel communication times. Notice that two different problem (global) sizes $120^2, 240^2$ are tested. As shown in both plots, speedup and scaleup performances are noticeably increased without considering the communication time. In particular, the scaleup test without the communication overhead indicates that an approximately similar efficiency is well achieved on a *four* times bigger problem size 240^2 with using *four* times more processors than on a smaller problem 120^2 using 4 processors. This is predicted in many ideal scaleup performances, yet we clearly observe in the plot that the communication time drops both speedup and scaleup performances. Note also that there are increases in both speedup and scaleup performances on a larger problem (e.g., 240^2) than on a smaller problem (e.g., 120^2). This is because each task should spend relatively more time in communication in a smaller problem than in a larger problem. Figure ?? shows averaged computation times t_c in μ seconds per each grid at each time step. The quantities are normalized by the problem sizes in consideration according to a relation $T_P = t_c N \times N$, where T_P is a total computation time with P processors for a problem size $N \times N$. On average USM-MHD algorithm calculates 8 primary variables, $\rho, u, v, w, B_x, B_y, B_z, p$, at each grid with double precision accuracy. This again confirms that a better efficiency can be observed on a relatively larger problem. Notice also that there are big gradients in the slope of t_c curves from a single-processor computation to a multiple-processor computation. This indicates that the relative efficiency rapidly drops in switching to a multiple-process architecture and slowly recovered thereafter.

Figure 3. Density plots of Orszag-Tang MHD vortex problem at $t = 0.159$ and $t = 0.5$, respectively.

3.2. Computational Results

We show in figure ?? a high resolution numerical result of Orszag-Tang's problem. The computation was performed using 16 processors on a global mesh size 800×800. Two plots are densities at an early time ($t = 0.159$) and the final time ($t = 0.5$). At time $t = 0.159$ the flow is transient becoming chaotic in parts while mostly smooth. In contrast, the density at $t = 0.5$ represents fully developed MHD turbulent flow, and discontinuities are dominant. A promising result was well predicted and obtained using USM-MHD algorithm, simulating full aspects of detailed MHD flows.

Figure ?? presents averaged amount of total times spent by each subroutine in our USM-MHD algorithm. Basically USM-MHD is subdivided into several different subroutines: predictor-corrector, MUSCL-Hancock, TVD slope limiter, Riemann solver, MHD eigenstructure, conversions between primitive and conservative variables, staggered mesh treatment of divergence-free magnetic fields, boundary condition (MPI communication), CFL stability, and some other minor subroutines. To analyze overall performance contributed by each individual subroutine we measured wall clock times taken by all different subroutines. It is found out that the predictor-corrector routines dominate most of computing times, and along with the Riemann solver and MHD eigensystem routines, these three subroutines take almost 75% of computing times at each time step. It is only 0.25% that the staggered mesh algorithm takes to handle $\nabla \cdot \mathbf{B} = 0$ constraint and it is very efficient in many practical purposes in numerical MHD simulations.

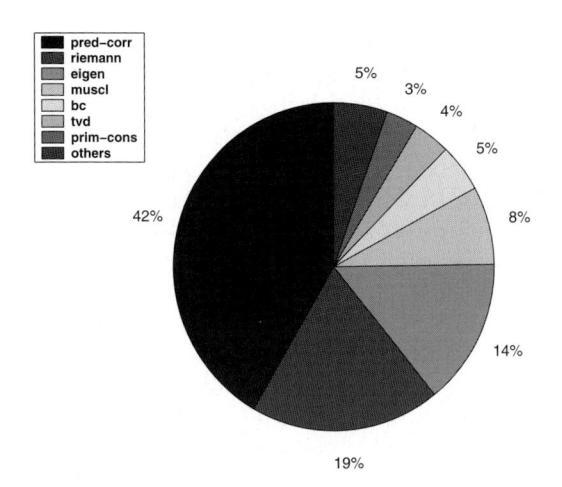

Figure 4. Averaged percentages of time stent by each subroutine for Orszag-Tang MHD vortex problem. In the legend, we denote "pred-corr (42%), riemann (19%), eigen (14%), muscl (8%), bc (5%), tvd (4%), prim-cons (3%), others (5%)" by predictor-corrector, Riemann solver, MHD eigenstructure, MUSCL-Hancock, boundary condition, TVD slope limiter, primitive-conservative conversion, and other subroutines, respectively.

4. Conclusion

The second-order accurate parallel, multidimensional unsplit MHD algorithm has been successfully implemented on the staggered grid to maintain $\nabla \cdot \mathbf{B} = 0$ numerically. The method preserves the numerical MHD constraint extremely well without any evidence of numerical instability or accumulation of unphysical errors. We have performed parallel performances of our USM-MHD algorithm and validated reliable results in simulating 2D Orszag-Tang MHD problem. We envision extending the scheme to 3D with AMR capability. The ultimate interest of our study includes space physics applications and the accuracy of the scheme, its robustness, and its divergence properties as demonstrated on the test problems of interest encourage us to pursue simulations relevant to solar wind configurations.

REFERENCES

1. D. S. Balsara, D. S. Spicer, A Staggered Mesh Algorithm Using High Order Godunov Fluxes to Ensure Solenoidal Magnetic Fields in Magnetohydrodynamics Simulation, J. Comp. Phys., 149 (1999), 270–292.
2. D. S. Balsara, Divergence-Free Adaptive Mesh Refinement for Magnetohydrodynamics, J. Comp. Phys., 174 (2001), 614–648.
3. P. Colella, Multidimensional Upwind Methods for Hyperbolic Conservation Laws, J.

Comp. Phys., 87 (1990), 171–200.
4. R. K. Crockett, P. Colella, R. T. Fisher, R. I. Klein, C. F. McKee, An Unsplit, Cell-Centered Godunov Method for Ideal MHD, *J. Comp. Phys.*, 203 (2005), 422–448.
5. A. E. Deane, Parallel Performance of An AMR MHD code, Parallel CFD, Egmond En Zee, Netherlands, (2001).
6. D. Lee, A. E. Deane, An Unslit Staggered Mesh Scheme for Magnetohydrodynamics with High-Order Godunov Fluxes, *In progress*, (2005).
7. C. R. Evans, J. F. Hawley, Simulation of magnetohydrodynamic flows: a constrained transport method, *Astrophys. J.*, 332 (1988), 659–677.
8. I. Foster, *Designing and Building Parallel Programs, Concepts and Tools for Parallel Software Engineering*, Addison Wesley, (1995).
9. T. A. Gardiner, J. M. Stone, An Unsplit Godunov Method for Ideal MHD via Constrained Transport, *Preprint submitted to Elsevier Science*, (2005)
10. D. Ryu, T. W. Jones, Numerical Magnetohydrodynamics in Astrophysics: Algorithm and Tests for One-dimensional Flow, *Astrophys. J.*, 442 (1995), 228–258.
11. K. G. Powell, P. L. Roe, J. T. Linde, T. I. Gombosi, D. L. De Zeeuw, A Solution-Adaptive Upwind Scheme for Ideal Magnetohydrodynamics, *J. Comp. Phys.*, 154 (1999), 284–309.
12. D. A. Roberts, M. L. Goldstein, A. E. Deane, S. Ghosh, Quasi-Two-Dimensional MHD Turbulence in Three-Dimensional Flows, *Phys. Rev. Lett.*, 82 (1999), 548–551.
13. M. S. Ruderman, M. L. Goldstein, D. A. Roberts, A. E. Deane, L. Ofman, Alfven wave phase mixing driven by velocity shear in two-dimensional open magnetic configurations, *J. Geophys. Res.*, 104 (1999), 17057.
14. J. M. Stone, The Athena Test Suite, <http://www.astro.princeton.edu/~jstone>
15. E. F. Toro, *Riemann Solvers and Numerical Methods for Fluid Dynamics, A Practical Introduction*, Springer, 1997.
16. G. Tóth, The $\nabla \cdot B = 0$ Constraint in Shock-Capturing Magnetohydrodynamics Codes, *J. Comp. Phys.*, 161 (2000), 605–656.

Coupled magnetogasdynamics – radiative transfer parallel computing using unstructured meshes[*]

V.A. Gasilov[a], S.V. D'yachenko[a], O.G. Olkhovskaya[a],
O.V. Diyankov[b], S.V. Kotegov[b], V.Yu. Pravilnikov[b]

[a]*Institute for Mathematical Modelling, Russian Ac.Sci., 4-A, Miusskaya Sq., 125047, Moscow, Russia*

[b] *Neurok Techsoft LLC, Troitsk, 142190, Moscow region, Russia*

Keywords: Computational fluid dynamics; unstructured mesh; parallel algorithms; distributed computations

Application of new RMHD numerical technologies to plasma physics studies

During 2000-2005, the work of a team from the Institute for Mathematical Modeling, Russian Ac. Sci. (IMM RAS) resulted in the development of a radiative magneto-hydrodynamic (RMHD) code *MARPLE* (Magnetically Accelerated Radiative Plasmas – Lagrangian-Eulerian). The predecessor of *MARPLE* is the Lagrangian-Eulerian code *RAZRYAD*, which was under development since 1979 up to 1992 on the basis of original methods developed in the Keldysh Institute of Applied Mathematics (KIAM RAS), and IMM RAS. The code was primarily designed for simulations pertinent to the two-dimensional pulsed-power problems. The main *RAZRYAD* version was oriented to simulations of plasma flows like Z-pinch, Theta-pinch and others.
Since its origin, *MARPLE* was applied to many problems in the field of plasma dynamics, especially to those related with modern pulsed-power systems. This experience has led us to the new *MARPLE* version based on new computational technologies and programming principles.
MARPLE performs calculations in terms of a cylindrical (r,z) frame of reference, (x,y) or (r,φ) geometry is also available. A flow of plasma is considered in one-fluid MHD approximation represented by the well-known Braginskii model. The MHD equations

---

[*] The study is supported by the State contract 10002-251/OMH-03/026-023/240603-806, ISTC project 2830, and RFBR project 04-01-08024-OFI-A.

are written in a so-called 2.5-dimensional fashion, i.e. the flowfield vectors are presented by all three components: velocity **V**=(u,w,v), magnetic inductance **B**=(B_r,B_φ,B_z), and electric intensity **E**=(E_r,E_φ,E_z). We take into account anisotropy of dissipative processes in presence of magnetic field. The energy balance is given by the two-temperature model describing the electron-ion relaxation. Accordingly, the total thermal pressure P is the sum of electron (P_e), and ion (P_i) components. The governing system incorporates a radiative transport equation for the spectral intensity and is completed by data tables of plasma state, coefficients of transport processes, spectral opacity and emissivity.

The radiative energy transfer is calculated by means of a characteristic-interpolation algorithm constructed on the base of a Schwarzschield-Schuster approximation.

Depending on the studied problem the governing system can be supplemented by the equation describing the electric current evolution for the whole electric circuit. For a typical pulsed-power problem the circuit includes an electric generator, current supply facilities and a discharge chamber.

The new RMHD code is based on unstructured grid technology. The versatility of unstructured grids makes *MARPLE* a convenient tool for investigations of plasma dynamics taking place in various pulsed-power facilities.

The main features of the innovative numerical tools are defined by a strongly nonlinear nature of physical problems which the new code must be capable to simulate. The developed application software should be compatible with different operating systems (MS Windows, Linux, UNIX) and different platforms (PC, powerful workstation, mainframe including parallel processing). The object-oriented programming by means of C++ language seems to be the most suitable technique for this project. Our experience in using C++ proved that the technique of derived classes and virtual functions is very effective for solution of CFD problems by the technology of irregular or unstructured meshes.

Governing system: 2 temperature MHD

Non-dissipative MHD

$$\frac{\partial}{\partial t}\rho + \nabla(\rho\vec{w}) = 0,$$

$$\frac{\partial}{\partial t}\rho w_i + \sum_k \frac{\partial}{\partial x_k}\Pi_{ik} = 0,$$

$$\Pi_{ik} = \rho w_i w_k + P\delta_{ik} - \frac{1}{4\pi}\left(B_i B_k - \frac{1}{2}B^2\delta_{ik}\right),$$

$$\frac{\partial}{\partial t}\vec{B} - \nabla\times\left(\vec{w}\times\vec{B}\right) = 0,$$

$$\frac{\partial}{\partial t}\left(\rho\varepsilon + \frac{1}{2}\rho w^2 + \frac{B^2}{8\pi}\right) + \nabla\vec{q} = 0,$$

$$\vec{q} = \left(\rho\varepsilon + \frac{1}{2}\rho w^2 + P\right)\vec{w} + \frac{1}{4\pi}\vec{B}\times\left(\vec{w}\times\vec{B}\right),$$

$$P = P(\rho\varepsilon).$$

Dissipative processes:
Magnetic field diffusion

$$\vec{E} = \frac{\vec{j}_\parallel}{\sigma_\parallel} + \frac{\vec{j}_\perp}{\sigma_\perp},$$
$$rot\ \vec{B} = \frac{4\pi}{c}\vec{j} = \frac{4\pi}{c}\hat{\sigma}\ \vec{E},$$
$$\frac{1}{c}\frac{\partial \vec{B}}{\partial t} = -rot\ \vec{E}$$

Heat conductivity, electron-ion relaxation, Joule heating, radiative cooling.

$$\frac{\partial(\rho\varepsilon_e)}{\partial t} = -div(\hat{\kappa}_e\ grad\ T_e) + Q_{ei} + G_J + G_R$$
$$\frac{\partial(\rho\varepsilon_i)}{\partial t} = -div(\hat{\kappa}_i\ grad\ T_i) - Q_{ei}$$
$$\varepsilon = \varepsilon_i + \varepsilon_e,\ P = P_i + P_e$$

Common notations for physical values are used here. The volumetric sources are: Q_{ei} – for electron-ion exchange, G_J – for Joule dissipation and G_R – for radiation.

The splitting scheme

A splitting scheme is applied to the governing system with the subsets of equations describing different physical processes being solved in sequence by appropriate program modules. The MHD system is solved by the generalized TVD Lax-Friedrichs scheme which was developed for the unstructured mesh applications [1]. A general monotonous reconstruction of mesh-defined functions is designed taking into account the dependence on two variables. For the case of a regular triangulation this scheme ensures the second order approximation to spatial derivatives (the third order is possible with a special choice of the antidiffusion limiters). The time integration is explicit, the second approximation order is reached due to the predictor – corrector procedure. The time step is restricted by the Courant criterion. The predictor and corrector steps are organized similarly but the numerical fluxes differ due to different reconstruction of the functions. Namely a nonmonotonic piecewise-linear continuous interpolation is used for the predictor, and special monotonic discontinuous reconstruction is created for the corrector. Thus the corrector scheme not only improves the time-advance resolution but also appears to be a stabilizing procedure. For the solution of parabolic equations describing the conductive heat transfer, we developed the new finite-volume (FV) schemes constructed by analogy with mixed finite-element method.
Radiative energy transfer is described by the equation for spectral radiation intensity. Practical calculations are done via multigroup spectral approximation. We solve the radiative transport equation by means of semi-analytical characteristic algorithm. The analytical solution along the characteristic direction is constructed by means of the backward-forward angular approximation to the photon distribution function [2], [3]. The two-group angular splitting gives an analytical expression for radiation intensity dependent on opacity and emissivity coefficients. The energy exchange between

radiation field and the gas is taken into account via a radiative flux divergence, which is incorporated into the energy balance as a source function.

Non-dissipative MHD	Dissipative processes	
	Magnetic field diffusion, heat transfer, electron-ion exchange, Joule heat	Radiative energy transport
Local processes	Quasi-local processes	Non-local processes
Explicit scheme	Implicit scheme	Explicit fractional-steps scheme
High-resolution TVD scheme with flux correction	Integro-interpolaton FV schemes	Characteristic scheme
2-nd order predictor-corrector time advanced scheme		

Grids and discretization

The solution of the RMHD system in the *MARPLE* program complex is carried out by use of unstructured triangular mesh. We utilize the node-centered (nonstaggered) storage of calculated values. The technique of finite volumes is used for approximation of the governing system. In the present program version the finite volumes are formed by modified Voronoi diagrams. A fragment of triangular mesh and finite volumes are shown at the Fig. 1.

A universal program tools were developed convenient for input and acquisition of geometrical/physical data as well as for storage and treatment of a discrete computational model. Cellular model is used for both the computational domain and the meshes described as geometric complexes. A formalized description of geometric and topological properties of meshes is thus provided. Topological complexes of various dimensionalities are suitable for representation of irregular continuum as well as discrete structures. 3D geometric complex includes: 0-order elements (nodes), 1-order elements (edges), 2-order elements (faces), and 3-order elements (cells). The boundary

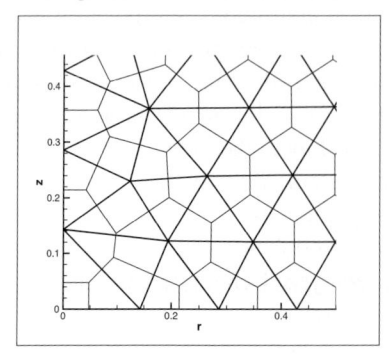

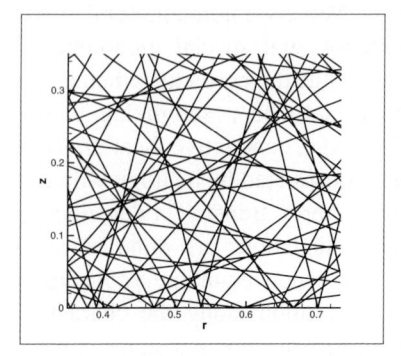

Fig.1. Triangular grid and finite volumes.   Fig.2. Grid of rays.

of each element is formed by the elements of lower order. Topology is described by relations between the elements. Two types of relations are used: incidence relations (between elements of different dimensions) and adjacency relations (between elements of the same dimension). Only a few base incidence relations are stored permanently, and other relations are calculated every time we need them.

Dynamic topology changes are supported. Each element is identified by its number. Having the number of an element, it is possible to find all the data assigned to this element. Special methods were developed that allow implementing for all data structures appropriate changes caused by variations in elements numeration (e.g. adding or removing the elements while grid construction or refinement).

A uniform description of both the computational domain topology and its discretization is offered. The basic structure used for spatial discretization is *Discretized domain* including computational domain geometry, problem statement and computational grids. Preprocessor tools providing problem statement, computational domain description and mesh generation are included.

This geometric data treatment technique is especially effective for unstructured grids, but may be applied for regular grids as well. It is also useful for handling subdomains processed by a distributed computer system.

Radiative transfer

Transport equation for quasistationary radiation field in cylindrical geometry:

$$\sin\theta\left(\cos\varphi\frac{\partial I_\omega}{\partial r}+\frac{\sin\varphi}{r}\frac{\partial I_\omega}{\partial\varphi}\right)+\cos\theta\frac{\partial I_\omega}{\partial z}=-\aleph_\omega I_\omega+j_\omega$$

A set of equations for the forward/backward intensity functions $I^{f/b}$:

$$\frac{\cos\theta_{n+1}-\cos\theta_n}{\Delta\theta_n}\left(\frac{\partial I^b_{n+1/2}}{\partial r}+\frac{I^b_{n+1/2}}{r}\right)+\frac{\sin\theta_{n+1}-\sin\theta_n}{\Delta\theta_n}\frac{\partial I^b_{n+1/2}}{\partial z}=-\aleph I^b_{n+1/2}+j$$

$$\frac{\cos\theta_n-\cos\theta_{n+1}}{\Delta\theta_n}\left(\frac{\partial I^f_{n+1/2}}{\partial r}+\frac{I^f_{n+1/2}}{r}\right)+\frac{\sin\theta_{n+1}-\sin\theta_n}{\Delta\theta_n}\frac{\partial I^f_{n+1/2}}{\partial z}=-\aleph I^f_{n+1/2}+j$$

The radiation energy density:

$$U=\frac{\pi}{c}\sum_{n=1}^{N}\left(I^f_{n+1/2}+I^b_{n+1/2}\right)\left(\cos\theta_n-\cos\theta_{n+1}\right)$$

The forward/backward intensities along a ray in the direction $\theta_{n+1/2}$:

$$I^{f/b}=(I^{f/b}_{i,j+1}-I^{eq}_{i,j})\exp(-\kappa_{i,j}\xi_{i,j})+I^{eq}_{i,j}$$

For the purpose of the radiation energy transport calculation a special grid of characteristics is constructed in the computational area. This grid represents a number of sets (families) of parallel right lines and is further referred to as the grid of rays. Each set of parallel lines is characterized by the angle of inclination to coordinate axes and spatial density of rays. The grid of rays introduces some discretization of the computational area in the plane (r,z) and with respect to the angle θ $(0 \leq \theta < \pi)$, which is required for numerical integration of the radiation transport equation according to the described above model. The grid of rays is superimposed on the initial computational grid intended for gas dynamics and heat transfer computations. A fragment of the grid of rays (12 angle sectors) is shown above at the Fig. 2.

Parallel implementation

The developed explicit difference scheme allows quite natural parallel implementation for distributed computer systems. Each processing node is associated with a section of the triangular computational grid (a subdomain). Each processor carries out the computations only inside this subdomain. Subdomains cover the entire computational grid and may have common nodes only at the boundaries. The data exchange between subdomains is organized through the «margins». We called so the layers of grid elements belonging to the neighboring subdomains. In that way some elements of a subdomain are included in the «margins» of the neighboring subdomains (see Fig.3). Basic (non-fictive) near-bound grid elements in the right are drawn with solid lines, and the artificial margin – fictive near-bound grid elements, associated with the pre-image (in the left) are drawn with dashed lines. Any data from a margin node are to be retrieved via high-speed network from the processor where this node is stored as an element of the related subdomain. We use MPI for distributed computations and ParMetis for mesh partition.

Implementation of complex physical models incorporating multi-scaled and essentially non-local processes requires more than one grid structure. For instance, an additional grid of rays (characteristics) is necessary for the simulation of radiative energy transfer by the method of characteristics. As a rule, the number of elements (and therefore the

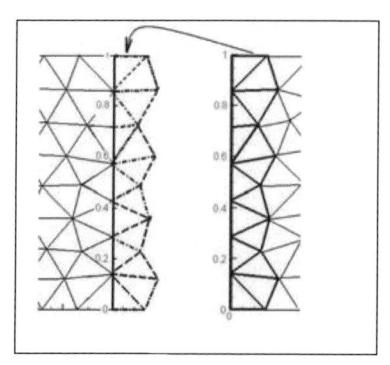

Fig. 3. Artificial margins. 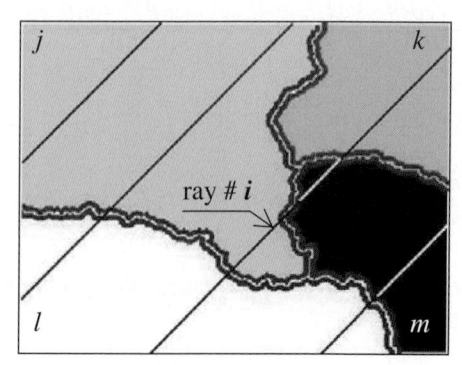   Fig. 4. Rays over grid partition.

resources consumption) of these grids are comparable. The multigroup model allows separated computations for each spectral group, so the spectral groups may be distributed for a several processors. It is important that an intensive non-local data exchange exists between the different grid structures. When the transport equation is solved along the characteristics each of them crosses several subdomains and both the internal and boundary cells are crossed. Thus additional conditions are introduced into the interprocessor communications balancing problem – we have to optimize the rays distribution over the triangular grid partition. Rays crossing the basic grid subdomains are schematically shown at the diagram Fig.4.

The basic triangular grid is partitioned to N domains located at N processors ($N=16$ in our case). We are to distribute M rays (grid of characteristics) over the same processors ($M=3845$). Each ray is divided into segments, associated with the basic grid cells (control volumes). Each ray is processed individually. The computations are carried out along a ray, and each segment utilizes the calculated data from the related cell (because the optical properties at a segment are dependent upon the temperature and the density in the cell). That means that if the ray i is processed by the processor j and the cells involved are located in the domains j, k, l, m, only the data from the domain j cells may be acquired directly, and those from the domains k, l, m require communications between the processors $j-k, j-l, j-m$. The number of ray segments varies significantly. Some rays may include only two segments and some a hundred or more. A lot of rays cross several domains. That's why an optimization of ray distribution is important.

Boolean linear programming problem statement.
Unknown variables:
$$x_{ij} = \begin{cases} 1 & \text{if the ray } \# i \text{ is at the processor } \# j \\ 0 & \text{otherwise} \end{cases}$$

Limitations: each ray should be included in the partition exactly once.
$$\forall i \quad \sum_j x_{ij} = 1$$

Weight factor a_{ij} is the number of ray i segments located in the domain j.
It is the number of segments acquiring the information from the processor j. In particular $a_{ij}=0$ if the ray i does not cross the domain j.

Criterion function 1. The number of interprocessor communications.
$$F_1 = \sum_i \sum_j a_{ij}(1-x_{ij}) \to min$$

Criterion function 2. The processor load balance.
$$F_2 = \sum_k \sum_l (\varphi_k - \varphi_l)^2 \to min$$

Here $\varphi_j = \sum_i b_i x_{ij}$ is the number of rays segments processed by the processor j.
$b_i = \sum_j a_{ij}$ is the total number of ray i segments.

The aggregate criterion function
$$F = \alpha F_1 + \beta F_2.$$

The factors α and β depend upon the estimations of the interprocessor communication time (α) and the ray processing time (β).
For practical computations a heuristic optimization algorithm was applied.

Numerical results

The studied benchmark problem concerns with the compression of the plasma shell by an azimuthal magnetic field. The problem formulation corresponds to the experiments carried out in Sandia National Laboratory with a multi-megaampere pulsed current generator "Z-facility" [4]. In these experiments the generator is loaded by the tungsten wire array (so-called "squirrel cage") which is heated up to the plasma state by a 20 megaampere-amplitude electric pulse thus producing an imploding Z-pinch. After the pinch collapse the soft x-ray energy yield reaches ~ 1.8 MJ [4].
In our study we chose the power source parameters approximating those of the Z machine. The simplified electrical circuit represents a voltage generator U, a resistance R, an inductance L, and the simulation domain (Z-pinch) all connected in series (Fig.5). The experimental generator voltage waveform is presented at the Fig.6.
The simulation starts at the time $t = 0.08\mu s$, when the voltage becomes significant. The shell height is 2cm. The shell mass is $4.108 \cdot 10^{-3}$g (constant and unperturbed density $\rho_0 = 0.002$g/cm^3). The shell initial position is between $r = 1.96$cm and $r = 2.04$cm. The initial temperature is 2eV. The rarefied background plasma density is 10^{-7}g/cm^3.
Boundary conditions are the following:
the left bound ($r=0$) is the symmetry axis, the upper and the lower bounds ($z=0$ or $z=2$) are metallic electrodes, the right bound ($r=2.5$) is the external electric circuit with L=11.44nH, R=0.12mOhm and the voltage waveform $U(t)$ (Fig.6).
The triangular computational grid in the plane (r,z) was refined near the shell initial position and near the axis. Radiation transport model includes 12 angle sectors and 20 spectral intervals for opacity and emissivity factors.
The numerical results are in good agreement with the experimental data, as concerns the liner acceleration time, density and temperature growth and the radiation yield. Fig. 7 (on the next page) demonstrates the evolution of the density ρ (solid line), the magnetic field intensity H_φ (dotted line), and the electron temperature T_e (dashed line). The upper left figure depicts the initial position of the liner.

Fig.5. Model electrical circuit.

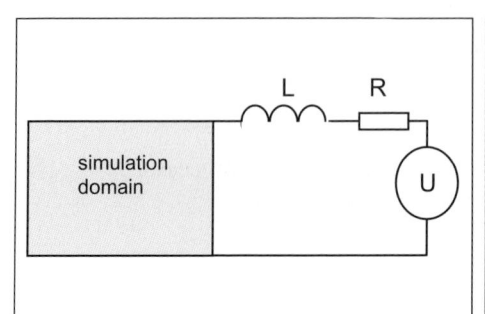

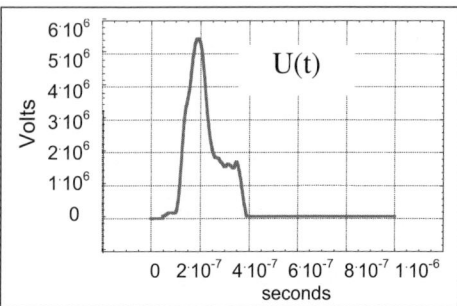

Fig.6. Voltage waveform.

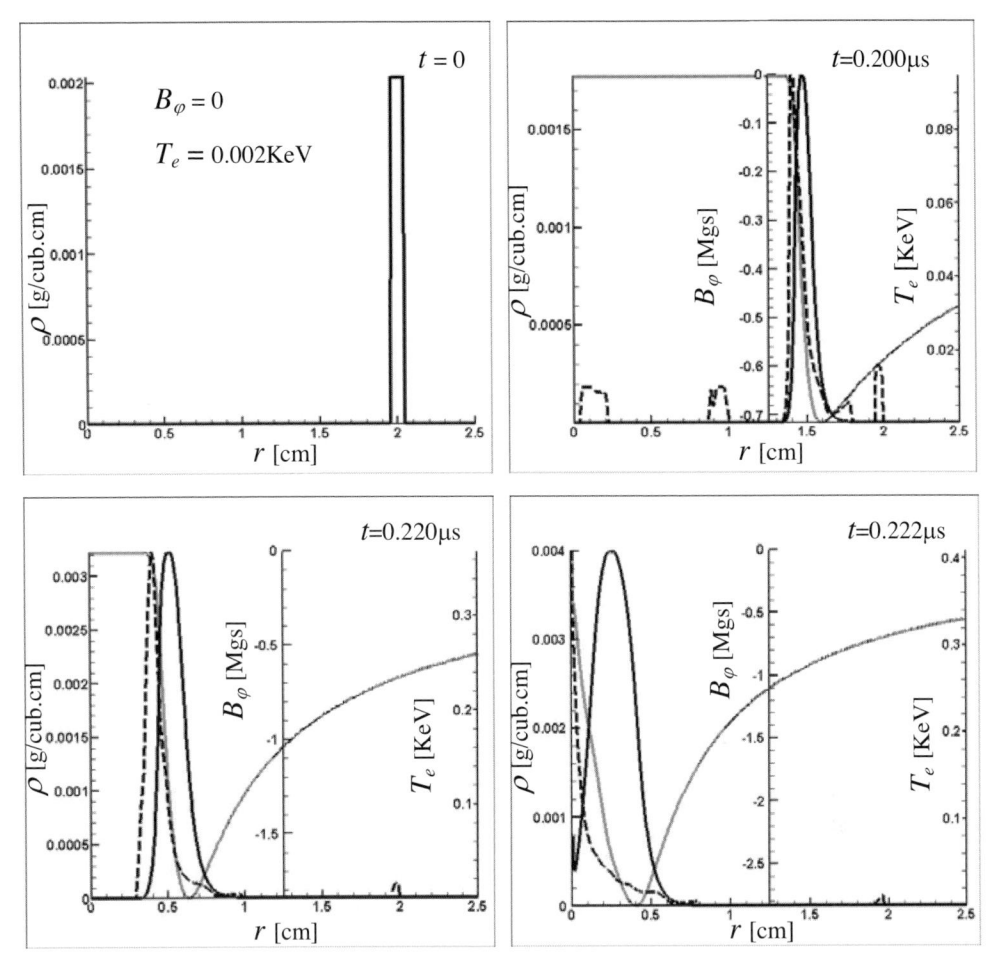

Fig.7. Liner compression by the magnetic field

The first tests of this new code showed promising results, which make us confident in good perspectives of using the unstructured grid technologies in pulsed-power simulations and give opportunity for future code developments.

REFERENCES
1. V.A.Gasilov and S.V.D'yachenko. Quasimonotonous 2D MHD scheme for unstructured meshes. Mathematical Modeling: modern methods and applications. Moscow, Janus-K, 2004, pp.108-125.
2. R.Siegel, J. R.Howell. Thermal radiation heat transfer. Tokyo, McGraw-Hill, 1972.
3. B.N.Chetverushkin. Mathematical modeling in radiative gasdymnamic problems. Moscow, Nauka publ., 1985.
4. C. Deeney, M.R.Douglass, R.B.Spielman, et. al. Phys. Rev. Lett., **81**, 4883, (1998).

Parallel performance of a UKAAC helicopter code on HPCx and other large-scale facilities

A. G. Sunderland[a], D. R. Emerson[b], C. B. Allen[c]

[a]*Advanced Research Computing Group, CSED, CCLRC Daresbury Laboratory, Warrington WA4 4AD, UK.*

[b]*Computational Engineering Group, CSED, CCLRC Daresbury Laboratory, Warrington WA4 4AD, UK.*

[c]*Department of Aerospace Engineering, University of Bristol, Bristol BS8 1TR, UK*

Keywords:Helicopter simulation; High Performance Computing; HPCx; Parallel performance; MPI.

Despite the major computing advances in recent years there has been a significant increase in the demand for more accurate, robust and reliable calculations of aerodynamic flows. New challenges are emerging that are pushing the limits of CFD calculations due to the complexity of the problems being proposed. Many of these problems are emerging from the aerospace industry and, in response to this, the UK has formed a new consortium that is focused on simulating challenges identified and driven by industrial aerospace needs. In this paper we analyse the performance of a helicopter rotor blade simulation code, developed by members of the consortium, on several large-scale high-end computer architectures available in the UK. Finally we also benchmark the PMB (Parallel Multi-Block) code on the HPCx supercomputer.

1. INTRODUCTION

The UK Applied Aerodynamics Consortium (UKAAC) [1] promotes the benefits of High Performance Computing (HPC) and brings together more than 30 researchers from 12 institutions and 3 industrial partners. The consortium has been awarded significant computational resources on the UK's flagship supercomputer HPCx to tackle several major aerodynamic challenges. The consortium is led by Dr. Ken Badcock (University of Glasgow) and Dr. David Emerson (Daresbury Laboratory). This 3 year programme started on June 1st, 2004.

Helicopter aerodynamics is one of the most scientifically challenging and industrially rewarding problems for CFD. Flow speeds range from subsonic to supersonic with vortices and wakes needing to be resolved and maintained over long times. Dynamic flow separation effects are also present. Part of the consortium's effort will target the understanding of multi-component helicopter interactions, including the main rotor/tail rotor interaction, responsible for noise and handling problems, and the fuselage influence on rotor flows. These simulations are going to have a significant impact on current efforts to feed full CFD simulations into simpler models, to enable improved design in the short term, and will be exploited to raise the profile of CFD within industrial companies for the longer term.

This paper will highlight the performance of such codes on large-scale massively parallel computing facilities. The machines considered involve a range of architectures with different processors, configurations and interconnects.

2. COMPUTING FACILITIES

Three recent procurements for centralised High Performance Computing (HPC) facilities in the UK have led to the establishment of an IBM system operated by the HPCx Consortium, an SGI-based solution at CSAR at the University of Manchester and an AMD Opteron-based compute cluster at the CCLRC e-Science Centre. These three facilities, described in more detail below, are primarily used by UK academic research groups.

HPCx is the UK's Flagship High Performance Computing Service. It is a large IBM p690+ cluster whose configuration is specifically designed for high-availability capability computing. HPCx is a joint venture between the Daresbury Laboratory of the Council for the Central Laboratories of the Research Councils (CCLRC) and Edinburgh Parallel Computing Centre (EPCC) at the University of Edinburgh [2]. The current phase of HPCx has 50 IBM p690+ Regatta nodes for computation with a total of 1.60 TBytes of memory and the High Performance Switch (HPS) for message passing, also formally known as "Federation". Each Regatta node houses 32 1.7 GHz IBM p690+ processors, making 1600 processors in total.

CSAR is the National High-Performance Computing Service run on behalf of the UK Research Councils by Computation for Science (CfS), a consortium comprising Computer Sciences Corporation (CSC), Silicon Graphics and the University of Manchester [3]. Their facilities now include "Newton"; a 512 processor SGI Altix 3700 which has 384 1.3 GHz Itanium2 processors and 128 1.5 GHz processors with 1 TByte of memory and the NUMAflex interconnect.

SCARF (Scientific Computing Application Resource for Facilities) [4] is a compute cluster run by the CCLRC e-Science Centre. It comprises of 128 dual CPU AMD Opteron 248 MSI 1000-D processors with 2.2 GHz clock speed. Message passing is achieved via a Myrinet M3F-PCIXD-2 interconnect.

The key challenge for the UK's flagship services is to deliver on the capability computing aspirations of the scientific and engineering community across a broad spectrum of disciplines, ranging from *ab initio* materials science through to applied aerodynamics and computational fluid dynamics.

3. HELICOPTER SIMULATIONS AND THE ROTORMBMGP CODE

Hover simulation requires the capture of several turns of the tip vortices to compute accurate blade loads, resulting in the requirement for fine meshes away from the surface, and a long numerical integration time for this wake to develop. For forward flight simulation, not only does the entire blade domain need to be solved, rather than a single blade for the hover simulation, but also the wake flow is now time dependent. The simulations are therefore computationally expensive due to the unsteady nature of the flow.

ROTORMBMGP (Rotor Multi-Block Multi-Grid Parallel) is a fixed- and rotary-wing flow simulation code developed by Dr. Chris Allen at the University of Bristol. The code uses an unsteady finite-volume upwind scheme to solve the integral form of the Euler equations. The code has been developed for structured multi-block meshes and incorporates an implicit temporal approach, with an explicit-type scheme within each real time-step. The code is written in Fortran95 and uses MPI for message passing. As a spatial stencil of five points in each direction is used, the point-to-point (send/receive) communication between processors represents the solution from two adjacent planes at each internal block boundary. A more detailed description of the problem, computational methodology and parallel code can be found in [5,6].

4. PROFILING ON HPCX AND CODE OPTIMIZATIONS

As reported in [6], the parallel performance of the original ROTORMBMGP code on HPCx is generally very good. However, after profiling the parallel code on HPCx with the Vampir tool it became evident that some further improvements could be made to the message passing efficiency within the code.

4.1. Profiling the code with Vampir & VampirTrace

VAMPIR (Visualization and Analysis of MPI Resources) [7] is a commercial postmortem trace visualization tool from Intel GmbH, Software & Solutions Group, the former Pallas HPC group. It uses the profiling extensions to MPI and permits analysis of the message events where data is transmitted between processors during execution of a parallel program. The tool comes in two components - VampirTrace and Vampir. VampirTrace is a library which, when linked and called from a parallel program, produces an event tracefile. The Vampir tool interprets the event tracefiles and represents the data in a graphical form for the user. Using the basic functionality of VampirTrace for MPI is straightforward: relink your MPI application with the appropriate VampirTrace -lVT library, add some environment variables to your batch script and execute the application as usual.

The standard view in Vampir is the global *timeline* display, with processors listed on the vertical axis and time represented by the horizontal axis. Figure 1 shows a zoomed view of the timeline display for one cycle of the solver in ROTORMBMGP on 16 HPCx processors.

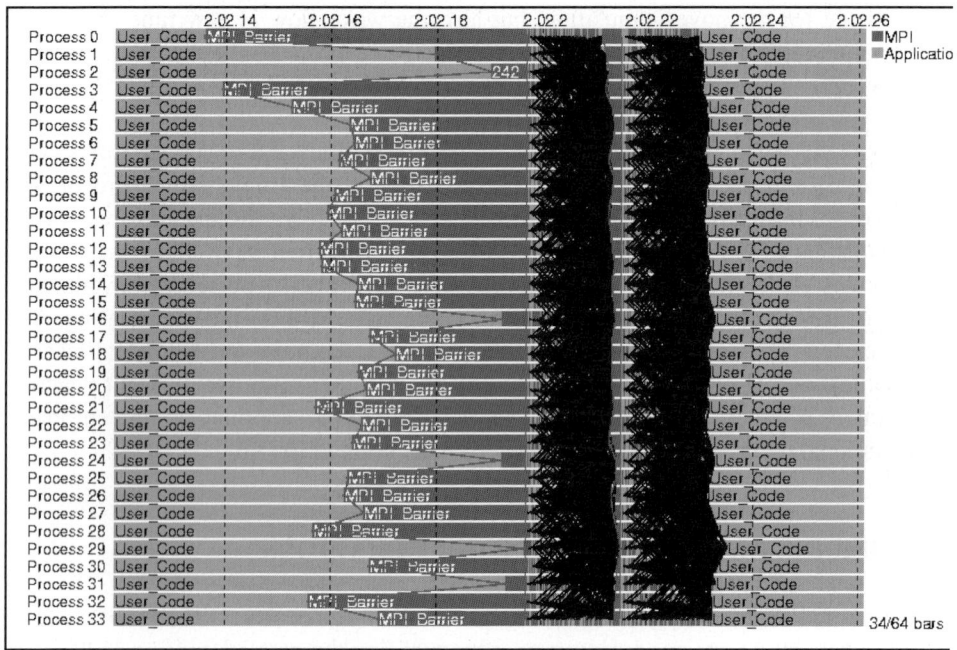

Figure 1. Vampir visualization of message passing across block boundaries on HPCx (original code).

Using a greyscale colour scheme, the dark grey areas represent time spent in communication (or time spent waiting for communication to take place) and light grey areas represent time spent in computation. The two distinct bands of black lines denote message passing between processors, representing the exchange of data across block boundaries for both planes of solution, separated by global calls to MPI Barrier synchronisations. This separation is unnecessary and, on HPCx, enforces multiple, relatively costly processor synchronizations (the cost may vary depending on the computer architecture). By removing these global synchronizations, more flexibility is introduced into the communication pattern and processors need wait only for their neighbours to send them data before proceeding with their computation for the next cycle. The detailed timeline view for the message passing associated with one cycle of the solver from the optimized version is shown in Fig. 2.

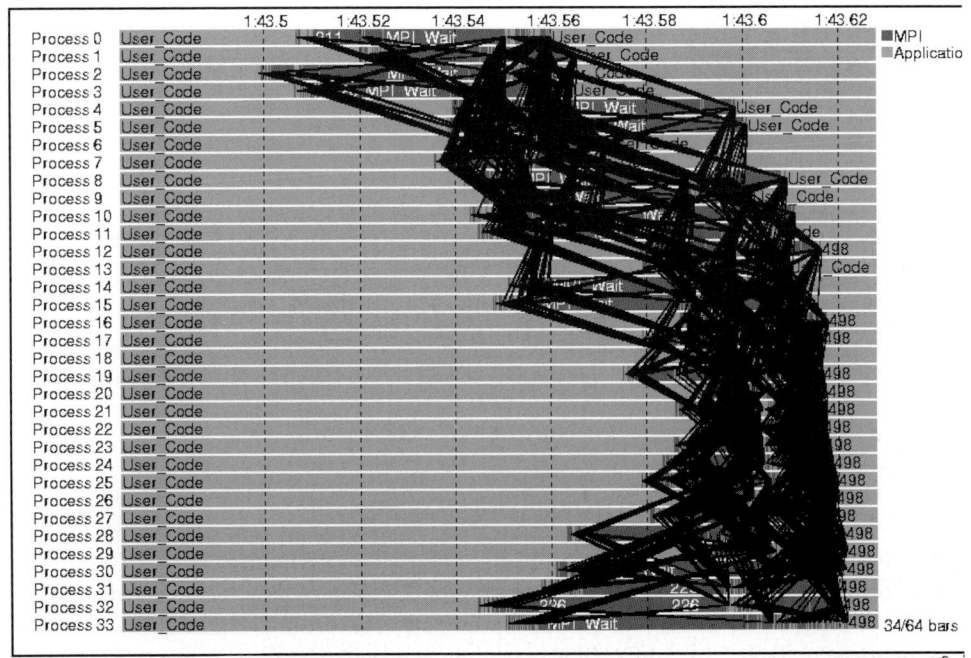

Figure 2. Vampir visualization of message passing across block boundaries on HPCx (optimized code).

5. BENCHMARK TIMINGS ON HIGH PERFORMANCE COMPUTING ARCHITECTURES

5.1. Parallel performance of ROTORMBMGP on HPCx

The benchmark runs used a multi-block multi-grid parallel scheme to model the unsteady flow about a 4-bladed rotor in forward flight. Simulations have been run using mesh densities of 20M points and the scaling performance on HPCx for both the original code and the optimized version of the code are shown in Fig. 3.

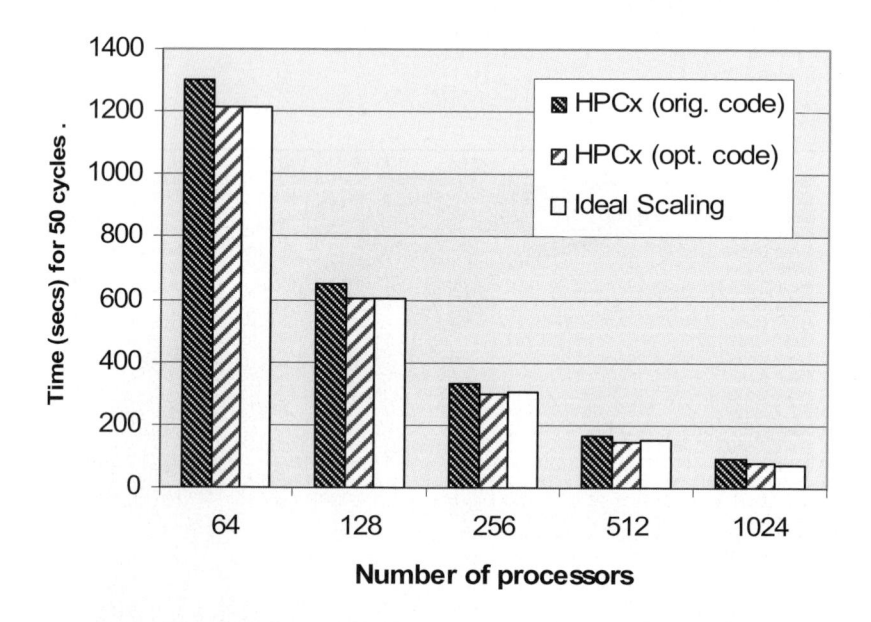

Figure 3. Parallel performance of ROTORMBMGP on HPCx.

The performance gain with the optimized version is more marked at large processor counts resulting in a reduction in compute time of 7.5% on 64 processors and 11.5% on 1024 processors. The speed-up for the optimized code from 64 processors to 1024 processors is a factor of 14.98 (linear scaling corresponds to a factor of 16). The performance on each processor averages 475 Mflop/s, which is around average for CFD codes on HPCx.

5.2. Parallel performance on other platforms

The performance of ROTORMBMGP on the three computing platforms decribed in Section 2 is shown in Fig. 4. It can be seen that the code scales well across all the platforms investigated. The code scales well on all three systems, but parallel performance is best on the Altix system (Newton).

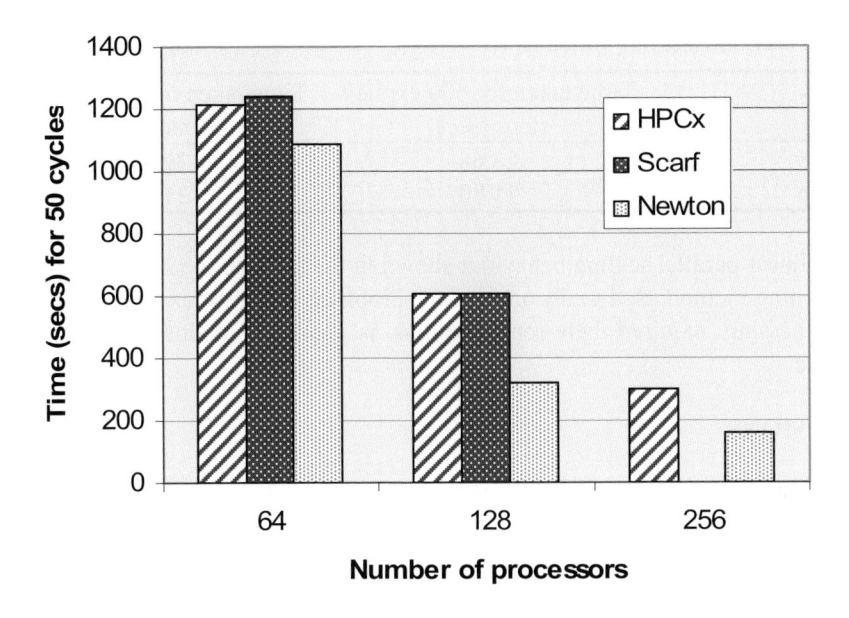

Figure 4. Parallel performance of the optimized ROTORMBMGP code on the three HPC architectures.

5.3. Capability incentive scheme

Developers and users of codes that demonstrate excellent scaling properties on HPCx can apply for "Seals of Approval" awards for their codes that qualifies them for discounted use of HPCx CPU resources. After undergoing these scaling tests, ROTORMBMGP was awarded a Gold Seal of Approval, thereby entitling users of this code to a 30% discount for CPU resources.

6. PMB FLOW SOLVER

Another UKAAC code benchmarked on HPCx is the PMB (Parallel Multi-Block) [8] code developed by the Computational Fluid Dynamics Group at the University of Glasgow. The code utilizes a control volume method to solve the Reynolds averaged Navier-Stokes (RANS) equations with a variety of turbulence models including k-ω, LES and DES-SA. The predominant numerical method in the code is a Krylov sub-space linear solver with pre-conditioning. As with the ROTORMBMGP code, information at boundaries is transferred using MPI-based message passing.

The benchmark case on HPCx is a cavity simulation with 8 million grid points. The model simulates a no-doors environment that involves a highly turbulent, unsteady separated flow field. Due to the high memory requirements this simulation cannot be run on fewer than 128 HPCx processors.

Table 1. Parallel performance of PMB on HPCx

Number of Processors	Time taken for 600 explicit steps (secs)	Time taken for 600 implicit steps (secs)
128	28.68	79.29
256	11.80	31.95

The super-linear parallel scaling behaviour shown in the table above is a consequence of the partitioning method used to decompose the problem. It transpires that the amount of message passing required between partitions is markedly reduced in the 256 processor case.

7. CONCLUSIONS

Both UKAAC codes scale very well on HPCx for the benchmark datasets tested here. In particular, the message passing characteristics of ROTORMBMGP have been profiled using specialized tools and the code has been modified accordingly in order to improve performance further. The optimized code scales near-linearly upto 1024 processors and has therefore been granted a Gold award on HPCx. Similarly on other HPC architectures the parallel scaling is excellent, with particularly impressive results demonstrated on the Altix machine.

ACKNOWLEDGEMENTS

The authors would like to thank EPSRC for their support of the UK Applied Aerodynamics Consortium (UKAAC) under grant GR/S91130/01 and the CCP12 programme.

REFERENCES
1. Information on the UKAAC is at http://www.cse.clrc.ac.uk/ceg/ukaac/ukaac.shtml
2. HPCx UoE Ltd, UK, http://www.hpcx.ac.uk.
3. Computer Services for Academic Research (CSAR), University of Manchester, UK, http://www.csar.cfs.ac.uk.
4. Scientific Computing Application Resources for Facilities (SCARF), CCLRC, UK, http://www.scarf.rl.ac.uk.
5. Allen, C. B., *"An Unsteady Multiblock Muligrid Scheme for Lifting Forward Flight Simulation"*, International Journal for Numerical Methods in Fluids, Vol. 45, No. 7, 2004.
6. Allen C. B., *" Parallel Simulation of Lifting Rotor Wakes"*, in Proceedings of Parallel CFD Conference, University of Maryland, 2005.
7. Pallas High Performance Computing products, http://www.pallas.com/e/products/index.htm.
8. Steijl, R., Nayyar, P., Woodgate, M. A.; Badcock, K. J., Barakos, G. N., *"Application of an implicit dual-time stepping multi-block solver to 3D unsteady flows"*, in Proceedings of Parallel CFD Conference, University of Maryland, 2005.

Parallel Computations of Unsteady Aerodynamics and Flight Dynamics of Projectiles

Jubaraj Sahu

U.S. Army Research Laboratory

Aberdeen Proving Ground, Maryland 21005-5066, USA

Keywords: Parallel computing; unsteady aerodynamics; free flight; coupled method

1. INTRODUCTION

As part of a Department of Defense High Performance Computing (HPC) grand challenge project, the U.S. Army Research Laboratory (ARL) has recently focused on the development and application of state-of-the art numerical algorithms for large-scale simulations [1-3] to determine both steady and unsteady aerodynamics of projectiles with and without flow control. One of the objectives is to exploit computational fluid dynamics (CFD) techniques on HPC platforms for design and analysis of micro adaptive flow control (MAFC) systems for steering spinning projectiles for infantry operations. The idea is to determine if the MAFC using synthetic jets can provide the desired control authority for course correction for munitions. Another goal is to perform real-time multidisciplinary coupled computational fluid dynamics/rigid body dynamics computations for the flight trajectory of a complex guided projectile system and fly it through a "virtual numerical tunnel" similar to what happens with the actual free flight of the projectile inside an aerodynamics experimental facility. Our initial attempt in the past has been to use a quasi-unsteady approach [4] in an effort to save computer time and resources. Recently progress has been made in coupling CFD and rigid body dynamics (RBD) to perform required time-accurate multidisciplinary simulations for moving body problems. Also, recent advances in computational aerodynamics and high performance computing make it possible to solve increasingly larger problems including these multidisciplinary computations at least for short time-of-flights. Complete time-accurate multidisciplinary simulations of entire trajectory for long-range guided munitions are still beyond the scope. A few highly optimized, parallelized codes [5,6] are available for such computations. Various techniques, including structured, unstructured, and overset methods can be employed to determine the unsteady aerodynamics of advanced guided projectiles and missiles. In the present study, a real time-accurate approach using unstructured methodology has been utilized to perform such complex computations for both a spinning projectile and a finned projectile. The maneuver of a projectile through flow control adds another complexity to the trajectory and the unsteady aerodynamics computations. Trajectory computations for the spinning projectile have been accomplished both without and with aerodynamic flow control through the use of synthetic jets.

Calculations for the spinning projectile as well as a finned configuration have been performed using a scalable parallel Navier-Stokes flow solver, CFD++ [7,8]. The method used is scalable on SGI Origin 3000, IBM SP3/Sp4, and the Linux PC Cluster and incorporates programming enhancements such as dynamic memory allocation and highly optimized cache management. It has been used extensively in the parallel high performance computing numerical simulations of projectile and missile programs of interest to the U.S. Army. The advanced CFD capability used in the present study solves the Reynolds-Averaged Navier-Stokes (RANS) equations [9] and incorporates unsteady boundary conditions for simulation of the synthetic jets [10,11]. Also, a hybrid RANS/ LES (Large eddy Simulation) turbulence model [12-15] was used for accurate numerical prediction of unsteady jet flows. Computed lift forces due to the jets using the hybrid RANS/LES model were found to match well the experimental data [16-18]. The present numerical study is also a big step forward which now includes numerical simulation of the actual fight paths of the projectile both with and without flow control using coupled CFD/RBD techniques.

2. COMPUTATIONAL METHODOLOGY

The complete set of three-dimensional (3-D) time-dependent Navier-Stokes equations [9] is solved in a time-accurate manner for simulations of unsteady flow fields associated with both spinning and finned projectiles during flight. The 3-D time-dependent RANS equations are solved using the finite volume method [7,8]:

$$\frac{\partial}{\partial t}\int_V \mathbf{W} dV + \oint [\mathbf{F} - \mathbf{G}] \cdot dA = \int_V \mathbf{H} dV \qquad (1)$$

where $\mathbf{W}$ is the vector of conservative variables, $\mathbf{F}$ and $\mathbf{G}$ are the inviscid and viscous flux vectors, respectively, $\mathbf{H}$ is the vector of source terms, V is the cell volume, and A is the surface area of the cell face.

Two-equation [15] and higher order hybrid RANS/LES [7] turbulence models were used for the computation of turbulent flows. The hybrid RANS/LES approach based on Limited Numerical Scales (LNS) is well suited to the simulation of unsteady flows. With this method a regular RANS-type grid is used except in isolated flow regions where denser, LES-type mesh is used to resolve critical unsteady flow features. The hybrid model transitions smoothly between an LES calculation and a cubic k-ε model, depending on grid fineness. Dual time-stepping was used to achieve the desired time-accuracy. In addition, special boundary conditions were developed and used for numerical modeling of synthetic jets. Grid was actually moved to take into account the spinning motion of the projectiles.

- **Grid Movement**

Grid velocity is assigned to each mesh point. For a spinning projectile, the grid speeds are assigned as if the grid is attached to the projectile and spinning with it. A proper treatment of grid motion requires careful attention to the details of the implementation of the algorithm applied to every mesh point and mesh cell so that no spurious numerical effects are created. For example, a required consistency condition is that free stream uniform flow be preserved for arbitrary meshes and arbitrary mesh velocities. Another important aspect deals with how boundary conditions are affected by grid velocities. Two significant classes of boundary conditions: slip or no slip at a wall, and far field boundary, are considered here. Both are treated in a manner which works seamlessly with or without mesh velocities. In both cases, the contravariant velocity which includes the effect of grid motion is used and appropriate boundary conditions are applied at the wall and the far field boundaries.

- **6-Degree-of-Freedom (6-DOF) Rigid Body Dynamics**

The coupling between CFD and RBD refers to the interaction between the aerodynamic forces/moments and the dynamic response of the projectile to these forces and moments. The forces and moments are computed every CFD time step and transferred to a 6DOF module which computes the body's response to the forces and moments. The response is converted into translational and rotational accelerations that are integrated to obtain translational and rotational velocities and integrated once more to obtain linear position and angular orientation. The 6-DOF rigid body dynamics module uses quaternions to define the angular orientations; however, these are easily translated into Euler angles. From the dynamic response, the grid point locations and grid point velocities are set.

3. PARALLEL COMPUTATIONAL ISSUES

The CFD++ code was designed from the outset to include unified computing. The "unified computing" capability includes the ability to perform scalar and parallel simulations with great ease. The parallel processing capability in CFD++ allows the code to run on a wide variety of hardware platforms and communications libraries including MPI, PVM, and proprietary libraries of nCUBE, Intel Paragon etc. Currently MPI is a de facto standard and CFD++ uses public domain and vendor-specific versions of MPI depending on the hardware platform. The code is compatible with and can provide good performance on standard Ethernet (e.g. 100Mbit, 1Gbit, 10Gbit) as well as high performance communications channels of Myrinet and Infiniband etc.

CFD++ code can be easily on any number of CPUs in parallel. The mesh files, restart and plot files) that are needed/generated for single CPU runs are identical to those associated with multi CPU runs for any number of CPUs. One can switch the use of an arbitrary number of CPUs at any time. The only extra file required is a domain-decomposition file, which defines the association between cell number and which CPU should consider that cell as its "native" cell. Depending on the number of CPUs being employed, the corresponding domain decomposition is utilized. Several domain decomposition tools are avaialble that are fully compatible with the METIS tool developed at the University of Minnesota. The code runs in parallel on many parallel computers including those from Silicon Graphics, IBM, Compaq (DEC and HP), as well as on PC workstation clusters. Excellent performance (see Figure 1 for the timings on a 4-million mesh) has been observed up to 64 processors on Silicon Graphics O3K (400 MHz), IBM SP P3 (375 MHz), IBM SP P4(1.7GHz), and Linux PC cluster (3.06 GHz). Computed results on the new Linux PC

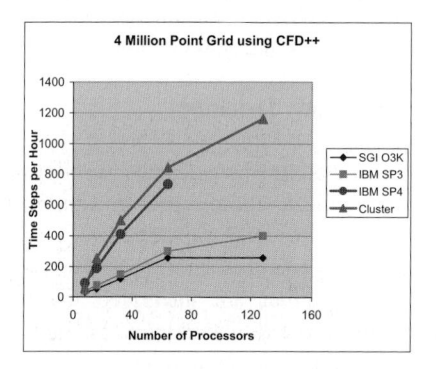

Figure 1. Parallel Speedups (4-million grid).

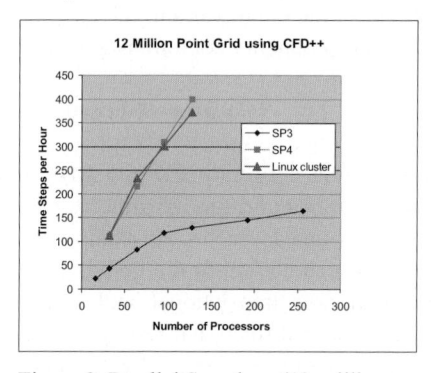

Figure 2. Parallel Speedups (12-million grid).

cluster seem to show 2 to 4-fold reduction in CPU time for number of processors larger than 16. The reason for somewhat poor performance with 8 processors is most likely due to memory requirements. Similar good performance is also achieved on the Linux PC cluster for a larger 12-million mesh (see Figure 2) up to 128 processors.

4. RESULTS

    o   **Spinning Projectile CFD with 6-DOF**

Time-accurate unsteady numerical computations were performed to predict the flow fields and the flight paths of a spinning projectile using coupled multi-disciplinary CFD/RBD technique. Numerical computations have been made at an initial Mach number, M = 0.39, initial angle of attack, $\alpha = 2°$, and an initial spin rate of 434 Hz with and without synthetic jet. The jet was located just upstream of the base [18]. The jet width was 0.32 mm and the peak jet velocity used was 110 m/s operating at a frequency of 1000 Hz.

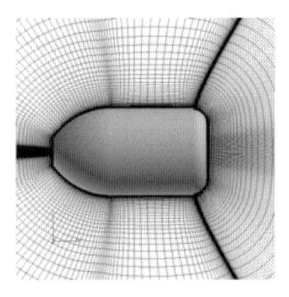

Figure 3. Computational grid near the projectile.

The projectile used in this study is a 1.8-caliber ogive-cylinder configuration. A computational grid expanded near the vicinity of the projectile is shown in Figure 3. The jet actuation corresponds to one-fourth of the spin cycle as shown in Figure 4. The jet is off during the remaining three-fourths of the spin cycle. For the part of the spin cycle with jet-on, the jet operated for approximately four cycles. Time-accurate CFD modeling of each jet cycle required over 20 time steps. The actual CPU time for one full spin cycle of the projectile was about 50 hours using 32 processors on an IBM SP3 system for a mesh size about four million grid points. Multiple spin cycles and, hence, a large number of synthetic jet operations were required in the trajectory simulations after the jet is activated.

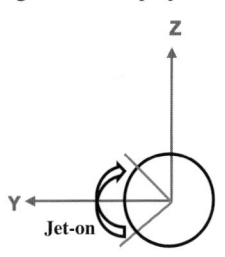

Figure 4. Schematic showing the jet actuation in one spin cycle.

Figure 5 shows the velocity magnitude contours at a given time in the trajectory. It clearly shows the orientation of the body at that instant in time and the resulting asymmetric flow field in the wake due to the body at angle of attack. Computed results have been obtained with unsteady RANS (URANS) as well as the hybrid RANS/LES (or LNS) for the jet-on conditions. The jet-off URANS calculations were first obtained and the jets were activated beginning at time, t = .28 sec or a distance of approximately 41 m of travel. The computed side force (F_Y) is shown as a function of distance (or range) in Figure 6. The mean amplitude of the side force is seen to increase with distance. The RANS approach

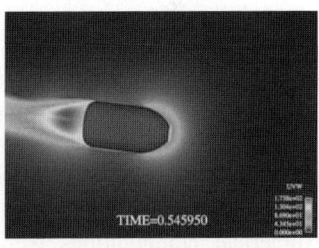

Figure 5. Computed velocity magnitudes at a given instant in time.

clearly shows when the jet is on and when it is off during each spin cycle. The effect due to the jet for the LNS case is not as easily seen. It is more oscillatory. In general, the levels of the force oscillations predicted by the LNS models are larger than those predicted by the RANS.

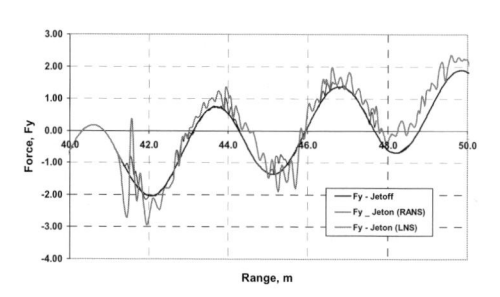

Figure 6. Comparison of computed side force, URANS and LNS, jet-off and jet-on).

Figure 7. Comparison of the computed distance, URANS and LNS, jet-off and jet-on.

Figure 7 shows the computed y-distance as a function of x or the range. Even though the jet was turned on at x = 41 m, the effect of the jet is not immediately felt for about 1.5 m. The RANS model predicts a small increase in y with increase in range. With the LNS model however, y decreases a little at first until x = 46.5 m and then increases more rapidly. These computed results strongly indicate that applying the jet in the +ve y-direction moves the projectile in the same positive direction.

o Finned Projectile CFD with 6-DOF

Time-accurate unsteady numerical computations were performed using Navier-Stokes and coupled 6-DOF methods to predict the flow fields and the flight paths of a ogive-cylinder-finned projectile (see Fig. 8) at supersonic speeds. Four fins are located on the back end of the projectile. Numerical computations were made for these cases at an initial velocities 1037 and 1034 m/s depending on whether the simulations were started from the muzzle or a small distance away from it. The corresponding initial angles of attack were, $\alpha = 0.5°$ or $4.9°$ and initial spin rates were 2800 or 2500 rad/s, respectively.

Figure 8. Computational Finned Configuration.

Figure 9 shows the computed z-distance as a function of x. The computed results are shown in solid lines and are compared with the data measured from actual flight tests. For the computed results the aerodynamic forces and moments were completely obtained through CFD. One simulation started from the gun muzzle and the other from the first station (4.9 m away from the muzzle) where the actual data was measured. Both sets of results are generally found to be in good agreement with the measured data, although there is a small discrepancy between the two sets of computed results.

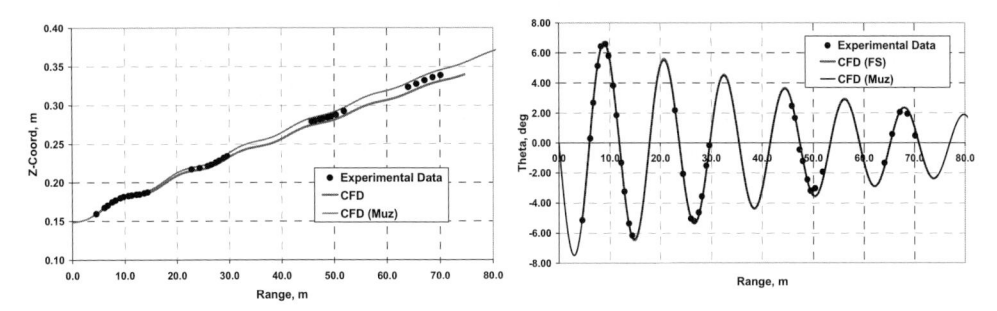

Figure 9. Computed z distance vs. range.   Figure 10. Euler pitch angle vs. range.

Figure 10 shows the variation of the Euler pitch angle with distance traveled. As seen here, both the amplitude and frequency in the Euler pitch angle variation are predicted very well by the computed results and match extremely well with the data from the flight tests. Both sets of computations, whether it started from the muzzle or the first station away from the muzzle, yield essentially the same result. One can also clearly see the amplitude damped out as the projectile flies down range i.e. with increasing x-distance.

5. CONCLUDING REMARKS

This paper describes a new coupled CFD/RBD computational study undertaken to determine the flight aerodynamics of both a spinning projectile and a finned projectile using a scalable unstructured flow solver on various parallel computers such as the IBM, and Linux Cluster. Advanced scalable Navier-Stokes computational techniques were employed to compute the time-accurate aerodynamics associated with the free flight of a finned projectile at supersonic velocities and a spinning projectile at subsonic speeds both with and without flow control. High parallel efficiency was achieved for the real time-accurate unsteady computations. In the coupled CFD/RBD calculations for the spinning projectile, the hybrid RANS/LES)results are found to be more unsteady than the RANS predictions. The hybrid RANS/LES approach predicted a larger increase in the side force and hence, a larger increase in the side distance when the jet was applied in that direction. For the finned configuration, computed positions and orientations of the projectile obtained from the coupled CFD/RBD calculations have been compared with actual data measured from free flight tests and are found to be generally in very good agreement.

- **REFERENCES**

    1. J. Sahu, Pressel, D., Heavey, K.R., and Nietubicz, C.J., "Parallel Application of a Navier-Stokes Solver for Projectile Aerodynamics", Parallel CFD'97 Meeting, Manchester, UK, May 1997.
    2. J. Sahu, Edge, H.L., Dinavahi, S., and Soni, B., "Progress on Unsteady Aerodynamics of Maneuvering Munitions" Users Group Meeting Proceedings, Albuquerque, NM, June 2000.
    3. J. Sahu, "Unsteady Numerical Simulations of Subsonic Flow over a Projectile with Jet Interaction", AIAA paper no. 2003-1352, Reno, NV, 6-9 Jan 2003.
    4. J. Sahu, H.L. Edge, J. DeSpirito, K.R. Heavey, S.V. Ramakrishnan, S.P.G. Dinavahi, "Applications of Computational Fluid Dynamics to Advanced Guided Munitions", AIAA Paper No. 2001-0799, 39th AIAA Aerospace Sciences meeting Reno, Nevada, 8-12 January 2001.
    5. Peroomian, O. and Chakravarthy S., "A `Grid-Transparent' Methodology for CFD," AIAA Paper 97-0724, Jan. 1997.
    6. Meakin, R. And Gomez, R., "On Adaptive Refinement and Overset Structured Grids", AIAA Paper No. 97-1858-CP, 1997.
    7. P. Batten, U. Goldberg and S. Chakravarthy, "Sub-grid Turbulence Modeling for Unsteady Flow with Acoustic Resonance", AIAA Paper 00-0473, Reno, NV, January 2000
    8. O. Peroomian, S. Chakravarthy, S. Palaniswamy, and U. Goldberg, "Convergence Acceleration for Unified-Grid Formulation Using Preconditioned Implicit Relaxation." AIAA Paper 98-0116, 1998.
    9. T. H. Pulliam and J. L. Steger, "98dOn Implicit Finite-Difference Simulations of Three-Dimensional Flow" *AIAA Journal*, vol. 18, no. 2, pp. 159–167, February 1982
    10. B. L. Smith and A. Glezer, "The Formation and Evolution of Synthetic Jets." *Journal of Physics of Fluids*, vol. 10, No. 9, September 1998
    11. M. Amitay, V. Kibens, D. Parekh, and A. Glezer, "The Dynamics of Flow Reattachment over a Thick Airfoil Controlled by Synthetic Jet Actuators", AIAA Paper No. 99-1001, January 1999
    12. S. Arunajatesan and N. Sinha, "Towards Hybrid LES-RANS Computations of Cavity Flowfields", AIAA Paper No. 2000-0401, January 2000
    13. R.D. Sandberg and H. F. Fasel, "Application of a New Flow Simulation Methodology for Supersonic Axisymmmetric Wakes", AIAA Paper No. 2004-0067.
    14. S. Kawai and K. Fujii, "Computational Study of Supersonic Base Flow using LES/RANS Hybrid Methodology", AIAA Paper No. 2004-68.
    15. U. Goldberg, O. Peroomian, and S. Chakravarthy, "A Wall-Distance-Free K-E Model With Enhanced Near-Wall Treatment" *ASME Journal of Fluids Engineering*, Vol. 120, 457-462, 1998
    16. C. Rinehart, J. M. McMichael, and A. Glezer, "Synthetic Jet-Based Lift Generation and Circulation Control on Axisymmetric Bodies." AIAA Paper No. 2002-3168
    17. McMichael, J., GTRI, Private Communications.
    18. Sahu, J., "Unsteady CFD Modeling of Aerodynamic Flow C ontrol over a Spinning Body with Synthetic Jet." AIAA Paper 2004-0747, Reno, NV, 5-8 January 2004.

Parallel Computational Fluid Dynamics – Theory and Applications
A. Deane et al. (*Editors*)
Published by Elsevier B.V.

Parallel Adaptive Solvers in Compressible PETSc-FUN3D Simulations[*]

S. Bhowmick,[a] D. Kaushik,[b] L. McInnes,[b] B. Norris,[b] P. Raghavan [c]

[a]Department of Applied Physics and Applied Mathematics, Columbia University,
200 S.W. Mudd Building, 500 W. 120th Street, New York, NY 10027,
E-mail: *sb2423@columbia.edu*

[b]Mathematics and Computer Science Division, Argonne National Laboratory,
9700 South Cass Avenue, Argonne, IL 60439-4844,
E-mail: *[kaushik,mcinnes,norris]@mcs.anl.gov*

[c]Department of Computer Science and Engineering, The Pennsylvania State University,
343K IST Building, University Park, PA 16802-6106, E-mail: *raghavan@cse.psu.edu*

Abstract. We consider parallel, three-dimensional transonic Euler flow using the PETSc-FUN3D application, which employs pseudo-transient Newton-Krylov methods. Solving a large, sparse linear system at each nonlinear iteration dominates the overall simulation time for this fully implicit strategy. This paper presents a polyalgorithmic technique for adaptively selecting the linear solver method to match the numeric properties of the linear systems as they evolve during the course of the nonlinear iterations. Our approach combines more robust, but more costly, methods when needed in particularly challenging phases of solution, with cheaper, though less powerful, methods in other phases. We demonstrate that this adaptive, polyalgorithmic approach leads to improvements in overall simulation time, is easily parallelized, and is scalable in the context of this large-scale computational fluid dynamics application.

1. INTRODUCTION

Many time-dependent and nonlinear computational fluid dynamics (CFD) applications involve the parallel solution of large-scale, sparse linear systems. Typically, application developers select a particular algorithm to solve a given linear system and keep this algorithm fixed throughout the simulation. However, it is difficult to select *a priori* the most effective algorithm for a given application. Moreover, for long-running applications in which the numerical properties of the linear systems change as the simulation progresses, a single algorithm may not be best throughout the entire simulation. This situation has motivated us to develop an adaptive, polyalgorithmic approach for linear solvers, which this paper discusses in the context of parallel, three-dimensional transonic Euler flow using PETSc-FUN3D [2]. This application employs pseudo-transient Newton-Krylov methods for a fully implicit solution. We

---

[*]This work was supported in part by the National Science Foundation through grants ACI-0102537, CCR-0075792, CCF-0352334, CCF-0444345, ECS-0102345 and EIA-022191 and by the Mathematical, Information, and Computational Sciences Division subprogram of the Office of Advanced Scientific Computing Research, Office of Science, U.S. Department of Energy, under Contract W-31-109-ENG-38 and DE-AC03-76SF00098.

present a technique for adaptively selecting the linear solver methods to match the numeric properties of the linearized Newton systems as they evolve during the course of the nonlinear iterations. Our approach combines more robust, but more costly, methods when needed in particularly challenging phases of the solution, with faster, though less powerful, methods in other phases. Our previous work focused on sequence-based adaptive heuristics in an uniprocessor environment [4,5,9]. In this paper, we extend our research to solve a more complicated parallel application, where we demonstrate that the adaptive polyalgorithmic approach can be easily parallelized, is scalable, and can lead to improvements in overall simulation time.

The remainder of this paper is organized as follows. Section 2 introduces our motivating application. Section 3 explains our approach to adaptive solvers, and Section 4 presents some parallel experimental results. Section 5 discusses conclusions and directions of future work.

2. PARALLEL COMPRESSIBLE FLOW EXAMPLE

FUN3D is an unstructured mesh code originally developed by W. K. Anderson of the NASA Langley Research Center [1] for solving the compressible and incompressible Navier-Stokes equations. This code uses a finite volume discretization with a variable order Roe scheme on a tetrahedral, vertex-centered mesh. Anderson et al. subsequently developed the parallel variant PETSc-FUN3D [2], which incorporates MeTiS [7] for mesh partitioning and the PETSc library [3] for the preconditioned Newton-Krylov family of implicit solution schemes.

This paper focuses on solving the unsteady compressible three-dimensional Euler equations

$$\frac{\partial u}{\partial t} + \frac{1}{V} \oint_\Omega \left(\vec{F} \cdot \hat{n} \right) d\Omega = 0, \tag{1}$$

where

$$u = \begin{bmatrix} \rho \\ \rho u_x \\ \rho u_y \\ \rho u_z \\ E \end{bmatrix}, \qquad \vec{F} \cdot \hat{n} = \begin{bmatrix} \rho U \\ \rho U u_x + \hat{n}_x p \\ \rho U u_y + \hat{n}_x p \\ \rho U u_z + \hat{n}_z p \\ (E+p)U \end{bmatrix},$$

$$U = \hat{n}_x u_x + \hat{n}_y u_y + \hat{n}_z u_z, \qquad p = (\gamma - 1)\left[E - \rho \frac{(u_x^2 + u_y^2 + u_z^2)}{2}\right].$$

Here ρ, E, and U represent the density, energy, and velocity in the direction of the outward normal to a cell face, respectively. The pressure field p is determined by the equation of state for a perfect gas (given above). Also the cell volume (V) is enclosed by the cell boundary (Ω).

We solve the nonlinear system in Equation (1) with pseudo-timestepping [8] to advance towards an assumed steady state. Using backward Euler time discretization, Equation (1) becomes

$$\frac{V}{\Delta t^\ell}(u^\ell - u^{\ell-1}) + f(u^\ell) = 0, \tag{2}$$

where $\Delta t^\ell \to \infty$ as $\ell \to \infty$, u is a vector of unknowns representing the state of the system, and $f(u)$ is a vector-valued function of residuals of the governing equations, which satisfies $f(u) = 0$ in the steady state.

This code employs Roe's flux-difference splitting to discretize the convective terms. Initially Equation (2) is discretized by using a first-order scheme. When the nonlinear residual sinks

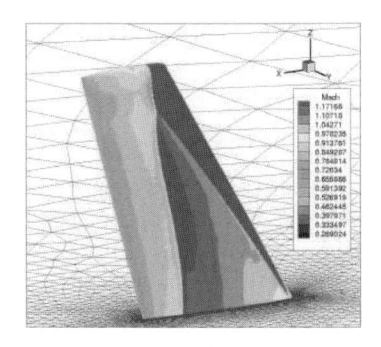

Figure 1. Mach contours on the ONERA M6 wing at freestream Mach number = 0.839.

below a given threshold, a second-order discretization is applied. The timestep is advanced toward infinity by a power-law variation of the switched evolution/relaxation (SER) heuristic of Mulder & Van Leer [10]. To be specific, within each of the first-order and second-order phases of computation we adjust the timestep according to the CFL number,

$$N_{CFL}^{\ell} = N_{CFL}^{0} \left(\frac{\|f(u^0)\|}{\|f(u^{\ell-1})\|} \right)^{\sigma},$$

where σ is normally unity but is damped to 0.75 for robustness in cases in which shocks are expected to appear.

At each timestep we apply a single inexact Newton iteration (see, e.g., [11]) to Equation (2) through the two-step sequence of (approximately) solving the Newton correction equation

$$(\frac{V}{\Delta t^{\ell}} I + f'(u^{\ell-1})) \delta u^{\ell} = -f(u^{\ell-1}), \qquad (3)$$

where I is the identity matrix, and then updating the iterate via $u^{\ell} = u^{\ell-1} + \delta u^{\ell}$. We employ matrix-free Newton-Krylov methods (see, e.g., [6]), with which we compute the action of the Jacobian on a vector v by directional differencing of the form $f'(u)v \approx \frac{f(u+hv)-f(u)}{h}$, where h is a differencing parameter. We use a first-order analytic discretization to compute the corresponding preconditioning matrix.

We explore the standard aerodynamics test case of transonic flow over an ONERA M6 wing using the frequently studied parameter combination of a freestream Mach number of 0.839 with an angle of attack of 3.06°. The robustness of solution strategies is particularly important for this model because of the so-called λ-shock that develops on the upper wing surface, as depicted in Figure 1. As mentioned earlier, the PDEs are discretized by using a first-order scheme; but once the shock position has settled down, a second-order discretization is applied. This change in discretization affects the nature of the resulting linear systems. The time for the complete simulation is dominated by the time to solve the linear systems generated at each nonlinear iteration, where this phase typically requires around 77 percent of the overall time. Moreover, changes in the numerical characteristics of the linear systems reflect the changing nature of the simulation. For example, the use of pseudo-transient continuation generates linear systems that become progressively more difficult to solve as the simulation advances. This situation is discussed further in Section 4; in particular, see the right-hand graph of Figure 2.

3. ADAPTIVE SOLVERS

Applications where the properties of the linear systems differ significantly throughout the simulation, such as PETSc-FUN3D's [2] modeling of compressible Euler flow, are an ideal case for exploring adaptive solvers. Our goal is to improve overall performance by combining more robust (but more costly) methods when needed in particularly challenging phases of solution with faster (though less powerful) methods in other phases. Adaptive solvers are designed with the goal of reducing overall execution time by dynamically selecting the most appropriate method to match the needs of the current linear system.

Adaptive solvers can be defined by the heuristic employed for method selection. The efficiency of an adaptive heuristic depends on how appropriately it determines *switching points*, or the iterations at which to change linear solvers. Adaptive heuristics monitor changes in *indicators* to detect switching points. We have observed that a combination of several indicators generally leads to better results. Some common examples of indicators are linear (nonlinear) solution time, convergence rate, change in nonlinear residual norm, etc.

In this paper we employ sequence-based adaptive heuristics, which rely on a predetermined sequence of linear solvers and then "switch up" to a more robust but more costly method or "switch down" to a cheaper but less powerful method as needed during the simulation. In this class of heuristics, only three methods are compared when making a given switching point decision – the current method and the methods directly preceding and succeeding it in the sequence. Adaptive heuristics are nonsequence-based when *all* the methods in the available set are compared. This class of adaptive methods requires more time for method selection than does the sequence-based class but has greater flexibility. Sequence-based methods are used in simulations where linear systems tend to become progressively difficult or easier; nonsequence-based strategies are used in applications where a monotonic pattern is missing. As discussed in Sections 2 and 4, the PETSc-FUN3D simulation falls in the former category.

We employed the following two indicators to construct the adaptive, polyalgorithmic solver:

- *Nonlinear residual norm:* The pseudo-transient continuation depends on the CFL number [8], which, as explained in Section 2 and shown in the left-hand side of Figure 2, increases as the nonlinear residual norm decreases. As the CFL number increases, the corresponding Newton correction equations (3) become more difficult to solve. Thus, in this case the nonlinear residual norm is a good indicator of the level of difficulty of solving its corresponding Newton correction equation: the lower the residual norm, the more difficult the linear system is likely to be. Based on trial runs of the application, we divided the simulation into four sections: (a) $||f(u)|| \geq 10^{-2}$, (b) $10^{-4} \leq ||f(u)|| < 10^{-2}$, (c) $10^{-10} \leq ||f(u)|| < 10^{-4}$, and (d) $||f(u)|| < 10^{-10}$. Whenever the simulation crosses from one section to another, the adaptive method switches up or down accordingly.

- *Average time per nonlinear iteration:* Our second indicator provides a rough estimate of the strength of the linear solver. The higher the value of the indicator, the more likely the solver is to effectively solve difficult linear systems. The base solvers are arranged in increasing order of their corresponding indicator values.

The parallelization of the adaptive scheme was straightforward; we invoked the linear solvers as determined by the heuristic, without redistributing the parallel data. No changes were needed to either the compressible Euler code or the base parallel solvers in PETSc [3] to accommodate this adaptive approach.

4. EXPERIMENTAL RESULTS

To perform numerical experiments on the compressible Euler application introduced in Section 2, we used the Jazz cluster at Argonne National Laboratory, which has a Myrinet 2000 network and 2.4 GHz Pentium Xeon processors with 1-2 GB of RAM. Our experiments focus on a problem instance designated as 1Grid (with 357,900 vertices and 2.4 million edges), which generates a Jacobian matrix of rank approximately 1.8×10^6 with 1.3×10^8 nonzeros. The large problem size makes it imperative to use a multiprocessor architecture. We ran the simulations on 4, 8, 16, and 32 processors using various Krylov methods and various subdomain solvers for a block Jacobi preconditioner with one block per processor. The relative linear convergence tolerance was 10^{-3}, and the maximum number of iterations for any linear solve was 30. The left-hand side of Figure 2 shows how the CFL number increases as the nonlinear residual norm decreases for the pseudo-transient Newton-Krylov algorithm; this situation was discussed in more detail in Section 2. The right-hand side of Figure 2 shows the time per nonlinear iteration for various solvers on four processors.

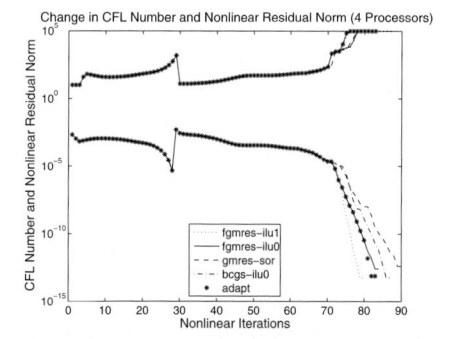

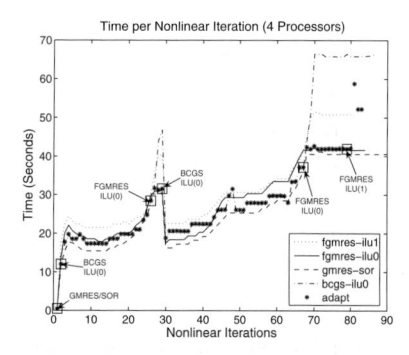

Figure 2. *Left:* Convergence rate (lower plot) and CFL number (upper plot) for the base and adaptive solvers on 4 processors. *Right:* Time per nonlinear iteration for the base and adaptive solvers on 4 processors. The labeled square markers indicate when linear solvers changed in the adaptive algorithm.

We employed four linear solvers: (1) GMRES with a block Jacobi (BJ) preconditioner that uses SOR as a subdomain solver, designated as GMRES-SOR; (2) bi-conjugate gradient squared (BCGS) with a BJ preconditioner that uses no-fill incomplete factorization (ILU(0)) as a subdomain solver, called BCGS-ILU0; (3) flexible GMRES (FGMRES) with a BJ preconditioner that uses ILU(0) as a subdomain solver, designated as FGMRES-ILU0; and (4) FGMRES with a BJ preconditioner that uses ILU(1) as a subdomain solver, called FGMRES-ILU1. We considered these as traditional base methods that remain fixed throughout the nonlinear solution process, and we also combined them in an adaptive scheme as introduced in Section 3. We ordered these methods for use in the adaptive solver as (1), (2), (3), (4), according to the average time taken per nonlinear iteration in the first-order discretization phase of the simulation, which can serve as a rough estimate of the strength of the various linear solvers for this application. The graphs in Figure 3 show the switching points among these methods in the adaptive polyalgorithmic approach. The simulation starts with method (1), then switches to method (2) at the next iteration. The switch to method (3) occurs at iteration 25. The discretization then shifts to second order at iteration 28, and the initial linear systems become easier to solve. The adaptive method therefore switches down to method (2). From this point onward, the linear systems become progressively more difficult to solve as the CFL number increases; the adaptive method switches up to method (3) in iteration 66 and method (4) in iteration 79. In the right-hand graph of Figure 2, this last change is accompanied by an increase in the time taken for the succeeding nonlinear iteration. This increased time is devoted to setting up the new preconditioner, which in this case changes the block Jacobi subdomain solver from ILU(0) to ILU(1) and consequently requires more time for the factorization phase.

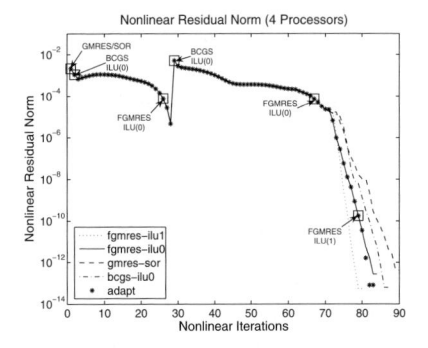

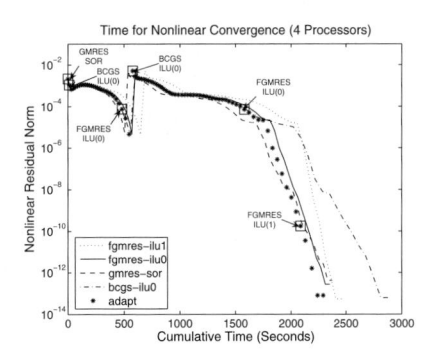

Figure 3. Performance of base and adaptive methods on 4 processors. *Left:* Residual norm vs. nonlinear iteration number. *Right:* Residual norm vs. cumulative simulation time. The labeled square markers indicate when linear solvers changed in the adaptive algorithm. The first solution method was GMRES-SOR, and switching occurred at iterations 1, 25, 28, 66, and 79.

The execution time of the adaptive polyalgorithmic scheme is 3% better than the fastest base method (FGMRES-ILU0) and 20% better than the slowest one (BCGS-ILU0). We also observe that the final nonlinear residual norm obtained by using the adaptive method is comparable to that obtained from the best base method ($O(10^{-14})$). In addition, the number of linear iterations required by the adaptive method for the overall simulation is smaller than that needed by any of the base methods. The overhead for switching among base methods in the adaptive scheme is minimal: approximately 0.02% − .06% of the total execution time.

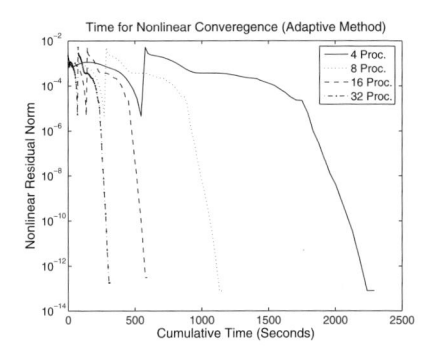

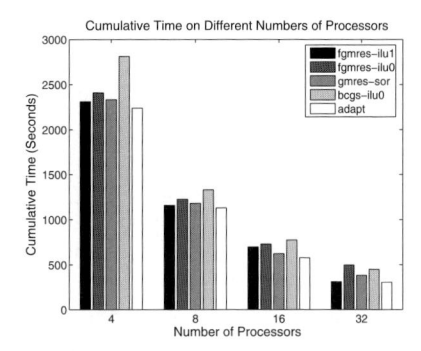

Figure 4. Performance of base and adaptive algorithms on 4, 8, 16, and 32 processors.

Figure 4 shows the performance of the adaptive method on 4, 8, 16, and 32 processors. As the number of processors increases, the simulation requires less time to converge. Although the best base method varies with the number of processors (FGMRES-ILU1 for 4, 8, and 32 processors, to GMRES-SOR for 16 processors), the adaptive method is always the one requiring the least time. The improvement of the adaptive scheme compared to the best base method varies from 2% to 7%. Moreover, as it is impossible to know *a priori* what particular linear solution algorithm will be fastest for a long-running nonlinear simulation, the adaptive, polyalgorithmic approach adjusts linear solvers according to the levels of difficulty encountered throughout a given run.

We consider the speedups of the adaptive and base solvers in Figure 5. Since this large-scale problem size requires a minimum of 4 processors, we use T_1 as the best estimate of the time for the full simulation on one processor using the fastest base method. We set T_1 to four times the time required with FGMRES-ILU1. We calculate the speedup $S = \frac{T_1}{T_p}$, where T_p is the observed time on p processors. The results show that the speedup of the adaptive method is almost ideal and as good as or better than any of the base methods.

5. CONCLUSIONS AND FUTURE WORK

In this paper we presented an adaptive, polyalgorithmic approach that dynamically selects a method to solve the linearized systems that arise in the PETSc-FUN3D application's modeling of compressible Euler flow using a pseudo-transient Newton-Krylov method. This approach reduced overall execution time by using cheaper though less powerful linear solvers for relatively easy linear systems and then switching to more robust but more costly methods for more difficult linear systems. Our results demonstrate that adaptive solvers can be implemented easily in a multiprocessor environment and are scalable. We are now investigating adaptive solvers in additional problem domains and considering more adaptive approaches, including a polynomial heuristic where the trends of the indicators can be estimated by fitting a function to the known data points. We also are combining adaptive heuristics with high-performance component infrastructure for performance monitoring and analysis, as described in [12, 13].

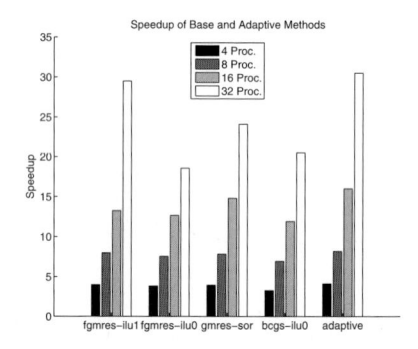

Figure 5. Scalability of the base and adaptive algorithms on 4, 8, 16, and 32 processors.

REFERENCES

1. W. K. Anderson and D. Bonhaus. An implicit upwind algorithm for computing turbulent flows on unstructured grids. *Computers and Fluids*, 23(1):1–21, 1994.
2. W. K. Anderson, W. D. Gropp, D. K. Kaushik D. E. Keyes, and B. F. Smith. Achieving high sustained performance in an unstructured mesh CFD application. In *Proceedings of Supercomputing 1999*. IEEE Computer Society, 1999. Gordon Bell Prize Award Paper in Special Category.
3. S. Balay, K. Buschelman, W. Gropp, D. Kaushik, M. Knepley, L. McInnes, Barry F. Smith, and H. Zhang. PETSc users manual. Technical Report ANL-95/11 - Revision 2.2.1, Argonne National Laboratory, 2004. http://www.mcs.anl.gov/petsc.
4. S. Bhowmick. Multimethod Solvers, Algorithms, Applications and Software, 2004. Ph.D. Thesis, Department of Computer Science and Engineering, The Pennsylvania State University.
5. S. Bhowmick, L. C. McInnes, B. Norris, and P. Raghavan. The role of multi-method linear solvers in PDE-based simulations. *Lecture Notes in Computer Science, Computational Science and its Applications-ICCSA 2003*, 2667:828–839, 2003.
6. P. N. Brown and Y. Saad. Hybrid Krylov methods for nonlinear systems of equations. *SIAM Journal on Scientific and Statistical Computing*, 11:450–481, 1990.
7. G. Karypis and V. Kumar. A fast and high quality scheme for partitioning irregular graphs. *SIAM Journal of Scientific Computing*, 20:359–392, 1999.
8. C. T. Kelley and D. E. Keyes. Convergence analysis of pseudo-transient continuation. *SIAM Journal on Numerical Analysis*, 35:508–523, 1998.
9. L. McInnes, B. Norris, S. Bhowmick, and P. Raghavan. Adaptive sparse linear solvers for implicit CFD using Newton-Krylov algorithms. *Proceedings of the Second MIT Conference on Computational Fluid and Solid Mechanics*, June 17-20,2003.
10. W. Mulder and B. Van Leer. Experiments with implicit upwind methods for the Euler equations. *Journal of Computational Physics*, 59:232–246, 1985.
11. J. Nocedal and S. J. Wright. *Numerical Optimization*. Springer-Verlag, 1999.
12. B. Norris, L. McInnes, and I. Veljkovic. Computational quality of service in parallel CFD. Argonne National Laboratory preprint ANL/MCS-P1283-0805, submitted to *Proc. of the 17th International Conference on Parallel CFD*, Aug 2005.
13. B. Norris and I. Veljkovic. Performance monitoring and analysis components in adaptive PDE-based simulations. Argonne National Laboratory preprint ANL/MCS-P1221-0105, Jan 2005.

Numerical Simulation of Transonic Flows by a Double Loop Flexible Evolution

Gabriel Winter, Begoña González, Blas Galván, Esteban Benítez

Institute of Intelligent Systems and Numerical Applications in Engineering (IUSIANI) Evolutionary Computation and Applications Division (CEANI). University of Las Palmas de Gran Canaria. Edif. Central del Parque Científico y Tecnológico, 2ª Planta, Drcha. 35017 Las Palmas de G.C., Spain

Keywords: Evolutionary Algorithms, Flexible Evolution, Genetic Algorithms, EDP non-linear boundary problem, transonic flow.

1. INTRODUCTION

The solution to the problem of simplifying the application to industry-related problems of the new developments in Evolutionary Algorithms (EAs) can be provided by the Artificial Intelligence concepts. This basically involves EA implementations having to incorporate capabilities to make decisions, analyse their results, identify the correct and wrong ones, memorize the correct decisions and use them in the following algorithm steps. The Artificial Intelligence based on Evolutionary Algorithms can bring together lessons learned by developers and decision tools into advanced software developments, which can facilitate their use to solve complex challenging problems in industry.

As the target of verification of a numerical scheme for nodal point is equivalent to minimizing a corresponding objective function, we use EAs as a numerical simulation method. We demonstrate the capability and applicability of the Evolutionary Computation for solving an EDP non-linear boundary problem, as is the case in obtaining a numerical solution for the stationary full potential flow problem, which corresponds to the calculation of the speeds for transonic flow regime in the compressible and isentropic flow within a nozzle. The difficulty of the problem is well known [1][2]. Our objective is to use the evolutionary method to obtain an approximated value for the speed components in a totally developed flow.

We specifically use a double-loop strategy that considers two nested evolutionary algorithms, the outer one is an Evolutionary Intelligent Agent-based software named

Flexible Evolution Agent (FEA) [3] [4] [5] and the inner one is a Genetic Algorithm with real encoding [6].

The latest results for the non-linear potential flow problem inside a nozzle are shown here, the first results were presented in [7] [8], where a GA was run a certain number of times before obtain an acceptable solution. With the double-loop strategy we try to automate this process in an optimal sense and parallel computation help us to obtain acceptable solution in a lesser time period because we can run the GA with different individual (of the FEA population) at the same time.

2. THE PROBLEM TO SOLVE

We have studied the velocities for transonic flow in the compressible and isentropic flow inside the nozzle shown in Figure 1. The upper wall of the nozzle is given by:

$$H(x) = \frac{3}{8} + \frac{1}{8}\sin\left[\pi\left(x+\frac{1}{2}\right)\right], 0 \leq x \leq 2 \quad (1)$$

The resulting differential equation for the speed components can be written as follows [8]:

$$\left[\frac{\gamma+1}{2}u'^2 + \frac{\gamma-1}{2}v'^2 - 1\right]\frac{\partial u'}{\partial x} + u'v'\frac{\partial u'}{\partial y} + u'v'\frac{\partial v'}{\partial x} + \left[\frac{\gamma+1}{2}v'^2 + \frac{\gamma-1}{2}u'^2 - 1\right]\frac{\partial v'}{\partial y} = 0 \quad (2)$$

where $(u', v') = (u/c_0, v/c_0)$, c_0 is the speed of sound in normal conditions and $\gamma = 1.4$ for the air.

Boundary conditions:
    Inlet and Outlet:
$$u'(0, y) = u'(2, y) = C \text{ (constant value 0.2)};$$
$$v'(0, y) = v'(2, y) = 0;$$
    Symmetry axis: $v'(x, 0) = 0$;
    Nozzle wall: $v'(x, H(x))/u'(x, H(x)) = dH(x)/dx$;

Also, in this problem, **curl**$(u', v') = 0$.

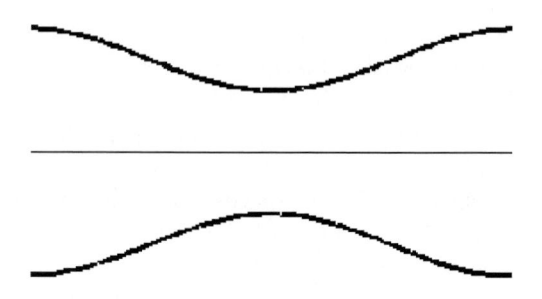

Figure 1. Lateral section of the nozzle for compressible and isentropic flow. The central line is the symmetry axis.

With the boundary conditions considered, the solution has axial symmetry and thus the area from the symmetry axis until the upper wall of the nozzle is sufficient to be considered as the domain of the problem.

3. PARTIAL GRID SAMPLING

For the application of EAs, the first step consists of establishing one solution-node set inside the nozzle, where the velocities are to be calculated, and one initial solution (u', v') for each solution-node. In particular, we have generated in the 2D domain, a regular set of solution-nodes, with variable steps hx, and hy. In this case, a numerical scheme that proved to be effective, for getting an acceptable evolution of the solution-nodes, was the centred scheme.

The partial grid sampling is based on choosing j different nodes, each fixed number of generations, in such a way that they are not neighboring. This condition is to guarantee that the schemes considered for the j nodes do not overlap. In successive generations the solution-nodes are chosen to complete the whole grid (several times) until the stop criterion is verified.

An advantage of this methodology is that in effect it works with a subset of nodes at each EA generation, instead of working with all of them at the same time.

4. EVOLUTIONARY ALGORITHMS AS A MESHLESS METHOD

The chromosome or individual, in each generation of the EA, consist on the adimensional values (u', v') associated to the j nodes chosen.

$$\left(u'_1, u'_2, \ldots, u'_j, v'_1, v'_2, \ldots, v'_j\right)$$

The fitness function is the sum of the numerical schemes associated to the j nodes chosen and their respective neighbors.

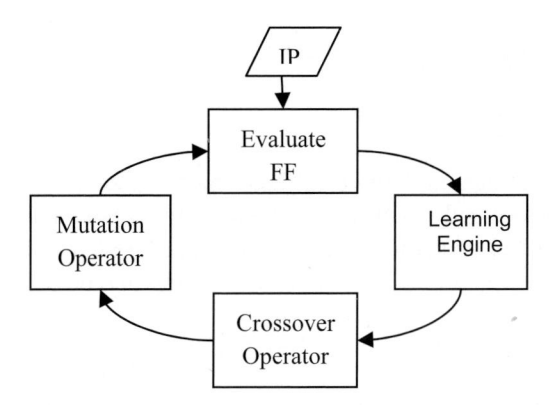

Figure 2. A general scheme of a Genetic Algorithm: IP is Initial Population and FF is Fitness Function.

We have considered a GA with real encoding to solve the problem with this partial grid sampling technique. This GA runs for 20 individuals, considers elitism and has as genetic operators the following ones:
- Selection Operator: A tournament selection operator between two candidates chosen randomly.
- Crossover Operator: Arithmetic crossover:
  Let $\alpha = U(-0.5, 1.5)$, then:
  $$Offspring_1 = \alpha\,Parent_1 + (1 - \alpha)\,Parent_2$$
  $$Offspring_2 = \alpha\,Parent_2 + (1 - \alpha)\,Parent_1$$
- Mutation Operator: Smooth mutation:
  Let $\alpha = U(0, 1)$ and $i \in \{1, 2, ..., j, j+1, j+2, ..., 2j\}$, then:
  If $(i \leq j)$
  $$u_i^{'} = u_i^{'} \pm 0.0005\,\alpha$$
  Else
  $$v_{i-j}^{'} = v_{i-j}^{'} \pm 0.0005\,\alpha$$

The crossover probability is 0.5 and the mutation probability is 0.1.

4.1. Two-stage optimisation

Due to the non-linearity of the differential equation (2), the optimization process was divided into two stages. The fist one is one-dimensional an the second one is bi-dimensional.

In a first stage of the algorithm, the part of the numerical scheme that only contains the function u' (ideal fluid problem with only component in the x-axis) is chosen as the objective function. The rest of the differential equation is regarded as a small perturbation.

After a number of generations, for example here 20,000 generations, the second stage starts and the full numerical scheme of the differential equation is used (20,000 generations).

The first stage is a learning stage. The solutions obtained at this first stage are a training set to begin the next stage, which is now in 2D context.

5. FLEXIBLE EVOLUTION AGENT

Many researchers have struggled to build different mending/adapting methods in order to improve EAs. Consequently, advanced methods to define, adapt, self-adapt or eliminate parameters and/or operators have been developed, tested and compared.

New work on both some kind of self-adaptation algorithms and on the use of agents has produced promising results.

However, the knowledge gained is still far from being really useful for many industry-related applications.

We have developed in CEANI a new idea in optimisation: the Flexible Evolution Agent (FEA) [3][4][5], that has the capacity to adapt the operators, the parameters and the algorithm to the circumstances faced at each step of every optimisation run and is able to take into account lessons learned from different research works in the adaptation of operators and parameters. So the FEA has been subdivided into several functions, called 'engines'. These subroutines have been designed to group the diverse actions that are to be executed during the optimisation depending on their objectives. In this way, all the learning tasks will be clustered in a 'learning engine', and something similar will happen with all the selection schemes, sampling strategies or decision mechanisms.

A general scheme of the Flexible Evolution Agent can be seen in Figure 3. Starting with the initial population IP, and for all the iterations of the population obtained earlier, we evaluate the fitness function of each candidate solution and from here, and for each generation, the Decision Engine acts over all the different stages of the algorithm. So the decision engine will decide what kind of learning, selection and/or sampling strategy will be used in every generation until the stop criterion is reached, which is also determined by the Decision Engine. The Learning Engine stores everything that could be useful afterwards, such as information about the variables or statistics. The intention is to use this information to learn about the process and even to establish rules that could be fruitful and will be included in the decision engine afterwards.

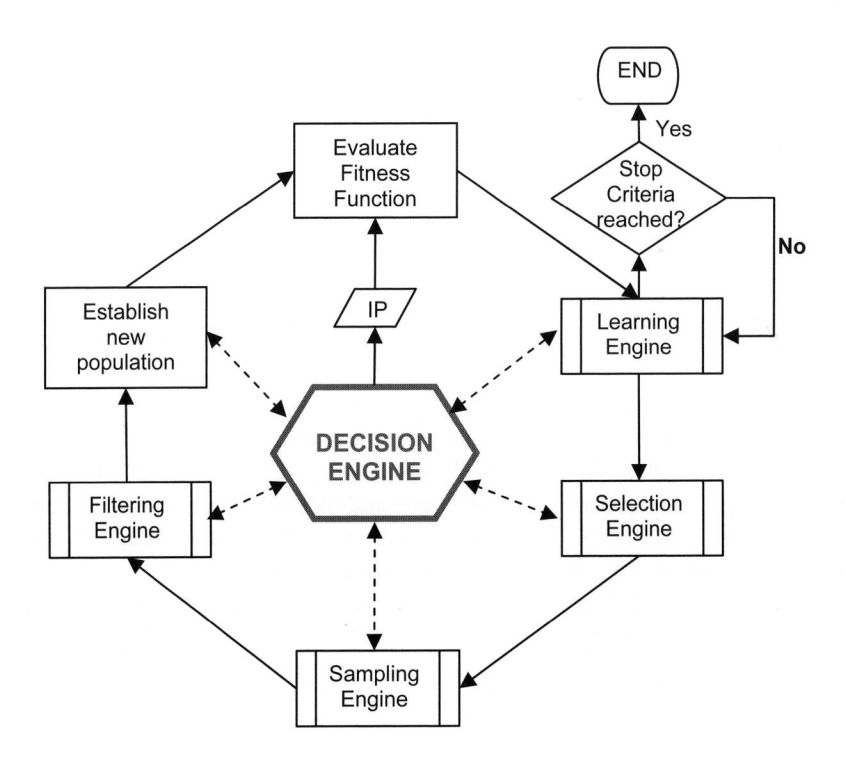

Figure 3. A general scheme of the Flexible Evolution Agent

6. DOUBLE LOOP STRATEGY WITH EVOLUTIONARY ALGORITHMS

In order to use the partial sampling of the solution nodes with EAs, it is necessary to begin with an initial solution for each of them that permits the algorithm to evolve to a near-optimum solution. Thus, this final solution will be conditioned by the initial one.

We propose here a double loop strategy that considers two nested EAs: the outer one is a Flexible Evolution Agent and the inner one is a Genetic Algorithm with real encoding. The chromosome or individual, in each generation of the FEA, consists of the adimensional values associated with all the solution nodes and the fitness function is the sum of the numerical schemes associated with all of them.

The algorithm associated to this double loop strategy is very similar to the FEA one but, in this occasion, for evaluating the fitness function, the GA with real encoding and the partial grid sampling technique is run for each individual (of the FEA population) and returns the best solution obtained from it, that will be used to evaluate its FEA fitness function.

7. PARALLEL IMPLEMENTATION

Parallel Genetic Algorithms can be classified in four main types :
- Global Master-Slave
- Island or Coarse-grained or Distributed Model
- Cellular or Fine-grained or Diffusion Model
- Hybrid or Hierarchical Models

There are articles which have dealt with comparison between these models, for example [9], and some surveys about parallel genetic algorithms have been developed, among which we cite [10].

Here we propose a Global Master-Slave algorithm to parallelize the double loop strategy with Evolutionary Algorithms (EAs). This Global Master-Slave algorithm considers a unique whole population and differs from the sequential one in that the evaluation of the fitness function is distributed among some processors to be evaluated in parallel with an asynchronous migration phase. In our case the master processor hosts the FEA and distributes, in an asynchronous way, the individuals among some processors that host the GA with real encoding and the partial grid sampling technique, and compute the fitness. In Figure 4 we can see the results obtained after 100 generations of the FEA and with 20 individuals per population.

The computed results are in acceptable agreement with the numerical results obtained from the linearization algorithm of Gelder [11] [12].

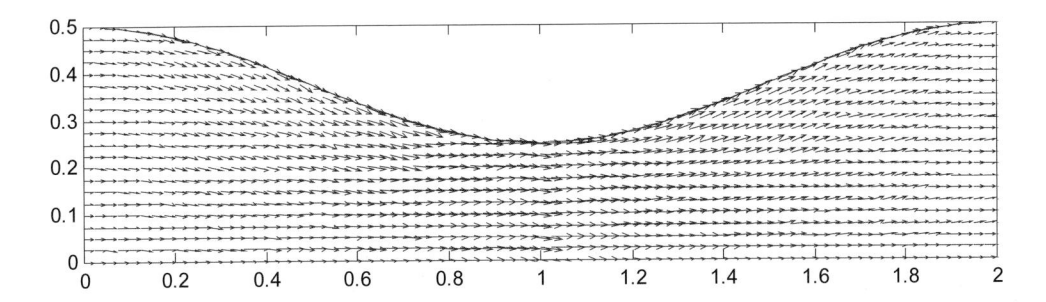

Figure 4. Solution obtained after 100 generations of the FEA and with 20 individuals per population.

8. CONCLUSIONS AND FUTURE WORK

- We have proposed a methodology that uses Evolutionary Algorithms to obtain fluid velocities for potential flows inside a nozzle.
- We have proposed a genetic algorithm with real encoding which, with the technique of partial sampling of the solution nodes, presents a number of advantages over other evolutionary algorithms. These include the avoidance of a rigid connectivity to discretize the domain, thus making it a meshless method.
- We have also proposed a double loop strategy with EAs which enables improvement of the above methodology.
- The amount of computer storage is low and convergence behaviour is good as the qualitative characteristics of the solution are taken into account in the algorithm.
- The method is easy to implement in parallel environments.
- However, it is a line of research which is very open in many directions as, in our proposed methodology, the use of higher order approximation schemes into the objective function has no strong effect on either the total computational cost or memory requirements.

REFERENCES

1. Landau, L.D., Lifshitz E.M. (1995). *Fluid Mechanics* 2º Ed. Butterworth-Heinemann.
2. Shapiro, A.H. (1953). *The dynamics and thermodynamics of Compressible Fluid Flow*. 2 Vol. Wiley. New York.
3. Winter, G., Galván, B., Cuesta, P.D, Alonso, S. (2001). *Flexible Evolution*. In the book: Evolutionary Methods for Design, Optimization and Control with Applications to Industrial Problems (Proceedings of the EUROGEN2001 Conference), pp. 436-441.

4. Winter, G., Galván, B., Alonso, S., González, B. (2002). *Evolving from Genetic Algorithms to Flexible Evolution Agents*. Late Breaking Papers GECCO 2002, pp 466-473. New York.
5. Winter, G., Galván, B., Alonso, S., González, B., Greiner, D., Jiménez, J.I. (2005). "Flexible Evolutionary Algorithms: cooperation and competition among real-coded evolutionary operators". *Soft Computing - A Fusion of Foundations, Methodologies and Applications*. **9** (4), pp. 299-323. Springer-Verlag.
6. Davis, L. (1991). *Handbook of Genetic Algorithms*. Van Nostrand Reinhold, New York.
7. Winter,G., Abderramán, J.C., Jiménez, J.A., González, B., Cuesta, P.D. (2001). Meshless Numerical Simulation of (full) potential flows in a nozzle by GAs. *STS-8: Test Cases Results Installed in a Database for Aerospace Design*, European Congress Computational Fluid Dynamics(CFD2001), Swansea.
8. Winter,G., Abderramán, J.C., Jiménez, J.A., González, B., Cuesta, P.D. (2003). Meshless Numerical Simulation of (full) potential flows in a nozzle by GAs. *International Journal for Numerical Methods in Fluids*, **43**: 10-11, pp. 1167-1176. John Wiley & Sons.
9. Gordon, V., Whitley, D. (1993). Serial and Parallel Genetic Algorithms as Function Optimizers, *The Fifth International Conference on Genetic Algorithms*, pp. 177-183, Morgan Kaufmann.
10. Cantú-Paz, E. (1997). A Survey of Parallel Genetic Algorithms. *Calculateurs Paralleles, Reseaux et Systems Repartis*, **10** (2), pp. 141-171.
11. Gelder D. (1987). Solution of the compressible flow equations. *International Journal for Numerical Methods in Engineering*, **3**, pp. 35– 43.
12. Pironneau O. (1988) *Methods des Elements Finis pour les Fuides*. Masson: Paris.

Numerical Simulation of 2D Radiation Heat Transfer for Reentry Vehicles

B.N. Chetverushkin, S.V. Polyakov, T.A. Kudryashova[1], A. Kononov, A. Sverdlin

Institute for Mathematical Modeling, Russian Academy of Science,
4-A Miusskaya Square, 125047 Moscow, Russia

    This is a highly relevant subject because planetary missions have been launched and indeed some of them are currently in progress such as Spirit or Dnepr.

    The design of a thermal protection system for the space vehicle is an important problem. It requires accurate prediction of the heating environments. This study uses diffusion approximation to investigate radiation heat transfer for reentry vehicles. The goals of the paper to present a parallel algorithm: to calculate a strongly radiating axisymmetric flow field over a blunt body; to introduce a database of absorption coefficients and use it in parallel computing; to compare results of flow with and without radiation influence.

1. INTRODUCTION

    The purpose of this paper is the development of parallel calculated algorithms for radiation gas dynamics problems adapted to architecture of multiprocessor systems with distributed memory. Investigations of flows over solid body coming into atmosphere is traditionally important direction of high temperature gas dynamics.

    The work is devoted the simulation of radiation heat transfer for reentry vehicles. Radiation heat flux become significant during the reentry of vehicles into Earth's atmosphere. When the space vehicle flies at Mach number $M_\infty > 10$ a shock wave is formed ahead of flying body. High temperature appears behind the shock wave and creates strong radiation. In such process radiation affects flow field and flow field affects radiation properties. Mutual influence means the join computing of the equation of energy for substance and the equation of radiation transport.

    For our investigations the model of diffusion approximation [1, 4] has been used. The model was coupled with the gas dynamical part by the including the radiation flux in the energy equation. The quasigasdynamics equations (QGD) [2, 3, 4] were performed to compute gas dynamics flows. A radiation database has been developed in order to use the values of the absorption coefficients for problems like that. The experimental data of the absorption coefficients for database have been taken from [5]. The measurements of radiation are presented for full spectrum.

---

[1] e-mail: kudryashova@imamod.ru

The radiation heat transfer problem is computed for two dimensional axisymmetric flow around of the blunt body [6, 7]. The calculations were carried out on rectangular grids. The mesh construction methods were developed in [9].

The present paper is intended to show evolution of previous work [4].

2. NUMERICAL PROCEDURE

Radiation gas dynamics (RGD) equations describe high temperature processes with essential influence of radiation transport. The governing equations were the quasigasdynamics equations (QGD) [2, 3, 10] and the system of diffusion equations [1, 4]. We offer splitting of the problem into subproblems.

Radiation gas dynamics code (RGDC) was created as block structure. It should include some blocks:
1) computing of gas dynamics flow fields
2) calculations for radiation fluxes
3) recalculations of the energy equation

As we have the separated gas dynamics block we can use any computation method for first block or gas dynamics parameters can be taken from external sources. Further the value of total energy will be recomputed taking into account the influence of the radiation.

The diffusion approach model was coupled with the gas dynamical block by the including the radiation flux in the energy equation [1]. The total radiation flux is obtained as the sum of the radiation fluxes for each frequency groups.

The computation finished when the steady-state solution is achieved according to the criterion:

$$\varepsilon = \frac{1}{N_r N_z} \sum_h \left| \frac{E^{j+1} - E^j}{E^j \Delta t} \right| \leq 10^{-6}, \tag{1}$$

here N_r - number of space steps on r-axis, N_z - number of space steps on z-axis, E - total

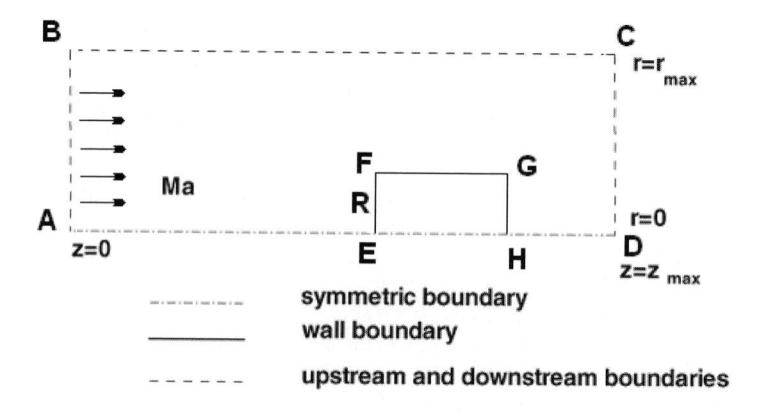

Figure 1. Computational domain

Table 1
Computational grids

Type of grids	1	2	3
Number of points	120*50	440*100	880*200
h_z / r	1.0	0.5	0.25

energy, Δt - time step, j is the time step index.

RGDC provides calculations for arbitrary number of frequencies groups.

The radiation heat transfer problem is computed for two dimensional axisymmetric flows around of the blunt body [6, 7]. A sample of computational domain is shown in Fig. 1.

Cartesian coordinates were used for computing. The space steps in the axial directions are uniform. The calculations were carried out for 3 computational grids given in Table 1. Here h_z and h_r are the steps of the computational grid in z and r directions, respectively.

The calculations were carried out with flow conditions. The parameters of flow are reported in the table below (Table 2).

3. DATABASE

Information from [5] have been used for design database. A radiation database has been developed in order to use the values of the absorption coefficients for problems like that.

The diffusion equations were evaluated for 600 frequency groups. Spectral interval is $250 cm^{-1}$ -$150000 cm^{-1}$ In the database the spectral step is $250 cm^{-1}$. Temperature step is 500K. Database supports approximately 10^5 meanings of the absorption coefficients. We can call data from database and use them in our parallel code. Well known limitations for values of the absorption coefficients for diffusion approach are checked.

$$\frac{l_v}{L} \langle\langle 1 \quad , l_v = \frac{1}{\chi_v}, \tag{2}$$

here L is characteristic size of the problem, χ_v - absorption coefficient.

We should consider these restrictions for uniform loading the computing system. At the certain temperatures some groups do not accept participation in calculations. For example for temperature **T=T1** we can use only **n1<600** groups. In the parallel program redistribution of groups on processors is organized. The scheme of this transformation is shown in the Table 3.

Table 2.
The initial conditions of the free stream.

Background gas temperature $T_0(K)$	266
Temperature of head shield $T_1(K)$	$2000 \div 3300$
Pressure $p_0(Pa)$	4.3*10
Density $\rho_0(kg/m^3)$	5.63*10^{-4}
Characteristic size of problem $L(m)$	5.035*10^{-1}
Mach number M_∞	12; 24

Table 3.
The scheme of the redistribution of groups on processors.

Npr	1	2	3	...	np
Nfg	n1	n2	n3	...	Nnp

⇓

Nfg	n	n	n	...	Nn

Npr – number of processor, *Nfg* – number of frequency groups for each processor.

We have 18 meanings for temperature and 10 meanings for pressure. To obtain intermediate values of the absorption coefficients the two dimensional logarithmic approximation has been used. The scheme of the approximation is shown in Figure 2. We consider the approximation on (P,T) cell. If we have value of the absorption coefficients inside of a cell (P_i,T_j), (P_i,T_{j+1}), (P_{i+1},T_{j+1}), (P_{i+1},T_j), the following algorithm is applied. For approximation we shall take logarithms from the absorption coefficients in nodes of the cell. The absorption coefficient is designated as $f_{i,j}$. Value in point ($P_{i+1/2}, T_{j+1/2}$) is equal $\frac{1}{4}\left(\ln(f_{i+1,j})\ln(f_{i,j})\ln(f_{i+1,j+1})\ln(f_{i,j+1})\right)$. Then it is defined in what triangle the point has got. Depending on it approach for definition of the absorption coefficients is under construction. Here we use linear interpolation on triangle.

For air radiation O_2, N_2, NO, N_2^+ are considered.

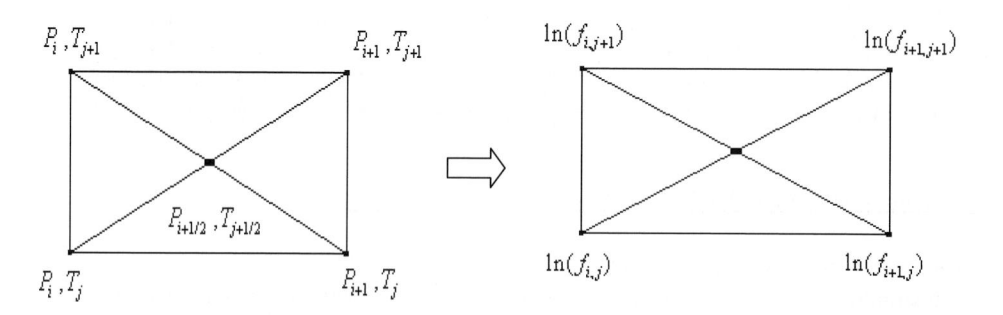

Fig.2 The scheme of the approximation

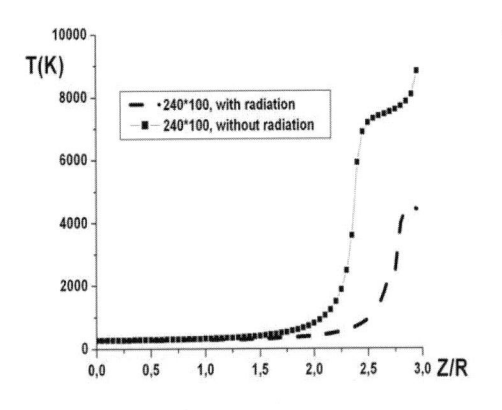

Fig. 3 Temperature distribution with and without radiation for grid (240*100) for M=12

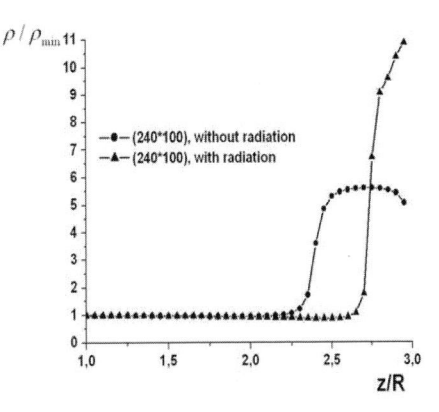

Fig. 4 Density distribution with and without radiation for grid (240*100) for M=12

4. PARALLEL COMPUTING

Two topologies were used to parallel algorithm: "pipe-line" and "grid". The grid-topology was used for parallelization of gas-dynamic equations. The pipe-line topology was used for solution of radiation equation system. Let us consider some details of parallelization.

The domain decomposition technique was applied for parallelization of the gas dynamics block. The computational domain is divided in the sub-domains in z-direction (or for "grid" topology in z-direction and r-direction).

For radiation part the parallelization in accordance with radiation frequencies groups and a parallel conjugate gradient method was implemented to obtain the radiation fluxes. Dynamics load-balancing procedure is used as you see in Table 3.

Parallelization for recalculating of the energy equation is similar to the parallelization of gas dynamics block.

Numerical method has been realised as a FORTRAN parallel code. The homogeneous parallel computer system MVS-5000 equipped with 336 - PowerPC970 (1.6GHz) microprocessors was employed. The total performance is over 2.1 Tflops and the fast communication links give up to 340 MB/sec (peak speed) data transmission rate.

5. NUMERICAL RESULTS

Radius of the blunt body is R=0.5035m.

Temperature distribution is shown in Figure 3. Radiation influence smoothes the shock wave. The maximum value of the temperature with radiation is lower than without radiation.

It is the opposite for density distribution. Peak value of density without radiation is lower than with radiation. Density distribution is shown in Figure 4.

Figures 5, 6 show distribution of Mach numbers with radiation and without radiation. Here we see the shock wave in front of the blunt body and contours of Mach numbers.

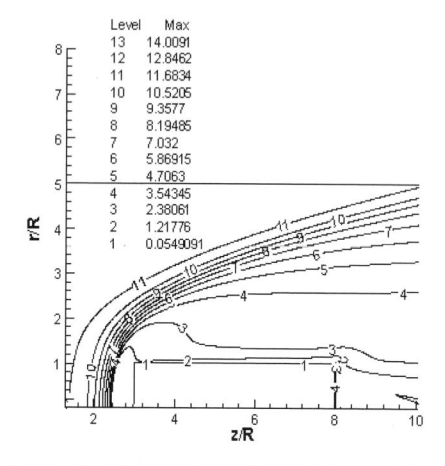

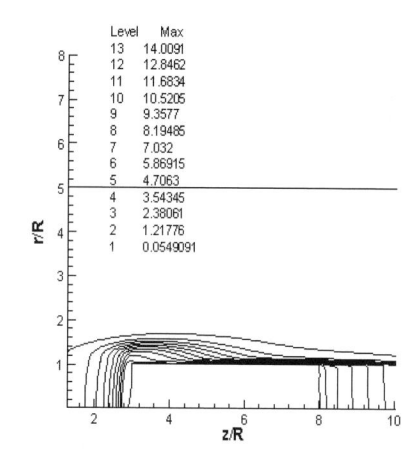

Fig. 5 Mach numbers for grid (240*100) for variant without radiation (initial M=12)

Fig. 6 Mach numbers for grid (240*100) for variant with radiation (initial M=12)

The radiation heat flux rises in presence of shock wave.

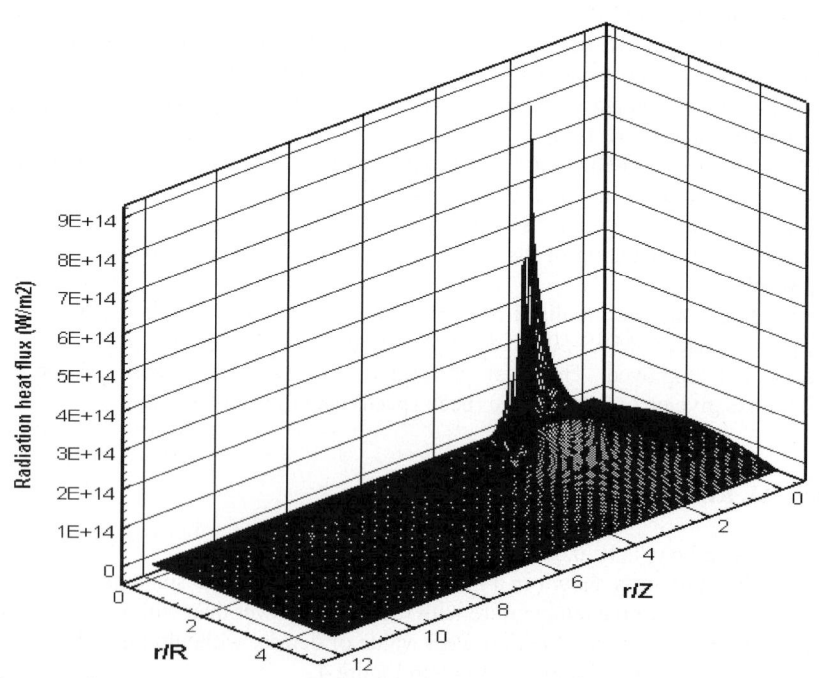

Fig. 7 Radiation heat flux distributions along the computational domain for grid 480*200, for M_∞=12

6. CONCLUDING REMARKS

The high performance parallel computing system was used here in order to reproduce a structure of flow for the radiation heat transfer processes in reasonable computation time.

Present numerical method has been realized as a parallel program for distributed memory computer systems. The parallel realization is based on the geometrical parallelism principle.

Database allows compute more exactly values of the absorption coefficients for the radiation problems.

Numerical results show that the radiation heat transfer processes are of primary importance for accurate prediction of gas dynamics fields for problems of this nature.

Parallel code is the combination of independent procedures.

REFERENCES

1. B.N. Chetverushkin . Dynamics of radiating gas. Moscow, Nauka, 1980 (in Russian).
2. T.G.Elizarova, Yu.V. Sheretov (2001) Theoretical and numerical investigation of quasigasdynamics and quasihydrodynamics equations, Comput. Mathem. And Mathem. Phys., 41, 219-234.
3. T.G.Elizarova, M. Sokolova Numerical algorithm for computing of supersonic flows is based on quasi-hydrodynamics equations. Vestnik of Moscow State University, Phisics & Astronomy, 2003 (in Russian).
4. I.A. Graur, T.A. Kudryashova, S.V. Polyakov, Modeling of Flow for Radiative Transport Problems, International Conference Parallel CFD 2004 (Las Palmas de Gran Canaria, SPAIN, May 24-27, 2004), A Collection of Abstracts. Published by Institute of Intelligent Systems and Numerical Applications in Engineering (IUSANI), University of Las Palmas de Gran Canaria, 2004, pp. 197-200.
5. Avilova I.V., Biberman L.M., Vorob'ev V.S., Zamalin V.M., Kobzev G.A. The optical characteristics of hot air. Moscow, Nauka, 320 p. (in Russian), 1970.
6. Park C., Milos F.S. Computational Equations for Radiating and Ablating Shock Layers. AIAA Paper 90-0356. June 1990.
7. Sakai T, Tsuru T., and Sawada K. Computation of Hypersonic Radiating Flowfield over a Blunt Body. Journal of Thermophysics and Heat Transfer. Vol. 15, №1, pp. 91-98, 2001.
8. Suzuki T., Furudate M., and Sawada K. Unified Caclulation of Hypersonic Flowfield for a Reentry Vehicle. Journal of Thermophysics and Heat Transfer. Vol. 16, №1, pp. 94-100, 2002.
9. I.V. Popov, and S.V. Polyakov. Construction of adaptive irregular triangular grids for 2D multiconnected non-convex domains. Mathematical modeling, 14(6), pp. 25-35, 2002 (in Russian).
10. B.N. Chetverushkin. Kinetic schemes and quasigasdynamics system of equations. Moscow, Maks Press, 2004 (in Russian).

Parallelism Results on Time Domain Decomposition for Stiff ODEs Systems

D. Guibert [*] and D. Tromeur-Dervout [†] [a]

[a]CDCSP/UMR5208, University Lyon 1, 15 Bd Latarjet, 69622 Villeurbanne, France

Keywords: ODE and DAE, Runge-Kutta and extrapolation, parallel computation, grid computing, parareal domain decomposition
MSC : 65L80, 65L06, 65Y05, 65Y10, 65N55.

1. Introduction

Modeling complex systems can lead to solve large ODEs, and/or DAEs systems to understand the dynamic behavior, eventually stiff, of the solutions. The main difficulties in term of parallel implementation of such systems of equations are the poor granularity of the computations, and the sequential nature of the computing. Previous time steps are required to compute current time step. This last constraint occurs in all the time integrator used in computational fluid dynamics of unsteady problems. From the numerical point of view the development of time domain decomposition methods for stiff ODEs problems allows to focus on the time domain decomposition without perturbation coming from the space decomposition. In other side, the classical advantages of the space decomposition on the granularity are not available. From the computer science point of view, especially, for the grid computing architecture, which is characterized by a large number of processors, the space domain decomposition is limited by the size of the problem and the time decomposition becomes necessary. In another side, the low cost of computing resources on the grid architecture change the concept of efficiency. The number of available processors allows to consider these computational resources to improve the confidence in the solution. One can combine several schemes of different orders and different discretisations to validate and to verify the results. Indeed, from the engineering point of view in the design with ODEs/DAEs, the efficiency of the algorithm is more focus on the saving time in the day to day practice.

In this paper, we investigate the time domain decomposition on the ODEs system. We first implement the main approach used in the past to obtain parallel solver for ODEs systems which consists in parallelizing "across the method". This method distributes to the processors the computation of steps of multi-step methods as Runge-Kutta method of order 5 [1,2]. We will show that the number of processors used is limited as it depends of

---

[*]This work is backward to the Région Rhône-Alpes thru the project: "Développement de méthodologies mathématiques pour le calcul scientifique sur grille".
[†]This author was partially supported thru the GDR MOMAS project:"simulation et solveur multidomaine"

the number of steps. Then we slightly modified the concept of Parareal[3] /Pita[4] domain decomposition that allows to introduce adaptivity in the time stepping an domain decomposition in order to solve stiff problems. Then we investigate two parallel implementations of these algorithms and finally we conclude with some development perspectives.

2. Parallelization across the method

The first approach to obtain a parallel solver for ODEs systems consists in parallelizing "across the method" which distributes to the processors the steps computation of multi-step methods as in [1,2]. Let us recall briefly the method

Definition 2.1. Let b_i, a_{ij} ($i = 1, \cdots, s$) be reals and $c_i = \sum_{j=1}^{i-1} a_{ij}$. Then the s-stages Runge-Kutta method writes:

$$k_i = f\left(x_0 + c_i h, y_0 + \sum_{j=1}^{s} a_{ij} k_j\right) \text{ and } y_1 = y_0 + h \sum_{i=1}^{s} b_i k_i \qquad (1)$$

Implicit Runge-Kutta with s-stages needs the solution of the nonlinear system of size $n \otimes s$. If $A = (a_{ij})$ is non-singular the non linear system $Z = (A \otimes I)hF(Z)$ is solved by a Newton iteration scheme ([5, page 118]). The work of K. Burrage consists in exploiting the tensorial product to compute the different stages of the method in parallel. The number of processors that can be used is limited as it depends of the number of stages. For example let us consider the Radau IIA method of order 5. It involves 3-stages that set to 3 the number of processors needed in this parallelizing approach. Moreover as the method computes numerically the Jacobian matrix several times, this computation has to be also parallelized, in order to not affect the parallel efficiency performances of the method. Table 1 shows the comparison on elapsed time between the sequential

Elapse time(s)	seq. code	parallel code	speed up
Jacobian computing		260	
Jacobian comm.		28.6	
total Jacobian	847	288.6	2.93
comput. stage		43.5	
comm. stage		44.7	
total stage	143	88.2	1.62
total elapse time	1082	436	2.48

Table 1
Comparison between the sequential version and the parallel version of Radau IIa (on 3 processors) on a engine injectors problem with 131 unknowns.

version and the parallel version of the 3-stages Radau IIa method for an engine injector problem. Numerical experiments have been performed a 4 nodes ALTIX350 with Itanium2 processors cadenced to 1.3 Ghz, with 3 MB L2-cache, 2 GB of RAM per node, and 1.2 Gb/s NUMAlink network. It exhibits that follows:

- the speed-up concerning the stages computing is not so good compared to those of the Jacobian computing.
- the communications take as much time as the stages computing.

To have a method which does not limit the number of processors, time domain decomposition methods are now investigated.

3. The Parareal algorithm

The principle of the parareal algorithm to solve

$$\frac{dy}{dt} = f(y,t), \quad \forall t \in \Omega_t =]T^0, T^f], \quad y(t^0) = y_0 \quad (2)$$

consists in splitting the time domain in m time-slices $\{S^i = [t^i, t^{i+1}]\}$ of different sizes with $t^0 = T^0$ and $t^m = T^f$. Let Y^i denote the values of the exact solution of problem (2) at the beginning of the time-slice S^i. The principle of the parareal algorithm consists in defining an approximation Y_k^i of these Y^i on a coarsest grid. Y_k^i known, the solution $y_k^i(t)$ on the m time-slices S^i can be computed as

$$\frac{dy_k^i}{dt} = f(y_k^i, t) \; \forall t \in S^i, \; y_k^i(t^i) = Y_k^i. \quad (3)$$

These solutions exhibit jumps $\Delta_k^i = y_k^{i-1}(t^i) - Y_k^i$, $1 \leq i \leq m-1$ at the time-instances t^i. A correction function c_k piecewise C^1 in Ω_t is introduced to update the Y_k^i values with a Newton-type linearized method around y_k, for $0 \leq i \leq m-1$,

$$\frac{dc_k}{dt} = f_y(y_k, t)c_k, \text{ with } c_k(t^0) = 0, \text{ and } c_k(t^{i+}) = c_k(t^{i-}) + \Delta_k^i \text{ for } i \geq 1 \quad (4)$$

We introduce adaptivity in the definition of the fineness of the grids based on the relative tolerance of the time integrator. The advantage is that adaptivity in the time step can occur in order to pass the strong nonlinearities

4. Adaptive Parareal: numerical and parallelism results

4.1. Numerical results for non-linear test problems:

Figure 1 shows the convergence of the method for the Lotka-Volterra problem. For 10 time subdomains, the method blows up for $rtol = 10^{-3}, 10^{-4}$ and finally converges at the 10th iterate. Even with $rtol = 10^{-7}$ convergence takes 7 iterates. For this number of subdomains the method has no interest. Nevertheless, if the number of equal size subdomains is increased to 200, the convergence is reached between 5 and 7 iterates for $rtol = 10^{-6}$ and 10^{-3}, providing speed up. For 1168 subdomains convergence is obtained between 2 and 7 iterates for $rtol = 10^{-6}$ to 10^{-3}. Let us notice that the correction can converge to 10^{-14} but the convergence on solution is limited by the fine grid solver (here $rtol = 10^{-7}$. It is not necessary to reach the machine accuracy for correction to have the effective convergence on fine grid.

304

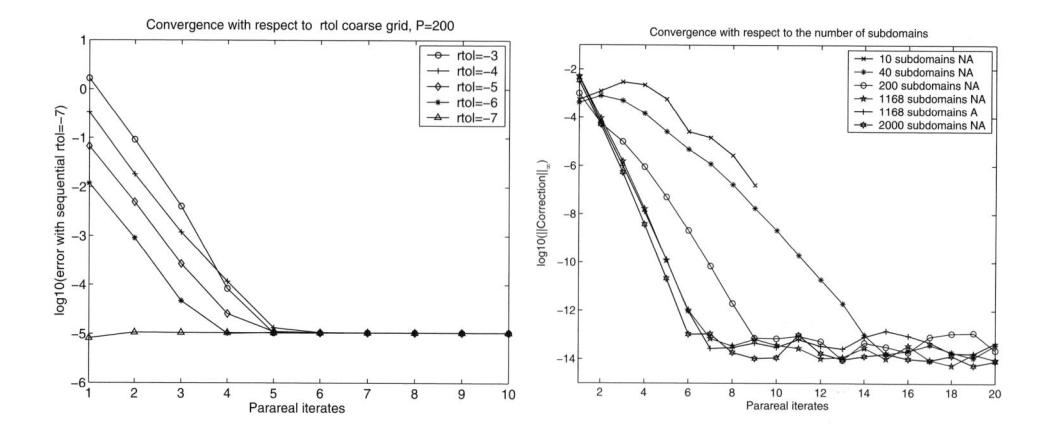

Figure 1. (Left) Modified parareal convergence for Lotka-Volterra Problem on $\mu = (1.5, 1, 3, 1)$ with 200 subdomains with respect to different rtol for the initialization and the correction. (Right) Convergence of the correction in parareal with respect to the number of subdomains with and without adaptivity and $rtol = 10^{-5}$ for the correction.

The size of the subdomain is adapted with respect to strong variations of the step size. Figure 1 shows that the number of subdomains 1168 defined by the adaptivity (A) gives better results. Nevertheless, this is not the optimal number of subdomains, 2000 regular subdomains (NA) lead to a faster convergence, and 1168 regular subdomains give quite the same convergence. For the Oregonator problem, the convergence blows up even with 1000 subdomains for $rtol = 10^{-6}$ to 10^{-3} and even with time decomposition adaptivity.

4.2. Parallelism results for the non-linear beam test problem:

We focus the parallelism results on the beam problem test case. It originates from mechanics and describes the motion of an elastic beam which is supposed inextensible, of length 1 and thin. Moreover, it is assumed that the beam is clamped at one end and a force $F = (F_u, F_v)$ acts at the free end. The semi-discretization in space of this equation leads to a stiff system of n non-linear second order differential equations which is rewritten in 1rst order form, thus providing a stiff system of ordinary differential equations of dimension $2n$. An accurate description of this classical test case can be found at [6].

The DASSL solver, based on backward differentiation formulae (BDF) and developed by Petzold [7] [8], is used as time integrator.

Numerical experiments presented below have been performed a Compaq Sierra-cluster with alpha ev67/600Mhz processors with 8MB L2-cache and a with 800Mb/s bandwidth communication network. The MPI communication library is used to manage the communications between processors.

The number of subdomains needed is higher than the available number of processors. Thus, each processor manages several subdomains. The first algorithm implements straightforward the parareal scheme. Let P be the number of available processors,

$nbdomloc$ be the number of time subdomains per processor, and n be the number of equations of the ODEs.

4.2.1. Implementation: Algorithm 1

S1 Initialization of subdomains IBV with a coarse time integrator tolerance. This part is a sequential computation. The solution of the last subdomains $[T_{i,m}^+, T_{i,m+1}^-]$ managed by processor i is sent into the IBV of the first time subdomain $[T_{i+1,1}^+, T_{i+1,2}^-]$ of processor $i+1$.

S2 until convergence do

S2.1 Parallel solve of the ODEs system on subdomains $[T_{i,j}^+, T_{i,j+1}^-]$, $j = 1, \ldots, m$

S2.2 Sequential correction process : the correction with IBV equal to $c_i(T_{i,m}^+) + \Delta_{i,m}$ is solved on subdomain m. In practice, the subdomain correction c is added to the $y_i(T_{i,j+1}^-)$ fine solution, then the result is copied to the next subdomain correction IBV where $y_i(T_{i,j+1}^-)$ is subtracted to the copied value. For the last subdomain of processors i the value $y_i(T_{i,m+1}^-) + c_i(T_{i,m+1}^-)$ is sent to the processor $i+1$ and $y_{i+1}(T_{i+1,1}^+)$ is subtracted to the received value in order to have the correction IBV $c_{i+1,1}(T_{i+1,1}^-) + \Delta_{i+1,1}$.

S2 end do

Let us emphasize on the difficulties that occur in the parallel implementation and execution of the parareal scheme for the test problem:

- As the time interval is divided in subdomains, the time integrator for the subdomains IBV initialization performs a cumulative number of time steps greater than the number of time steps used in sequential code. This is mainly due to the adaptation of the time step to reach the end time of the subdomain. As the initializing is sequential, its computational cost is great.

- For complex problem, one have to compute numerically the Jacobian matrix of the correction problem. This implies $n+1$ matrix-vector product of n^2 complexity. Consequently, the correction function evaluation is about n times more costly than the evaluation function for the initialization.

- As the correction solution is small in absolute value, the time step for the time integration is smaller than for the initialization. Consequently the time integrator performs more time steps for this correction.

- As the algorithm step [S2.2] is sequential, only one processor is active at the same time. This correction step is the most time consuming part of the implementation. A solution is needed to introduce parallelism in this correction part.

We propose a relaxation of the transmission condition between the last subdomain of a processor and the first subdomain of the next processor. We define **Algorithm 2** where step S2.2 of Algorithm 1 is replaced by :

4.2.2. Implementation: Algorithm 2

S2.2b Instead of sending the $y_i(T^-_{i,m+1}) + c_i(T^-_{i,m+1})$ to the next processor, we first send only the $y_i(T^-_{i,m+1})$ and perform the classical transmission of step S2.2 for the other subdomains belonging to the processor. Then before to performing a new parareal iterate, the modified solution $y_i(T^-_{i,m+1}) + \tilde{c}_i(T^-_{i,m+1})$ generated starting from $y_{i-1}(T^-_{i-1,m+1})$ is sent to $y_{i+1}(T^-_{i+1,1})$ as the new IBV of the first subdomain of processor $i+1$ for the next parareal iteration.

4.2.3. Comparisons between the two algorithms

The parareal iterate of Algorithm 2 becomes totally parallel with this approach. Nevertheless, it introduces some error between subdomains as a non-sharp corrected value of the solution is exchanged between adjacent subdomains of different processors. Moreover this error is propagated on the time interval. Figure 2 tests for the Lotka-Volterra problem the convergence of the two algorithms with the adaptive dassl time integrator. Algorithm 1 performs better numerically in three iterates, Algorithm 2 needs more parareal iterates to converge to the same value. The convergence depends on the number of processors for Algorithm 2 due to the S2.2b step. Nevertheless, Algorithm 2 is fully parallel and reduces effectively the most time consuming correction part of the parareal method. The first communication of step S2.2b can be overlapped by the Jacobian linearizing on each subdomain. We plan to investigate domain decomposition coarse grid correction techniques in order to make the convergence independent of the number of processors.

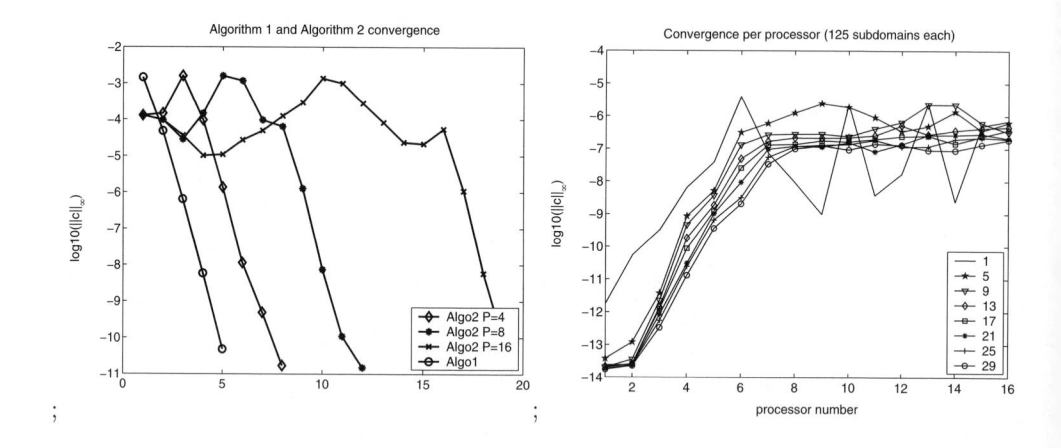

Figure 2. (left) Comparison of numerical convergence on beam problem of the Algorithms 1 and 2 with respect to the parareal iterates. (right) Numerical convergence in maximum norm of the correction on beam problem in each processor (125 subdomains per processors) with respect to the number of parareal iterates.

Figure 2 exhibits the convergence in the maximum norm of the correction on the beam problem for each processor (125 subdomains per processor) with respect to the number of parareal iterates. The convergence is faster for the firsts processors that manage the firsts subdomains. This can allow to dynamically free the processors where convergence is reached. Let us notice, that the solution is of order $O(1)$ in the last subdomains, and the relative tolerance is set to 10^{-5}. Consequently when the correction is under 10^{-6} the reachable accuracy for the solution is obtained.

Table 2 gives the elapse time and the parallel efficiency of the parareal algorithm 2 for the beam problem with $n = 200 \times 2$ unknowns to perform 3 parareal iterates.

Beam problem with 200×2 unknowns				
elapse time (s)	# processors			
	2	4	8	16
Initializing $rtol = 10^{-2}$	1394	1366	1364	1366
Fine grid $rtol = 10^{-5}$	2526	1080	587	312
correction $rtol = 10^{-3}$	11260	4265	2184	1149
Total	15180	6711	4135	2827

Table 2
Parallel efficiency of Algorithm 2 on the beam problem.

The parallel efficiency is limited here with the small number of available processors. Nevertheless, the performance trends of the algorithm make it attractive to actually reduce the elapse time. The elapse time of Algorithm 2 depends strongly on the number of processors available and is balanced with the parareal iterates needed to reach a given tolerance.

Beam problem 200×2 unknowns			
elapse time (s)	# processors		
	4	8	16
Initializing $rtol = 10^{-2}$	1135	2271	4550
Fine grid $rtol = 10^{-5}$	1484	1531	294 (1it)
correction $rtol = 10^{-3}$	5946	6012	1198 (1it)

Table 3
Scalability of Algorithm 2 on the beam problem.

Table 3 gives the scalability of the Algorithm 2. Each processor manages 250 subdomains and 3 parareal iterates are performed. The 16 processors run converges only in one parareal iteration with the number of subdomains taken. It shows that the initializing does not scale while the correction and the fine grid solution steps scales quite well.

5. Conclusions and future works

Two concepts of parallel methods to solve stiff ODEs systems have been investigated. The parallelizing across the method showed some limitation on the number of processors to be use and on the speed up achieved. The modified Parareal/Pita time domain decomposition seems to be able to achieved good reducing on the elapsed time on the grid architecture concepts. Nevertheless, these methods for stiff problems are very sensitive to the computing of the Jacobian matrix for the correction. Moreover, there are the same difficulties as the parallelizing of the coarse grid problem in the multigrid methods. The correction equation in the studied cases is the more time consuming part of the algorithm due to its sequential nature. We proposed a slight relaxation of this constraint with the Algorithm 2 that saved time without to much penalty on the number of parareal iterates. however, our current developments focus on adaptive parallel extrapolations based on the behavior of the time integrator solver that use the same concept of time domain decomposition and avoid the linearizing of the Jacobian function. First numerical investigations [9] seem promising and will be extended by parallel implementations in a next paper.

REFERENCES

1. K. Burrage, Parallel and sequential methods for ordinary differential equations, Oxford University Press.
2. P. Van Der Houwen, S. B.P., Parallel iteration of high-order runge-kutta methods with stepsize control, J. Comput. Appl. Math. 29 (1990) 111–117.
3. J.-L. Lions, Y. Maday, G. Turinici, Résolution d'edp par un schéma en temps "pararéel", CRAS Sér. I. Math. 332 (7) (2000) 661–668.
4. C. Farhat, M. Chandesris, Time-decomposed parallel time-integrators: theory and feasibility studies for fluid, structure, and fluid-structure applications, Internat. J. Numer. Methods Engrg. 58 (9) (2003) 1397–1434.
5. E. Hairer, S. Norsett, W. G., Solving Ordinary Differential Equations II: Stiff and Differential-Algebraic Problems, 2nd Edition, Series in Computational Mathematics, Springer, Berlin, 1991.
6. CWI, Tests set for ivp solvers, http://pitagora.dm.uniba.it/~testset.
7. L. Petzold, A description of dassl: A differential and algebraic system solver, SAND82-8637.
8. K. Breman, S. Campbell, L. Petzold, Numerical Solution of initial-Value Problems in Differential-Algebraic Equations, Elsevier Science Publishing Company, New York, 1989.
9. D. Guibert, D. Tromeur-Dervout, Adaptive parareal for system of odes, in: Proc. Int. Conf. DD16, 2005, to appear.

Application of an implicit dual-time stepping multi-block solver to 3D unsteady flows

R. Steijl[a], P. Nayyar[a]*, M.A. Woodgate[a], K.J. Badcock[a] and G.N. Barakos[a]

[a]CFD Laboratory, Department of Aerospace Engineering, University of Glasgow, Glasgow, G12 8QQ, United Kingdom

This work discusses the application of a parallel implicit CFD method to challenging 3D unsteady flow problems in aerospace engineering: transonic cavity flows and the flow field around a helicopter rotor in forward flight. The paper discusses the computational details of simulations using the HPCx supercomputer (1600 processors) of Daresbury Lab., U.K. and a Beowulf cluster (100 processors) of the CFD Laboratory of the University of Glasgow. The results show that accurate simulations based on Large-Eddy Simulation (LES) and Detached-Eddy Simulation (DES) at realistic Reynolds numbers require impractical run times on the Beowulf cluster. A simulation of a full helicopter geometry is similarly beyond the limits of the 100-processor Beowulf cluster.

1. INTRODUCTION

Many of the present CFD applications in aerospace engineering involve unsteady three-dimensional aerodynamic problems. In contrast to steady state flows, which can typically be tackled in a matter of hours on a multi-processor machine or on a Beowulf cluster, unsteady flows require days of CPU time.

This paper presents the application of a parallel, unfactored, implicit method for the solution of the three-dimensional unsteady Euler/Navier-Stokes equations on multi-block structured meshes [1]. For time-accurate simulations, dual time-stepping is used. The solver and its performance on Linux clusters was previously discussed in refs.[2] and [3].

The application examples presented here are for three-dimensional unsteady aerodynamic problems: transonic cavity flow and flow around a helicopter rotor in forward flight. The simulations were carried out on the HPCx supercomputer of the Daresbury Lab. in the UK[6] and the local Beowulf cluster (comprising 100 Pentium 4 processors).

The CFD method and its parallelization are described in Sections 2 The application to 3D unsteady flow problems is described in Section 3, while conclusions are drawn in Section 4.

---

*Present address: Aircraft Research Association Ltd., Manton Lane, Bedford, Bedfordshire, England MK41 7PF

2. CFD METHOD AND PARALLELISATION

The unsteady Navier-Stokes equations are discretised on a curvilinear, multi-block, body conforming mesh using a cell-centred finite volume method. The convective terms are discretised using Osher's upwind scheme [4] and MUSCL variable extrapolation is used to provide second-order accuracy. The Van Albada limiter is used to prevent spurious oscillations around shock waves. Central differences are used for the viscous terms. The solver includes a range of one- and two-equation turbulence models as well as LES based on the Smagorinsky model and DES Spalart-Almaras model, as described by[9]. A dual-time stepping method is employed for time-accurate simulations, where the time derivative is approximated by a second-order backward difference [5]. The resulting non-linear system of equations is solved by integration in pseudo-time using a first-order backward difference. In each pseudo-time step, a linearisation in pseudo-time is used to obtain a linear system of equations, which is solved using a Generalised Conjugate Gradient method with a Block Incomplete Lower-Upper (BILU) pre-conditioner. The method is detailed in ref.[1]. Regarding parallelisation of the above method few changes were necessary:

- The flux Jacobians resulting from the linearisation in pseudo-time are employed in an approximate form that reduces the number of non-zero entries and as a result the size of the linear system. The use of the approximate Jacobian also reduces the parallel communication since only one row of halo cells is needed by the neighbouring process in the linear solver instead of two in the case of an 'exact' Jacobian.

- The communication between processes is minimised by decoupling the BILU factorisation between blocks.

- On each processor a vector is allocated that contains all the halo cells for all grid blocks.

- Inter-process communication is performed by sending a series of messages between the respective processes, each corresponding to a block connection, containing the halo cell data. The messages are sent in chunks of 10,000 double precision numbers using non-blocking send and receive MPI functions.

The parallel implementation was presented previously in refs. [2] and [3]. and the solver has been used on a range of platforms, including Beowulf clusters consisting of various generations of Pentium processors and multi-processor workstations. Recently, the solver was ported to the HPCx computer at Daresbury Laboratory. The HPCx system comprises 50 IBM Power4+ Regatta nodes, i.e. 1600 processors, delivering a peak performance of 10.8 TeraFlops[6].

3. EXAMPLES OF 3D UNSTEADY APPLICATIONS

3.1. Transonic cavity flows

This section presents results from a computational study of transonic cavity flows, in which the formation of highly unsteady turbulent flow structures and the resulting noise production is the main interest[7]. In cavity flows, the flow separates at the sharp edge at the front of the cavity while further downstream two flow patterns may be encountered.

For the first pattern, the shear layer formed by the separation at the cavity front spans the entire cavity and re-attaches at the rear of the cavity. This is referred to as *open* cavity. In contrast, the shear layer in a *closed* cavity re-attaches at the cavity floor, then separates from the cavity floor further downstream, forming a shear layer that re-attaches at the rear cavity edge. The conditions considered here result in an *open* cavity flow.

Three approaches for the turbulence modelling were used: unsteady RANS (URANS) using the $k - \omega$ model, Large-Eddy Simulation (LES) and Detached-Eddy Simulation (DES). LES works by filtering the flow structures in terms of scale size, with the larger scales explicitly resolved and the smaller ones modelled using a sub-grid scale (SGS) model. Pure LES can still be expensive, however, and recent endeavours have looked at developing hybrids of Unsteady Reynolds-Averaged Navier-Stokes (URANS) and LES to compromise the best of both methods. One example of such developments is the DES method introduced by Spalart et al. [9].

Here, a clean rectangular cavity with a length-to-depth ratio (L/D) of 5 and a width-to-depth ratio (W/D) of 1 is considered for two cavity configurations: one with doors-on and another with doors-off. The free-stream Mach number is 0.85 and a Reynolds number of one million based on the cavity length. These conditions result in an *open* cavity flow for both configurations.

Pressure traces and visualisation of the flow-field inside the cavity from DES simulation for the doors-on configuration and LES simulation for the doors-off configuration are illustrated in Figure 1. Experimental pressure signals (provided by Ross et al.[10] and sampled at 31.25 kHz for doors-on and 6 kHz for doors-off) and numerical results with URANS for both doors-on (Figure 1(a)) and doors-off (Figure 1(b)) are also included for reference. The high Reynolds number considered here requires the use of high density grids. Small time-steps are required as a result of the high frequency unsteady flow features, with frequencies as high as 1 kHz. This results in an overall number of time-steps of approximately 50,000 required to simulate 0.2 seconds of the flow, which is just enough for gathering the flow statistics required for LES. The high density grids combined with the large number of time steps makes this flow computation very demanding, which is why the HPCx super-computing facility was exploited.

At a Reynolds number of 1 million, the flow in the cavity is turbulent. Combined with the presence of walls and the presence of a shear layer that separates the external (fast-flowing) fluid with the internal (slow-moving) cavity fluid, high levels of dissipation exist signifying that a large number of turbulent length scales are present. Good resolution of this turbulent spectrum is important in order to understand the function of turbulent processes and the source of acoustics inside the cavity. Without the use of massively parallel computers such as the HPCx, simulation of such turbulent flow-fields within realistic run times becomes impossible.

Table 1 shows details of three calculations: A DES on a grid with 4.5 million points on 320 processors (HPCx), an LES on a grid with 4.5 million points (24 processors on Beowulf cluster) and an URANS ($k - \omega$) simulation on a 1.5 million point grid on 19 processors of the local Beowulf cluster. A proper spectral decomposition of the DES and LES flow-field requires the calculation to run for long durations (at least 0.1s) to obtain sufficient samples (after sampling at either 31.25 kHz (for doors-on) and 6 kHz (for doors-off)) for analysis of the frequency content inside the cavity. Even after approximately

312

(a) Doors-On Pressure (DES)

(b) Doors-Off Pressure (LES)

(c) Doors-On flow field (DES)

(d) Doors-Off flow field (LES)

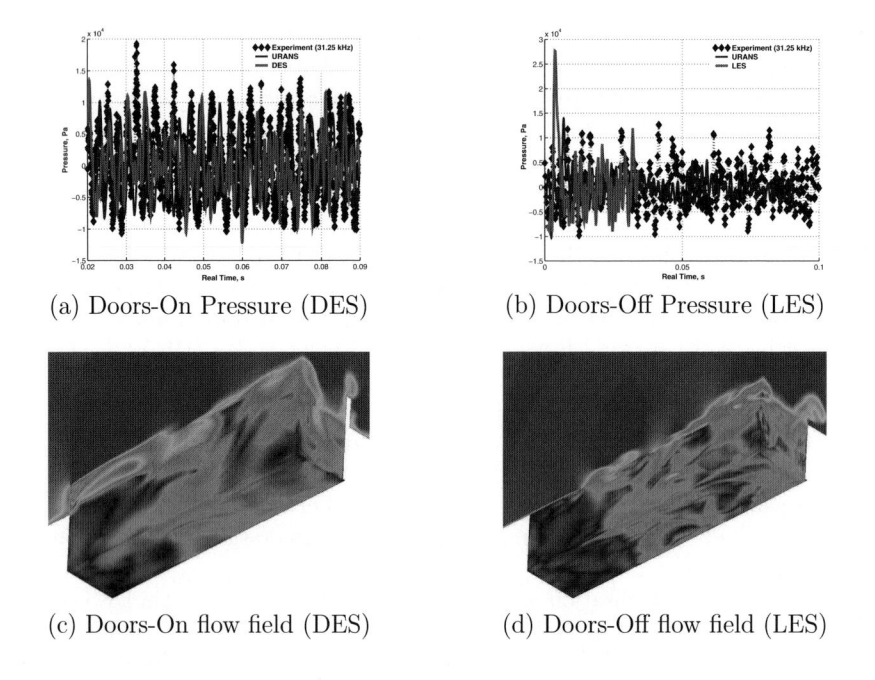

Figure 1. Pressure traces and visualisation of the flow features inside the 3D, L/D=5, W/D=1 cavity with doors-on using DES and doors-off using LES. Pressure traces contain experimental signal (black with diamond symbols) and DES results (red). For reference, results from URANS are also included (blue). Flow-field plot consist of Mach contours normalised by free-stream Mach number of 0.85.

Power spectral density

Sound presssure level

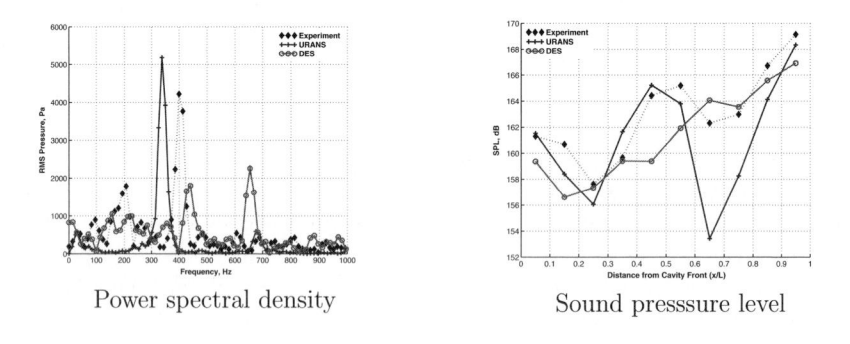

Figure 2. Power spectral density and sound pressure level versus distance along cavity floor. Doors-on case, L/D=5, W/D=1 cavity, free-stream Mach number 0.85.

40,000 CPU hours of run-time, the 4.5 million LES calculation is far from complete as shown in the pressure signal in Figure 1(b).

For the doors-on cavity configuration, the power spectral density and the sound pressure level on the cavity floor versus distance from the cavity front are presented in Figure 2. The CFD results are compared with the experimental data of Ross et al.[10]. The comparison shows a good agreement for the DES results. The unsteady RANS results show poor correlation with experimental data, especially for the high-frequency modes. The poor predictions for power spectral density and the sound pressure level from the URANS simulation can be explained by the failure of this simulation to resolve the break-up of the shear layer that spans the cavity. This shear layer break-up is resolved in the LES and DES simulations.

Table 1
DES, LES and URANS calculation details on HPCx and Beowulf cluster

Calculation Details	DES	LES	URANS
Platform	HPCx	Beowulf cluster	Beowulf cluster
Cavity Configuration	Doors-On	Doors-Off	Doors-On
Grid Size	4.5×10^6	4.5×10^6	1.5×10^6
Processors	320	24	19
Time-Step (s)	1.81×10^{-6}	1.81×10^{-6}	1.81×10^{-5}
Pseudo-Steps/Time-step	6	4	39
Time-Steps/min.	9.72	0.723	0.425
Total Time-Steps	50,200	50,000	5,506
Total CPU Hours	28,100	1,565	3,121
Signal Duration	0.1 s	0.1 s	0.1 s
Total Run-time	3.46 days	48 days	9 days

3.2. Helicopter rotor in forward flight

This example combines a complex geometry with a flow field rich in fluid mechanics phenomena including strong interacting vortices, the formation of a vortex wake that spirals down below the rotor disk, transition to turbulence and a wide variation of the Mach and Reynolds numbers in the radial direction and around the azimuth[11]. An additional difficulty, is the strong link between the aerodynamics and the aeromechanics, i.e. to achieve a level flight, the rotor requires a blade pitch that changes periodically during the rotor revolution. The forward flight velocity leads to one side of the rotor disk with high blade-normal velocities (advancing side) and one with lower blade-normal velocities (retreating side). Using a lower blade pitch on the advancing side and an increased blade pitch on the retreating side, the rotor revolution-averaged roll and pitching moments can be canceled out. Furthermore, the rotor blades are hinged to allow for flapping (blade motion normal to vertical plane) and a lead-lag deflection (motion in the

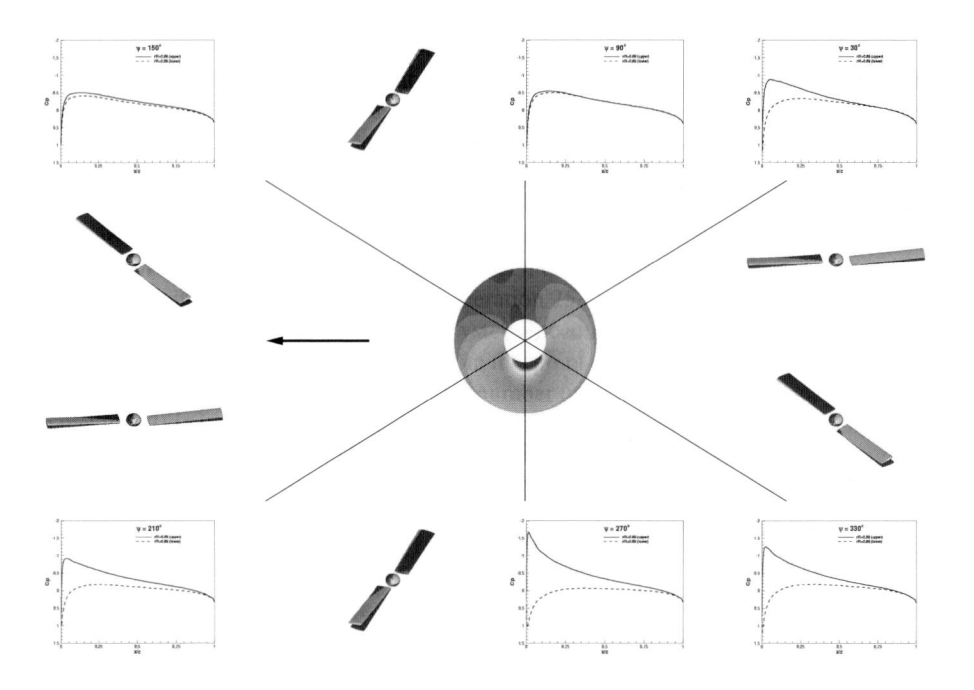

Figure 3. Geometry and chordwise pressure distribution for a fully articulated 2-bladed rotor in forward flight. The grey shade shows the rotor surface with periodic changes in pitch, flapping and lead-lag deflection. The blue shade shows the original blade position.

horizontal plane). The required control input (the blade pitch) and the resulting blade motion form part of the solution. This is known as the *trimming problem* and good CFD investigations are described in refs.[12], [13] amongst others.

The test case considered here is a two-bladed rotor with low-aspect ratio blades. The tip and forward flight Mach numbers are 0.6 and 0.09, respectively. The simulation involves periodic blade pitching, flapping and lead-lag motions. Figure 3 shows how the rotor geometry changes during a rotor revolution. The grey shadings show the geometry at various azimuthal positions compared to an equivalent rotor geometry without blade motions (blue). Also shown is the chordwise surface pressure distribution for a radial station at 89% of the rotor radius. Table 2 shows details of the forward flight case shown in Figure 3 and estimates of CPU times for forward flight cases currently underway. As shown, an affordable Beowulf cluster (≤ 100 processors) does not provide the capability to simulate the viscous flow around a full helicopter configuration in forward flight (mesh size $15\text{-}30 \cdot 10^6$ points) within realistic time. Supercomputing facilities, such as the HPCx may thus be required for simulations of such flows.

Table 2
Computational requirements for simulations of helicopter rotor in forward flight

Calculation Details	2-bladed rotor	2-bladed rotor	4-bladed rotor	4-bladed rotor	4-bladed rotor + fuselage
	inviscid (measured)	RANS (estimate)	inviscid (measured)	RANS (estimate)	RANS (estimate)
Grid Size	$1.2 \cdot 10^6$	$4.0 \cdot 10^6$	$4.0 \cdot 10^6$	$10 \cdot 10^6$	$15 \cdot 10^6$
Processors	20	40	40	100	200
Pseudo-Steps/Time-step	40	40	40	40	40
Total Time-Steps	5000	5000	5000	5000	5000
Total CPU Hours	1,440	4,000	4,000	10,000	20,000
Total Run-time	72 hrs	100 hrs	100 hrs	100 hrs	100 hrs

4. CONCLUSIONS

The application of a parallel implicit multi-block CFD method to two challenging problems in aerodynamics is presented. The computer platforms used here are a Beowulf cluster and the HPCx supercomputer. Obtained results indicate that the parallel implementation of the solver is both robust and efficient on both platforms.

The problems considered were the transonic cavity flow and the flow around a helicopter rotor in forward flight. Using a Beowulf cluster (comprising 100 Pentium 4 processors), these flow can be analysed using CFD simulations. The run times, however, become excessive if the details of the turbulent flow must be resolved using LES or DES.

Acknowledgements
The financial support of the Engineering Physical Sciences Research Council (EPSRC) and the UK Ministry of Defense (MoD) under the Joint Grant Scheme is gratefully acknowledged for this rotorcraft project. This work forms part of the Rotorcraft Aeromechanics Defense and Aerospace Research Partnership (DARP) funded jointly by EPSRC, MoD, DTI, QinetiQ and Westland Helicopters. For the cavity work, the financial support from BAE Systems is gratefully acknowledged.

REFERENCES

1. K. Badcock, B. Richards and M. Woodgate, Elements of Computational Fluid Dynamics on Block Structured Grids Using Implicit Solvers. (2000) Progress in Aerospace Sciences, 36(5-6): 351-92.
2. M. Woodgate, K. Badcock, B. Richards and R. Gatiganti. A parallel 3D fully implicit unsteady multiblock CFD code implemented on a Beowulf cluster. In *Parallel CFD 1999*, Williamsburg, USA, 1999.

3. M. Woodgate, K. Badcock and B. Richards. The solution of pitching and rolling delta wings on a Beowulf cluster. In *Parallel CFD 2000*, Trondheim, Norway, 2000.
4. S. Osher and S. Chakravarthy. Upwind schemes and boundary conditions with applications to Euler equations in general geometries. (1983) Journal of Computational Physics, 50:447–481.
5. A. Jameson. Time dependent calculations using multigrid, with applications to unsteady flows past airfoils and wings. AIAA Paper 91-1596, 1991.
6. HPCx capability computing. http://www.hpcx.ac.uk
7. P. Nayyar, G. Barakos, K. Badcock and B. Richards. Analysis and Control of Weapon Bay Flows. NATO RTO AVT-123 Symposium on "Flow Induced Unsteady Loads and the Impact on Military Applications", Budapest, 24-28 April, 2005.
8. D. Rizzetta and M. Visbal. Large-Eddy Simulation of Supersonic Cavity Flowfields Including Flow Control. 32nd AIAA Fluid Dynamics Conference, 2002, AIAA Paper 2003-0778.
9. P.R. Spalart. Strategies for Turbulence Modelling and Simulations. (2000) International Journal of Heat and Fluid Flow, 21:252–263.
10. J. Ross. Cavity Acoustic Measurements at High Speeds. Technical Report DERA/MSS/MSFC2/TR000173, QinetiQ, March 2000.
11. R. Steijl, G. Barakos and K. Badcock. A CFD Framework for Analysis of Helicopter Rotors. AIAA Paper 2005-5124, 17th AIAA CFD Conference, Toronto, 6-9 June, 2005.
12. H. Pomin and S. Wagner. Aeroelastic Analysis of Helicopter Rotor Blades on Deformable Chimera Grids. (2004) J. Aircraft 41(3):577-584.
13. M. Potsdam, W. Yeo and W. Johnson. Rotor Airloads Prediction Using Loose Aerodynamic/Structural Coupling. American Helicopter Society 60th Annual Forum. Baltimore, MD, June 7-10, 2004.

Improving the Resolution of an Elliptic Solver for CFD Problems on the Grid

B. Hadri [a], M. Garbey [a] and W. Shyy [b]

[a]Department of Computer Science,
University of Houston, Houston, TX 77204, USA

[b]Department of Aerospace Engineering,
University of Michigan, Ann Arbor, MI 48109, USA

This paper foccuses on the fast resolution of elliptic problems generated by algorithms to solve CFD problems. A typical application is a pressure solver in an incompressible Navier-Stokes flow code. We describe a domain decomposition method that is numerically efficient, scales well on a parallel computer, and is highly tolerant to the high latency and low bandwidth of a slow network.Our method is based on the implementation of a domain decomposition technique in parallel and also on the performance tuning of the linear solver depending on the subdomain and the processor architecture.

1. Introduction and Motivation

There are many methods to solve a linear problem, such as a direct solver with some decomposition of the operator(e.g, LU decomposition) or an iterative method such as Krylov methods or multigrid. Each of these methods have their own advantages and disadvantages. If we consider only the elapsed time for the resolution, some of these methods are faster for small sizes and others are faster for large size problems. So the wrong choice of the solver can slow down dramatically a computation since the elapsed time of the resolution of the linear system can take a major part of a code [1].

The goal is to present a fast parallel solver for elliptic equations as follows, complemented by appropriate boundary conditions,

$$-div(\rho \nabla p) = f, \quad \rho \equiv \rho(x) \in \mathbf{R}^+, \ p \equiv p(x) \in \mathbf{R}^+ \ x \in \Omega \subset \mathbf{R}^3, \tag{1}$$

The specificity of our solver is that it is designed to combine numerical efficiency and parallel efficiency on a grid of parallel computers.

We recall that a grid is a complex system with a large number of distributed hardware and software components and no centralized control [2, 3, 4, 5]. Communication latency between some pairs of nodes might be high, variable, and unpredictable. The parallel computers of the grid are heterogeneous and have variable performances. One may also expect an unusually high unreliability of computing nodes. We assumed that the discretization grid is topologically equivalent to a Cartesian grid, or can be embedded to such a spatial discretization grid. This assumption holds for subdomains only, when we use an overset method [6]. We look at second order finite volume or finite difference discretization of the equation in 2D that provides a sparse linear operator with regular bandwidth structure. The structure of the matrix and its storage scheme is therefore fairly simple.

After this introduction, we present in Section 2 the subdomain solver tuning, then Section 3 describes the domain decomposition technique, and finally, Section 4 gives some results.

2. The subdomain solver tuning

We have written an interface software [7] to reuse a broad variety of existing linear algebra softwares for each subdomain such as LU factorization, a large number of Krylov methods with incomplete LU preconditioner, and Geometric or Algebraic multigrids solvers. Lapack, Sparskit [8] and Hypre [9] were implemented in this interface. These libraries are representative of the state of the art for the resolution of a linear system. However many options are available for each iterative solver, and each of these softwares has a different language and/or style of coding. This interface is helpful for the user to speed up his code without cumbersome programming, it has the simplicity of Matlab command with the call of the subroutine with its parameters and it keeps the original Fortran or C. The complexity of the choice of the argument in the solver is hidden by the fact that the software can determine the best method case by case by exhaustive experimentation on the grid. The least square quadratic polynomial approximation for elapsed time with a set of test problems [10], we can predict with a good accuracy the elapsed time for the resolution of a linear solver with LU or a Krylov solver with a relative error of the prediction less than 3% . For Hypre, the prediction did not give good aproximation because the elapsed time is very sensitive to the size due to the coarsing of the grid. So for the Krylov methods and the LU domain decomposition, we did a surface response of the elapsed time of the fastest solver depending only on the grid size, the number of points in x direction and in y direction. The automatic tuning of the solver helps us to choose wisely the fastest solver for each subdomain when we do a domain decomposition technique.

3. Domain Decomposition Technique

We decompose the domain into overlap domain and use the general framework of the additive Schwarz algorithm [11].The additive Schwarz algorithm is particularly attractive for grid computing, because we have *a priori* only local communications of interfaces between neighbor subdomains at each iteration step. In addition, the algorithm is memory scalable and very easy to code. The convergence of the additive Schwarz method for linear operators is obtained if the spectral radius of the trace transfer operator is less than one. However this method is numerically inefficient unless one uses, for example, a coarse grid preconditioner or some optimized boundary conditions.

We have introduced recently an acceleration technique that in the simplest case resumes to a generalized Aitken acceleration method. The interfaces generated by the Schwarz algorithm must be written in a linear basis, such that the scalar Aitken acceleration can be efficiently applied to each individual component of the iterative sequence of vectors.

This method was called the Aitken-Schwarz algorithm [12]. The salient feature of the method is that it requires only a post-processing procedure of the interface obtained by an existing parallel code. There is no code rewriting involved.

The Aitken-Schwarz method is numerically efficient provided that one can get a good approximation of the main eigenvectors associated to the trace transfer operator. Main eigenvectors are those which correspond to the largest eigenvalues. The acceleration technique is applied to the eigenvector components that are responsible for the slow convergence of the original Schwarz method. We refer to [12, 13] for a description of the method in the general case.

4. Results

4.1. Subdomain solver performance

First, we will show results on a single processor.

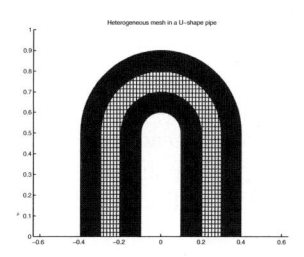

Figure 1. Composite Mesh for the curved pipe flow problem

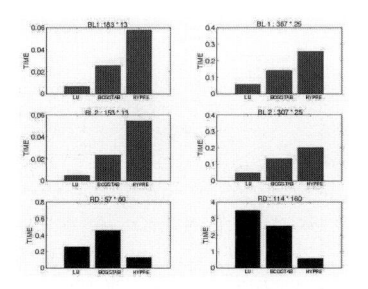

Figure 2. Comparison of the elapsed time for each subdomain with preconditioning for the curved pipe flow problem

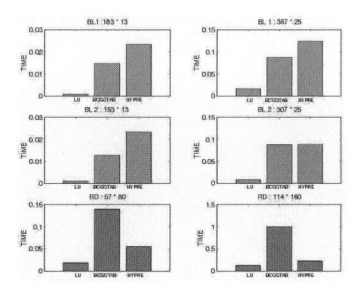

Figure 3. Comparison of the elapsed time for each subdomain with a precomputed preconditioner for the curved pipe flow problem

Let us illustrate the performance of subdomain solvers with an incompressible flow in a curved pipe. The model uses Navier Stokes equation with no slip boundary condition on the wall of the pipe and prescribed flow speed at the inlet and outlets. For large Reynolds number we use an overlapping domain decomposition method with non matching grids. Two thin subdomains, denoted *BL1* and *BL2* fit the wall and have orthogonal meshes to approximate the boundary layer. The central part of the pipe denoted *RD* is polygonal and it is overlapping the boundary subdomains by few mesh cells. This is basically a Chimera approach that is convenient to compute fluid structure interaction. In Figures 2 and 3, the performance of different solvers in *RD* and boundary layer subdomains are represented. Figure 2 corresponds to the elapsed time for each family of solver and three grid sizes of the problem. This elapsed time includes the computation of the factorization of the matrix, or the computation of the preconditioner. In Figure 3, we report the elapsed time for the same problem, excluding the time spent to construct the factorization and/or the preconditioner. We observe first that LU decomposition performs best for small problems. The performance of LU deteriorates when the size of the problem growth while Hypre provides the best performance if the problem is large enough. The boundary layer problem is rather special because of the large aspect ratio on grid dimensions. With appropriate ordering the bandwidth of the corresponding matrix is much smaller than the dimension of the linear system, and LU has very low arithmetic complexity. If we exclude the time spent in the factorization or in the preconditioner, we can notice that LU is always the fastest even for large grid size.

So, the optimum choice of the solver for each subdomain, i.e., the solver that processes the subdomain in the shortest elapsed time, depends on the type of subdomain, the fact that we use or not the same preconditioner or decomposition of the operator, the architecture of the processor and the size of the problem.

We are going now to present the performance of the Aitken Schwarz method on a Beowulf cluster.

4.2. Performances of Aitken-Schwarz

Figure 4. Speedup of the Aitken-Schwarz method solved by LU in each subdomain

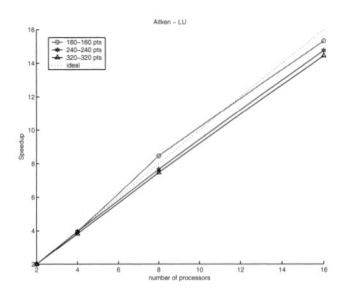

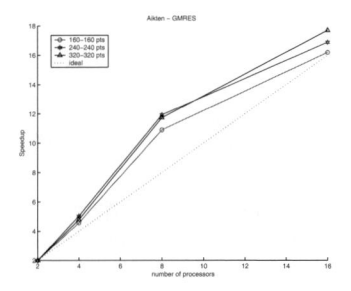

Figure 5. Speedup of the Aitken-Schwarz method solved by krylov in each subdomain

Let us look at the parallel performances of the Aitken-Schwarz method and consider the Poisson equation in a square domain with Dirichlet boundary conditions.

We note from Figures 4 and 5 that Aitken Schwarz performs very well on small problems. Further, the Krylov method seems to be more sensitive to the cache effect, since we have a superlinear speedup.

4.3. Limit of parallel subdomain solver performance

We are going to study the impact of the choice of the subdomain solver on the overall domain decomposition performance. The goal here is not to obtain the best subdomain solver from all existing methods, but rather to use the best solver from Sparskit (Krylov solver tool developed by Saad [8]) or Lapack only. As a matter of fact from all existing computational algorithms a full multigrid algorithm should be, in theory, the optimum.

The performance modeling on a single processor of the Poisson subdomain solver gives Figure 6. We checked that the elapsed time with four successive runs stays consistent. Horizontal axis and vertical axis of the figure corresponds to the dimension of the subdomain in each space direction. The dark region indicates where Sparskit (Bicg) is faster than Lapack (LU decomposition) while the light shows where Lapack is faster.

Let us compare this prediction model on a single processor and the performance of the Aitken-Schwarz algorithm with a fixed size of the subdomain per processor that corresponds to the points in Figure 6. We test the scalability of the Aitken-Schwarz algorithm, that is: we fix the size of the subdomain per processor once for all, and let the number of processors grow. The global size of the problem grows linearly with the number of processors. We use up to 16 processors of a Beowulf cluster (Dual AMD Athlon 1800) equipped with a Gigabit Ethernet switch.

The following table gives the total elapsed time to solve the problem and the bold number gives the best elapsed from both options i.e., Lapack with LU or Sparskit with Bicgstab solver .

	2 processors		4 processors		8 processors		16 processors	
points per processors	LU	Bicg	LU	Bicg	LU	Bicg	LU	Bicg
100 x 160	**1.44**	2.26	**1.58**	1.87	1.43	1.43	**1.79**	1.82
120 x 160	**2.08**	3.19	**2.45**	2.54	2.29	**1.98**	2.75	**2.50**
140 x 160	**3.28**	4.17	3.61	**3.36**	3.54	**2.75**	3.89	**3.45**
160 x 160	**4.67**	5.21	4.81	**4.36**	4.95	**3.49**	5.37	**4.37**
180 x 160	6.46	**6.39**	6.43	**5.63**	6.52	**4.31**	7.72	**5.45**
200 x 160	8.24	**8.18**	8.89	**6.69**	8.27	**5.25**	9.69	**6.54**

We can notice that the prediction of the best subdomain solver according to Figure 6, is that one should uses Sparskit (respectively Lapack) for the first space dimension above 160 (respectively below 160). This prediction is correct for the 2 processors computation. However as the number of processors grows, this prediction is incorrect, and one should favor the Krylov solver. Overall, we observe that the optimum choice of the subdomain solver should use a surface response model that includes as a third dimension the number of processors.

Indeed Figures 7, 8 and 9 of the surface response depending on the grid size and the number of processors confirm our statement. For the next section, we use Sparskit for

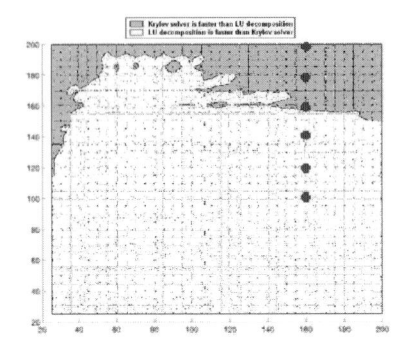

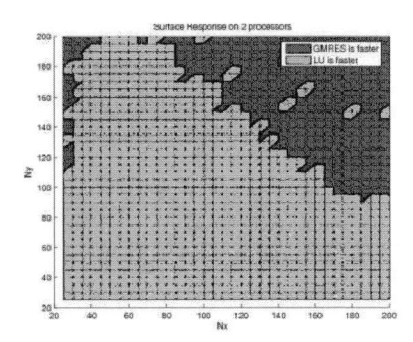

Figure 6. Comparison between Krylov Solver and LU.

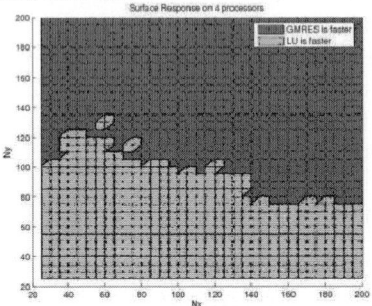

Figure 7. Surface response with 2 processors

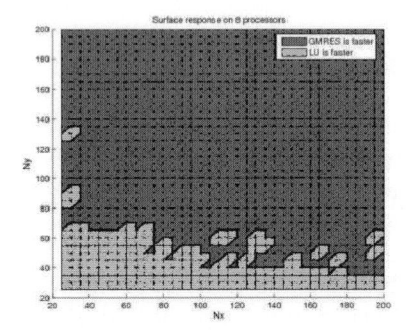

Figure 8. Surface response with 4 processors

Figure 9. Surface response with 8 processors

subdomain solver only, since this solver gives us the best performance for large number of processors. Let us compare the performance of our Aitken-Schwarz code with the Portable, Extensible Toolkit for Scientific Computation (PETSc).

4.4. Comparison with PETSc

PETSc [14] is an excellent general purpose software for comparison purpose. PETSc consists of a variety of libraries which include many linear solvers such as Lapack, Krylov solver and algebraic multigrid solver [15].

In Figure 10 , we report the speedup performance of PETSc and Aitken-Schwarz on the same graphic, while in Figure 11, we give the elapsed time. We choose to run PETSc using V-cycle multigrid with 3 levels which is the fastest method to solve the Poisson problem. PETSc, as expected, is faster than Aitken-Schwarz with two processors and also for 3 processors. However as the number of processors increases, one can observe

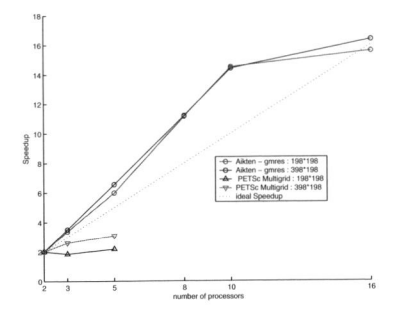

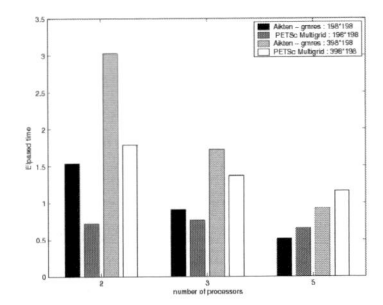

Figure 10. speedup between Aitken solved with GMRES and PETSc multigrid

Figure 11. Elapsed times for Aitken solved with GMRES and PETSc multigrid

that the multigrid solver does not speed up well, while our method performs better. Eventually for more than three processors, Aitken-Schwarz gives a better elapsed time than the multigrid solver. PETSc does not have a good speedup, this is explained by the fact that the results are obtained with a high latency network and PETSc is very sensitive to the performance of the network.

This is by no means a general conclusion because this test case is particularly simple. But it is rather a demonstration that the Aitken-Schwarz algorithm is tolerant for high latency network, while traditional optimum servers are not.

We should have used PETSc as a subdomain solver, and Aitken-Schwarz for the domain decomposition method in this specific case. This is part of our ongoing software development.

5. Conclusion

We have presented a domain decomposition for elliptic solver that is interesting for distributed computing with high latency network which provides good scalability results. We have demonstrated the impact of the choice of the subdomain solver and presented a methodology with surface response that can help to tune the solver in an optimal way. While generating such a model is cumbersome and time consuming, there are advantages to generating them in an automatic manner using the resources of the grid. We also show that for slow network and a simple Poisson problem, Aitken-Schwarz is efficient and eventually out-performs the multigrid solver of PETSc for a large number of processors. But, we can improve the performance of the code by implementing PETSc library for our subdomain solver.

REFERENCES

1. D.E. Keyes, Letting Engineers be Engineers and Other Goals of Scalable Solver R & D , Parallel CFD 2005
2. I.Foster and C.Kesselman, The Grid: Blueprint for a New Computing Infrastructure, Morgan Kaufmann, 1998
3. I. Foster and N.Karonis, A Grid-Enabled MPI: Message Passing in Heterogeneous Distributed Computing Systems, IEEE/ACM SC98 Conference , p46, 1998
4. I. Foster and C. Kesselman and J. M. Nick and S. Tuecke, Grid Services for Distributed System Integration, Computer ,35, p37-46, 2002
5. I. Foster, The Grid: A New Infrastructure for 21st Century Science, Physics Today,55,p42-47,2002
6. Overture: Object-Oriented Tools for Overset Grid Applications D. L. Brown, W.D. Henshaw and D.J. Quinlan, AIAA conference on Applied Aerodynamics, UCRL-JC-134018, 1999
7. M.Garbey, W.Shyy, B.Hadri and E.Rougetet, Efficient Solution Techniques for CFD and Heat Transfer, ASME Heat Transfer/Fluids Engineering Summer Conference, 2004
8. Y. Saad, Iterative Methods for Sparse Linear Systems,SIAM, edition 2,2003
9. R.D. Falgout and U. Meier Yang, Hypre: A Library of High Performance Preconditioners, Lecture Notes in Computer Science, Springer-Verlag, editors C.J.K. Tan. J.J. Dongarra, and A.G. Hoekstra, 2331, p632-641, 2002
10. D.C. Montgomery and R.H. Myers, Response Surface Methodology: Process and Product Optimization Using Designed Experiments, Wiley, 2nd edition, 2002.
11. B. Smith and P. Bjorstad and W. Gropp, Domain Decomposition, Parallel Multilevel Methods for Elliptic Partial Differential Equations, Cambridge University Press, 1996
12. M.Garbey and D.Tromeur Dervout, On some Aitken like acceleration of the Schwarz Method, Int. J. for Numerical Methods in Fluids, 40, p1493-1513, 2002
13. M.Garbey, Acceleration of the Schwarz Method for Elliptic Problems, SIAM, J. SCI COMPUT., 26, 6, p 1871-1893, 2005.
14. S. Balay, W. D. Gropp, L.C. McInnes and B.F. Smith, PETSc Users Manual,ANL-95/11 - Revision 2.1.5, Argonne National Laboratory,2003
15. W.L. Briggs ,V.E. Henson and S.F. McCormick, A Multigrid Tutorial, SIAM Books, 2000

FLASH: Applications and Future

K.B. Antypas,[a] * A.C. Calder,[a] A. Dubey,[a] J.B. Gallagher,[a] J. Joshi,[a] D.Q. Lamb,[a]
T. Linde,[a] E.L. Lusk,[a] O.E.B. Messer,[a] A. Mignone,[b] H. Pan,[a] M. Papka,[a] F. Peng,[a]
T. Plewa,[a] K.M. Riley,[a] P.M. Ricker,[c] D. Sheeler,[a] A. Siegel,[a] N. Taylor,[a] J.W. Truran,[a]
N. Vladimirova,[a] G. Weirs,[d] D. Yu,[a] J. Zhang[a]

[a]ASC/Flash Center, University of Chicago, Chicago, IL 60637

[b]Via Osservatorio, 20, 10025 Pino Torinese (TO), Italy

[c]Dept of Astronomy, University of Illinois, Urbana, IL 61801

[d]Sandia National Laboratories, Albuquerque, NM 87185 Mailstop 0196

FLASH is a publicly available, modular, parallel, adaptive mesh application code capable of simulating the compressible, reactive flows found in many astrophysical environments. It is a collection of inter-operable modules which can be combined to generate different applications. Such component based architectures have historically met with varying degrees of success. FLASH is unique in that it started out as a more traditional scientific code and evolved into a modular one as insights were gained into manageability, extensibility, and efficiency. Hence, the development of the code has been, and continues to be, driven by the dual goals of application requirements and modularity. In this paper we describe the architecture of the latest released version of the code, give insights into the code development process, and discuss future directions of code development. We also describe some of the applications generated by scientists at the Flash center.

1. Introduction

FLASH is a publicly available, modular, parallel, adaptive mesh application code capable of simulating the compressible, reactive flows found in many astrophysical environments (1). It has been developed primarily at the ASC/Alliances Flash Center, founded in 1997 under the auspices of the Department of Energy Advanced Simulation and Computing program, with the purpose of examining thermonuclear flashes, *i.e.* events of rapid or explosive thermonuclear burning, that occur on the surfaces and in the interiors of compact stars. Scientists at the Flash Center have applied FLASH to a wide variety of astrophysical problems, including X-ray bursts (2), magnetized galactic bubbles (3), classical novae (4), and Type Ia supernovae (5). Simulations of the relevant basic physics problems include the cellular structures of detonations (6), wind/wave interactions (7), and the Rayleigh-Taylor instability (8). Today the FLASH code has a wide user base

---

*This work is supported by the U.S. Department of Energy under Grant No. B523820 to the Center for Astrophysical Thermonuclear Flashes at the University of Chicago

and is unique among astrophysics simulation codes in that it has been subjected to a formal verification and validation program (9; 10). The FLASH code was awarded the 2000 Gordon Bell Award in the Special Category (11).

The FLASH code uses the PARAMESH library (12) to manage a block-structured adaptive grid. The compressible Euler equations of hydrodynamics are solved using an explicit, directionally split version of the piecewise-parabolic method that allows for general equations of state (13; 14). FLASH contains magnetohydrodynamics solvers useful in problems of magnetic reconnection typical of solar physics, as well as problems of magnetically driven outflows characteristic of many accreting objects such as neutron stars or black holes. FLASH also contains a special relativistic solver (15) that is useful for accretion problems. Gravitational potential and acceleration can be calculated using either a multigrid scheme, or, for more symmetrical problems, a faster multipole scheme. FLASH supports active particles used to calculate the gravitational field in cosmological simulations, as well as passive tracer particles for simulation diagnostics. There are functions in FLASH to define the material properties of the constituent species and the equations of state. FLASH also has support for a variety of source terms, including ionization, heating, cooling, and nuclear burning. In addition to the astrophysics, routines for cosmology provide a numerical framework required for solving problems in co-moving coordinates. These include the evolution of the scale factor and the redshift assuming a Friedmann-Robertson-Walker solution.

Large-scale simulations have significant performance requirements, and much of the challenge of developing FLASH has been balancing the need for performance with the need for a maintainable and extensible code. As the complexity of the code and the number of developers have grown, code verification and management of the software development process have become increasingly important to the success of the project.

FLASH development is currently focused on the next version of the code, FLASH3, which better incorporates modern software engineering practices, including unit testing and encapsulation, by eliminating lateral communication between code modules and by implementing well-defined rules for inheritance, interoperability, and data movement. While some of these practices were incorporated in earlier versions, they are integral to the design of FLASH3 and enhance the interchangeability of code modules with the same functionality but different implementations. In earlier versions of FLASH, this capability was limited to physics kernels, while in FLASH3 it is further formalized and extended to infrastructure such as the mesh package. The objective of this effort is to create a community code to which new functionality may easily be added.

2. Architecture

The FLASH architecture is characterized by three entities: the modules, the configuration layer, and the problem setups.

Each module provides a well-defined functionality and conforms to a structure that facilitates interactions with other modules. A module can have interchangeable implementations of varying complexity, as well as sub-modules that inherit from and override the functionality in the parent module. Physics modules such as *hydro* and *gravity* solve specific equations, infrastructure modules such as *mesh* and *IO* handle the housekeeping of the simulation grid and check-pointing, and utility modules manage runtime parameters, logging, and profiling. The driver module defines the evolution in time and the interaction between different modules.

The FLASH code is not a single application (or a single binary). Rather, it is a collection of modules combined by the configuration layer into a single application for specific simulations. The configuration layer consists of text files called *Config* files and the *setup script*. The Config files reside with modules, and they specify the module requirements, such as physical variables, runtime parameters, and other modules that must be included or excluded in the simulation. The setup script parses the Config files to find the set of modules, physical variables, runtime parameters and required libraries to create an application.

The problem setup directory plays a very important role in the configuration process. Its Config file is the starting point for the setup script and provides the initial set of modules and variables for the new application. A user can replace any native FLASH routine by providing an alternative implementation in the problem setup directory. For example, users can provide their own criteria for refining and derefining the mesh by including the appropriate files with compatible function interfaces in the setup directory.

2.1. Database

The biggest challenge in modularizing any application software, especially one that includes large sections of legacy code, is managing the data movement therein. Traditionally, FORTRAN codes have relied on common blocks to tame the argument lists in passing data. This practice, however, makes it extremely difficult to determine the source of data and its ownership. FLASH2 solved this problem by adopting a centralized database. The global scope data, including the solution grid, physical variables, and the simulation state variables (*e.g.* the timestep) were owned by a central database that provided functions to access and modify the data. A similar functionality covered physical constants and properties of multiple species of materials. Thus, instead of passing data as arguments or using common blocks between interacting modules, each module accessed data directly through the database interface. This database and accessor function approach streamlined data management, but the drawback was that it gave all modules equal access to data, which could sometimes lead to conflicts. Improving this situation was one of the issues addressed in the design of the next version of the code, as described in the next section.

2.2. FLASH3 Architecture

The Flash Code Group is currently in the processes of designing the next generation architecture for the FLASH code. The primary goal of this effort is to prepare FLASH to become the astrophysical community code of choice. To achieve this, it is necessary to further simplify and streamline the data management, rigorously define module architecture, and clarify the rules of interoperability between different modules. Better data management in FLASH3 is achieved by decentralizing the database. The centralized database of previous versions of FLASH has two major disadvantages. The first is the conflict in ownership of data between the database and the mesh packages. The database owns the solution grid data, but the mesh package assumes unrestricted access to it. Additionally, IO requires unrestricted access to the grid data for check-pointing, thereby violating the basic assumption of modularity. The second problem with the centralized database is its inability to distinguish between modules that should and should not be allowed to access and modify data. This inability can lead to errors if one module attempts to modify data which it should not be allowed to access.

Avoiding such errors is already a challenging problem for FLASH's internal developers, and one that becomes even more challenging when the code is opened to external collaborators. Decentralizing the database resolves these problems by allowing individual modules to own the data most relevant to them and to control external access to their data by using accessor functions. For example, the driver module owns the timestep variable. While any module may read the value of this variable, the driver alone reserves the right to modify it. This is an improvement over the earlier scheme in which all modules were allowed equal write-access to any writable variable. FLASH3 also formalizes namespaces, interfaces, and inheritance of the modules, and eliminates lateral interactions between modules. These abstractions allow greater flexibility in adding new modules, or in adding a new implementation of an existing module.

3. Applications

FLASH has been applied to a variety of problems, but its principal applications are astrophysical thermonuclear flashes, events of rapid or explosive thermonuclear burning. These events occur as the result of the transfer of combustible stellar material from a normal star onto the surface of a nearby compact star, either a white dwarf or a neutron star. Accretion, *i.e.* the buildup of material on the surface of the compact star, leads to the thermonuclear explosion. There are three main classifications of thermonuclear flashes. The first may result from the steady accretion of matter onto the surface of a neutron star. The accreted matter may explosively burn or "flash", an event observed mostly as a burst of X-rays and referred to as a Type I X-ray burst. The second type of flash, called a nova, can occur when material accretes on the surface of a white dwarf and explodes. The third and most spectacular event can occur if the accretion rate of material onto the surface of the white dwarf is sufficiently high. If this is the case, the ever-accumulating stellar material will begin to compress the white dwarf. As the interior of the star is compressed it will heat and, if circumstances are right, a thermonuclear runaway will result which will consume the entire star and produce a violent explosion known as a Type Ia supernova.

The challenge of modeling these complex phenomena lies in understanding and including all of the relevant physics. The wide range of length scales of these problems necessitates exploration of detailed microphysics independent of simulations of the full problems with the goals of a better understanding and development of sub-grid-scale models. Consequently, FLASH has been applied to problems on a variety of relevant length scales, including centimeter-scale nuclear flames (16), meter-scale turbulent mixing processes (7), flashes occurring on the surfaces of compact stars (2), and whole-star supernova simulations (17).

Figure 1 shows a rising bubble of nuclear ash from a whole-star simulation of a Type Ia supernova. In this case, the nuclear flame was initiated slightly off-center deep in the core of the star. The simulation indicates that a slightly off-center ignition leads to a rapidly

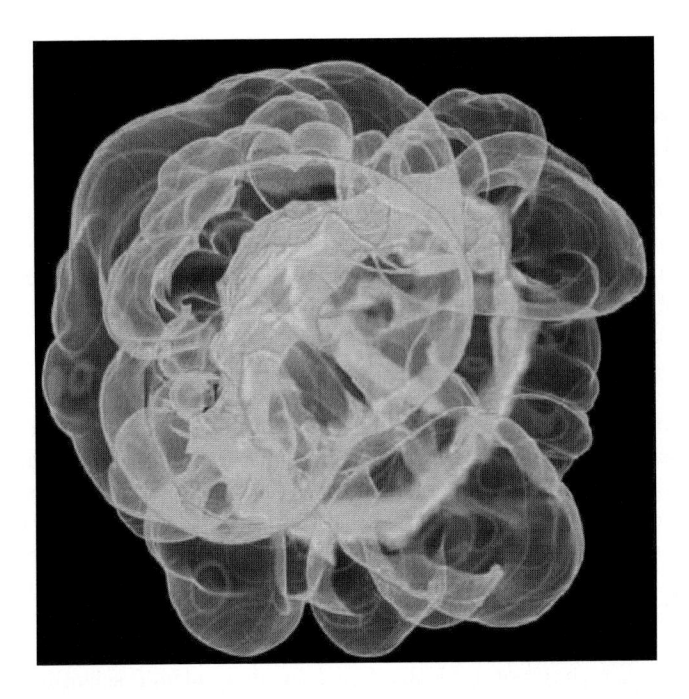

Figure 1. Image of the rising bubble of magnesium resulting from an initially off-center carbon deflagration in a Chandrasekhar-mass carbon/oxygen white dwarf (17). Shown is a volume rendering of magnesium abundance. Image produced by the ANL Futures Laboratory.

rising bubble of burning material that is one part of a proposed mechanism for Type Ia supernova explosions (5).

4. Software Process

Developing and managing FLASH is a complex issue for a variety of reasons. Some of this complexity arises from the non-linear equations evolved by FLASH. The solvers often do not lend themselves readily to algorithm separation, and encapsulation is correspondingly difficult. Performance is another challenge. Modularity, and hence maintainability, of the code is often achieved at the expense of computational efficiency, and balancing these two conflicting requirements is an ongoing concern. Another issue is the difficulty of ensuring that the code continues to run without errors while undergoing simultaneous modification by multiple developers. Finally, an often overlooked complication in scientific computing is the loss of expertise that follows when contributors move on to new positions, leaving parts of the code unsupported. For all these reasons, good design and management of the software development process and thorough documentation are critical to maintaining and enhancing the code.

Good software practices demand substantial effort in the areas of maintenance, documentation, and code enhancement policies. Perhaps the most important maintenance policy is regression testing. Each night the FLASH test suite is run on multiple platforms. The suite consists of a collection of applications that exercise all major parts of the code. With many developers at work, the test suite has become a vital tool for implementation verification. The code documentation includes a detailed user's guide, which is part of the released code, and extensive online descriptions of major functions. Any significant code modification requires a proposal which includes the specifications and an analysis of its impact on ongoing work. A detailed plan is required for any modification before changes are made to the code, and extensive documentation and rigorous testing are implemented at every step of development.

5. Summary and Conclusions

FLASH has been successful in simulating a wide variety of astrophysical problems both within the Flash Center and in the external community. The code has steadily gained acceptance since its initial release for the following reasons. First, it is easily ported to a variety of computer architectures and the distribution includes support for many standard machines. Second, creation of new applications is straightforward and well-documented, and the suite of tools provided allows users to quickly process and interpret simulation results. Third, the modularity of the code allows users to readily add functionality.

The unusually large scale of FLASH necessitates the adoption of practices already in wide commercial use, but not typically found in an academic scientific project. The principal lesson learned during the development of FLASH is the importance of good software practices. These include thorough documentation, detailed design specifications for the framework and components, and especially regression testing, which is essential in a project with multiple developers.

FLASH3 builds upon the success of earlier versions and promises a more modular architecture that will be easy to augment and maintain without sacrificing computational efficiency.

References

[1] B. Fryxell, K. Olson, P. Ricker, F.X. Timmes, M. Zingale, D.Q. Lamb, P. MacNeice, R. Rosner, J.W. Truran, H. Tufo., ApJS,**131**, 273 (2000)

[2] M. Zingale, J.W. Truran, F.X. Timmes, B. Fryxell, D.Q. Lamb, K. Olson, A.C. Calder, L.J. Dursi, P. Ricker, R. Rosner, P. MacNeice & H. Tufo, ApJS, **133**, 195 (2001)

[3] K. Robinson, L.J. Dursi, P.M. Ricker, R. Rosner, T. Linde, M. Zingale, A.C. Calder, B. Fryxell, J.W. Truran. F.X. Timmes, A. Caceres, K. Olson, K. Riley, A. Siegel & N. Vladimirova, ApJ, **601**, 621 (2004)

[4] A. Alexakis, A.C. Calder, A. Heger, E.F. Brown, L.J. Dursi, J.W. Truran, R. Rosner, D.Q. Lamb, F.X. Timmes, B. Fryxell, M. Zingale, P.M. Ricker & K. Olson, ApJ, **602**, 931 (2004)

[5] T. Plewa, A.C. Calder & D.Q. Lamb, ApJ, **612**, L37 (2004)

[6] F.X. Timmes, M. Zingale, K. Olson, B. Fryxell, P. Ricker, A.C. Calder, L.J. Dursi, J.W. Truran, H. Tufo, P. MacNeice & R. Rosner, ApJ, **543**, 938 (2000)

[7] A. Alexakis, A.C. Calder, L.J. Dursi, R. Rosner, J.W. Truran, F.X. Timmes, B. Fryxell, M. Zingale, P.M. Ricker & K. Olson, Phys. Fluids, **16**, 3256 (2004)

[8] G. Dimonte, D. Youngs, A. Dimits, S. Weber, M. Marinak, S. Wunsch, C. Garasi, A. Robinson, M.J. Andrews, P. Ramaprabhu, A.C. Calder, B. Fryxell, J. Biello, L. Dursi, P. MacNeice, K. Olson, P. Ricker, R. Rosner, F. Timmes, H. Tufo, Y.-N. Young & M. Zingale, Phys. Fluids, **16**, 1668 (2004)

[9] A.C. Calder, B. Fryxell, T. Plewa, R. Rosner, L.J. Dursi, V.G. Weirs, T. Dupont, H.F. Robey, J.O. Kane, B.A. Remington, R.P. Drake, G. Dimonte, M. Zingale, F.X. Timmes, K. Olson, P. Ricker, P. MacNeice & H.M. Tufo, ApJS, **143**, 201 (2002)

[10] G. Weirs, V. Dwarkadas, T. Plewa, C. Tomkins & M. Marr-Lyon, ApSS, **298**, presented at the 5th International Conference on High Energy Laboratory Astrophysics, Tucson, AZ, March 10-13, 2004, pp. 341-346

[11] A.C. Calder, B.C. Curtis L.J. Dursi, B. Fryxell, G. Henry, P. MacNeice, K. Olson, P. Ricker, R. Rosner, F.X. Timmes, J.W. Truran, H.M. Tufo & M. Zingale, in Proc. Supercomputing 2000, IEEE Computer Soc. 2000, http://sc2000.org (Gordon Bell Prize)

[12] P. MacNeice, K.M. Olson, C. Mobarry, R.de Fainchtein & C. Packer, Comp. Phys. Comm., **126**, 330 (2000)

[13] P. Colella & P. Woodward, J. Comp. Phys., **54**, 174 (1984)

[14] P. Colella & H.M. Glaz, J. Comp. Phys., **59**, 264 (1985)

[15] A. Mignone, T. Plewa, & G. Bodo, ApJ, in press, available as astro-ph/0505200, 2005.

[16] L.J. Dursi, M. Zingale, A.C. Calder, B. Fryxell, F.X. Timmes, N. Vladimirova R. Rosner, A. Caceres, D.Q. Lamb, K. Olson, P.M. Ricker, K. Riley, A. Siegel, & J.W. Truran, ApJ, **595**, 955 (2003)

[17] A.C. Calder, T. Plewa, N. Vladimirova. E.F. Brown, D.Q. Lamb, K. Robinson, J.W. Truran. BAAS, **35** 1278, 2004.

An Adaptive Cartesian Detonation Solver for Fluid-Structure Interaction Simulation on Distributed Memory Computers

R. Deiterding[a]

[a]California Institute of Technology, Mail Code 158-79
1200 East California Blvd., Pasadena, CA 91125, USA

Time-accurate fluid-structure interaction simulations of strong shock and detonation waves impinging on deforming solid structures benefit significantly from the application of dynamic mesh adaptation in the fluid. A patch-based parallel fluid solver with adaptive mesh refinement in space and time tailored for this problem class is presented; special attention is given to the robustness of the finite volume scheme with embedded boundary capability and to a scalable implementation of the hierarchical mesh refinement method.

1. Introduction

The Center for Simulating the Dynamic Response of Materials at the California Institute of Technology has constructed a virtual test facility (VTF) for studying the three-dimensional dynamic response of solid materials subject to strong shock and detonation waves propagating in fluids. While the fluid flow is simulated with a high-resolution Cartesian finite volume upwind method that considers the solid as an embedded moving body represented implicitly with a level set function, Lagrangian finite element schemes are employed to describe the time-accurate material response subject to the current hydrostatic pressure loading. A loosely coupling temporal splitting method is applied to update the boundary's positions and velocities between time steps. The Cartesian finite volume scheme is incorporated into a parallel structured dynamic mesh adaptation algorithm that allows very fine local resolutions to capture the near-body fluid-structure interaction (FSI) and incoming waves in the fluid at minimal computational costs.

In this paper, we describe the dynamically adaptive solver for compressible flows, including shock and detonation waves with one-step reaction model, that enables highly efficient FSI simulations in the VTF on distributed memory machines. After introducing the governing fluid equations, we explain our specific implementation of the ghost fluid approach [11]. In Sec. 4, we outline the structured adaptive mesh refinement (SAMR) algorithm of Berger and Collela [3], and in particular the locality-preserving rigorous domain decomposition paradigma under which the method has been parallelized [7]. Section 5 details the extension of the SAMR implementation to loosely coupled FSI problems. The final section, Sec. 6, gives a numerical example in which a detonation wave propagating through a high explosive interacts with a surrounding solid cylinder. The enormous savings in compute time from mesh adaptation and parallelization demonstrate the efficiency of the approach.

2. Governing Equations

In order to model detonation waves we utilize the single-phase model proposed by Fickett and Davis [13], which has also been used by Clarke et al. [4] to evaluate numerical methods for detonation simulation. We assume a single chemical reaction $A \longrightarrow B$ that is modelled by a progress variable λ corresponding to the mass fraction ratio between the density of the product B and the total density ρ, i.e. $\lambda = \rho_B/\rho$. The governing equations of the model read

$$\begin{aligned} \partial_t \rho + \nabla \cdot (\rho \mathbf{u}) &= 0, & \partial_t(\rho \mathbf{u}) + \nabla \cdot (\rho \mathbf{u} \otimes \mathbf{u}) + \nabla p &= 0, \\ \partial_t(\rho E) + \nabla \cdot ((\rho E + p)\mathbf{u}) &= 0, & \partial_t \lambda + \mathbf{u} \cdot \nabla \lambda &= \psi. \end{aligned} \quad (1)$$

Herein, $\mathbf{u}$ is the velocity vector and E the specific total energy. The hydrostatic pressure p is given by $p = (\gamma - 1)(\rho E - \frac{1}{2}\rho \mathbf{u}^T \mathbf{u} + \rho \lambda q)$ with γ denoting the ratio of specific heats and q the heat release due to the chemical reaction per unit mass. System (1) together with above pressure equation is a valid model both for detonations in combustible gases and high energetic solid materials. As our focus in this paper is on the latter, we use the simple rate function $\psi = \frac{2}{T_R}(1-\lambda)^{1/2}$ proposed by Fickett for detonations in solids in the following. Herein, T_R denotes a typical time associated with the reaction in which the depletion from A to B is complete.

3. Cartesian Finite Volume Scheme with Embedded Boundaries

Following Clarke et al. [4], we apply the method of fractional steps to decouple the chemical reaction and hydrodynamic transport numerically. The *homogeneous* system of (1) and the scalar ordinary differential equation $\partial_t \lambda = \psi(\lambda)$ are solved successively with the data of the preceding step as initial conditions. As the homogeneous system (1) is a hyperbolic conservation law that admits discontinuous solutions, cf. [4], we use a time-explicit finite volume discretization that achieves a proper upwinding in all characteristic fields. The scheme is based on a straightforward generalization of the Roe scheme for the purely hydrodynamic Euler equations and is extended to a multi-dimensional Cartesian scheme via the method of fractional steps, cf. [22]. To circumvent the intrinsic problem of unphysical total densities and internal energies near vacuum due to the Roe linearization, cf. [10], the scheme has the possibility to switch to the simple, but extremely robust Harten-Lax-Van Leer (HLL) Riemann solver. The occurrence of the disastrous carbuncle phenomena [20], a multi-dimensional numerical crossflow instability that affects every simulation of strong grid-aligned shocks or detonation waves, is prevented by introducing a small amount of additional numerical viscosity in a multi-dimensional way [21]. This hybrid Riemann solver is supplemented with the MUSCL-Hancock variable extrapolation technique of Van Leer [22] to achieve second-order accuracy in regions where the solution is smooth.

Geometrically complex moving boundaries are considered by utilizing some of the finite volume cells as ghost cells to enforce immersed moving wall boundary conditions [11]. The boundary geometry is mapped onto the Cartesian mesh by employing a scalar level set function ϕ that stores the signed distance to the boundary surface and allows the efficient evaluation of the boundary outer normal in every mesh point as $\vec{n} = -\nabla \phi / |\nabla \phi|$.

In coupled Eulerian-Lagrangian simulations, ϕ is updated after every boundary synchronization step by calling the closest-point-transform algorithm developed by Mauch [18]. A cell is considered to be a valid fluid cell within the interior, if the distance ϕ in the cell *midpoint* is positive and is treated as exterior otherwise.

For system (1), the boundary condition at a rigid wall moving with velocity $\vec{w}$ is $\vec{u}\cdot\vec{n} = \vec{w}\cdot\vec{n}$. Enforcing the latter with ghost cells, in which the discrete values are located in the cell centers, requires the mirroring of the primitive values ρ, $\vec{u}$, p, λ across the embedded boundary. The normal velocity in the ghost cells is set to $(2\vec{w}\cdot\vec{n} - \vec{u}\cdot\vec{n})\vec{n}$, while the mirrored tangential velocity remains unmodified. Mirrored values are constructed by calculating spatially interpolated values in the point $\vec{\tilde{x}} = \vec{x} + 2\phi\vec{n}$ from neighboring interior cells. For instance, in two space dimensions, we employ a bilinear interpolation between (usually) four adjacent cell values, but directly near the boundary the number of interpolants needs to be decreased, cf. Fig. 1. It has to be emphasized that for hyperbolic problems with discontinuities like detonation waves, special care must be taken to ensure the monotonicity preservation of the numerical solution. Figure 1 highlights the necessary reduction of the interpolation stencil for some exemplary cases. The interpolation locations are indicated by the origins of the arrows normal to the complex boundary (dotted).

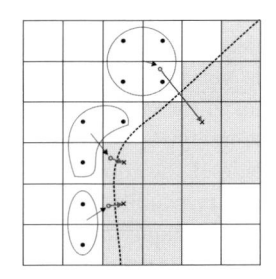

Figure 1: Construction of mirrored values to be used in interior ghost cells (gray).

After the application of the numerical scheme, cells that have been used to impose internal boundary conditions are set to the entire state vector of the nearest cell in the interior. This operation ensures proper values in case such a cell becomes a regular interior cell in the next step due to boundary movement. The consideration of $\vec{w}$ in the interior ghost cells ensures that the embedded boundary propagates at most one cell further in every time step.

4. Structured Adaptive Mesh Refinement

Numerical simulations of detonation waves require computational meshes that are able to represent the strong local flow changes due to the reaction correctly. The shock of a self-sustained detonation is very sensitive to changes in the energy release from the reaction behind and the inability to resolve all reaction details usually causes a considerable error in approximating the correct speed of propagation. In order to supply the required temporal and spatial resolution efficiently, we employ the structured adaptive mesh refinement (SAMR) method of Berger and Colella [3]. Instead of replacing single cells by finer ones, as it is done in cell-oriented refinement techniques, the Berger-Colella SAMR method follows a patch-oriented approach. Cells being flagged by various error indicators (shaded in Fig. 2) are clustered with a special algorithm [2] into non-overlapping rectangular grids. Refinement grids are derived recursively from coarser ones and a hierarchy of successively embedded levels is thereby constructed (cf. Fig. 2). All mesh widths on level l are r_l-times finer than on level $l-1$, i.e. $\Delta t_l := \Delta t_{l-1}/r_l$ and $\Delta x_{k,l} := \Delta x_{k,l-1}/r_l$ with $r_l \geq 2$ for $l > 0$ and with $r_0 = 1$, and a time-explicit finite volume scheme will (in principle) remain stable on all levels of the hierarchy.

The numerical scheme is applied on level l by calling a single-grid routine in a loop over all subgrids. The subgrids get computationally decoupled by employing additional ghost cells around each computational grid. Three different types of ghost cells have to be considered: Cells outside of the root domain are used to implement physical boundary conditions; ghost cells overlaid by a grid on level l have a unique interior cell analogue and are set by copying the data value from the grid, where the interior cell is contained (synchronization). On the root level no further boundary conditions need to be considered, but for $l > 0$ internal boundaries can also occur. They are set by a conservative time-space interpolation from two previously calculated time steps of level $l-1$.

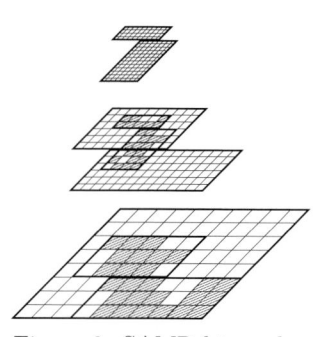

Figure 2: SAMR hierarchy.

The regularity of the SAMR data allows high performance on vector and super-scalar processors that allow cache optimizations. Small data arrays are effectively avoided by leaving coarse level data structures untouched when higher level grids are created. Values of cells covered by finer subgrids are overwritten by averaged fine grid values subsequently. This operation leads to a modification of the numerical stencil on the coarse mesh and requires a special flux correction in cells abutting a fine grid. The correction replaces the coarse grid flux along the fine grid boundary by a *sum* of fine fluxes and ensures the discrete conservation property of the hierarchical method (at least for purely Cartesian problems without embedded boundaries). See [3] or [6] for details.

In our SAMR solver framework AMROC (Adaptive Mesh Refinement in Object-oriented C++) [8], we follow a rigorous domain decomposition approach and partition the SAMR hierarchy from the root level on. A careful analysis of the SAMR algorithm uncovers that the only parallel operations under this paradigm are ghost cell synchronization, redistribution of the data hierarchy and the application of the previously mentioned flux correction terms. Interpolation and averaging, but in particular the calculation of the flux corrections remain strictly local [6]. Currently, we employ a generalization of Hilbert's space-filling curve [19] to derive load-balanced root level distributions at runtime. The entire SAMR hierarchy is considered by projecting the accumulated work from higher levels onto the root level cells. Figure 3 shows a representative scalability test for a three-dimensional spherical shock wave problem for the computationally inexpensive Euler equations for a single polytropic gas without chemical reaction. The test was run on a Linux Beowulf cluster of Pentium-4-2.4 GHz dual processor nodes with Quadrics Interconnect. The base grid had 32^3 cells and two additional levels with refinement factors 2 and 4. The adaptive calculation used approx. 7.0 M cells in each time step instead of 16.8 M cells in the uniform case. Displayed are the average costs for each root level time step. Although we utilize a single-grid update routine in Fortran 77 in a C++ framework with full compiler optimization, the fraction of the time spent in this Fortran routine are 90.5 % on four and still 74.9 % on 16 CPUs. Hence, Fig. 3 shows a satisfying scale-up for at least up to 64 CPUs.

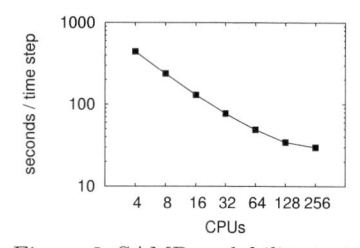

Figure 3: SAMR scalability test.

```
advance_level( l )
  repeat r_l times
    if time to regrid
      regrid( l )
    level_set_generation( ϕ^l, I )
    update_fluid_level( Q⃗^l, ϕ^l, w⃗|_I, Δt_l )
    if level l+1 exists
      advance_level(l+1)
      Correct Q⃗^l(t+Δt_l) with Q⃗^{l+1}(t+Δt_l)
    if l = l_c
      send_interface_data( p(t+Δt_l)|_I )
      receive_interface_data( I, w⃗|_I )
    t := t + Δt_l
return
```

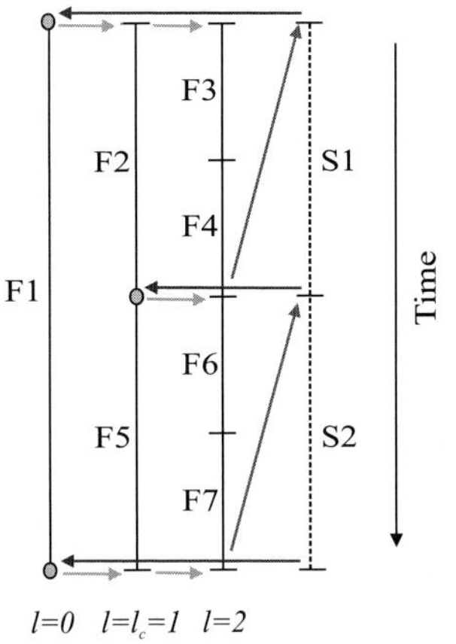

Figure 4. Left: SAMR algorithm for fluid-structure coupling. Right: data exchange between advance_level() and a conventional solid solver throughout one level 0 time step.

5. Fluid-Structure Coupling with SAMR

In the VTF, we apply a loosely coupled, partitioned approach and use separated solvers to simulate the fluid and solid sub-problem. Fluid-structure interaction is assumed to take place only at the evolving interface between fluid and solid and is implemented numerically by exchanging boundary data after consecutive time steps. The Eulerian fluid solver with embedded boundary capability (cf. Sec. 3) receives the velocities and the discrete geometry of the solid surface, while only the hydrostatic pressure is communicated back to the Lagrangian solid solver as a force acting on the solid's exterior [1,16]. As the inviscid Euler equations can not impose any shear on the solid structure, cf. [12], the fluid pressure is sufficient to prescribe the entire stress tensor on the solid boundary. An efficient parallel communication library has been implemented to support the boundary data exchange between (dedicated) fluid and solid processes on distributed memory machines, see [9] for details on this.

While the implementation of a loosely coupled FSI method is straightforward with conventional solvers with consecutive time update, the utilization of the recursive SAMR method is non-apparent. In the VTF, we treat the fluid-solid interface I as a discontinuity that is a-priori refined at least up to a coupling level l_c. The resolution at level l_c has to be sufficiently fine to ensure an accurate wave transmission between fluid and structure, but will often not be the highest level of refinement, cf. Sec. 6. We formulate the corresponding extension of the recursive SAMR algorithm of Berger and Collela [3] in the routine advance_level() outlined in pseudo-code on the left side of Fig. 4. The algorithm calls the routine level_set_generation() to evaluate the signed distance $ϕ$ for the actual level l based on the currently available interface I. Together with the recent solid velocity on the interface $w⃗|_I$, the discrete vector of state in the fluid $Q⃗$ is updated for the entire level with the scheme detailed in Sec. 3. The method then proceeds recursively to higher levels and utilizes the (more accurate) data from the next higher level to correct cells overlaid by refinement. If level l is the coupling level l_c, we use the updated fluid data to evaluate the pressure on the nodes of $\mathcal{I}$ to be sent to the solid and to receive updated mesh positions and nodal velocities. The recursive order of the SAMR algorithm automatically ensures that updated interface mesh information is available at later time steps on coarser levels and to adjust the grids on level l_c dynamically before the current surface mesh, i.e. the level set information derived from it, is actually used to again advance level l_c.

The data exchange between the solid and advance_level(), is visualized in the right graphic of Fig. 4 for an exemplary SAMR hierarchy with two additional levels with $r_{1,2} = 2$. Figure 4 pictures the recursion in the SAMR method by numbering the fluid update steps (F) according to the order determined by advance_level(). The order of the solid update steps (S) on the other hand is strictly linear. The red arrows correspond to the sending of the interface pressures $p|_{\mathcal{I}}$ from fluid to solid at the end of each time step on level l_c. The blue arrows visualize the sending of the interface mesh $\mathcal{I}$ and its nodal velocities $\vec{w}|_{\mathcal{I}}$ after each solid update. The modification of refinement meshes is indicated in Fig. 4 by the gray arrows; the initiating base level that remains fixed throughout the regridding operation is indicated by the gray circles.

6. HMX Detonation in a Tantalum Cylinder

As computational example we present the three-dimensional dynamic interaction of a detonation wave in the high explosive HMX ($C_4H_8N_8O_8$) with the walls (thickness 0.01 m) and the closed end of a cylinder made of Tantalum. The cylinder has the length 0.10 m and an outer radius of 0.0185 m. An inner combustion chamber of depth 0.055 m opens at its left end. A non-adaptive tetrahedral structure mechanics finite element solver with special artificial viscosity formulation for capturing dilatation and shear waves [15] is employed for the solid update. The Tantalum is assumed to obey J2-flow theory of plasticity and Vinet's thermal equation equation of state with parameters derived from first-principle calculations [14]. The shown computation used a solid mesh of 56,080 elements.

For the fluid initial conditions, we assume a fully developed one-dimensional steady Chapman-Jouguet detonation with its front initially located at $x = 0.01$ m that we prescribe according to the theory of Zeldovich, Neumann, and Döring (ZND) (see [13] or [7] for detailed derivations). The detonation is propagating into the positive direction, which allows the prescription of constant inflow boundary conditions at the open left end (cf. Fig. 5). No deformations are allowed in the entire solid for $x < 0.01$ m to model a fully rigid material downstream of the initial wave. Further, no deformations are possible on the outer hull of the cylinder for $0.01\,\text{m} \le x \le 0.03\,\text{m}$.

Unreacted HMX has a density of $\rho_0 = 1900\,\text{kg/m}^3$ and gets depleted by a Chapman-Jouguet detonation with propagation speed $\approx 9100\,\text{m/s}$ resulting in an energy release of $q \approx 5176\,\text{kJ/kg}$ [17]. The hydrodynamic flow can be described with reasonable accuracy with a constant adiabatic exponent of $\gamma = 3$ [17]. We assume atmospheric pressure $p_0 = 100\,\text{kPa}$ in the unreacted material and set the unknown rate factor to $T_R = 1\,\mu s$. Fig. 5 displays the steadily propagating

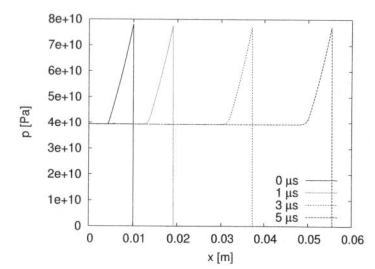

Figure 5: One-dimensional simulation of the detonation in HMX.

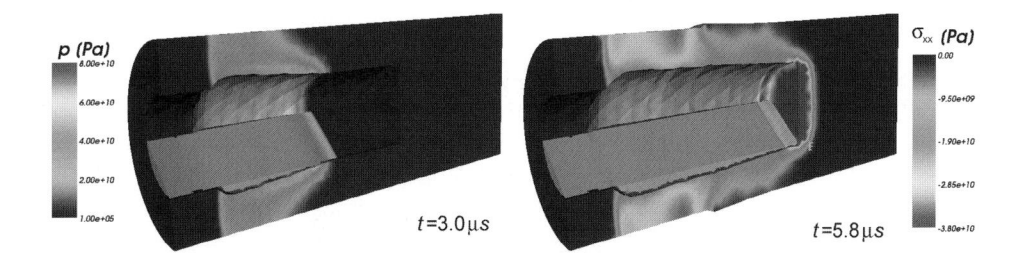

Figure 6. Left: compression of the wall material next to the combustion chamber due to the detonation passage. Right: outward movement of the unconstrained walls and the strong compression in axial direction due to the impact event.

pressure distribution of the ZND wave in a one-dimensional computation on a uniform mesh with 960 finite volume cells. At considerably coarser resolutions, the head of the detonation wave is not approximated to sufficient accuracy leading to an incorrect speed of propagation and a significant reduction of the maximal pressure value. Hence, this FSI problem benefits considerably from the application of dynamic mesh adaptation in the fluid.

Two snapshots of the simulation displaying a cut through the hydrodynamic pressure distribution and the normal stress in the axial direction are shown in Fig. 6. The left graphic shows the initiation of stress waves at the head of the detonation (the slight delay in the solid is due to its coarser mesh resolution). The right graphic at later time exhibits the influence of the boundary conditions: While the material gets strongly compressed initially, no normal stresses arise at the outer surface in the unconstrained section with $x \geq 0.03$ m. At $t = 5.8\,\mu$s, the HMX is fully depleted and the impact of the detonation wave at the closed end has caused a very strong compression wave in the solid in the axial direction. The reflected hydrodynamic shock wave is visible.

The fluid sub-problem has been run on a Cartesian domain of $0.03\,\text{m} \times 0.03\,\text{m} \times 0.06\,\text{m}$ and was discretized with $60 \times 60 \times 120$ cells at the SAMR root level. While the solid boundary is fully refined at the coupling level $l_c = 1$ with $r_1 = 2$, level 2 is only used to capture the head of detonation wave accurately ($r_2 = 4$). The SAMR mesh increases from initially approx. 706 k cells on level 1 and 6.5 M on level 2 to about 930 k and 10.0 M cells at later times. The number of grids on both levels varies between 400 and 1000. Compared with a uniform fluid mesh of $480 \times 480 \times 960 \simeq 221$ M cells, the enormous saving from mesh adaptation is apparent. Figure 7 displays the adapted fluid mesh in the mid plane for $t = 3.0\,\mu$s by overlaying a schlieren plot of the fluid density onto regions covered by level 1 (blue) and 2 (red). The simulation ran on 4 nodes of a Pentium-4-2.4 GHz dual processor system connected with Quadrics interconnect for about 63 h real time. Six processes were dedicated to the adaptive fluid simulation, while two were used for the significantly smaller solid problem.

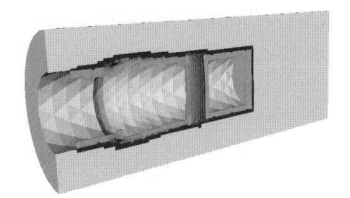

Figure 7: Adaptation at $t = 3.0\,\mu$s.

7. Conclusions

An efficient Cartesian finite volume scheme for the simulation of detonation waves in FSI problems has been presented. The method is patch-based and considers embedded complex boundaries with the ghost-fluid approach [11]. Distributed memory parallelization is provided by a parallel variant of the SAMR method of Berger and Collela [3] that utilizes a rigorous domain-decomposition strategy [7]. An algorithmic extension of the recursive SAMR method to loosely coupled FSI simulation with time-explicit solvers has been described. The approach allows the accurate capturing of near-body fluid-structure interaction, while resolution in space *and* time can be reduced effectively in the fluid far field. As computational example, a detonation wave in a solid high energetic material impinging on a Tantalum cylinder has been discussed. The example demonstrates the enormous savings in the computational costs that can be obtained through structured dynamic mesh adaptation in the fluid for this problem class: While the parallel calculation required only 504 h CPU (63 h real time), the costs of a simulation with an equivalent fluid unigrid mesh can be expected to be in the range of 10^5 h CPU.

REFERENCES

1. M. Aivazis, W.A. Goddard, D.I. Meiron et al., Comput. Science & Eng. 2(2) 2000 42.
2. J. Bell, M. Berger, J. Saltzman, M. Welcome, SIAM J. Sci. Comp. 15(1) (1994) 127.
3. M. Berger and P. Colella, J. Comput. Phys. 82 (1988) 64.
4. J. F. Clarke, S. Karni, J. J. Quirk et al., J. Comput. Phys. 106 (1993) 215.
5. J. C. Cummings, M. Aivazis, R. Samtaney et al., J. Supercomput. 23 (2002) 39.
6. R. Deiterding, Parallel adaptive simulation of multi-dimensional detonation structures, PhD thesis, Brandenburgische Technische Universität Cottbus, 2003.
7. R. Deiterding, in Notes Comput. Science & Eng. 41, Springer, New York, (2005) 361.
8. R. Deiterding, AMROC, available at http://amroc.sourceforge.net (2005).
9. R. Deiterding, R. Radovitzky, S. P. Mauch et al., Engineering with Computers, Springer, (2005) submitted.
10. B. Einfeldt, C. D. Munz, P. L. Roe, and B. Sjögreen, J. Comput. Phys. 92 (1991) 273.
11. R. P. Fedkiw, T. Aslam, B. Merriman, S. Osher, J. Comput. Phys. 152 (1999) 457.
12. R. P. Fedkiw, J. Comput. Phys. 175 (2002) 200.
13. W. Fickett, W. .C. Davis, Detonation, Univ. Cal. Press, Berkeley, 1979.
14. D. E. Johnson, E. Cohen, in Proc. IEEE Conf. Robotics & Automation (1998) 3678.
15. A. Lew, R. Radovitzky, and M. Ortiz, J. Comput-Aided Mater. Des. 8 (2002) 213.
16. R. Löhner, J.D.Baum et al. in Notes Comput. Science 2565, Springer,Berlin (2003) 3.
17. C. L. Mader, Numerical modeling of detonations, Univ. Cal. Press, Berkeley, 1979.
18. S. P. Mauch, Efficient Algorithms for Solving Static Hamilton-Jacobi Equations, PhD thesis, California Institute of Technology, 2003.
19. M. Parashar and J. C. Browne, in Proc. 29th Hawaii Int. Conf. System Sciences, 1996.
20. J. J. Quirk. Int. J. Numer. Meth. Fluids 18 (1994) 555.
21. R. Sanders, E. Morano, M.-C. Druguett, J. Comput. Phys. 145 (1998) 511.
22. E. F. Toro, Riemann solvers and numerical methods for fluid dynamics, Springer, Berlin, Heidelberg, 1999.

PARAMESH: A Parallel, Adaptive Grid Tool

Kevin Olson[a]

[a]Goddard Earth Science and Technology Center,
University of Maryland Baltimore County,
NASA/GSFC,
Code 610.6, Greenbelt, MD 20771, USA

PARAMESH is a portable high performance software toolkit that enables parallel, adaptive-mesh, computer simulations. It was originally conceived to support hydrodynamic and MHD applications of the Solar atmosphere. It has been so successful that it is now being used in a much broader range of applications. These applications include General Relativistic models of colliding compact objects in preparation for the LISA mission, space weather models of the sun, inner heliosphere and magnetosphere, and models of radiation hydrodynamics in novae, supernovae, and gamma-ray bursts. These different applications stress the PARAMESH package in different ways. As a result, we are currently working to extend the functionality of PARAMESH and rationalize its internal structure to best serve our user community.

This paper gives a general description of PARAMESH and the algorithms it uses. Secondly, an overview of some of the applications for which it is being used with will be given.

1. INTRODUCTION TO PARAMESH

PARAMESH[1] was written primarily by Kevin Olson and Peter MacNeice (Drexel Univesity) at NASA's Goddard Space Flight center as part of the NASA/ESTO-CT project (formally HPCC). Other contributors to date include C. Mobary, R. deFainchtein, C. Packer, J. vanMetre, M. Bhat, M. Gehrmeyer (NASA/GSFC), R. DeVore (NRL), M. Zingale, J. Dursi, A. Siegel, K. Riley (U. of Chicago), and R. Loy (Argonne Lab).

1.1. PARAMESH: What is it ?

PARAMESH is a set of subroutines which are written in Fortran 90. The package is fully parallel and communications between processes are handled using calls to the MPI communications library. PARAMESH has been tested using a number of different Fortran 90/95 compilers and different computer architectures, with an emphasis on using Beowulfs. Some of these include the Portland Group compiler, the NAG compiler, the Intel compiler, and the Lahey/Fujitsu compiler. Architectures which have been used to run PARAMESH include the IBM SP, SGI, HP-Compaq, and Cray. The current released version of PARAMESH is version 3.3. We expect to release version 3.4 before the end of August, 2005.

The kind of application developers we are targeting with PARAMESH are those who already have a pre-existing, uniform-mesh, serial code. PARAMESH is designed to enable such users to easily incorporate both parallelization and dynamic adaptive mesh refinement (AMR) into their application. We provide templates that help them to do this. Further, we distribute the source code, allowing users to modify it for their specific application. Alternately, we collaborate with users to make modifications to PARAMESH for their specific application if they request this.

PARAMESH is a subset of the block-adaptive technique described by Burger and Oliger[2] and Burger and Collela[3]. In our case we bi-sect the region of space covered by the computational domain, forming a set of child blocks. Child blocks can themselves be bi-sected, their children bi-sected, and so on. This process is carried out recursively until the desired resolution in a certain region of space is reached. Only jumps in refinement of a factor of 2 in resolution are allowed. This naturally leads to the use of a tree data structure to organize the blocks and their relationships to one another.

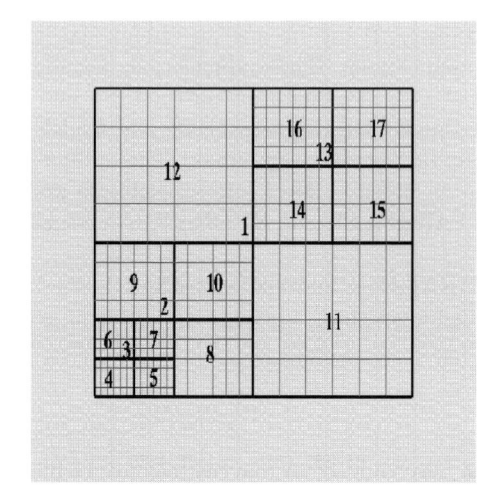

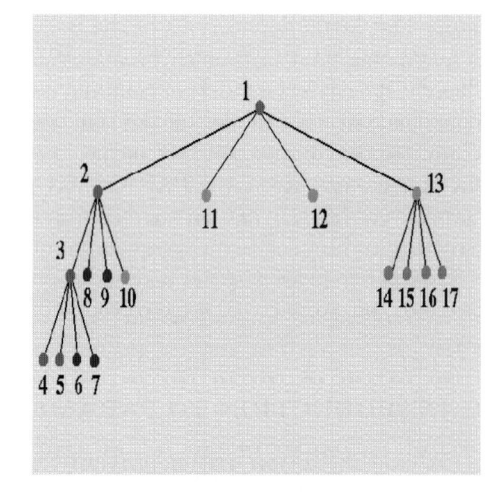

Figure 1. PARAMESH blocks are created by recursively bisecting space as shown here.

Figure 2. PARAMESH blocks are arranged according to a tree data structure. We show in the figure above the tree relationships of the blocks originally shown in Figure 1. The numbers shown at the 'nodes' of the tree refer to the numbers of the blocks shown in Figure 1.

For the purposes of parallelization, the blocks are ordered according to a morton space-filling curve[4]. This curve is 'cut' into a number of pieces equal to the number of processors. The length of each piece is determined by a user selectable work weighting which can be specified at runtime and enables some control over the load balance of the algorithm.

This type of curve has the property that blocks which are nearby in physical space also tend to be nearby in the sorted morton curve.

The refinement-derefinement process is very simple from the users' point of view. Blocks are simply marked with logical flags which indicate whether the blocks are to refine or derefine. Calling the PARAMESH routine for refinement-derefinement then results in the appropriate set of new blocks being created (or destroyed), the tree being reconstructed, the blocks being redistributed according to the morton space filling curve, and all internal control information being reconstructed and stored for later use by other PARAMESH functions.

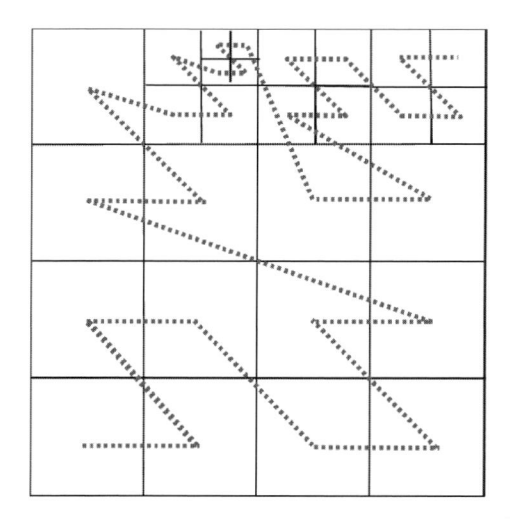

Figure 3. Morton Space Filling Curve. This figure shows an example of a 'morton space filling' curve threading a set of PARAMESH blocks.

Each PARAMESH block is itself a Cartesian mesh of cells. Each block is surrounded by a 'guard' or 'ghost' cell region. The guardcells are filled by either copying data from neighboring blocks at the same refinement level or by interpolation in the case that there is a jump in refinement. The PARAMESH package provides a routine which takes care of this interpolation procedure using Lagrange polynomials to a user chosen, arbitrary order. The user is also free to modify this routine to satisfy any special interpolation requirements they might have for their application.

The user has the option of storing cell, face, edge, or corner data, or any combination of them, on a single computational cell within a PARAMESH block.

We provide support for 'flux' conservation or 'refluxing' at refinement jumps where fluxes (or any quantity) that are computed at cell faces are averaged (or summed) from the fine cells to course cell faces where two blocks of different refinement abut one another.

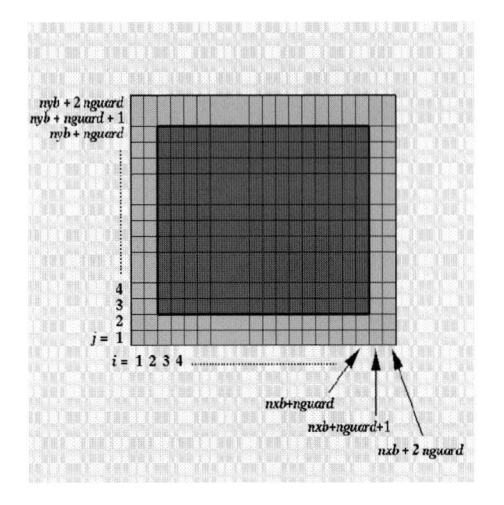

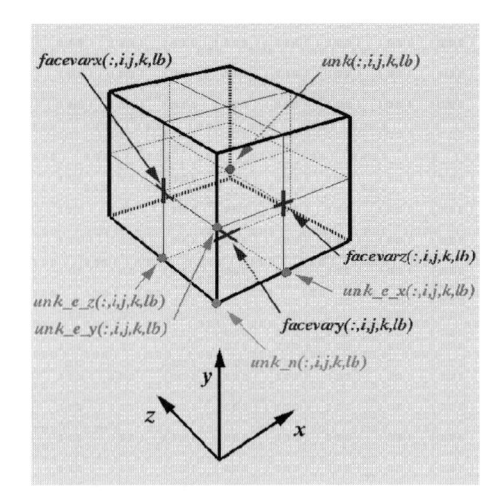

Figure 4. Each block in PARAMESH is a logically Cartesian mesh of cells. This figure shows an example of a single PARAMESH block. Interior cells of the block are shown in red and the 'guardcells' (cells which overlap with neighboring blocks) are shown in blue. A user of PARAMESH is free to choose the number of guardcells and the number of interior cells in a block.

Figure 5. This figure shows a single cell within a PARAMESH block. Data can be located in any combination at cell centers, cell edges, cell faces, or cell corners.

We also provide a similar procedure which can be applied to values computed at cell edges such that circulation integrals around cell faces are consistent at refinement jumps.

2. APPLICATIONS

2.1. NRL ARMS

The Naval Research Lab's MHD code, known as ARMS, was the application for which PARAMESH was originally developed. Many of the choices we made in designing PARAMESH were influenced by the requirements of this code and the requirement that it get a certain level of performance on the Cray T3E.

It is a Flux Corrected Transport MHD code, with magnetic fields stored at cell faces[5]. It uses 3 guardcells around each block. Electric fields are computed at cell edges and the edge averaging described earlier is used to ensure a consist update of the magnetic fields. All other variables are stored at the cell centers. It is mostly being used for Solar physics applications.

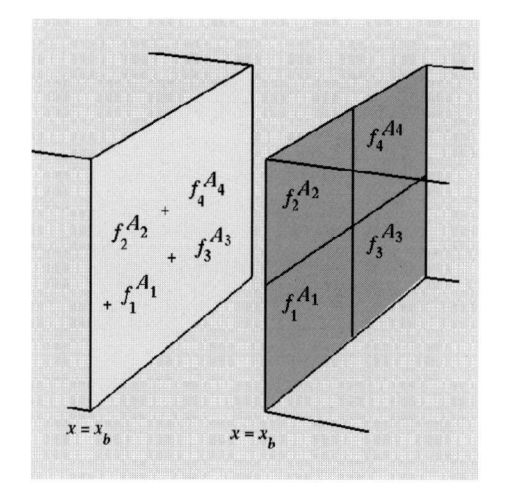

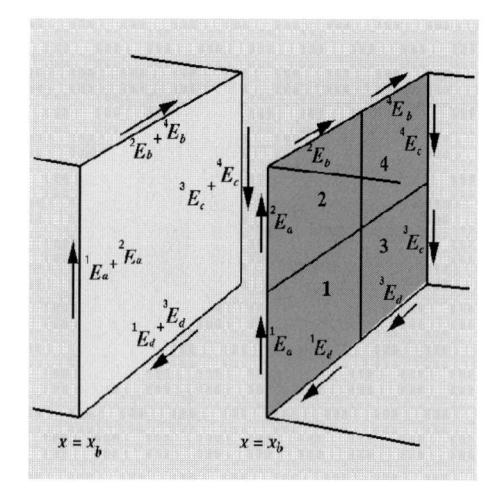

Figure 6. PARAMESH supports 'refluxing' or flux averaging at jumps in refinement.

Figure 7. PARAMESH supports averaging of edge based quantities which can be useful for algorithms which require the circulation integral of that quantity to be continuous at refinement jumps.

2.2. GENERAL RELATIVITY

A group lead by Dr. Joan Centrella at Goddard Space Flight Center is developing an application which solves the Einstein equations in order to simulate the collision of super-massive Black Holes colliding in the centers of galaxies. They are interested in simulating events of this type since it is fundamental to predicting the amplitude of the propagating gravitational wave-forms which would result. Such simulations are therefore critical to the design and interpretation of measurements from the Laser Interferometer Space Antenna (LISA) to be launched in 2011[8].

AMR is important for this application since one needs to model the details of the collision (which occur on small, relative size scales) at the same time a large volume around the collision is modeled to simulate the gravitational wave forms which result from the collision of the black holes and then propagate away from the collision site. This group is using finite difference techniques and multigrid.

2.3. FLASH

The FLASH code[7,9] is an astrophysics code which uses PARAMESH as its main AMR package. This code implements a number of CFD schemes, an MHD algorithm, a nuclear reaction network with energy feedback, stellar equations of state, self-gravity using multigrid, and particles (which can be passive or gravitationally interacting). It uses 4 guardcells around each block with all variables stored at cell centers. The main purpose for this code is the simulation of astrophysical thermonuclear flashes which occur

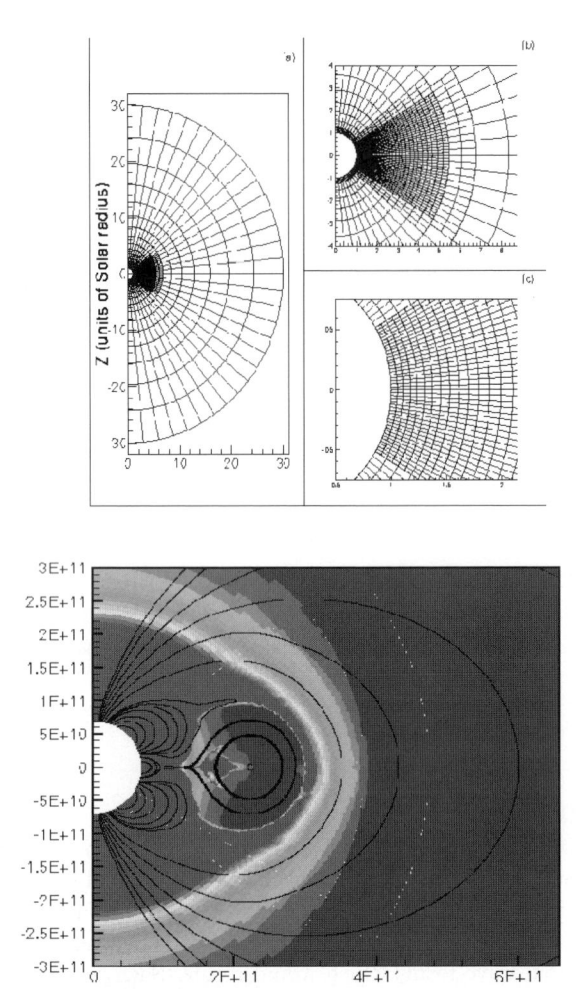

Figure 8. Results of simulation using a variant of the NRL ARMS code [6]. It uses FCT in the same way as previously described except that it uses spherical coordinates and works in 2.5 dimensions (carries azimuthal B field). The mesh is refined near the inner boundary representing the solar surface as shown in panels a, b and c above. The initial field is a combination of the sun's dipole field superposed with an octopole field near the equator. Shear along the model solar surface is introduced near the equator. As result, a flux rope forms and propagates away from the surface. This is shown in the bottom panel where the magnetic field lines are shown superposed over the density.

in close binary star systems where one of the stars' sizes has exceeded its Roche limit and some of its material is accreting onto the companion star. Depending on conditions, it is believed that such a situation can give rise to observed X-ray bursts, classical novae, and Type I supernovae.

2.4. IBEAM

IBEAM [10] is a project funded by NASA/ESTO-CT to build a modern software framework for Astrophysics. They are using as their starting point the FLASH code and are collaborating with them and are adding functionality which is compatible with the FLASH code as well as their own framework. The physics they are attempting to model is radiation, hydrodynamics and they have developed relativistic hydro modules, Multigrid for radiation diffusion (single group), BiCGstab algorithm (for radiation). GMRES and Boltzmann transport are currently under development. Their target application is the simulation of Gamma Ray bursts.

2.5. CASIM

Written by Benna et al.[11]. A multifluid MHD code which models the a neutral particles, ions, and electrons and their interations using a chemical reaction network. Uses a TVD Lax-Friedrich's method with a second order Runge-Kutta time integrator. Time step is chosen using the Courant condition local to each cell. The algorithm is then integrated until an equilibrium is reached. Uses up to 28 levels of refinement. Used to simulate the interaction of the solar wind with expanding cometary atmospheres.

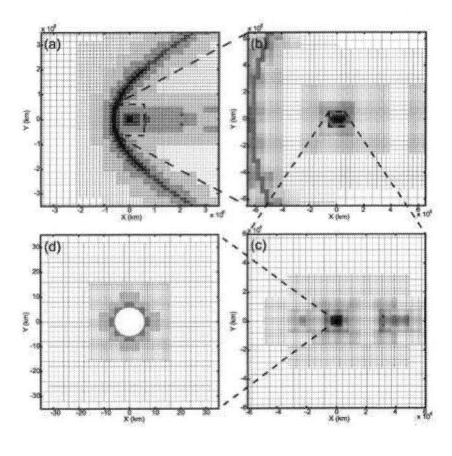

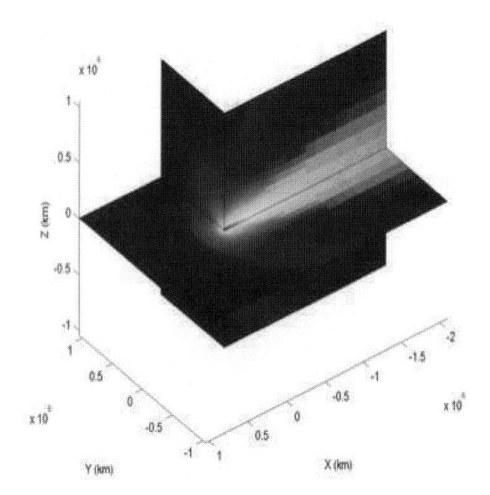

Figure 9. A CASIM mesh.

Figure 10. Casim results. The density structure of a Halley-like comet.

2.6. Zeus AMR

PARAMESH has been incorporated into several other MHD applications similar to those discussed earlier. Yet another MHD app. using PARAMESH was developed by a group from the University of California at Berkley[12]. They have combined PARAMESH with the Zeus code[13] and are using it for studying the emergence of magnetic flux from the solar surface.

REFERENCES

1. P. MacNeice, K. M. Olson, C. Mobarry, R. deFainchtein, and C. Packer, Comput. Phys. Comm., 126 (2000) 330.
2. M. J. Burger and J. Oliger, J. Comp. Phys., 53 (1984) 484.
3. M. J. Burger and P. Colella, J. Comp. Phys., 82 (1989) 64.
4. M. S. Warren and J. K. Salmon, in Proc. Supercomputing (Washington DC: IEEE Comput. Soc.) (1993) 12.
5. C. R. DeVore, J. Comp. Phys., 92 (1991) 142.
6. P. MacNeice, S. K. Antiochos, P. Phillips, D. S. Spicer, C. R. DeVore , and K. Olson, ApJ, 614 (2004) 1028.
7. B. Fryxell, K. Olson, P. Ricker, F. X. Timmes, M. Zingale, D. Q. Lamb, P. MacNeice, R. Rosner, J. W. Truran, and H. Tufo, ApJS, 131 (2000) 273.
8. http://lisa.jpl.nasa.gov
9. http://www.flash.uchicago.edu
10. http://www.ibeam.org
11. M. Benna, P. Mahaffy, P. MacNeice, and K. Olson ApJ, 617 (2004) 656.
12. Abbett, W. P. Ledvina, S. A., Fisher, G. H., and P. MacNeice, American Geophysical Union, Fall Meeting, (2001) abstract #SH11C-0727
13. J. M. Stone and M. L. Norman, ApJS, 80 (1992) 791.

Embarrassingly Parallel Computations of Bubbly Wakes

Andrei V. Smirnov, Gusheng Hu, Ismail Celik

West Virginia University
Morgantown, WV 26506, U.S.A.

Abstract

An statistical approach for simulating turbulent particle-laden wake flows is implemented and tested on the case of a bubbly ship wake. A zero communication overhead provides linear scale-up of parallel computations. The approach can be used in grid computing environments.

Key words: CFD, domain decomposition, multiphase flows, LES, wake flows, Lagrangian particle dynamics

1 Introduction

A major drawback in the Lagrangian particle simulation in dispersed two-phase flows, in terms of the computational cost and machine capacity, is the limitation on the number of particles, or particle clouds whose trajectories are to be tracked parallel to the solution of the continuous flow field. The insufficient number of particles commonly leads to inaccurate statistics.

Earlier work of the authors on parallelization of a LES solver by means of domain decomposition [1,2] provided the possibility to simulate large scale turbulent structures of typical ship-wakes on computer clusters. The next logical step is to extend the pure turbulence model with important multi-phase features of the wakes such as bubble dynamics. The algorithm for particle tracking and population dynamics developed earlier demonstrated

---

Email addresses: andrei.smirnov@mail.wvu.edu (Andrei V. Smirnov,), ghu@mix.wvu.edu (Gusheng Hu,), ismail.celik@mail.wvu.edu (Ismail Celik).

the ability to efficiently simulate large populations of particles including coalescence effects with even modest computer resources [3–6]. However, parallel implementation of a discrete particle dynamics algorithm and a continuum flow solver by means of domain decomposition technique commonly leads to large communication overheads and load balancing problems. In this study we pursued a simple embarrassingly parallel strategy, which enabled us to avoid these two problems. At the same time this strategy opens the possibility of using emerging grid computing infrastructures for massively parallel computations of bubbly wakes.

The idea of "embarrassingly" parallel computations presumes a complete absence or a small degree of inter-processor communications during parallel execution of a multi-processor problem. In applications to continuum mechanics it usually means that the domain decomposition is either not used at all or there is a very loose coupling between the parts of the decomposed domain. This idea is attractive because of low communication overhead and linear scale-up of the computations. However, this kind of parallelism can only be realized for a limited classes of problems. One such class involves statistical modeling, where the results are obtained as an average over a set of independent statistical ensembles. In this case each ensemble can be simulated on a separate processor and no communication between the processors is necessary. Unfortunately, most problems of fluid dynamics don't fall under this category. Moreover, parallel CFD computations are notoriously hard to realize because of the inherent strong coupling between the variables and sub-domains.

For these reasons it is especially interesting to exploit the idea of completely uncoupled CFD simulations, in cases where statistical modeling can still be applied. In this work we performed such simulations for the case of a bubbly wake. This is a specially interesting case, because statistical models can be used to represent the dynamics of both the bubbles and the fluid.

2 Method

Turbulent bubbly wake is a two phase system. In our model we use large eddy simulation technique (LES) [3] as a turbulence model for the fluid phase and a Lagrangian particle dynamics (LPD) approach with a stochastic random flow generation method (RFG) [7] as a bubble dynamics model. The RFG method like almost any stochastic particle method is well suited for parallel implementation, since it does not require inter-processor communication. At the same time the LES technique belongs to the category of statistical turbulence models, and as such is also well suited for parallel implementation, provided the domain decomposition is not used. It should be noted that the main reason for using domain decomposition for LES lies in large memory needed for adequate resolution of the large-scale flow turbulence. However, with the growing memory capacities, it is becoming possible to conduct simulations of reasonably big problems on a single processor. Indeed, with computer memories up into several gigabyte range one can easily put several million grid nodes on a single processor. With the grids this large computing times become a bigger issue. This situation is exacerbated by the fact that computing time is already one of the main limiting factors of the LES method, where long runs are needed to collect statistics on turbulent eddies. In this situation the "embarrassing" parallelism may come to help, drastically reducing the computation time. Thus, the straightforward approach is to perform the simulation of statistically independent realizations of the flow-field and particle ensembles, with each realization assigned to one cluster node. This strategy completely excludes any communication between the computing nodes, at the same time achieving the perfect load balance.

Behind the seeming simplicity of this parallelization strategy there lies a problem of ensuring statistical independence of ensembles generated on different computing nodes. This is achieved by imposing independent random realizations of three important flow and particle conditions: (1) initial conditions on the flow-field, (2) inlet flow conditions, (3) particle injection distribution. All three conditions are subjected to randomized time-dependent change, which nevertheless follows a predefined statistical distribution.

It should be noted, that while generating independent ensembles of injected particles is a simple matter of Monte-Carlo sampling with a given distribution, the generation of randomized inflow and initial conditions for the flow field should be subjected to certain restrictions imposed by the laws of continuum dynamics. For example, the continuity relation will generally not be satisfied for any random distribution of flow velocity vectors, obeying Gaussian or other valid statistics. In this respect the RFG technique [7] provides a continuity-compliant time varying random velocity field. This technique was applied successfully in this case to generate statistically independent and divergence-free flow-field ensembles.

The iterations of the discrete bubble solver were sub-cycled inside flow iterations following Lagrangian particle dynamics (LPD) in a dilute dispersed two-phase flow with one way coupling, i.e. where the particles motion is determined by the continuous phase flow, but the continuous flow is not influenced by the particles. The number of particles tracked in each simulation can vary, depending on the machine capacity and task load. The realizations from different runs on different nodes were collected and analyzed to produce histograms of bubble distribution. The global particle statistics was obtained by averaging over all the ensembles.

3 Results

The combined LES/LPD solver was set up to run on a Beowulf cluster, with 1GHz 1GB computing nodes. A set of simulations of a turbulent bubbly ship wake flow was performed, in which a total of about 254000 particles were tracked on nine different computing nodes. The combined statistics is compared to the statistics from a single-run, indicating qualitatively equivalent, but better results.

Figure 1 shows a snapshot of bubbles in the computational domain of length L, which is approximately equal to the length of the ship. The restriction on L was dictated by available computer memory. The cumulative cross-sectional distribution of bubbles close to the inlet is given in Fig.2. The post-processed bubble distributions are presented as histograms in Fig.3, showing the results from a single processor and multiple processors. Some degree of smoothing is evident in the results, however, increasing the data sample would be appropriate. The total bubble decay computed on the basis of the simulations (Fig.4) is consistent with the experimental data [8].

One interesting observation concerns the usage of different interpolation schemes for fluid-bubble coupling. A commonly used bi-linear interpolation:

$$V(x) = \left(\sum_i^N V(x_i)/d_i\right) / \left(\sum_i^N 1/d_i\right) \quad (1)$$

computes variable V at particle position x, by interpolating from N vertexes with coordinates, x_i ($i = 1..N$), and distances d_i between the particle and the node i. Fig.5(a) shows the velocity of the particle and the surrounding fluid computed using (1). Discontinuities in particle velocity are evident. These are attributed to the discontinuities in the interpolation (1) itself, which can occur at cell faces, as well as to the fact that particles with density lower than the fluid, such as bubbles, tend to overshoot their velocities compared to that of the surrounded fluid, as dictated by bubbles equation of motion [3]. Switching to a strictly piecewise linear interpolation formula remedied the problem (Fig.5(b)). The trade-off is in additional operations required for an exact linear interpolation as compared to (1).

4 Conclusions

In conclusion we would like to mention that along with the relative simplicity of implementation, and parallel efficiency of the approach, this *embarrass-*

ingly parallel strategy is especially suitable for the newly emerging grid computing infrastructure, which is still not well adapted for most CFD computations. Extending this type of simulations to grid computing environments will be the subject of future work.

Acknowledgments. This work has been performed under a DOD EPSCoR project sponsored by the Office of Naval Research, Grant N00014-98-1-0611, monitored by Dr. Edwin P. Rood and Patrick L. Purtell.

References

[1] A. Smirnov, I. Yavuz, C. Ersahin, I. Celik, Parallel computations of turbulent wakes, in: Parallel CFD 2003, Russian Academy of Sciences, Moscow, Russia, 2003.

[2] A. Smirnov, Domain coupling with the DOVE scheme, in: B. Chetverushkin (Ed.), Parallel Computational Fluid Dynamics: Advanced numerical methods, software and applications, Elsevier, North-Holland, Amsterdam, 2004, pp. 119–127.

[3] A. Smirnov, I. Celik, S. Shaoping, Les of bubble dynamics in wake flows, Journal of Computers and Fluids 34 (3) (2005) 351–373.

[4] S. Shi, A. Smirnov, I. Celik, Large eddy simulations of particle-laden turbulent wakes using a random flow generation technique, in: ONR 2000 Free Surface Turbulence and Bubbly Flows Workshop, California Institute of Technology, Pasadena, CA, 2000, pp. 13.1–13.7.

[5] A. Smirnov, S. Shi, I. Celik, Random Flow Simulations with a Bubble Dynamics Model, in: ASME Fluids Engineering Division Summer Meeting, no. 11215 in FEDSM2000, Boston, MA, 2000.

[6] A. Smirnov, I. Celik, A Lagrangian particle dynamics model with an implicit four-way coupling scheme, in: The 2000 ASME International Mechanical Engineering Congress and Exposition. Fluids Engineering Division, Vol. FED-253, Orlando, Fl, 2000, pp. 93–100.

[7] A. Smirnov, S. Shi, I. Celik, Random flow generation technique for large eddy simulations and particle-dynamics modeling, Trans. ASME. Journal of Fluids Engineering 123 (2001) 359–371.

[8] M. Hyman, Modeling ship microbubble wakes, Tech. Rep. CSS/TR-94/39, Naval Surface Warfare Center. Dahlgren Division. (1994).

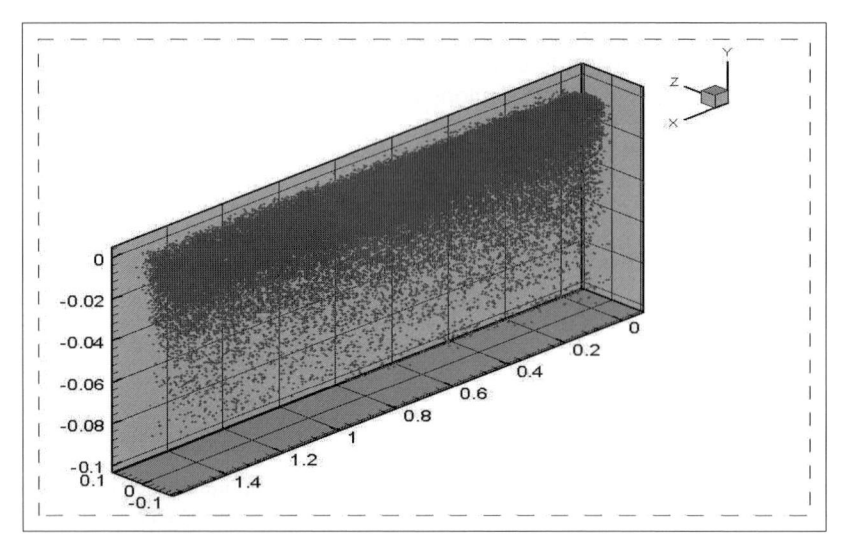

Fig. 1. Snapshot of about 50000 bubbles from 9 processors.

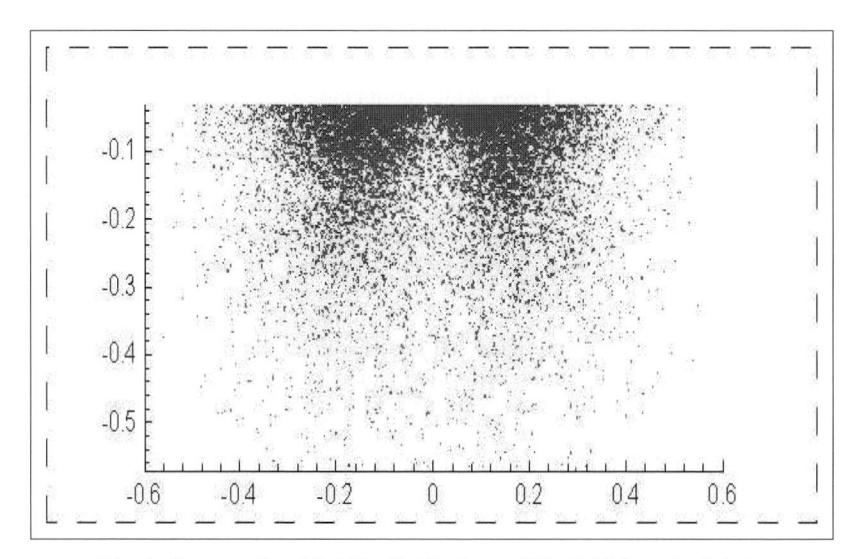

Fig. 2. Cross-sectional bubble distribution at X/L=0.25 from the inlet.

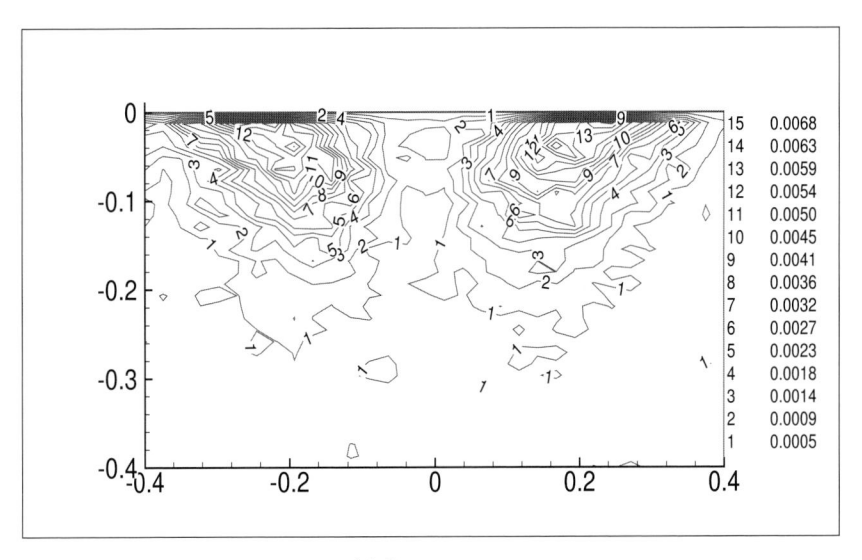

(a) One processor

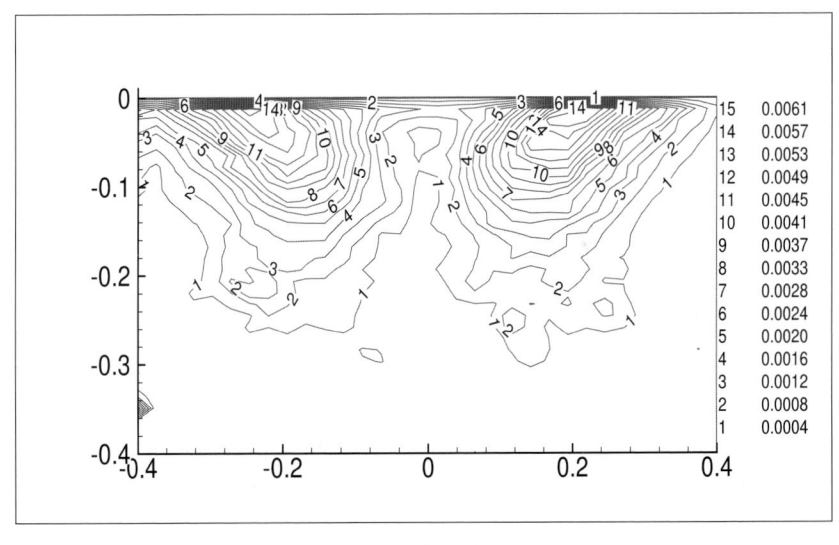

(b) 9 processors

Fig. 3. Particle distribution histograms in cross-sectional planes

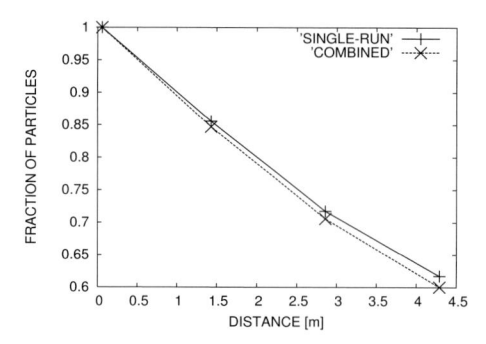

Fig. 4. Bubble decay

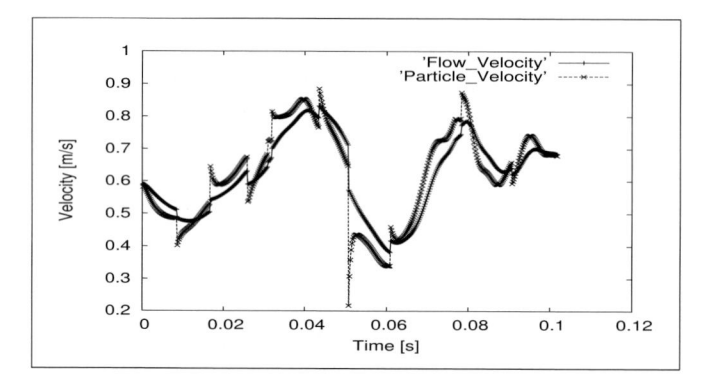

(a) Bilinear interpolation

(b) Linear interpolation

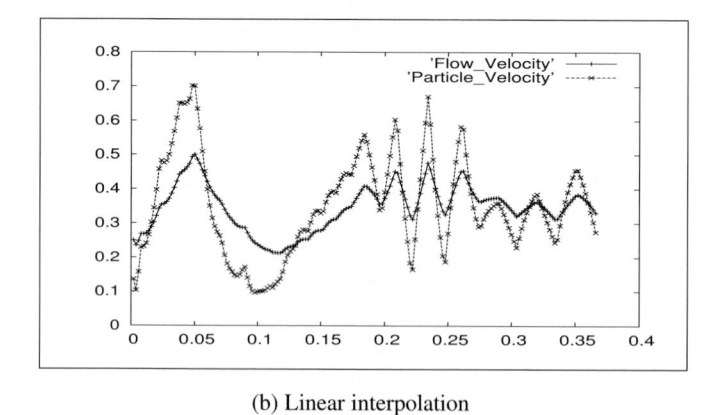

Fig. 5. Interpolation effects on bubble dynamics

Parallelisation of inundation simulations

S.C. Kramer[a] and G.S. Stelling[a]

[a] Environmental Fluid Mechanics Section,
Faculty of Civil Engineering,
Delft University of Technology,
P.O. Box 5048, 2600 GA Delft, The Netherlands.

In this paper the parallelisation of an inundation model is presented. The focus lies on the parallelisation of the matrix solver that combines a direct Gaussian elimination method with the iterative conjugate gradient algorithm. It is similar to the reduced system conjugate gradient method, but is applicable not only to problems that allow a red-black ordering of the equations, but to a more general class of unstructured sparse matrices. These arise in the discretisation of the flow equations on the combined 1D-2D grid of channel networks and flooded land, that is used in the inundation model. With a simple parallelisation strategy, the elimination process remains strictly local, and good scaling behaviour is achieved.

1. Introduction

In recent years there has been growing interest in accurate modelling of inundations caused by for example dike or dam breaks, tsunamis or other natural hazards. This is a result of frequent high water and flooding events in the past decade, and expected climatological changes and their impact in the future. The modelling of inundations over cultivated, especially urban areas is a complex matter due to the presence of all kinds of man made constructions and other obstacles that influence the flow. Nowadays detailed digital terrain models (DTM's) are available, allowing for a very accurate description of the flow geometry. Incorporating all such details in a numerical simulation leads to large computations that may easily exceed the capacity of a single computer. This paper describes the parallelisation, based on the MPI message passing technique, of a model[1] for the simulation of inundations. The scheme of the model is applicable in a wide range of situations, from flooding of rural or urban land due to a dike break to rainfall runoff simulations in a catchment area under extreme weather conditions.

The most important part of the parallelisation is that of the applied matrix solver. The solver combines a minimum degree ordered Gaussian elimination procedure with the conjugate gradient algorithm. It is similar to the reduced system conjugate gradient (RSCG, see e.g. [1]) method, that can be applied on problems that allow a red-black ordering of the matrix. Various references[2–4] show that good scalability can be reached

---
[1] the scheme of this model is implemented in SOBEK, an integrated 1D2D inundation modelling program, developed by Delft Hydraulics, http://www.wldelft.nl

for parallel implementations of this method on distributed memory architectures. The benefit of our approach is that it can be applied on a wider class of unstructured sparse matrices. This enables the application of unstructured 2D or 3D grids, or, as in our inundation model, a combination of 1D grids (channel networks) with 2D grids.

In the next section we will first explain the numerical scheme of the model, the special properties that make it suitable for simulating inundations in various situations. Also the need for the integration of one-dimensional channel networks and the two-dimensional grid of the flooded domain is explained. Section 3 deals with the parallelisation and performance of the matrix solver. Section 4 explains the parallelisation of the complete model, and discusses the load-balancing issues that arise when parts of the grid become (in)active during the computation due to flooding and drying.

2. Numerical modelling of inundations

An important aspect of flooding simulations is the presence of large gradients in the flow. These are present in the shock waves of tsunamis, dike or dam breaks, but also near steep changes in the bathymetry. Both subcritical and supercritical flow regimes are present with sudden transitions between them in hydraulic jumps. The modelling of such rapidly varied flows puts special demands on the numerical scheme. Conservation properties are important to get accurate shock solutions.

Numerical techniques for shallow water flow in coastal regions and estuaries are often based on the very efficient semi-implicit (see e.g. [5]) or ADI methods (see e.g. [6]) and the application of staggered grids. Most of them however do not give accurate results in the presence of large gradients. Godunov methods give correct shock solutions on smooth bottom topographies, but may still be inaccurate near steep bed slopes. Moreover they usually apply explicit time integration leading to strict time-step restrictions in the presence of both deep and shallow water. The scheme of our model[7] combines the efficiency of (semi-)implicit methods on staggered grids with local conservation considerations, and is thus able to simulate rapidly varied flows in an accurate and efficient way.

The model is based on the depth-integrated two-dimensional shallow water equations. As in [5] it combines a semi-implicit time integration with a spatial discretisation on a staggered grid, with water levels in the cell centres and the velocity normal to each cell face stored in the middle of the face(figure 1). A special discretisation of the advective velocity[7] gives the model the desired conservation properties. An orthogonality requirement on the computational grid enables a simple pressure gradient discretisation. This makes it possible to eliminate the velocities out of the system of equations. The resulting system, with only the water levels as unknowns, is symmetric positive definite(SPD) and easily solved by means of the conjugate gradient algorithm.

The inundation process often not only consists of a two-dimensional flow over land, but also involves one-dimensional channel flow. The channels, ditches and sewers of the normal hydraulic system may initially play an important role in the spreading of the water through the domain. Because the cross sections of these one-dimensional water ways are in general small compared to the two-dimensional length scales, it is important to model them separately in a one-dimensional grid. As the one-dimensional network starts flooding the surrounding land a coupling between the one and two-dimensional grids is necessary.

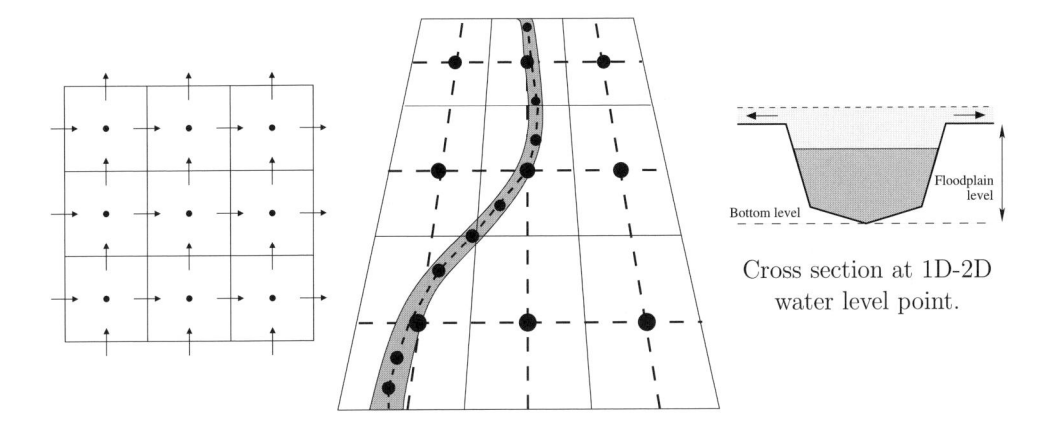

Figure 1. Staggered grid with water levels in the cell centres and face normal velocities on the edges of the cell(left picture). A 1D channel through a 2D Cartesian grid (middle picture). As soon as the water rises above the floodplain level (right picture) the connection with adjacent 2D cells become active.

Because of stability it is necessary to couple the equations in an implicit way. To do this both grids are combined in a 1D-2D grid. At those points where the channel can inundate the surrounding land, some of the water level points of the channel are identified with the nearest water level point of the 2D grid(figure 1). As long as the water stays below the level of the surrounding land, it acts as a normal 1D point. When the water rises above this level its connections with the adjacent 2D cells become active, and the water starts inundating the surrounding land.

Even if the 2D grid itself is Cartesian, the complete 1D-2D grid is unstructured in nature. Because the 1D and 2D water levels are solved together in a single linear system of equations, we also get to solve an unstructured sparse matrix. Thus the 1D-2D coupling already requires an unstructured setup of the administration of the model. Therefore the application of unstructured 2D grids is an interesting option, especially since the complicated flow geometry of flooded urban areas requires a lot of flexibility of the grid.

3. Parallelisation of the solver

The solution of the SPD system of equations, performed each time-step, is the most time consuming part of the calculation and therefore the most important step in an efficient parallelisation. The distribution of the work load over the processes is done according to a decomposition of the unstructured graph of the matrix, that minimises the edge-cut. Fast graph-based decomposition algorithms are readily available (Metis, Jostle). The decomposition of the graph of the matrix, which is, as we saw, equal to the adjacency graph of the 1D-2D grid, also prescribes the distribution of the work load for the rest of the computation. Thus the distribution of the work load corresponds to a decomposition of the flow domain in local subdomains.

Parallel Gaussian elimination and conjugate gradient

The one-dimensional part of the grid often contains long stretches of consecutive channel segments. The corresponding tri-diagonal blocks in the system of equations are most efficiently solved by a direct method. The two-dimensional part is better solved using the iterative conjugate gradient(CG) method with a suitable preconditioner. This is why we combine both methods for the 1D-2D grid. First we apply symmetric Gaussian elimination with a minimum degree ordering. The elimination process is stopped after a certain degree is reached. The remaining equations of higher degree, together still form a SPD matrix block that can be solved with CG. Finally this result is substituted in the eliminated equations. The stretches of the one-dimensional grid, that have a low degree, are thus automatically eliminated by the direct method.

The Gaussian elimination process is also used to improve the performance of the solution algorithm for the 2D part of the grid. For example if we have a 2D Cartesian grid and we eliminate up until a degree of 4, about halve of the 2D Cartesian cells are eliminated in a red-black manner. In this case the method is equal to the well known *reduced system conjugate gradient* (RSCG) algorithm[1]. The reduction of the problem size speeds up the vector operations in the CG algorithm. Moreover the condition number of the matrix is improved so that fewer iterations are needed to converge. If we combine this with a simple diagonal scaling in the CG algorithm, no further preconditioning is needed.

The reduction in number of iterations in our computations ranges between 25% and 50%, depending on the specific flow situation and the diagonal dominance of the problem. This is much less reduction than what could be achieved using other preconditioning techniques, such as incomplete Cholesky or polynomial preconditioners. However this further reduction of iterations hardly makes up for the high cost of performing preconditioner matrix inversions each iteration. This becomes more prominent in parallel computations, as the preconditioner inversions are in general hard to parallelise.

A scalable parallelisation of matrix factorisation by Gaussian elimination is in general not trivial (see e.g. [8]). As the fill-in of the matrix increases with each elimination step, the locality of the problem decreases. However as long as we only eliminate those nodes in the graph that do not lie along the boundary of the local domains, the operations can be performed within the local domains. Thus by excluding the local boundary nodes, the elimination process is a strictly local operation. Since the parallel reduction algorithm does not perform exactly the same elimination procedure as the serial version, we expect some decline in the conditioning of the reduced system, leading to slower convergence in the subsequent CG iterations.

Testing the parallel solver

To investigate the influence of a restricted reduction on the convergence of the conjugate gradient algorithm, we study the following test problem(similar to those in [2,9]). It has the same 5-point structure as the system of equations associated with the 2D part of the grid in the inundation model. Consider the Poisson equation

$$-\frac{\partial}{\partial x}\left(a(x,y)\frac{\partial u}{\partial x}\right) - \frac{\partial}{\partial y}\left(a(x,y)\frac{\partial u}{\partial y}\right) = f(x,y), \tag{1}$$

on a 6×6 square with a 1200×1200 Cartesian grid and define a and f by

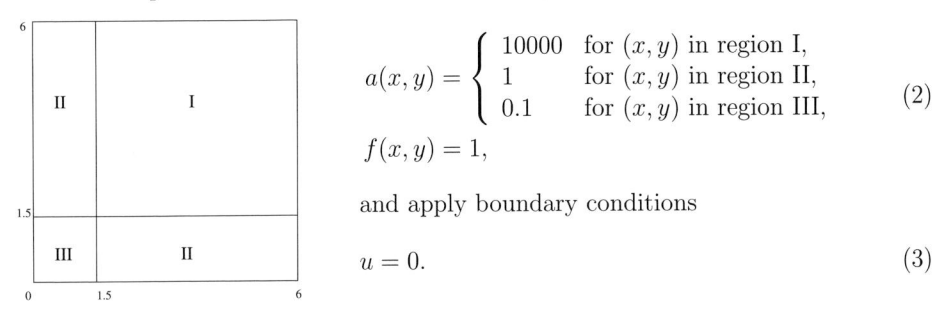

$$a(x,y) = \begin{cases} 10000 & \text{for } (x,y) \text{ in region I,} \\ 1 & \text{for } (x,y) \text{ in region II,} \\ 0.1 & \text{for } (x,y) \text{ in region III,} \end{cases} \quad (2)$$

$$f(x,y) = 1,$$

and apply boundary conditions

$$u = 0. \quad (3)$$

The CG algorithm with diagonal scaling applied on the corresponding matrix equation, converges in 2682 iterations. If we first reduce the system with the minimum degree ordered Gaussian elimination until a degree of 4, we only need 1488 iterations for convergence of the reduced system. In the domain decomposed parallel computation, the nodes along the local boundaries are excluded from elimination, and the number of iterations increases. In figure 2 we see that after an initial increase of 10% at $np = 2$, the number of iterations only varies slightly according to the different domain-decompositions for the different number of processors. At NP= 1024 the increase is still only 15%.

The number of extra iterations is small as long as the number of excluded points is small with respect to the problem size. This means the local domains need to be sufficiently large. In figure 3 we also increase the size of the problem as we increase the number of processors. We do this by keeping the local domains at a fixed size of 100×100 cells. Thus the grid size grows linearly with the number of processors, but since the local domain sizes do not grow, the ratio of points excluded from elimination to the total number of points stays the same. As we can see the number of extra iterations stays relatively small, between 10 and 20%.

The extra iterations give an efficiency loss on top of the communication cost in the parallel CG algorithm. The communication cost can stay relatively small in a distributed memory architecture[3], as long as the communication network provides sufficient bandwidth and the number of connections between two different local domains in the matrix graph is small. So with sufficiently large local domains we can keep both the number of extra iterations and the communication cost small. This is visible in figure 4. We see the usual good scaling of the conjugate gradient algorithm for a large enough problem. For the combination of Gaussian elimination and CG, the speed up, T_1/T_n, is slightly less but still scaling satisfactory. Here T_1 corresponds to unrestricted elimination and T_n to elimination with the local boundary nodes excluded. Thus the combined algorithm achieves good scalability and a simple yet efficient preconditioning of the matrix.

4. Parallelisation of the inundation model

The parallelisation of the rest of the model is fairly straightforward. We make use of ghost points, i.e. local copies of those grid points that lie just outside the local domain, in a domain assigned to another processor. The variables in those points are updated each time step. In this way the changes that have to be made in the serial code are limited.

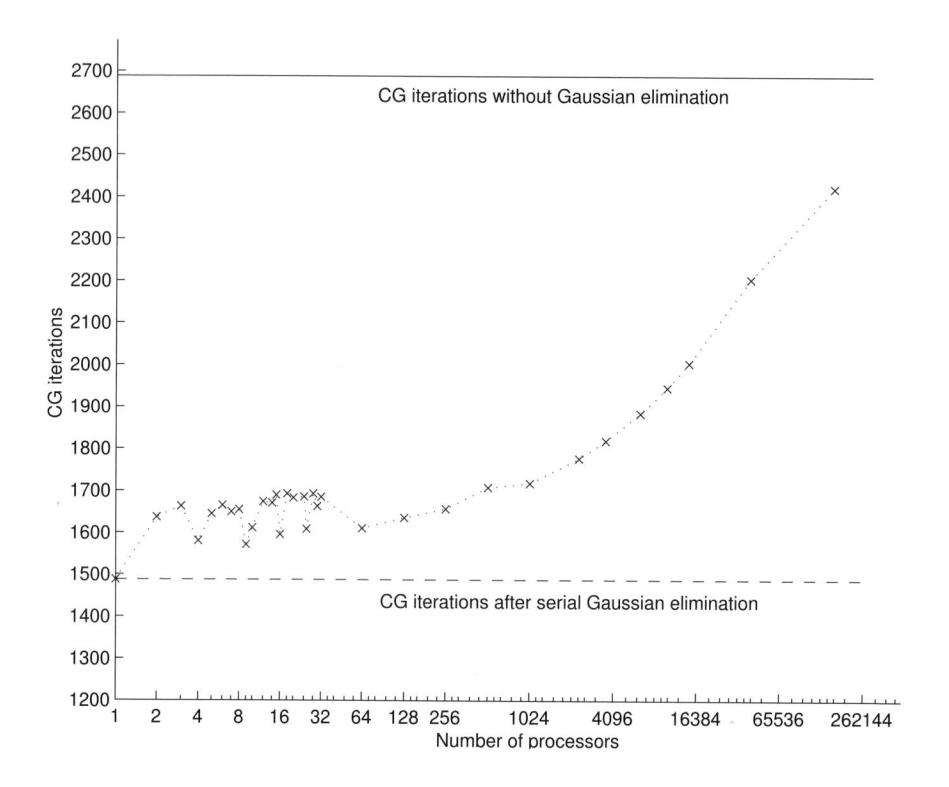

Figure 2. Number of iterations needed for convergence after Gaussian elimination on a varying number of processors for a fixed problem size of 1200 × 1200.

The performance of the parallel model has been studied with several test cases. Depending on what time the model spends in the solver routine, the scaling behaviour is comparable with that of the solver in the previous section. In typical serial computations 66 to sometimes 80% of the time is spent in the solver. Because the solver algorithm is the only part of the model that scales non-linearly with the problem size, this percentage only increases for the much larger computations that are possible on parallel computing systems. For small time steps however, sometimes necessary in drying and flooding situations (for instance the test case in figure 4), only a small number of iterations are needed and still a fair amount of time is spent in the coefficient computation outside the solver. This part is highly scalable; apart from updating the ghost points, the computation is strictly local. Moreover this part typically jumps very fast through large arrays of variables associated with cells or edges, and therefore suffers most from cache latency. Because these problems are diminished for smaller domains, the subdivision in local domains gives a relative advantage to parallel computation, sometimes leading to super-linear scaling.

A specific problem for parallel inundation models, is that due to flooding and drying different parts of the grid become active or inactive during the computation. With a

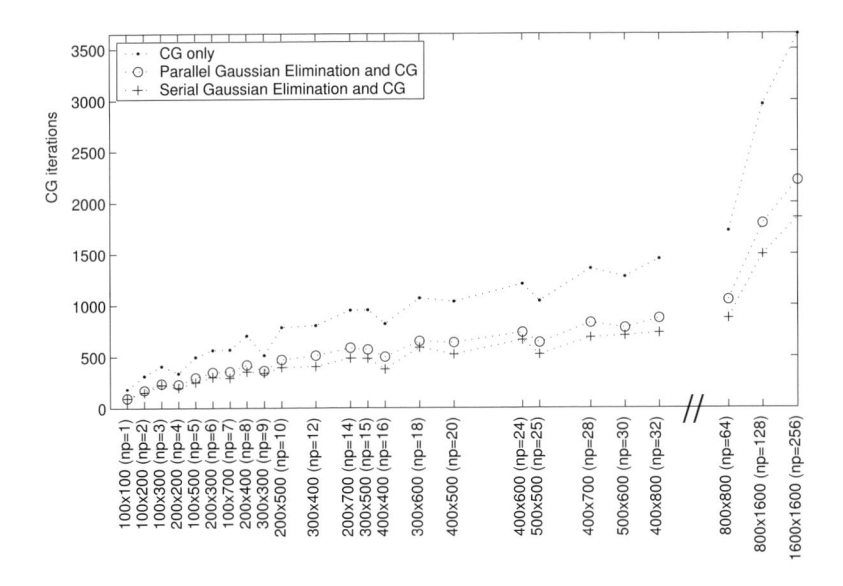

Figure 3. Number of iterations needed for convergence after Gaussian elimination on varying number of processors and varying problem size.

static domain decomposition this leads to load imbalances. Dynamic load balancing is a widely studied subject (see e.g. [10] and references therein) usually applied to adaptive grids. For our model a quasi-dynamic approach is chosen. If after some time steps the load imbalance, measured by a suitable function, exceeds a chosen limit, the computation is interrupted. The flooded part of the entire grid is partitioned from scratch, and the model is restarted after a redistribution of the data. A practical problem of partitioning only the active part of the grid, is that the inactive cells still need to be stored in memory and checked whether they need to be reactivated each time-step. If the inactive part is not reasonably well distributed among the processors, this can lead to a bottleneck for the reactivation check and memory problems in case of very large computations with a large number of processors.

5. Summary and conclusions

In this paper the parallelisation of a model for simulating inundations was presented, with a focus on the matrix solver. It is known that the RSCG algorithm gives well scalable performance on distributed memory architectures for similar problems on Cartesian grids. A more general approach based on minimum degree ordered Gaussian elimination allows the unstructured integrated 1D-2D grids used in this model. A simple parallelisation strategy, that exempts a small number of grid points from elimination, gives good results provided the local domains assigned to the individual processors are not too small. In this way the parallel model can provide the computational power that is necessary for a detailed and accurate modelling of inundations in urban environments.

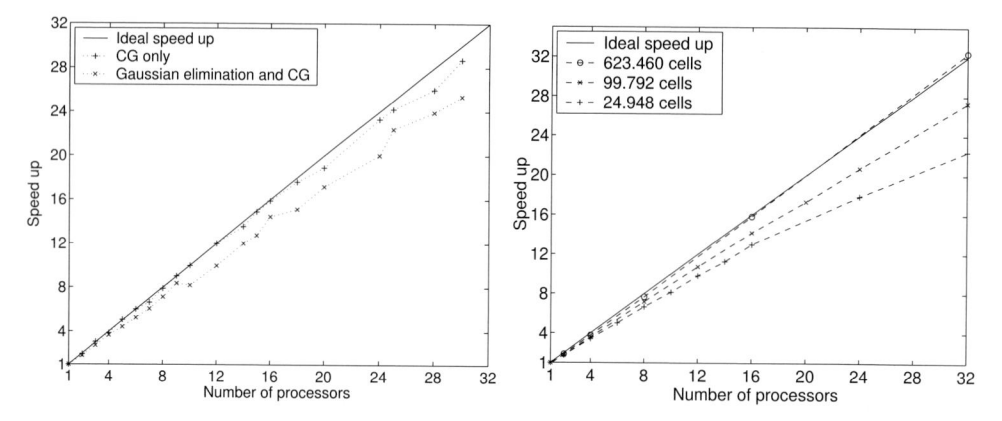

Figure 4. Parallel speed up T_1/T_n of the matrix solver for Poisson test problem (2) (left picture), and of the complete model for a dam break test case(see [7]) with varying resolution (right picture). All tests were performed on a 32 processor cluster (Intel Xeon 2.00 Mhz.) with Myrinet interconnect.

REFERENCES

1. L.A. Hagemand and D.M. Young, *Applied iterative methods*, Academic Press, New York 1981.
2. U. Meier and A. Sameh, *The behavior of conjugate gradient algorithms on a multivector processor with a hierarchical memory*, J. Comput. Appl. Math. 24 (1998) pp. 13–32.
3. L. Freitag and J.M. Ortega, *The RSCG Algorithm on Distributed Memory Architectures:*, Numerical Linear Algebra with Applications 2 (1995) pp. 401-414.
4. E.F. D'Azevedo and P.A. Forsyth and Wei-Pai Tang, *Ordering Methods for Preconditioned Conjugate Gradient Methods Applied to Unstructured Grid Problems*, SIAM J. Matrix Anal. Appl. 13 (1992), pp. 944–961.
5. V. Casulli, *Semi-implicit Finite-Difference Methods for the 2-dimensional Shallow-Water Equations*, J. Comput. Phys. 86 (1990) pp. 56–74.
6. J.J. Leendertse, *Aspects of a Computational Model for Long-Period Water Wave Propagation*, Memorandum RM-5294-PR, Rand Corporation, Santa Monica 1967.
7. G.S. Stelling and S.P.A. Duinmeijer, *A Staggered Conservative Scheme for Every Froude Number in Rapidly Varied Shallow Water Flows*, Int. J. Numer. Methods Fluids 43 (2003) pp. 1329–1354.
8. A. Gupta, G. Karypis and V. Kumar, *Highly Scalable Parallel Algorithms for Sparse Matrix Factorization*, IEEE Trans. Parallel Distrib. Syst. 8 (1997) pp. 502–520.
9. H.A. van der Vorst, *(M)ICCG for 2D problems on vector computers*, in Supercomputing, A.Lichnewsky and C.Saguez, eds., North-Holland 1988, pp. 321-333.
10. N. Touheed and P. Selwood and P.K. Jimack and M. Berzins, *A comparison of some dynamic load-balancing algorithms for a parallel adaptive flow solver*, Parallel Computing 26 (2000), pp. 1535-1554.

Towards numerical modelling of surface tension of microdroplets

X. J. Gu, D. R. Emerson, R. W. Barber and Y. H. Zhang

Computational Science and Engineering Department,
CCLRC Daresbury Laboratory, Daresbury, Warrington WA4 4AD, UK

Keywords: Droplet; Curvature; Surface tension; Volume-of-fluid; Free-surface flow

A parallel Volume-of-Fluid code has been developed to simulate interfacial flows with and without surface tension. The formulation follows a hybrid approach that uses a tangent transformation to capture the advected interface while surface tension effects are modelled by the Continuum Surface Force method. Numerical results are presented for several test problems and show good agreement with experimental data and theory.

1. INTRODUCTION

Free-surface flows involving droplets and bubbles are frequently encountered in Micro-Electro-Mechanical-Systems (MEMS). The Volume-of-Fluid (VOF) [1] and the level set method [2] are the two most popular approaches for the simulation of free-surface flows. The advantage of these methods is their ability to handle arbitrarily shaped interfaces with large deformations, including interface rupture and coalescence, in a natural way. The interface in the VOF method is implicitly expressed by a volume fraction, f. However, the step-like behaviour of f makes computing interface derivatives (normals and curvature) difficult. In contrast, the interface in the level set method is defined by the zero level set of a smooth function and interfacial derivatives can be readily calculated. A disadvantage of the level set method is that mass is not rigorously conserved. The VOF approach, however, has the advantage that mass is conserved, provided the discretisation scheme is conservative. In the present study, the difficulty in capturing interfacial derivatives has been avoided by using a hybrid approach [3] which transfers the step-like behaviour of the volume fraction, f, into a smooth function.

The Continuum Surface Force (CSF) approach, developed by Brackbill *et al.* [4], has previously been combined with the Volume-of-Fluid method to simulate flows where surface tension plays an important role. However, when large interfacial forces are present, the CSF technique has been shown to generate undesirable spurious currents near the interface. Renardy and Renardy [5] have shown that there are two main sources which generate these spurious currents. One arises from the coupling of the surface tension force with the flow field, while the other is caused by inaccuracies in the calculation of surface curvature. Both of these issues are addressed in this paper.

The present study combines the hybrid method for advecting the interface [3] and the CSF model [4] within a finite-volume, collocated grid scheme. A number of validation tests have been performed that involve flows under gravity, and problems where surface tension is important. The free-surface test involves the initial stages of a dam break and the results are compared with the experimental data of Stansby et al. [6]. The second test calculates the pressure rise due to surface tension inside a cylindrical liquid rod and the final test involves the simulation of a gas bubble rising in mineral oil. The computed results for the terminal velocity of the bubble are then compared to the experimental observations of Hnat and Buckmaster [7]. Surface tension can be a significant force at the micro-scale and the two final tests are a key step towards reliable modelling of surface tension phenomena in microsystems.

2. GOVERNING EQUATIONS

Consider two incompressible fluids, A and B, separated by an interface S. The continuity equation is given by

$$\frac{\partial u_i}{\partial x_i} = 0, \tag{1}$$

where u_i is the velocity and x_i is the spatial direction. The flow is governed by the incompressible Navier-Stokes equations:

$$\frac{\partial u_i}{\partial t} + \frac{\partial u_i u_j}{\partial x_j} = -\frac{1}{\rho}\frac{\partial p}{\partial x_i} + \frac{\mu}{\rho}\left(\frac{\partial u_i}{\partial x_j} + \frac{\partial u_j}{\partial x_i}\right) + \frac{F_i}{\rho} + g_i, \tag{2}$$

in which p, g_i and F_i are the pressure, gravity vector and the interfacial surface tension force, respectively. The local density, ρ, and viscosity, μ, are defined as

$$\rho = f\rho_A + (1-f)\rho_B \quad \text{and} \quad \mu = f\mu_A + (1-f)\mu_B, \tag{3}$$

where the subscripts denote the different fluids, and f is the volume fraction with a value of unity in fluid A and zero in fluid B. The volume fraction is governed by

$$\frac{\partial f}{\partial t} + u_i \frac{\partial f}{\partial x_i} = 0. \tag{4}$$

The Continuous Surface Force (CSF) method of Brackbill et al. [4] is employed to calculate the surface tension force:

$$F_i = \sigma \kappa \frac{\partial f}{\partial x_i}, \tag{5}$$

where σ is the surface tension and κ is the curvature of the interface. The CSF method converts the surface force into a volumetric continuous force, F_i, instead of a boundary condition on the interface. Equations (1-5) are discretised using a finite-volume method and an implicit second-order temporal scheme. The pressure and velocity fields are solved on a collocated grid using the SIMPLE algorithm coupled through Rhie and Chow interpolation [8]. Equation (4) is essential for capturing the motion of the fluid interface but accurate discretisation of its step-like behaviour is not straightforward. Following Yabe and Xiao [3], a tangent transformation is used to convert the volume fraction, f, into a smooth function, as detailed in Section 3.

3. TANGENT TRANSFORMATION

The solution of Eq. (4) on a fixed grid introduces numerical diffusion into the scheme and the initial sharpness of the interface will be smeared. To minimise this smearing, the volume fraction, f, is transformed to a tangent function, C, given by:

$$C(f) = \tan\left[(1-\varepsilon)\pi\left(f-\tfrac{1}{2}\right)\right] \Leftrightarrow f = \frac{\tan^{-1} C}{(1-\varepsilon)\pi} + \frac{1}{2}, \tag{6}$$

where ε is a small positive constant that can be used to tune the steepness of the transition layer. If Eq. (6) is substituted into Eq. (4), it can be shown that C obeys an equation analogous to f, i.e.

$$\frac{\partial C}{\partial t} + u_i \frac{\partial C}{\partial x_i} = 0. \tag{7}$$

Although f undergoes a rapid transition from 0 to 1 at the interface, C exhibits a smooth behaviour. The transformation function improves the local spatial resolution near the interface, allowing the discontinuity to be described easily, but results in a slight modification of the advection speed. However, the function guarantees the correct advection speed along the $f = 0.5$ surface and tends to produce a solution that counters the smearing across the transition layer with intrinsic anti-diffusion (Brackbill's analysis) [3]. This method does not involve any interface reconstruction procedure and is computationally economic. It is also more attractive for 3D computations.

4. CALCULATION OF INTERFACIAL CURVATURE

In the CSF approach, the value of the volumetric surface tension force is proportional to the interfacial curvature and it is therefore important to evaluate this curvature accurately. This is particularly true for flows involving microdroplets. Following Brackbill et al. [4], the curvature of the surface can be expressed as

$$\kappa = -\frac{\partial n_i}{\partial x_i} \tag{8}$$

where n_i is the unit normal to the surface. In the context of the Volume-of-Fluid method, the unit normal can readily be obtained from

$$n_i = \frac{\partial f / \partial x_i}{\sqrt{(\partial f / \partial x_k)^2}}. \tag{9}$$

Substituting Eq. (6) into Eq. (9), the unit normal to the surface can also be expressed by

$$n_i = \frac{\partial C / \partial x_i}{\sqrt{(\partial C / \partial x_k)^2}}. \tag{10}$$

As C is a smooth function, its derivatives can be accurately replaced by finite difference approximations. If we denote

$$G_i = \frac{\partial C}{\partial x_i} = (1-\varepsilon)\pi(1+C^2)\frac{\partial f}{\partial x_i} \quad \text{and} \quad T_{ij} = \frac{\partial G_i}{\partial x_j} = \frac{\partial^2 C}{\partial x_i \partial x_j}, \tag{11}$$

the curvature of the surface can be evaluated from

$$\kappa = -\left(\sqrt{G_k^2}\right)^{-3}\left(G_k G_k T_{ll} - G_l G_k T_{lk}\right). \tag{12}$$

Substituting G_i from Eq. (11) and κ from Eq. (12) into Eq. (5) allows the surface tension force to be evaluated as

$$F_i = -\sigma\left(\sqrt{G_k^2}\right)^{-3}\left(G_k G_k T_{ll} - G_l G_k T_{lk}\right)\frac{G_i}{(1-\varepsilon)\pi(1+C^2)}. \tag{13}$$

5. THE SURFACE TENSION FORCE IN A COLLOCATED GRID

As part of the pressure in the momentum equation is balanced by the volumetric surface tension, the pressure is split into two components, p_1 and p_2. Following Renardy and Renardy [5], p_1 is chosen such that

$$\frac{F_i}{\rho} - \frac{1}{\rho}\frac{\partial p_1}{\partial x_i} \tag{14}$$

is divergence free i.e.

$$\frac{\partial}{\partial x_i}\left(\frac{F_i}{\rho} - \frac{1}{\rho}\frac{\partial p_1}{\partial x_i}\right) = 0. \tag{15}$$

In discretising Eq. (15), the values of F_i and $\partial p_1 / \partial x_i$ have been defined at the faces of the control volume. For consistency, the values of F_i in the momentum equation are also calculated at the control volume faces. Figure 1(a) illustrates how spurious velocity vectors can be generated at an interface between two fluids. Evaluating the surface tension force, F_i, at the face of the control volume leads to a significant reduction in the magnitude of the spurious currents, as shown in Fig. 1(b).

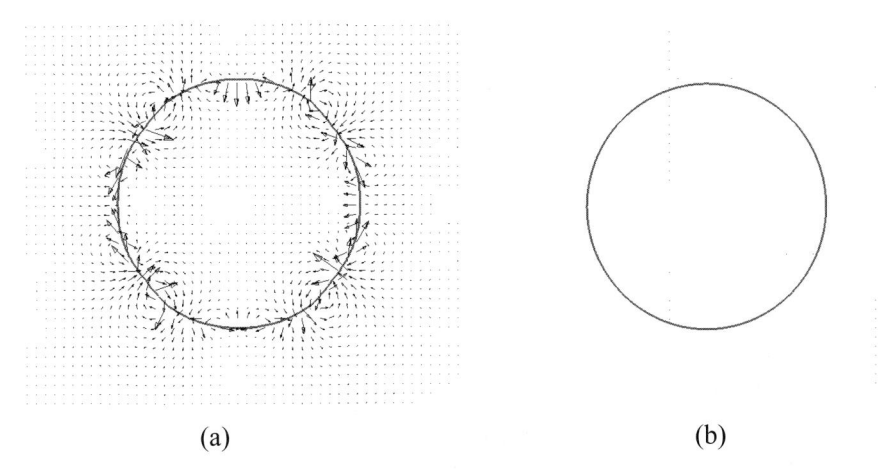

Figure 1. Computed velocity vectors for an initially static cylindrical water rod of radius 2.5 mm after 1 ms: (a) F_i evaluated at the cell centre using central differencing (b) F_i evaluated at the control volume interface. The velocity vectors are plotted at the same scale.

6. PARALLELISATION STRATEGY

The Centre for Microfluidics and Microsystems Modelling at Daresbury Laboratory has developed a 3-D Navier-Stokes code known as THOR. The code can deal with highly complex geometries through the use of multi-block grids and body-fitted coordinates. To accelerate the convergence of the numerical algorithm, a multigrid approach is employed based on the full approximation scheme. The code has been parallelised using standard grid partitioning techniques and communication is through the use of the MPI standard. At least one block must be allocated to each processor, but any processor may contain more than one block. Large blocks can therefore be divided into smaller blocks so that a balanced work load can be easily achieved. The scalability of the code has been tested on a range of platforms, as reported previously [9].

7. TEST CASES

Three studies have been performed to test the numerical implementation. The first case considers the initial stages of a classic dam break problem to test the accuracy of the interface advection. For this case, surface tension effects can be ignored. To test the CSF model, the second problem involves a static water rod. If gravitational effects are ignored, the difference in pressure across the water surface is purely related to the surface tension and results can be compared with theoretical values. Finally, a gas bubble rising in mineral oil has been simulated and the predicted terminal velocity compared against experimental observation.

7.1. Initial stages of a dam break

Experiments have been conducted by Stansby et al. [6] to investigate the early stages of dam-break flows. The experimental set-up employed a thin metal plate that separated water at two different levels. This plate was withdrawn impulsively in the vertical direction to create a travelling wave. The experiments were carried out in a horizontal flume with the following dimensions: 15.24 m long, 0.4 m wide, and 0.4 m high. The side walls were constructed from clear Perspex for observation of the free surface and the plate was located 9.76 m from the upstream end of the flume.

Figure 2. Simulation of a developing bore during the initial stages of a dam break.

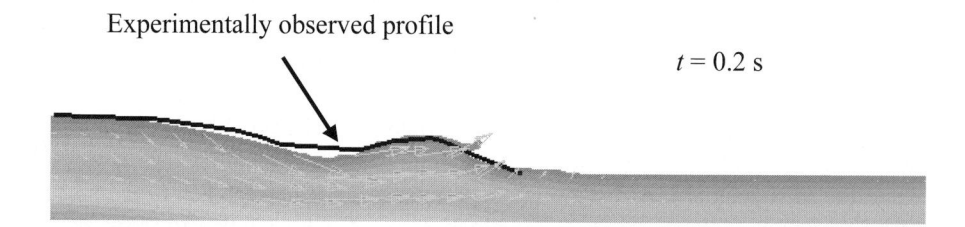

Figure 3. Comparison of the computed water level at $t = 0.2$ s against the experimentally observed surface profile reported by Stansby et al. [6].

The test case involved an upstream water depth of $\eta_0 = 0.1$ m and a downstream water depth of $0.45\eta_0$. The flume was modelled in 2D using 3840x100 equi-spaced grid points. The simulations were performed on HPCx at Daresbury Laboratory, which is an IBM Power4 platform, and required approximately 12 hours using 32 processors. A bore develops downstream of the plate as a result of highly complex flow interactions and is clearly captured by the simulation, as shown in Fig. 2. Figure 3 shows that the simulated water level and the experimentally observed surface profile are in very close agreement.

7.2. Modelling the pressure distribution in a static water rod

In the absence of gravity, Laplace's formula for the pressure rise in an infinite liquid cylinder is given by

$$\Delta P = \sigma \kappa = \frac{\sigma}{R}, \qquad (16)$$

where R is the radius of the cylindrical rod. Figure 4 shows the computed pressure distribution for a static water rod of radius $R = 1$ mm using the CSF model. The pressure rise is due solely to the surface tension force. The computed pressure jump across the interface is captured without any numerical oscillations. Figure 5 compares the predicted pressure rise for water rods of different radii against the theoretical solution and shows very good agreement, even for a radius as small as 0.1 mm.

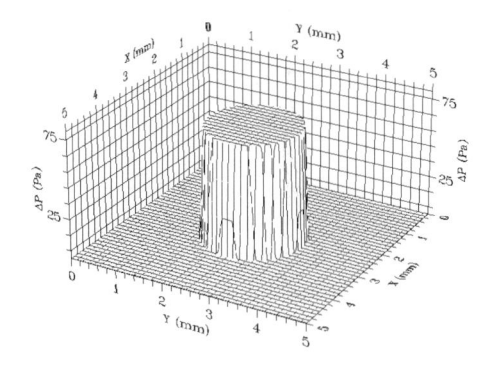

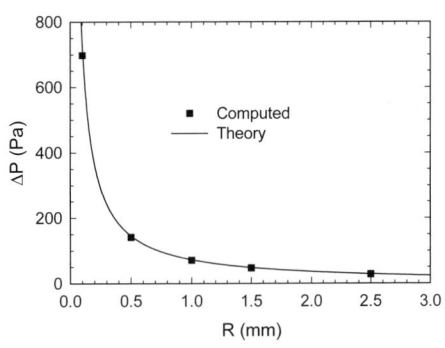

Figure 4. Computed pressure distribution in a water rod of radius $R = 1$ mm.

Figure 5. Predicted pressure rise in a water rod in comparison with theory, Eq. (16).

7.3. Interfacial dynamics of a rising bubble

The shape and behaviour of gas bubbles rising in mineral oil have been investigated experimentally by Hnat and Buckmaster [7]. However, simulating gas bubbles rising in a column of liquid presents a significant modelling challenge due to the complex balance of viscous, surface tension, and dynamic forces. The present test case involves a rising gas bubble with the following properties: initial radius $R = 6.1$ mm, $\rho_{oil} = 875.5$ kg/m^3, $\rho_{gas} = 1.0$ kg/m^3, $\mu_{oil} = 0.118$ Ns/m^2, $\mu_{gas} = 0.001$ Ns/m^2, $\sigma = 32.2 \times 10^{-3}$ N/m, and acceleration due to gravity, $g = 9.8$ m/s^2. The simulation was performed using 512x150 equi-spaced grid points in a 12Rx3R axisymmetric domain. The computations were carried out using 32 processors on Newton, an Itanium2 SGI Altix system located at the University of Manchester. Figure 6 shows the computed bubble shape while Fig. 7 illustrates that the predicted terminal velocity approaches the value observed by Hnat and Buckmaster [7]. Making the assumption that the flow is axisymmetric reduces the computational cost significantly. However, this particular simulation still required a total of 40x32 processor hours on Newton, indicating that parallelisation is essential.

Figure 6. Computed bubble shape and velocity field for a rising gas bubble.

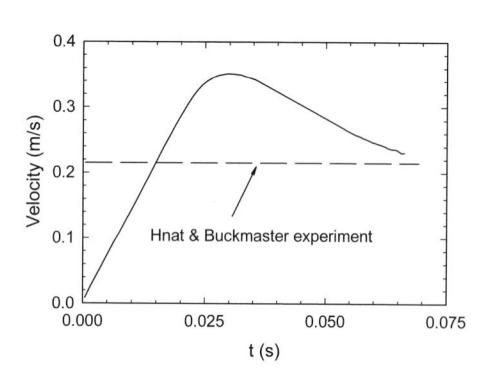

Figure 7. Predicted velocity of a gas bubble rising in mineral oil.

8. CONCLUSIONS

A parallel, hybrid Volume-of-Fluid and tangent transformation method has been combined with the Continuous Surface Force (CSF) model and successfully implemented in a collocated grid framework. The formulation ensures that the calculation of the volumetric surface tension force is consistent in the pressure and momentum equations, reducing spurious currents significantly. Predicted results from problems involving a dam break flow, static cylindrical water rods, and a gas bubble rising in mineral oil have all been shown to be in close agreement with available experimental data and theory.

ACKNOWLEDGEMENTS

The authors would like to thank the UK Engineering and Physical Sciences Research Council (EPSRC) for their support of this research under Grant No. GR/S83739/01. Additional support was provided by the EPSRC under the CCP12 programme.

REFERENCES

1. C. W. Hirt and B. D. Nichols, J. Comput. Phys. 39 (1981) 201.
2. J. A. Sethian and P. Smereka, Ann. Rev. Fluid Mech. 35 (2003) 341.
3. T. Yabe and F. Xiao, Computers Math. Applic. 29 (1995) 15.
4. J. U. Brackbill, D. B. Kothe, and C. Zemach, J. Comput. Phys. 100 (1992) 335.
5. Y. Renardy and M. Renardy, J. Comput. Phys. 183 (2002) 400.
6. P. K. Stansby, A. Chegini, and T. C. D. Barnes, J. Fluid Mech. 374 (1998) 407.
7. J. G. Hnat and J. D. Buckmaster, Phys. Fluids 19 (1976) 182.
8. C. M. Rhie and W. L. Chow, AIAA J. 21 (1983) 1525.
9. X. J. Gu, R. W. Barber, and D. R. Emerson, in Proc. Parallel CFD (2004) 497.

DNS Simulation of Sound Suppression in a Resonator with Upstream Flows

I. Abalakin [a], A. Alexandrov [a], V. Bobkov [a] and T. Kozubskaya [a]

[a] *Institute for Mathematical Modelling of Rus.Ac.Sci., 4-A, Miusskaya Sq., Moscow 125047,*

E-mail: tolik@imamod.ru - Web page: http://aeroacoustics.imamod.ru/

Keywords: DNS, CAA, parallel computing;

1. INTRODUCTION

Computational aeroacoustics (CAA) is one of the most advancing research fields. Such interest is caused by a variety of engineering problems that are studied in CAA. Among these problems there are jet noise, sonic boom, cavity tones, fan noises, aeolian tone etc. One of the interesting CAA problems of current importance is the process of noise suppression in acoustic resonator type liners – specially constructed noise absorbing panels. Resonator type liners are one of the most popular ways of engine noise damping.

Basically, the liner panel represents a screen with holes ended by resonator boxes (usually, of honeycomb form) figure 1. Generally, the liner may consist of a few panels of this kind. An acoustic wave falls on to the upper screen and penetrates into a system of resonators. An optimal design of acoustic liners for maximum suppression requires exhaustive understanding of mechanisms of noise absorption in such a construction. But due to the small size of the construction holes (the typical dimension of one used in aircraft engines is about 1mm) the experimental investigation of noise suppression in a real liner element is rather difficult. There are a few experimental works in which the mechanism of dissipation in such elements is investigated by means of natural experiments [1].

The physical mechanism of acoustic noise suppression for low intensity waves in a liner construction is studied well enough. Acoustic characteristics of resonance sound absorbing systems are constant for such waves. It is generally believed that in case of low sound intensity the friction at the hole entrance is in charge of the dissipation of acoustic energy and plays the dominant role in the noise suppression. At a high level of noise, non-linear effects become much more important.

In this paper direct numerical simulation (DNS) is used for investigation of mechanisms of acoustic noise absorption in a resonator which is the determining part of liner. The problem formulation under study is similar to the one given in [2,3], but in contrast to that one, in the present paper the incoming acoustic wave is propagating within the upstream subsonic flow. This formulation better corresponds to the real engineering problems. At the same time, it brings an additional computational difficulty.

In case of such formulation, the regimes with high gradients of gas dynamic parameters especially at the initial stage of the process may occur. In practice the problems with such high gradients of solution are predicted by the codes designed for solving gas dynamic problems which, in general case, is not well applicable to CAA (from here let us denote them as CFD codes in contrast to CAA codes).

In this paper the two stage algorithm is used. At first stage, the quasi-stationary flow field is predicted with the help of CFD code using the coarse mesh. At this stage, only the upstream flow and no acoustic waves are considered. Then at the second stage, the produced solution is re-interpolated to the fine mesh and is used as an initial state for the CAA code (WHISPAR 0.906). At this stage, the incoming monochromatic acoustic wave is added to the upstream flow at the inflow boundary conditions.

DNS of acoustic liner problem is expensive enough especially in case of high Reynolds number so the use of high performance parallel computer systems looks quite natural. In the paper presented all the CAA predictions on the refined meshes are carried out on the massively parallel computer system MVS 1000M [4].

2. PROBLEM FORMULATION

Geometrical parameters of the liner under consideration are shown in figure 1. The length of the computational domain is the length of total construction and comes to L=189mm. The computational domain width is equal to the duct diameter of D=2R=23.5mm. The hole radius is r=1.25mm and the screen thickness is δ=2mm. The length of duct part of the construction is h=L-H-δ=100mm. The resonator depth H=87mm is chosen equal to the quarter of the incoming wave length.

The upstream flow of subsonic Mach numbers (from 0.1 to 0.5) is excited by the incoming acoustic wave of frequency $f = 987$ Hz and different intensity (from 100 to 160 Db). The excited incoming flow falls on the screen and penetrates into a resonator.

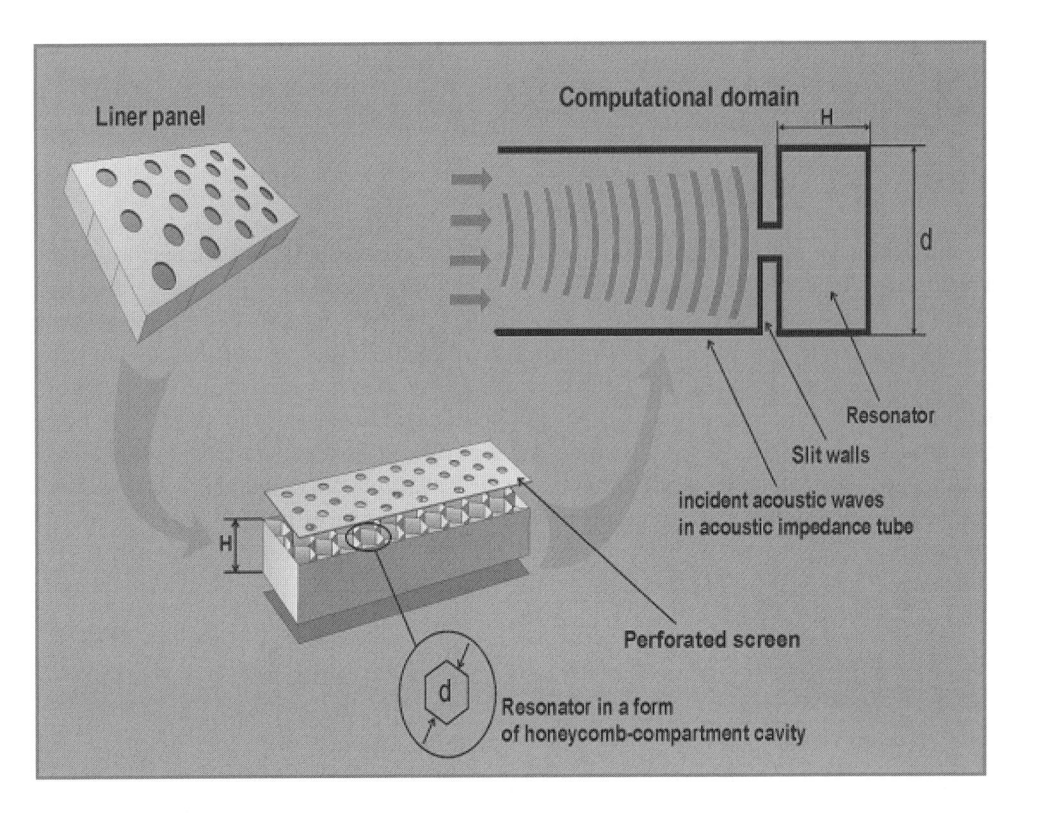

Fig. 1. Geometric parameters of liner under consideration

3. NUMERICAL METHOD

In accordance with the above-mentioned two-stage algorithm, two different codes are used. The CFD code used for solving compressible Navier–Stokes equations implements the finite volume solver for unstructured triangular meshes. It is based on the higher order Roe upwind scheme.

At second stage, the calculations presented in the paper are carried out with help of parallel program package WHISPAR 0.906 intended for solving CAA problems.

At this stage, the 2D unsteady compressible Navier-Stokes equations are solved within the finite difference approach. For the approximation of time and space derivatives the Dispersion Relation Preserving (DRP) finite difference scheme of 4th accuracy order is used [5].

The characteristic boundary conditions are used at the free (an open part of the duct) boundary of computational domain. The special modification of non-slip condition [6] is applied on the solid walls.

To avoid the high frequency non-physical oscillations, the filters proposed in [5] are used.

More details on the numerical algorithms in use are given in [2]. As distinct from the techniques described there, the present predictions are carried out with the help of DRP scheme adapted to the non–uniform meshes.

4. PARALLEL IMPLEMENTATION

At the first stage all the calculations are made on the coarse mesh and are performed on the non-parallel computer.

Then at the second stage the produced solution should be interpolated on the refined mesh and inherited by parallel CAA code (WISPAR 0.906). The calculations are held on 768-processor MVS-1000M MPP computer system equipped with 667MHz 64-bit 21164 EV67 Alpha processors [4]. All the predictions at the second stage have been performed on the base of explicit numerical algorithms. So the parallelization is based on the geometrical domain partitioning (in accordance with a number of processor available) in a way that each subdomain is served by one processor unit.

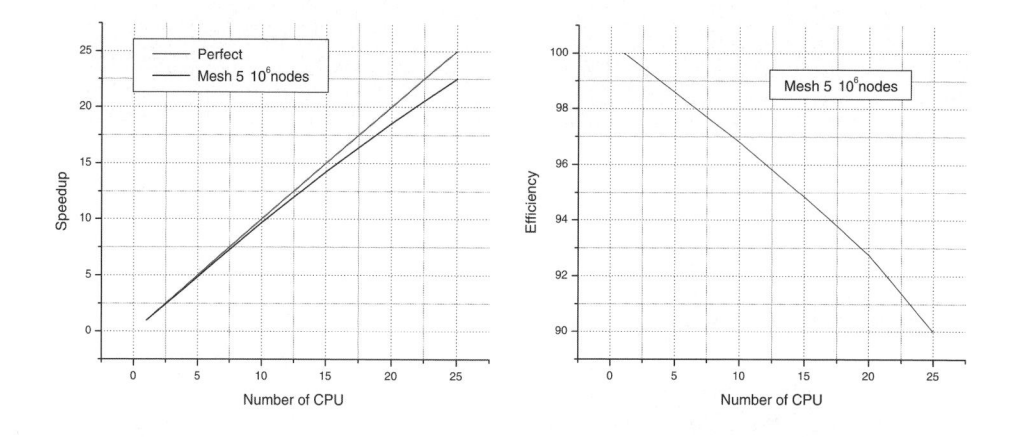

Fig. 2. Efficiency of parallelization and speedup

The requirement on the processors load balancing is provided automatically. Data exchange is based on the MPI standard. This way of doing results in a good scalability and portability for an arbitrary number of processorsn (figure 2).

5. NUMERICAL RESULTS

As a result of predictions of acoustic wave damping in a liner element several regimes of process have been detected. Depending on the magnitude of the intensity of incident wave there are laminar regime, the regime with synchronously separating and propagating couples of vortices and completely chaotic regime were. This set of regimes is similar to the one obtained in case of former formulation (figure 3,4.). But new formulation in addition to the detected regimes let us illustrate the influence of incoming flow on the development of the process.

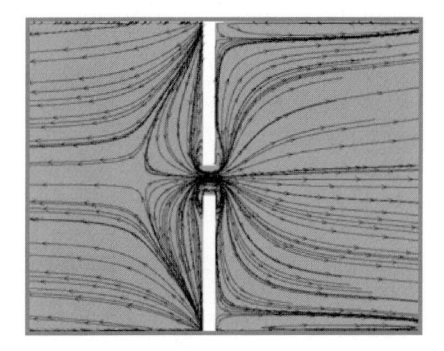

Fig. 3. laminar regime

Fig. 4. Chaotic motion and reflection of vortices.

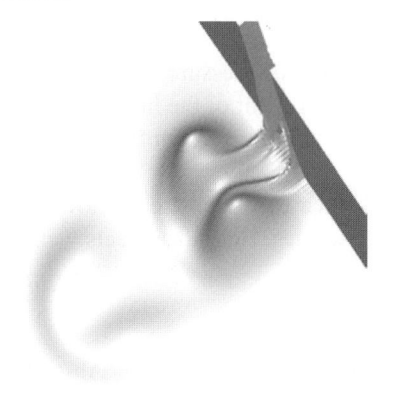

Fig. 5. Separation of a couple of vortices.

At figures 6-9 the pressure oscilation and it's spectra at the right point of the hurdle are shown for both cases (with and without upstream flow).

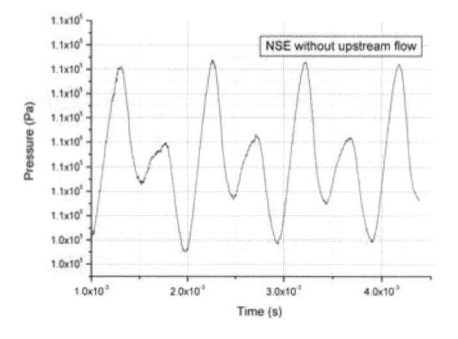

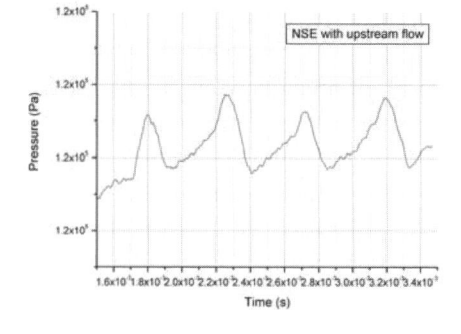

Fig. 6. Pressure oscilation at the right point of the hurdle (without upstream flow)

Fig. 7. Pressure oscilation at the right point of the hurdle (with upstream flow)

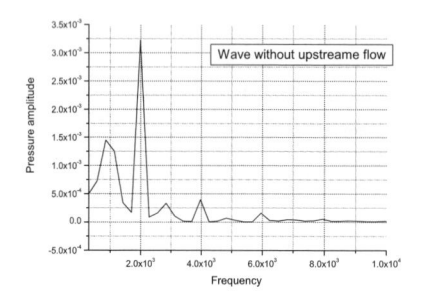

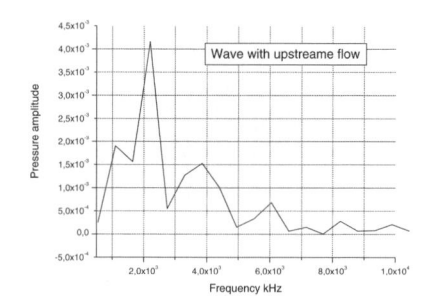

Fig. 8. Spectra of pressure oscilation at the right point of the hurdle (without upstream flow)

Fig. 9. Spectra of pressure oscilation at the right point of the hurdle (with upstream flow)

The influence of the upstream flow leads to the widening of peaks and to the shift of their frequencies. Both these facts well corresponds to the experimental data.

REFERENCES

1. Dragan S.P., Lebedeva I.V. "Absorption of intensive sound on the hole in the screen" Acoustic Journal, Vol 44 N 2, pp206-211.
2. A.V.Alexandrov, V.G.Bobkov, T.K.Kozubskaya . S Simulation of Acoustic Wave Damping in a Liner Element. Parallel CFD 2004 Conference, Las Palmos, Canary Islands, Gran Canaria, Spain, May 24-27

3. A.V.Alexandrov, V.G.Bobkov, T.K.Kozubskaya. Numerical Study of Sound Suppression Mechanisms in a Liner Element. ECCOMAS 2004 Conference, Jyvaskyla, Finland, July 24-28, 2004.
4. V.E. Fortov, V.K. Levin, G.I. Savin, A.V. Zabrodin, et al. Inform.-Analit. Zh. Nauka Prom-st Rossii, No 11(55)(2001) 49.
5. Tam C.K.W. and Webb J. C. "Dispersion-Relation-Preserving Schemes for Computational Aeroacoustics" , Journal of Computational Physics, Vol.107, 1993, pp262-281.
6. Tam C.K.W. , Shen H.,"Direct computation of nonlinear acoustic pulses using high order finite difference schemes". AIAA paper 93-4325.

Performance Characterization and Scalability analysis of a Chimera Based Parallel Navier-Stokes solver on commodity clusters

A. Hamed[a], D. Basu[a], K. Tomko[b], Q. Liu[b]

[a]*Department of Aerospace Engineering, University of Cincinnati, USA*

E-mail[a] : a.hamed@uc.edu, basud@email.uc.edu

[b]*Department of Electrical and Computer Engineering and Computer Science, University of Cincinnati, USA*

E-mail[b] : ktomko@ececs.uc.edu, liuqu@ececs.uc.edu

Keywords: Domain Connectivity, PEGASUS, Navier-Stokes, Implicit, Upwind Scheme, Cache Performance and Relative Speedup

1. Abstract

The present work focuses on the performance characterization and scalability analysis of a chimera based parallel Navier-Stokes (N-S) solver on a commodity cluster. The parallel code; FDL3DI; is MPI-based and the interconnectivity between the zones is obtained through the Chimera approach. In this strategy, the computational domain is decomposed into a number of overlapped zones and each of the zones is assigned to a separate processor. Each zone is solved independently and communication between them is accomplished through the interpolation points in the overlapped region by explicit message passing using MPI libraries. The code's performance is assessed on a commodity cluster of AMD Athlon processors in terms of relative speed up, execution time and cache analysis with different grids and domain sizes. Turbulent flow over a flat plate is considered as the benchmark problem for the current simulations. Three variants of the upwind biased Roe scheme, a second order central difference scheme as well as a high order compact differencing scheme with high order pade-type filtering are used for spatial discretization and they are compared with each other for relative execution time. Computed results indicate that the number of zones as well as the type of spatial discretization scheme used significantly affects the execution time. The relation between cache performance and the number of zones as well as the numerical scheme employed is also explored in the present analysis.

2. Introduction and Methodology

Parallel computational fluid dynamics (CFD) has evolved significantly over the past few years[1,2,3,4]. With the advent of high performance clusters, and the resulting increase in availability of computing cycles, large-scale parallel CFD computations are now realizable. Most of the commercial CFD codes are available in a parallel version. However, the performance of a parallel code is a combination of many parameters, including the code's scalability across an increasing number of nodes[5]. Also the number of zones, overlapped points and the numerical scheme influence the performance of the solver.

In the present analysis, an extensive performance characterization as well as scalability analysis of a Chimera based parallel Navier-Stokes solver is being carried out using commodity clusters. The main focus of the current work is to study the scalability of the solver, the cache analysis, its performance with varying number of zones and also its performance with different numerical schemes. The parallel time accurate three-dimensional solver FDL3DI was originally developed at AFRL[6]. In the Chimera based parallelization strategy[4] used in the solver, the computational domain is decomposed into a number of overlapped zones as shown in figure 1[7]. An automated pre-processor, PEGSUS[8], is used to determine the zone connectivity and interpolation function between the decomposed zones. In the solution process, each zone is assigned to a separate processor and communication between them is accomplished through the interpolation points in the overlapped region by explicit message passing using MPI libraries. The solver has been validated and proven to be efficient and reliable for a wide range of high speed and low speed; steady and unsteady problems[3,9,10]. Basu et al.[2] implemented different hybrid turbulence models in the solver and carried out extensive simulations of unsteady separated turbulent flows.

The Navier-Stokes equations written in strong conservation-law forms are numerically solved employing the implicit, approximate-factorization, Beam-Warming algorithm[11] along with the diagonal form of Pullinam and Chaussee[12]. Newton subiterations are used to improve temporal accuracy and stability properties of the algorithm. The solver has three variants of upwind biased Roe schemes and a second order central difference scheme for spatial discretization. Apart from that, high order compact differencing scheme and high order pade-type filtering schemes are also available in the code. The details of the different numerical schemes, as well as their relative merits and demerits are explained in references 3, 6-9, 10. The execution efficiency of the numerical schemes is assessed in the current study. In a prior investigation[4], the code's performance was analyzed using two supercomputers, namely the IBM SP3 and the Silicon Graphics Origin 2000. However, in the current analysis, the performance study and the scalability analysis is being carried out in a commodity cluster with AMD Athlon 2000+ processors. The 2nd order central difference scheme was used for spatial discretization in the earlier analysis[4]. In the current work, results are being obtained using different variants of the upwind biased Roe scheme (1st, 2nd and the 3rd order), 2nd order central scheme as well as a high order compact scheme in conjunction with high order pade type filtering. They are used for the spatial discretization of the governing equations. An extensive comparison is presented for the study of the different schemes including an investigation of memory cache utilization. Detailed analysis of cache usage is carried out for the different schemes as well as different numbers and sizes of zones. Level 1 and Level 2 cache effectiveness is characterized using the PAPI performance monitoring API[13]. A prior cache analysis of the 2nd order version of the solver is described in reference 14.

3. High order compact Scheme and Filtering

High order compact schemes[9] (schemes with an accuracy of fourth order and higher) are used in DNS (Direct Numerical Simulations) and LES (Large Eddy Simulation) of turbulent flows. High order compact schemes are non-dissipative and due to their superior resolution power, they represent an attractive choice for reducing dispersion, anisotropy and dissipation errors associated with low-order spatial discretization. However, high order compact-difference schemes are susceptible to numerical instability non-diffusive nature and approximate high order filters are used in conjunction to overcome their susceptibility to unrestricted growth of spurious perturbations. The filter has been proven superior to the use of explicitly added artificial dissipation (as used in upwind biased and central schemes) for maintaining both stability and accuracy. Compact schemes do however incur a moderate increase in computational expense over the non-compact schemes. A schematic of the formulation aspects of the compact scheme and the filtering technique is shown in figure 2[3,7]. In the current simulations, a fourth order compact scheme in conjunction with a sixth-order non-dispersive filter is used. The compact scheme used in the computations henceforth will be designated as C4F6.

4. Parallel Algorithm

A detailed description of the parallelization algorithm employed in the solver is given in references 15. The single Program Multiple Data (SPMD) parallel programming style is used for the parallelization strategy. The flow equations for each grid are solved independently in parallel and the interpolated boundary values are also updated in parallel. The boundary data is exchanged between processors and after that, on each processor, the interpolated Chimera boundary values and the physical boundary conditions are applied to the assigned grid. The code running on each processor is identical and the processor identification number is used to determine which grid is assigned to each processor. The number of processors utilized is equal to the number of zones/sub domains, which ensures that each zone has been assigned to an individual processor. The MPI message-passing library is used for inter-processor communication. Point-to-point communication using send and receive calls are used to exchange the Chimera boundary data between processors.

5. Results and Discussions

Computed results are presented for the relative execution times for the different schemes, for the relative speedup for the different domain sizes and for the cache analysis of the different schemes. Simulations are carried out for a flat plate geometry. The geometry and flow conditions correspond to the experimental upstream flow conditions of the backward facing step flow by Driver and Seegmiller[16]. The Mach number for the present simulations is 0.128 and the Reynolds number is 0.83×10^6/ft. The computational domain consists of 200 grid points in the streamwise direction, 100 points in the wall normal direction and 20 grids in the spanwise direction. At the inflow plane, the pressure is extrapolated from the interior and the other flow variables were prescribed. The pressure at the outflow boundaries was set equal to the free stream value and the other variables were extrapolated from the interior through a first order extrapolation. The top of the computational domain is set as outflow boundary. In the span-wise direction, symmetric boundary conditions are applied.

For the present study, three cases are considered. The computational domain is divided into 8-zone, 16 zone and 20 zone domains respectively. Figure 3 shows the computational grid with the different zones and the overlapped regions. Figure 4 shows the computed boundary layer profile and its comparison with the available experimental data. It can be seen that there is an excellent agreement between the computed solution and the experimental data[15]. The computations were carried out using 3rd order Roe scheme. Figure 5 shows the comparison of the execution time (in seconds) with the number of processors for the different numerical schemes. The execution time is based on 1000 iterations for each scheme. The schemes include the 2nd order central scheme, as well as 1st, 2nd and 3rd order Roe schemes and the C4F6 scheme (4th order compact scheme with 6th order filtering). It can be seen from figure 3 that for all of the schemes, there is a significant reduction in the execution time with the increase in the number of processors. However, among the four schemes, the 2nd order central scheme takes the lowest execution times. This is because the computation in Roe schemes involves additional limiter calculations that are not present in the central scheme. Among the Roe schemes, the 3rd order scheme takes the highest execution time, followed by the 2nd and the 1st order scheme. This is expected since the upwind biased Roe schemes have more computation compared to the central scheme and among the Roe schemes, the 3rd order Roe scheme deals with a larger stencil in all the directions compared to the 1st and 2nd order Roe schemes. Comparing the C4F6 scheme with the other schemes it can be observed that the execution time is approximately 45% greater for the C4F6 schemes. This is due to the fact that the compact scheme in conjunction with filtering involves significantly more computation compared to the upwind type schemes. However, the compact scheme is commonly used with a coarser grid discretization.

Figure 6 shows the comparison of the normalized execution time per grid point for all the different schemes. This is obtained by dividing the total execution time by the total number of grid points for the 8 zones, 16 zones and the 20 zones cases. The total number of grid points takes into account the effect of overlapping and also the interpolated boundary data points. It can be seen that for all the schemes, there is a significant reduction in the normalized execution time with the increase in the number of processors and there is also considerable difference in the normalized execution time between the different schemes. The trend is similar to that seen in figure 5.

Figure 7 shows the relative speedup of the computation with the number of processors. This is based on the 3rd order Roe scheme. The blue line gives the ideal/linear speedup. The green line shows best achievable speedup given the redundant calculation incurred due to grid overlap at Chimera boundaries, while the black line shows the present simulation case for the 3rd order Roe scheme. The results show both the redundant work as well as the inter-zone communications adversely affect the speed-up. Figure 8 through 10 illustrate the memory performance of different schemes with regard to the L1 data cache and L2 data cache. The data for the Roe schemes (1st order, 2nd order and 3rd order), Central scheme (2nd order) and the 4th order compact scheme are plotted by profiling the main subroutine for solving flow equations (the subroutine that deals with the bulk of the computation) in the solver. The compact scheme involves many more variables, which results in higher cache accesses. In figure 8, the miss rates of L1 & L2 cache are given on the top of corresponding bars. It shows that L1 data cache miss rate is lower than 4% for all the schemes, which means that the L1 cache can satisfy most of the data lookup. Figure 9 enlarges the cache performance for 1st order Roe scheme with 8 zones. A low L1 data cache miss rate of 3.6% is achieved and the L1 data cache misses are taken care of by the L2 cache, which has a hit rate of 66%. Combining with figure 8, it can be easily seen that all other schemes have similar performance. Figure 10 shows the cache behaviors with different sub grid sizes. Since a higher number of zones imply a smaller sub-zone size, both the cache access number and the cache miss number decrease accordingly. However, the miss rates of two levels of caches do not show any significant variation.

6. Conclusions

Computed results are presented for the performance characterization and scalability analysis of a chimera based parallel N-S solver on a commodity cluster. Comparison of the performance of the schemes shows that the relative execution time and the normalized execution time per grid point is similar for the 2nd order central scheme and the three Roe schemes but significantly greater for the C4F6 scheme. The execution times for all schemes can be effectively improved by increasing the number of chimera zones and hence processors used for a simulation but at some loss of efficiency due to the redundant work imposed by the grid over-lap and inter-zone communication. The cache analysis shows that the compact scheme results in higher cache accesses compared to the Roe schemes and the central scheme but still achieves a low miss rate. All schemes are effectively utilizing the 1st level cache and have miss rates less that 4%. Cache behaviors with different sub grid sizes indicate that both the cache access number and the cache miss number decrease with increasing number of zones.

7. Acknowledgements

The computations were performed in the AMD Athlon Cluster at UC. The authors would like to acknowledge Mr. Robert Ogden at the Aerospace Engineering Department at UC for setting up the cluster and for installing the profiling software PAPI in the cluster. The authors would like to thank Dr. Don Rizzetta and Dr. Philip Morgan at AFRL for their help and useful discussions regarding the FDL3DI code.

8. References

1) Juan J. Alonso, Andrey Belov, Scott G. Sheffer, Luigi Martinelli and Anthony Jameson, "Efficient simulation of three-dimensional unsteady flows on distributed memory parallel computers", 1996, Computational Aerosciences Workshop.
2) Basu, D., Hamed, A., and Das, K., "DES and Hybrid RANS/LES models for unsteady separated turbulent flow predictions", 2005, AIAA-2005-0503.
3) Morgan, P., Visbal, M., and Rizzetta, D., "A Parallel High-Order Flow Solver for LES and DNS", 2002, AIAA-2002-3123.
4) Morgan, P. E., Visbal, M. R., and Tomko, K., "Chimera-Based Parallelization of an Implicit Navier-Stokes Solver with Applications", 2001, AIAA-2001-1088.
5) R.P. LeBeau, Kristipati, P., Gupta, S., Chen, H., and Huang, G., "Joint Performance Evaluation and Optimization of two CFD Codes on Commodity Clusters", 2005, AIAA-2005-1380.
6) Gaitonde, D., and Visbal, M. R., "High-Order Schemes for Navier-Stokes Equations: Algorithm and Implementation into FL3DI", 1998, AFRL-VA-TR-1998-3060.
7) Visbal, M. and Rizzetta, D., "Large-Eddy Simulation on Curvilinear Grids Using Compact Differencing and Filtering Schemes", 2002, ASME Journal of Fluids Engineering, Vol. 124, No. 4, pp. 836-847.
8) Suhs, N. E., Rogers, S. E., and Dietz, W. E., "PEGASUS 5: An Automated Pre-processor for Overset-Grid CFD", June 2002, AIAA -2002-3186.
9) Visbal, M. R. and Gaitonde, D., "Direct Numerical Simulation of a Forced Transitional Plane Wall Jet", 1998, AIAA -1998-2643.
10) Rizzetta, D. P., and Visbal, M. R., 2001, "Large Eddy Simulation of Supersonic Compression-Ramp Flows", AIAA-2001-2858.
11) Beam, R., and Warming, R., "An Implicit Factored Scheme for the Compressible Navier-Stokes Equations," 1978, AIAA Journal, Vol. 16, No. 4, pp. 393-402.
12) Pullinam, T., and Chaussee, D., "A Diagonal Form of an Implicit Approximate-Factorization Algorithm," 1981, Journal of Computational Physics, Vol. 39, No. 2, pp. 347-363.
13) London, K., Moore, S., Mucci, P., Seymour, K., and Luczak, R., "The PAPI Cross-Platform Interface to Hardware Performance Counters," Department of Defence Users' group Conference Proceedings, June 18-21, 2001.
14) Katherine Hagedon, "Cache Performance and Tuning of an Application that Solves the Three-dimensional Navier-Stokes Equations", Master's Thesis, Wright State University, 1998.
15) Tomko, K. A., "Grid Level Parallelization of an Implicit Solution of the 3D Navier-Stokes Equations," Final report, Air Force Office of Scientific Research (AFOSR) Summer Faculty Research program, September 1996.
16) Driver, D.M. and Seegmiller, H.L., "Features of a Reattaching Turbulent Shear Layer in Divergent Channel Flow," AIAA Journal, Vol. 23, No. 2, Feb. 1985, pp. 163-171.

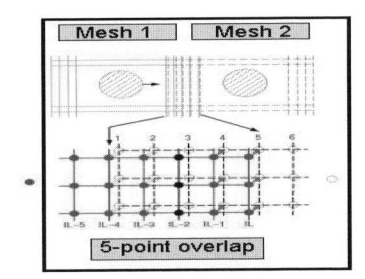

Figure 1 Schematic of the domain connectivity for the parallel solver (7)

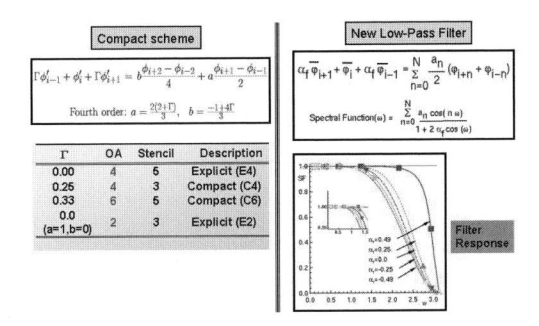

Figure 2 Schematic of compact scheme and filtering (7)

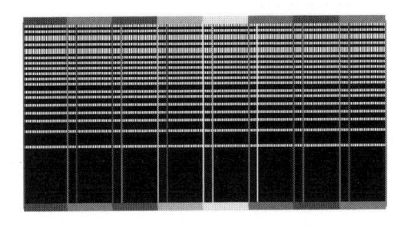

Figure 3 Computational grids for the flat plate with 8 zones

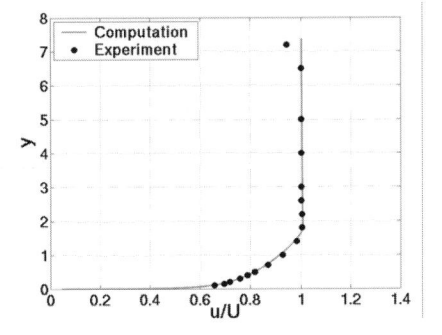

Figure 4 Comparison of computed boundary layer profile with experimental data

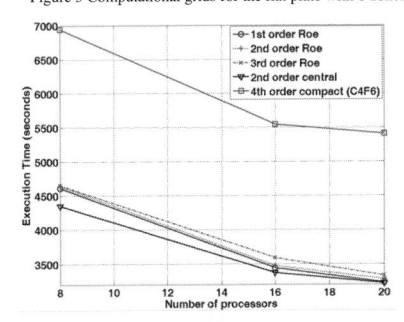

Figure 5 Effect of numerical scheme on the execution time and number of processors

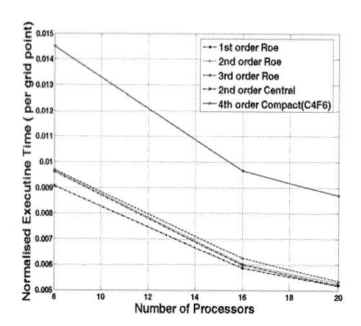

Figure 6 Normalized execution time (per grid point) for different schemes

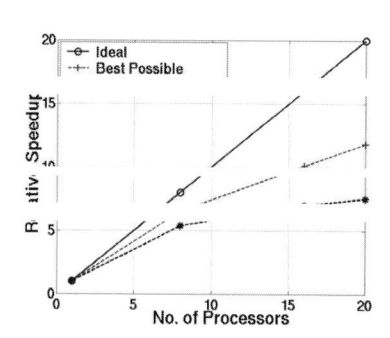

Figure 7 Comparison of relative speedup with increasing number of processors (3rd order Roe scheme)

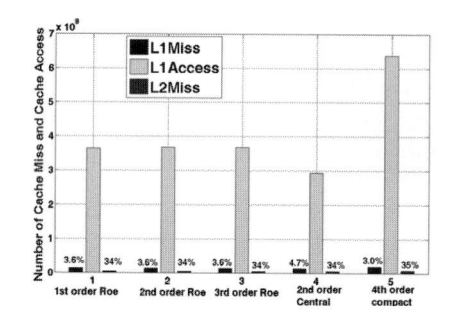

Figure 8 Number of Cache miss and Cache access for different schemes (8-zones)

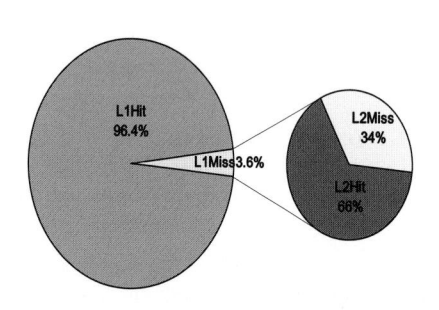

Figure 9 % Cache Miss for 1st order (8 zones)

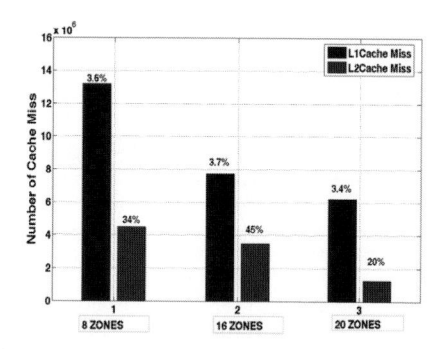

Figure 10 Number of cache miss for different zones (3rd order Roe scheme)

A Parallel Unstructured Implicit Solver for Hypersonic Reacting Flow Simulation

I. Nompelis T.W. Drayna & G.V. Candler[a]

[a]University of Minnesota,
Minneapolis, MN, USA

A new parallel implicit solver for the solution of the compressible Navier-Stokes equations with finite rate chemistry on unstructured finite volume meshes is presented. The solver employs the Data-Parallel Line Relaxation (DPLR) method for implicit time integration along lines of cells that are normal to the wall. A point-implcit method is used in regions where surface-normal lines are not constructed. The new method combines the robustness and efficiency of the implicit DPLR method with the flexibility of using unstructured discretizations. The solver employs a low-dissipation pure-upwind numerical scheme based on the Steger-Warming split flux method, as well as a MUSCL-type scheme designed for unstructured discretizations. Partitioning and load balancing of the computational mesh is automatic, and speed-up tests show that the code scales very well on commodity hardware.

1. Introduction

Simulations of high Reynolds number hypersonic flows require the use of large and highly stretched grids in order to accurately resolve boundary layers and shock waves. Additionally, performing such simulations may not be possible without making use of an implicit method of solution. The use of highly stretched grids often results in slow convergence rates even for implicit methods, as is the case with the popular LU-SGS method [1], which converges more and more slowly as the grid is stretched.[2] Other implicit methods converge rapidly with any level of grid stretching, for example the Gauss-Seidel Line Relaxation method[3] which solves the linearized equations exactly along grid lines, and the parallelizable extension of GSLR, the Data-Parallel Line Relaxation method.[2] The DPLR method also solves the linearized equations along grid lines, but the off-diagonal terms are relaxed through a series of parallelizable subiterations. The DPLR method is superior to other parallel methods, such as matrix-based point-implicit methods designed for chemically reacting hypersonic flow simulations.

In recent work[4] we introduced a generalized form of the DPLR method which was implemented on hybrid unstructured meshes, and is combined with the Full-Matrix Point Relaxation method. In this new approach, the concept is to use hexahedral and prismatic structured-like grids where possible, and use tetrahedral cells only when necessary. Our approach of combining the two implicit methods in a fully unstructured discretization is driven by our previous experience with hypersonic flows. It is generally desired to have sufficient grid resolution and high grid stretching in regions of the flow that exhibit large gradients, for example the boundary layers at solid walls, free shear-layers and shock waves. In these regions, the solution has a physical strong coupling in a particular direction, and that is where the DPLR method is most useful. In regions of the domain where the flow is generally "isotropic" the full-matrix point relaxation method is used.

Despite the potential that the hybrid discretization promises, there are several outstanding issues with regard to the solution accuracy. In particular, it is difficult to obtain second-order accurate inviscid flux reconstructions in regions with skewed tetrahedra. Furthermore, in these regions, the higher order flux reconstruction may require a special stencil that has an arbitrary number of cells for its support. This has an important consequence on the parallelization of the solver that is fully unstructured. These two issues are discussed in the present work. The paper focuses on these two aspects of the solver, as well as the parallel performance of the hybrid implicit method.

2. Solver Implementation

We use a cell-centered finite volume scheme for the solution of the compressible Navier-Stokes equations. In the finite volume method, the equations are discretized by splitting the domain into small volumes (cells), which can be arbitrary polyhedra. Flow variables are stored at cell centers, and the equations are integrated by discretizing the transient term and using the divergence theorem to write the rate of change of the conserved quantities U averaged over each cell as the sum of the fluxes over all faces of the cell. This is written as

$$\frac{\partial \bar{U}}{\partial t} = -\frac{1}{V} \sum_{\text{faces}} [(\vec{F}_c - \vec{F}_v) \cdot \hat{n} S] + \bar{W}$$

where the bar indicates a cell average of the quantity, V is the cell volume, $\hat{n}$ is the unit outward-pointing normal, S is the area of the face, and the sum is taken over the total number of faces of the polyhedron.

In our implementation, all variables that are associated with a particular cell are stored as a one-dimensional array. The size of the array is dictated by the number of cells in the computational domain and the number of ghost cells used for imposing boundary conditions. In the remaining of this paper, the subscript i always indicates that a quantity is associated with a cell in the discretization. Similarly, variables that are associated with a particular face are stored in appropriately sized arrays. For example, we store $\hat{n}_j$, the unit vector normal to face j. The index j always refers to a variable that is associated with a particular face. Additional storage is needed for the implicit solver.

3. Flux Evaluation Methods

The inviscid fluxes are computed using the modified Steger-Warming method [5]. This method uses an exact linearization of the fluxes, and upwind biasing is done on the sign of the eigenvalues. In the original Steger-Warming method, the flux at a face is computed according to $F = A_+ U_L + A_- U_R$, where A_+, A_- are flux Jacobians and U_L, U_R refer to the vector of conserved variables to the left and to the right of the face respectively. Upwinding is done when computing the Jacobians.

The Steger-Warming method is very dissipative, but a simple modification reduces the amount of numerical dissipation of the original method. In the modified method, the Jacobian of the flux is computed at each face using pressure weighted quantities of the adjacent cells. The weight takes the form

$$w_j = 1 - \frac{1}{2}\left[\frac{1}{(g\ \delta p)^2 + 1}\right], \quad \text{with} \quad \delta p = \frac{p_L - p_R}{\min(p_L, p_R)},$$

where p_L, p_R are the pressures left and right of the face, and g is a factor controlling the sensitivity of the weight – the typical value is 5. The method smoothly switches to true Steger-Warming in regions of large pressure gradients.

The viscous fluxes are computed exactly, provided that the spatial derivatives of the quantities involved are known at each face. Computing the gradient at each face would potentially involve building a special stencil with an arbitrary number of cells for its support. However, if the gradient of a quantity is known at each cell, we can simply average the gradient at adjacent cells component-by-component to get the gradient of the quantity at a face. To calculate the gradients at each cell we use a weighted least-squares fit method.

Second order accurate inviscid fluxes are calculated using the MUSCL scheme or a pure upwind extrapolation. The connectivity information required for this operation is provided by the pre-processor and is used directly in the solver. When MUSCL extrapolation similar to what is done on structured meshes is not possible due to the isotropicity of the cells, extrapolation is performed using gradients of primitive quantities to project variables to faces for the calculation of the flux. The gradients are calculated similar to those for the viscous fluxes. The extrapolation is performed by projecting the gradient to the face with a dot-product between the gradient $(\nabla \phi)$ and the vector that connects the upwind cell with the face of interest $(\vec{dr})$. This is analogous to the connectivity-based MUSCL scheme.

This, however, may lead to spurious oscillations of the solution due to the error-prone gradient employed, or in the presence of a discontinuity. To suppress such oscillations we employ a combination of a pressure-based smoothing function and the MUSCL scheme with a non-linear limiter $\psi(\cdot, \cdot)$. One of the arguments of ψ (that on the side of the gradient-based extrapolation) is multiplied by a dimensionless function f. The other argument is the difference of ϕ across the face. The dimensionless function is chosen to be $f(\tilde{p}) = \exp(-\frac{1}{2}\tilde{p}^2)$, where $\tilde{p}$ is the extrapolated pressure difference normalized by the smallest pressure on either side of the face.

4. Implicit Method of Solution

The implicit method requires that the right hand side of the equations is evaluated at the future time level. We can evaluate the inviscid fluxes at the future time level using the linearization

$$F_j'^{n+1} = F_j'^n + \left(\frac{\partial F'}{\partial U}\right)_j^n (U^{n+1} - U^n) = F_j'^n + A_j'^n \delta U^n$$

where $F'_j = (\vec{F}_c \cdot \hat{n})_j$ is the surface-normal inviscid flux at the face j, and is given in terms of the modified Steger-Warming Jacobians as:

$$F'^{n+1}_j = A'^n_{+j}(U^n_L + \delta U^n_L) + A'^n_{-j}(U^n_R + \delta U^n_R)$$

where A'_{+j}, A'_{-j} are the flux Jacobians at the face. This formulation is first-order accurate in space. In the second-order accurate formulation, the vectors U_L, U_R are computed based on a variable extrapolation method, and appropriate multipliers to δU_L, δU_R are introduced for consistency. We include the implicit approximation to the viscous flux by defining $\delta F'_v$ such that

$$F'^{n+1}_v = F'^n_v + \delta F'^n_v = F'^n_v + (R^{-1} M^n R)\frac{\partial}{\partial \eta}(N^n \, \delta U^n),$$

where M, R are the approximate viscous Jacobian and a rotation matrix respectively. The matrix N transforms from conserved variables to primitive variables and the derivative in η is simply a difference across a face. Similarly to the viscous and inviscid fluxes, the source term is linearized in time and a Jacobian for each cell is obtained.

The implicit equation represents a linear system of equations for the unknowns δU_i and is written as

$$\delta U_i + \frac{\Delta t}{V_i} \sum_{\text{faces}} \left[A'_+ \delta U_i + A'_- \delta U_o - (R^{-1} M R) N (\delta U_o - \delta U_i) \right]_j S_j - \Delta t \left(\frac{\partial W}{\partial U}\right)_i \delta U_i = \Delta U_i,$$

where ΔU_i is the right-hand side of the explicit equation times Δt, and we have assumed that the unit vector normal at each face is outward-pointing. The resulting system of equations for a structured discretization in 3D is a hepta-diagonal block-banded matrix. For such a system we can employ the Data-Parallel Line-Relaxation method to solve the system in parallel very efficiently. However, for general unstructured discretizations, constructing the system generally results in having no apparent banded-ness of the operator on the left-hand side, and solving the system directly is much more difficult.

4.1. Line-Relaxation

The implicit equation for each line of cells requires the solution of a block tri-diagonal system. The off-diagonal terms are relaxed by a series of relaxation sweeps. This procedure is straightforward on structured grids. Our implementation of the hybrid implicit method is more general in the sense that any type of element can be part of a particular line solution. Therefore, we are not limited to using the line-relaxation in regions where there are regular layers of cells.

4.2. Full-Matrix Data-Parallel Relaxation

The point-relaxation method is used to implement the line-relaxation method in the hybrid discretization. We use this method in regions of the domain where no line solutions are performed. The implicit operator for these cells consists only of the diagonal block-band, which is stored in an array. The operator is factored and stored on a per cell basis, which can be done efficiently. All off-diagonal terms are relaxed by a series of sweeps.

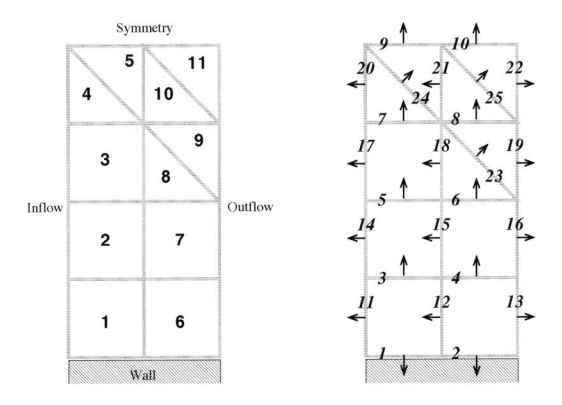

Figure 1. Example of a conceptual unstructured discretization.

4.3. Combined DPLR-FMDP Solution

We illustrate the use of the two methods with an example of a conceptual unstructured discretization as shown in Fig. 1. This simple two-dimensional example consists of 11 unstructured cells and has 4 different types of boundary conditions (Fig. 1a). The unit normal vector at each face is shown; normal vectors are outward-pointing at boundary faces (Fig. 1b). For simplicity, we combine all of the matrices in the implicit equation that are operating on δU_i with $\tilde{A}_i$ and those that operate on δU_o with $\tilde{B}_j^{\pm}$, where j corresponds to a face and the superscripted sign indicates the appropriate Jacobian based on the direction of the face normal. Then, the implicit equation at each time step n for this discretization is:

$$\begin{pmatrix} \tilde{A}_1 & \tilde{B}_3^- & & & \cdot & & \tilde{B}_{12}^+ & \cdot & & & \\ \tilde{B}_3^+ & \tilde{A}_2 & \tilde{B}_5^- & & & & & \tilde{B}_{15}^+ & & & \\ & \tilde{B}_5^+ & \tilde{A}_3 & \tilde{B}_7^- & & & & & \tilde{B}_{18}^+ & & \\ & & \tilde{B}_7^+ & \tilde{A}_4 & \tilde{B}_{24}^- & & & & & & \\ & & & \tilde{B}_{24}^+ & \tilde{A}_5 & & & & & \tilde{B}_{21}^+ & \\ \cdot & \tilde{B}_{12}^- & & & & \tilde{A}_6 & \tilde{B}_4^- & & & & \\ \cdot & \tilde{B}_{15}^- & & & & \tilde{B}_4^+ & \tilde{A}_7 & \tilde{B}_6^- & & & \\ & & \tilde{B}_{18}^- & & & & \tilde{B}_6^+ & \tilde{A}_8 & \tilde{B}_{23}^- & & \\ & & & & & & & \tilde{B}_{23}^+ & \tilde{A}_9 & \tilde{B}_8^- & \\ & & & & & & & & \tilde{B}_8^+ & \tilde{A}_{10} & \tilde{B}_{25}^- \\ & & & & \tilde{B}_{21}^- & & & & & \tilde{B}_{25}^+ & \tilde{A}_{11} \end{pmatrix} \begin{pmatrix} \delta U_1 \\ \delta U_2 \\ \delta U_3 \\ \delta U_4 \\ \delta U_5 \\ \delta U_6 \\ \delta U_7 \\ \delta U_8 \\ \delta U_9 \\ \delta U_{10} \\ \delta U_{11} \end{pmatrix} = \begin{pmatrix} \Delta U_1 \\ \Delta U_2 \\ \Delta U_3 \\ \Delta U_4 \\ \Delta U_5 \\ \Delta U_6 \\ \Delta U_7 \\ \Delta U_8 \\ \Delta U_9 \\ \Delta U_{10} \\ \Delta U_{11} \end{pmatrix}$$

Note that the implicit boundary conditions are folded into the diagonal band of the operator. In this very simple example, all cells with the exception of cell 10, which does not share faces with any boundary, have at least one type of boundary condition folded into their $\tilde{A}_i$ matrix. The terms in the implicit operator that are highlighted correspond to the two block tri-diagonal systems for each of the two line solutions. Note that line 1 could have easily been expanded to include cells 4 and 5, and similarly line 2 could have included cells 10 and 11, however in this example we use the FMDP method on these

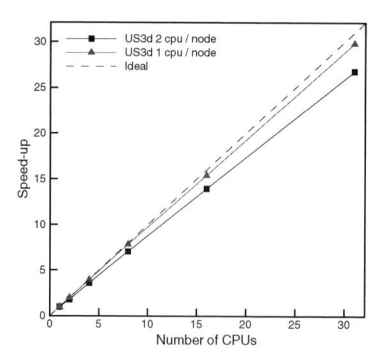

Figure 2. Parallel speed-up of the implicit solver versus the number of processors.

cells. Therefore, the shaded areas of the operator are factored together into an LU, and similarly the remaining $\tilde{A}_i$ blocks along the diagonal are factored individually. All the remaining off-diagonal ($\tilde{B}$) blocks are relaxed in a series of sweeps in exactly the same fashion for both line and point solutions. By applying this method of solution, the system is factored and no additional storage is required.

5. Partitioning and Parallelization

The parallelization of the solver is based on domain decomposition. The physical domain is split into a number of subdomains, which do not necessarily contain an equal number of finite volume cells of the unstructured discretization. An automatic procedure is needed in order to decompose the domain. Additionally, the amount of information that needs to be communicated across processors must be minimized and the overall amount of computation across processing elements must be balanced. The amount of communication across processors is directly related to the stencil used in the numerical method of solution, and it can be very large due to the arbitrariness of the stencil's support.

For unstructured solvers, typically, a graph partitioning algorithm (or library) is used for decomposing the domain based on certain criteria. In the case of the present solver, decomposition of the domain via graph partitioning is directly related to the choice of time integration method. Partitioning is performed in a way such that lines of cells that are part of a line solution in the DPLR method are not distributed across processors. We decompose the domain by making use of an existing graph partitioning library, Metis. [6]

6. Parallel Performance of the Solver

The parallel performance of the DPLR method for structured grids is very good, and simulations have shown that nearly perfect scaling can be achieved using this method on certain types of low-latency hardware.[2] However, because of the unstructured nature of the present solver the parallel performance of the code employing the DPLR method may degrade considerably. Thus, we performed a study to assess the parallel performance of the solver. The mesh used for the parallel performance test was partitioned in one grid direction. The simulations were run on a cluster of dual Pentium Xeon processors connected with a Myrinet fast interconnect. Figure 2 shows that the solver scales very well despite having some small load imbalance. We also see that when two CPUs per node are used, the solver runs about 8% slower. This is expected because the two CPUs share a single memory bus and performance is degraded due to doubling of the per-board memory bandwidth usage. Performance is greatly improved when only one processor per node is used.

REFERENCES

1. S. Yoon, and A. Jameson, "A Lower-Upper Symmetric Gauss-Seidel Method for the Euler and Navier-Stokes Equations," *AIAA Journal*, 26(8) (1987) 1025-1026.
2. M.J. Wright, G. V. Candler, and D. Bose, "Data-Parallel Line Relaxation Method for the Navier-Stokes Equations," *AIAA Journal,* 36(9) (1998) 1603-1609.
3. R.W. MacCormack, and G.V. Candler, "The Solution of the Navier-Stokes Equations Using Gauss-Seidel Line Relaxation," *Computers and Fluids*, 17(1) (1989) 135-150.
4. I. Nompelis, T.W. Drayna, and G.V. Candler, "Development of a Hybrid Unstructured Implicit Solver for the Simulation of Reacting Flows Over Complex Geometries," AIAA Paper No. 2004-2227, June 2004.
5. G.V. Candler, and R. W. MacCormack, "The Computation of Hypersonic Ionized Flows in Chemical and Thermal Nonequilibrium," *Journal of Thermophysics and Heat Transfer,* 5(3) (1991) 266-273.
6. G. Karypis, and V. Kumar, "Unstructured Graph Partitioning and Sparse Matrix Ordering System," http://www.cs.umn.edu/~metis, 1998.

Toward a MatlabMPI Parallelized Immersed Boundary Method

M. Garbey[a], F. Pacull[b]

[a]Dept of Computer Science, University of Houston

[b]Dept of Mathematics, University of Houston

The Immersed Boundary Method (IBM), originally developed by C.S. Peskin [1], is a very practical method of simulating fluid-structure interactions. It combines Eulerian and Lagrangian descriptions of flow and moving elastic boundaries using Dirac delta functions to distribute the force density. Incompressible Navier-Stokes and Elasticity theory can be unified by the same set of equations to get a combined model of the interaction.

There are numerous applications of the IBM in Bio-Engineering or in more general Computational Fluid Dynamics applications, which all are computationally expensive, especially in 3D. There are different ways to lower this cost and improve the method [2–4]. We have developed new methods [5] to improve the stability and the convergence order of the IBM. In this paper we will focus on domain decomposition techniques applied to the IBM in order to decrease the computation time. We will present our results in the context of parallel Matlab.

Matlab is a high-level technical computing language and interactive environment for algorithm development, data visualization, and numerical computation that is widely used by the Bio-Engineers. MPI, the library specification for message-passing, is a standard for a broadly based committee of implementers. MatlabMPI [6] is a set of Matlab scripts that implements a subset of MPI and allow any Matlab program to be run on a parallel computer.

Our goal is to be able to compute large scale simulations of fluid-structure interactions on a computer network, taking advantage of the interactivity of Matlab and the simplicity for coding of the IBM.

1. Algorithm

We started from our Matlab sequential implementation of the IBM, with finite differences and a uniform staggered mesh [7]. While we are using different semi-implicit or fully implicit codes, let us recall the basic projection scheme [8] at time step $n+1$ for the incompressible Navier-Stokes equations:

1- *Prediction*

$$\rho\left[\frac{V^* - V^n}{\Delta t} + (V^n.\nabla)V^n\right] - \nu\Delta V^n + \nabla P^n = F^n, \quad V_\Gamma^* = V_\Gamma^{n+1}, \tag{1}$$

2- *Pressure correction*

$$\Delta(\delta P)^n = \frac{\rho}{\Delta t}\nabla.V^*, \quad \left(\frac{\partial(\delta P)^n}{\partial \eta}\right) = 0, \tag{2}$$

$$P^{n+1} = P^n + (\delta P)^n,$$

3- *Correction*

$$\rho\left[\frac{V^{n+1} - V^*}{\Delta t}\right] + \nabla(\delta P)^n = 0. \tag{3}$$

Notations: V, P, ρ and ν are the velocity, pressure, density coefficient and viscosity coefficient of the fluid. F is the force term that contains that information about the moving immersed boundary. Δt is the time step. η is the normal outside vector to the domain boundary.

Now let us describe briefly the IBM, that is explicit here too. Let us call Ω the domain and Γ the immersed boundary, the IBM is based on two equations:

1- *Repartition of the force density on the Cartesian mesh*

$$F^n = \int_\Gamma f^n(s)\delta(x - X^n(s))\,ds, \tag{4}$$

2- *Motion of the immersed boundary*

$$\frac{X^{n+1}(s) - X^n(s)}{\Delta t} = \int_\Omega V^{n+1}\delta(x - X^n(s))\,dx. \tag{5}$$

The first equation is the evaluation of the force term that is plugged into the Navier-Stokes equations. The second one states that Γ is constrained to move at the same velocity V as the neighboring fluid particles.

Notations: δ is the shifted Dirac delta function. s is the curvilinear coordinate along Γ. $f(s)$ is the force density along Γ. $X(s)$ is the position of the moving boundary in the Cartesian mesh.

Now let us study which steps of this algorithm are the most expensive computationally, in order to target the parallelization on them.

2. Computational cost

In the projection scheme for the Navier-Stokes equations described above, only the pressure correction step (2) is computationally expensive, since a linear system has to be solved. In the boundary treatment process (4,5), we use numerous discrete Dirac delta functions distributed along the moving boundary as interpolating tools to switch from an Eulerian to a Lagrangian vector field or vise-versa. The arithmetic complexity of these two steps is proportional to the number of discrete points along the immersed boundary. We evaluated the CPU time spent on the pressure equation compared to the boundary treatment in a sequential 2D IBM test-case, the so called 2D "bubble" test-case: a closed elastic moving boundary is immersed inside an incompressible fluid. The domain is a closed square box. At the beginning the fluid velocity is null and the elastic boundary is stretched, with a large potential energy, as seen on figure 1. As one can observe on figure 2, the pressure equation is the most time consuming. It is important to specify that the program is written in Matlab, an interpreted language. We used the *cputime* command to measure the elapsed time. If we call N, the size of a side of the discrete 2D square mesh, the CPU time of (2) is of order N^3, (1) and (3) are of order N^2 while (4) and (5) are of order N.

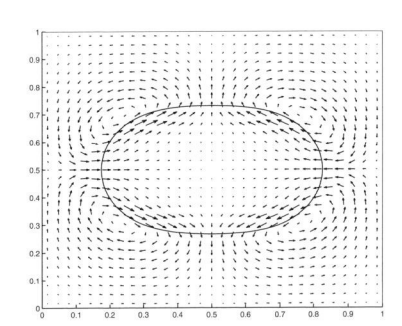

Figure 1. "Bubble" test-case. A closed elastic moving boundary is immersed inside an incompressible fluid

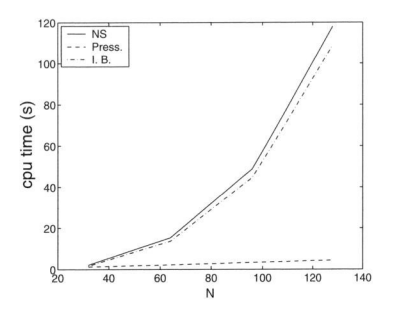

Figure 2. Maltlab CPU time in seconds with respect to N, the number of discrete points in each direction, for the fully explicit IBM/Navier-Stokes algorithm. NS: Navier-Stokes solver, Press: LU pressure solver, IB: Immersed boundary computations.

Consequently we will concentrate now on the parallelization of the pressure equation, that takes most of the CPU time.

3. The Parallel algorithm

The problem to solve is the following homogeneous Neumann problem :

$$\Delta P = RHS \text{ in } \Omega = [0,1]^2, \quad \frac{\partial P}{\partial \eta}|_{\partial \Omega} = 0. \tag{6}$$

We notice at first that the solution is defined up to a constant shift and that the right-hand side needs to satisfy a compatibility condition. We use the analytic additive Aitken-Schwarz algorithm [9], which is an excellent candidate to allow efficient distributed computing with slow networks. The rectangular uniform mesh is decomposed into an unidirectional partition of overlapping strip domains. The method is a post-process of the standard Schwarz method with an Aitken-like acceleration of the sequences of interfaces produced with the block-wise Schwarz relaxation. For simplicity, we restrict ourselves in this brief description of the algorithm to a decomposition of Ω into two overlapping subdomains: $\Omega = \Omega_1 \cup \Omega_2$ where $\Omega_1 = [0, x_r] \times [0,1]$ and $\Omega_2 = [x_l, 1] \times [0,1]$, $x_l < x_r$. The additive Schwarz algorithm is:

$$\Delta p_1^{n+1} = RHS \text{ in } \Omega_1, \quad \Delta p_2^{n+1} = RHS \text{ in } \Omega_2, \quad p_{1|\Gamma_1}^{n+1} = p_{2|\Gamma_1}^n, \quad p_{2|\Gamma_2}^{n+1} = p_{1|\Gamma_2}^n. \tag{7}$$

If we use a cosine expansion to describe the solution on the interface:

$$p(y)_{|\Gamma_i} = \sum_k \hat{p}_{|\Gamma_i}^k \cos(k\pi y) \; \forall k, \; i = 1 \text{ or } 2, \tag{8}$$

we observe that the cosine functions expansions of the solution on the interfaces Γ_1, Γ_2 provide a diagonalization of the trace transfer operator:

$$\left(p_{1|\Gamma_1}^n, p_{2|\Gamma_2}^n\right) \xrightarrow{T} \left(p_{1|\Gamma_1}^{n+1}, p_{2|\Gamma_2}^{n+1}\right). \tag{9}$$

As a matter of fact, $\Delta P = RHS$ decomposes onto a set of independent ODE problems:

$$\frac{\partial^2 \hat{p}_k(x)}{\partial x^2} - \mu_k \hat{p}_k(x) = \widehat{RHS}_k, \; \forall k. \tag{10}$$

Let us denote T_k the trace operator for each wave component of the interface:

$$\left(\hat{p}_{1|\Gamma_1}^{n,k} - \hat{P}_{\Gamma_1}^k, \hat{p}_{2|\Gamma_2}^{n,k} - \hat{P}_{\Gamma_2}^k\right) \xrightarrow{T_k} \left(\hat{p}_{1|\Gamma_1}^{n+1,k} - \hat{P}_{\Gamma_1}^k, \hat{p}_{2|\Gamma_2}^{n+1,k} - \hat{P}_{\Gamma_2}^k\right), \; \forall k. \tag{11}$$

The operators T_k are linear and the sequences $\{\hat{p}_{1|\Gamma_1}^n\}$ and $\{\hat{p}_{2|\Gamma_2}^n\}$ have linear convergence. If we call $\hat{\delta}_k^1$ and $\hat{\delta}_k^2$ the damping factors associated to the operators T_k and in the respective subdomains Ω_1 and Ω_2, we have:

$$\hat{p}_{1|\Gamma_2}^{n+1,k} - \hat{P}_{\Gamma_2}^k = \hat{\delta}_k^1 \left(\hat{p}_{2|\Gamma_1}^{n,k} - \hat{P}_{\Gamma_1}^k\right), \quad \hat{p}_{2|\Gamma_1}^{n+1,k} - \hat{P}_{\Gamma_1}^k = \hat{\delta}_k^2 \left(\hat{p}_{1|\Gamma_2}^{n,k} - \hat{P}_{\Gamma_2}^k\right). \tag{12}$$

These damping factors are computed analytically from the eigenvalues of the operators. We apply then the generalized Aitken acceleration separately to each wave coefficient in order to get the exact limit of the sequence on the interfaces based on the first Schwarz iterate. The solution at the interface can then be recomposed in the physical space from its discrete trigonometric expansion. After we get these limits we use the LU block solver to compute the exact solution over Ω. In order to get rid of the problem of the non-uniqueness of the solution we treat the zero mode aside and solve at first the equation with appropriate boundary conditions:

$$\frac{\partial^2 \hat{p}_0(x)}{\partial x^2} = \widehat{RHS}_0. \tag{13}$$

To summarize, the algorithm writes:

- step 1: solve the mode zero one dimensional system and subtract this mode from the right-hand side,
- step 2: compute analytically each damping factor for each wave number,
- step 3: perform one additive Schwarz iterate in parallel (a LU solver is used as a block solver),
- step 4: apply the generalized Aitken acceleration on the interfaces,
    - 4.1: compute the cosine expansion of the traces of p on the artificial interfaces for the initial condition and the first Schwarz iterate,
    - 4.2: apply the generalized Aitken acceleration separately to each wave coefficients in order to get the limit expressed in the cosine functions vector basis,
    - 4.3: transfer back the interface limit values into the physical space,
- step 5: compute the solution for each subdomain in parallel.

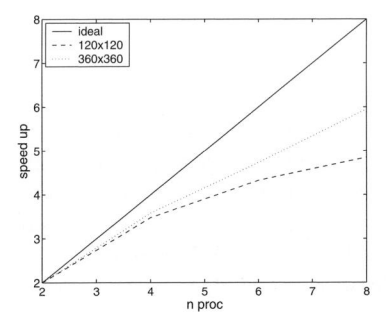

Figure 3. Parallel speed up for the pressure solver using an Aitken Schwarz method and a LU block solver and MatlabMPI. 8 way AMD Opteron (2GHz), 64 bit, 32 GB RAM. Two different problem sizes: 120 and 360.

Only two steps (3) and (5) of the algorithm are done in parallel, which are the most expensive ones since they consists in solving the system over the subdomains. As one can see on figure 3, the global performance of the method is encouraging. We can see that the results are better with the larger problem size since the ratio between communication and computation is then reduced. While the CPU time on figure 3 is measured between the beginning of step 2 and the end of step 5, let us look at the CPU time spent only on step 3 and step 5, respectively on figures 4 and 5.

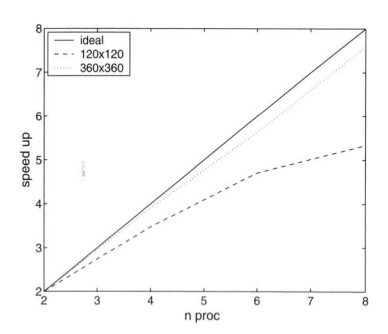

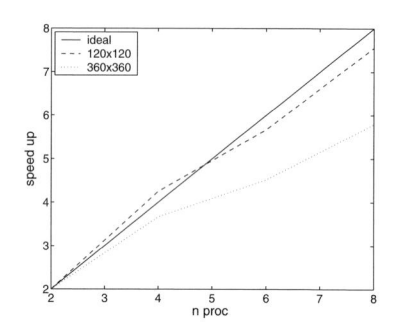

Figure 4. Matlab CPU time spent on step 3 of the algorithm. 8 way AMD Opteron (2GHz), 64 bit, 32 GB RAM.

Figure 5. Matlab CPU time spent on step 5 of the algorithm. 8 way AMD Opteron (2GHz), 64 bit, 32 GB RAM.

We can see that with a reasonably large problem size, the implementation of the third step of the algorithm has a satisfying speed up (figure 4). Step 1,2 and 4 are not parallelized. Step 4 CPU time is actually increasing with the number of subdomains since it contains a gathering of the interface data by one process and then a distribution of the updated interface data to the respective processes. The computational loads of these three steps are light, however, since they correspond to interface and not subdomain treatments. The results for the fifth step (figure 5) require some more investigation: they should be similar or even better than the ones from the first step. A better efficiency is expected with a larger problem size (360×360 compare to 120×120). Another aspect of the parallelization is that we gain computational time in the decomposition of the operator process: the cpu time is divided by the number of subdomains since the complexity of the decomposition is proportional to the bandwidth of the subdomains.

Now we study the numerical accuracy of this pressure solver in the IBM.

4. The IBM case

We use the parallel pressure solver code with a right-hand side similar to the ones we find in the IBM, and more specifically in the 2D "bubble" described in section 2. The singular part of the right-hand side in the pressure equation comes form the divergence of the force term, which is a collection of weighted shifted discrete Dirac delta functions along a closed curve. One can observe on figure 6 that even if we use regularized Dirac delta functions, we get a very stiff right-hand side, to which corresponds an irregular solution that you can see on figure 7.

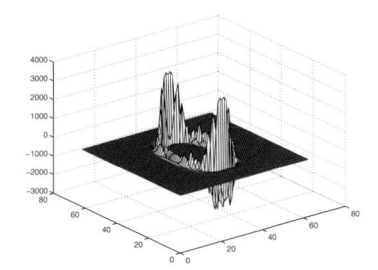

Figure 6. Typical right-hand side of a pressure equation in the 2D "bubble" test-case. Elasticity coefficient of the immersed boundary $\sigma = 10000$.

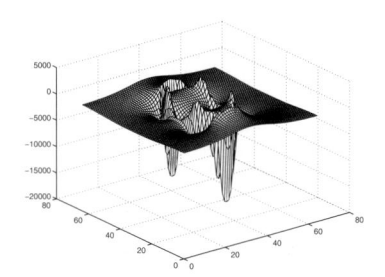

Figure 7. Discrete solution of the pressure equation at a time step in the 2D "bubble" test-case. Elasticity coefficient of the immersed boundary $\sigma = 10000$.

We note at first that the elapsed time and parallel efficiency are the same with this stiff right-hand side as with the smooth one used before. The problem is that the accuracy is deteriorated when the immersed boundary crosses the domain interfaces. In the Aitken-Schwarz algorithm, the solution at the interface is described with a cosine expansion, which introduces an error [10] at the membrane location due to the discontinuity of the pressure field shown on figure 7. For example in the "bubble" test-case with an elastic coefficient $\sigma = 10000$ stretched all across Ω and a problem size of 50×50, you can see the behavior of the error with respect to the number of subdomains on figure 8.
Remark: the error typically stays of the order of 10^{-14} if the "bubble" does not intersect the interfaces, no matter the number of interfaces.

5. conclusion

Some parallel efficiency is achieved by the Aitken-Schwarz algorithm on the pressure solver. The Matlab language allows a fast prototyping while the MatlabMPI toolbox provides the ability to solve large problems thanks to the parallelism.
There is an accuracy problem when the immersed boundary crosses the artificial interfaces. The solution contains a jump due to the singular right-hand side: it contains derivatives of discrete Dirac delta functions. This jump is hard to describe with the cosine expansion used in the Aitken acceleration technique.

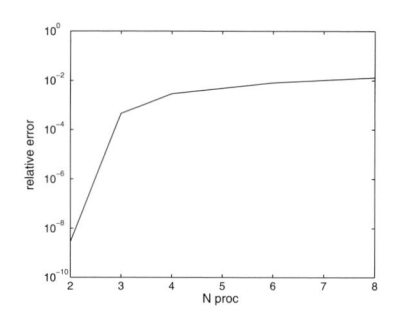

Figure 8. Relative error on the pressure solution using the Aitken-Schwarz algorithm with respect to the discrete solution. The Immersed boundary is stretched all over the domain so that each artificial interface crosses it. Size of the problem: 50×50. Elasticity coefficient of the immersed boundary $\sigma = 10000$.

REFERENCES

1. C. S. Peskin. The Immersed Boundary Method. Acta Numerica (2002), pp. 1-39, 2002.
2. E. Uzgoren, W. Shyy and M. Garbey. Parallelized domain decompositon techniques for multiphase flows. Proceedings of 2004 ASME Heat Transfer/Fluids Engineering Summer Conference, July 11-15 2004, Charlotte, North Carolina, 2004.
3. R. Leveque and Z. Li. The immersed interface method for elliptic equations with discontinuous coefficients and singular sources. SIAM Journal on Numerical Analysis, Volume 31 , Issue 4 (August 1994), pp.1019-1044, 1994.
4. Long Lee. Immersed Interface Methods for Incompressible Flow with Moving Interfaces, PhD Thesis. University of Washington, 2002
5. F.Pacull and M.Garbey. A Numerical Experimental Study of the Immersed Boundary Method. Proceeding of the Sixteenth International Domain Decomposition Conference, New York University January 12-15th 2005, 2005
6. J.Kepner and S. Ahalt, MatlabMPI. Journal of Parallel and Distributed Computing, vol.64, Issue 8,August 2004, pp.997-1005, 2004.
7. F. H. Harlow and E. Welch. Numerical calculation of time-dependent viscous incompressible flow of fluids with free surface. Phys. Fluids, vol. 8, pp. 2182-2189, 2004.
8. A.J. Chorin. Numerical solution of the Navier-Stokes equations. Math. Comp., vol. 22, pp. 745-762, 1968.
9. M.Garbey and D.Tromeur Dervout, On some Aitken like acceleration of the Schwarz Method. Int. J. for Numerical Methods in Fluids, 40 (12), pp. 1493-1513, 2002.
10. B. Fornberg, A Practical Guide to Pseudospectral Methods. Cambridge University Press, 1998.

Dynamics of multiphase flows via spectral boundary elements and parallel computations

Yechun Wang[a], Walter R. Dodson[a] and P. Dimitrakopoulos[a] * †

[a]Department of Chemical and Biomolecular Engineering,
University of Maryland, College Park, Maryland 20742, USA

We present the efforts of our research group to derive an optimized parallel algorithm for the efficient study of interfacial dynamics in Stokes flow and/or gravity. Our approach is based on our high-order/high-accuracy spectral boundary element methodology which exploits all the advantages of the spectral methods and the versatility of the finite element method. Our numerical code has been parallelized on shared-memory multiprocessor computers and thus we are able to utilize the computational power of supercomputers. Our parallel interfacial algorithm facilitates the study of a wide range of problems involving three-dimensional interfaces in Stokes flow and/or gravity.

1. INTRODUCTION

The dynamics of droplets and bubbles in infinite media or in restricted geometries under low-Reynolds-number flows and gravity is a problem of great technological and fundamental interest. These systems are encountered in a broad range of industrial, natural and physiological processes. Chemical engineering applications include enhanced oil recovery, coating operations, waste treatment and advanced materials processing. Pharmaceutical applications include emulsions which serve as a vehicle for the transport of the medical agent through the skin. An additional application is the blood flow in microvessels.

The industrial and biological applications of multiphase flows motivate the development of a parallel algorithm which can be employed to efficiently study the dynamics of three-dimensional interfacial problems. In this study we present the efforts of our research group to derive an optimized parallel algorithm for this problem based on a high-order/high-accuracy spectral boundary element methodology [1,2]. The main attraction of this approach is that it exploits all the benefits of the spectral methods (i.e. exponential convergence and numerical stability) and the versatility of the finite element method (i.e. the ability to handle the most complicated geometries) [3,4]. In addition, it is not affected by the disadvantage of the spectral methods used in volume discretizations; namely, the requirement to deal with dense systems, because in boundary integral formulations the resulting systems are always dense, independent of the form of discretization. The code

---

*Corresponding author, email: dimitrak@eng.umd.edu
†This work was supported in part by the National Science Foundation and the National Center for Supercomputing Applications (NCSA) in Illinois. Acknowledgment is made to the Donors of the American Chemical Society Petroleum Research Fund for partial support of this research.

has been parallelized on shared-memory supercomputers, such as the SGI Origin 2000, using OpenMP directives for the calculation of the system matrix, and highly optimized routines from the LAPACK system library for the solution of the dense system matrix. These properties result in a great computational efficiency which facilitates the study of a wide range of problems involving interfaces in Stokes flow.

2. MATHEMATICAL FORMULATION

We consider a three-dimensional fluid droplet in an infinite surrounding medium or a confined geometry; the droplet may be free-suspended or attached to the solid boundaries. The droplet size is specified by its volume V_0 or equivalently by the radius a of a spherical droplet of volume $(4/3)\pi a^3 = V_0$. The droplet (fluid 1) has density ρ_1 and viscosity $\lambda\mu$, while the surrounding medium (fluid 2) has density ρ_2 and viscosity μ. The gravitational acceleration is g, while the surface tension γ is assumed constant. Far from the droplet, the flow approaches the undisturbed flow $\boldsymbol{u}^\infty$ (e.g. shear or Poiseuille flow) characterized by a shear rate G.

Excluding inertial forces, the governing equations in fluid 2 are the Stokes equations and continuity,

$$\boldsymbol{\nabla}\cdot\boldsymbol{\sigma} = -\boldsymbol{\nabla} p + \mu\nabla^2 \boldsymbol{u} = 0 \tag{1}$$

$$\boldsymbol{\nabla}\cdot\boldsymbol{u} = 0 \tag{2}$$

while in the droplet, the same equations apply with the viscosity replaced by $\lambda\mu$.

The boundary conditions on the solid walls and at infinity give

$$\boldsymbol{u} = 0 \quad \text{on} \quad \text{solid walls} \tag{3}$$

$$\boldsymbol{u} \to \boldsymbol{u}^\infty \quad \text{as} \quad r \to \infty \tag{4}$$

At the interface, the boundary conditions on the velocity $\boldsymbol{u}$ and surface stress $\boldsymbol{f}$ are

$$\boldsymbol{u}_1 = \boldsymbol{u}_2 \tag{5}$$

$$\Delta \boldsymbol{f} = \boldsymbol{f}_2 - \boldsymbol{f}_1 = \gamma(\boldsymbol{\nabla}\cdot\boldsymbol{n})\boldsymbol{n} + (\rho_2 - \rho_1)(\boldsymbol{g}\cdot\boldsymbol{x})\boldsymbol{n} \tag{6}$$

Here the subscripts designate quantities evaluated in fluids 1 and 2, respectively. The surface stress is defined as $\boldsymbol{f} = \boldsymbol{\sigma}\cdot\boldsymbol{n}$, and $\boldsymbol{n}$ is the unit normal which we choose to point into fluid 2. The pressure as defined in $\boldsymbol{\sigma}$ is the dynamic pressure, hence the gravity force is absent from the Stokes equations and appears in the interfacial stress boundary condition.

For *transient* problems, the velocity field must satisfy an additional constraint - the kinematic condition at the interface

$$\frac{d\boldsymbol{x}}{dt} = (\boldsymbol{u}\cdot\boldsymbol{n})\boldsymbol{n} \tag{7}$$

For *equilibrium shapes* under flow conditions and/or gravity, the kinematic condition at the interface becomes

$$(\boldsymbol{u}\cdot\boldsymbol{n})\boldsymbol{n} = 0 \tag{8}$$

The magnitude of interfacial deformation due to viscous stress or gravity is given by the capillary number Ca and Bond number B_d, respectively, which are defined by

$$Ca = \frac{\mu G a}{\gamma} \qquad B_d = \frac{(\rho_1 - \rho_2)g a^2}{\gamma} \qquad (9)$$

Note that the capillary number Ca represents the ratio of viscous forces to surface tension forces while the Bond number B_d represents the ratio of gravitational forces to surface tension forces. The problem also depends on the viscosity ratio λ and geometric dimensionless parameters in the case of solid boundaries. For fluid volumes in contact with solid substrates, additional conditions are required to prescribe the interface shape in the vicinity of the contact lines as discussed in our earlier publication [2]; these conditions introduce additional dimensionless parameters.

We emphasize that, although the governing equations and boundary conditions are linear in $\boldsymbol{u}$ and $\boldsymbol{f}$, the problem of determining interfacial shapes constitutes a nonlinear problem for the unknown interfacial shape; i.e. the velocity $\boldsymbol{u}$, stress $\boldsymbol{f}$ and curvature $\boldsymbol{\nabla} \cdot \boldsymbol{n}$ are nonlinear functions of the geometrical variables describing the interface shape. For fluid volumes in contact with solid boundaries, the boundary conditions at the contact line involve the contact angle and thus constitute nonlinear functions of the unknown interfacial shape as well [2].

3. PARALLEL INTERFACIAL ALGORITHM

To solve the interfacial problem described in section 2, we transform the partial differential equations, Eqs.(1) and (2), which are valid in the system volume, into boundary integral equations valid on the surface of the volume [5,2]. This transformation results in a great reduction in CPU time, since a fully three-dimensional problem can be described and solved using only two (curvilinear) coordinates. For the case of a free-suspended droplet in an infinite medium, the velocity at a point $\boldsymbol{x}_0$ on the drop surface S_B is given by

$$4\pi\mu(1+\lambda)\,\boldsymbol{u}(\boldsymbol{x}_0) - 4\pi\mu\,\boldsymbol{u}^\infty(\boldsymbol{x}_0) =$$
$$-\int_{S_B} [\boldsymbol{S} \cdot (\Delta \boldsymbol{f} - \boldsymbol{f}^\infty) - \mu \boldsymbol{T} \cdot ((1-\lambda)\boldsymbol{u} - \boldsymbol{u}^\infty) \cdot \boldsymbol{n}](\boldsymbol{x})\,\mathrm{d}S \qquad (10)$$

where $\boldsymbol{S}$ is the fundamental solution for the three-dimensional Stokes equations and $\boldsymbol{T}$ the associated stress defined by

$$S_{ij} = \frac{\delta_{ij}}{r} + \frac{\hat{x}_i \hat{x}_j}{r^3} \qquad T_{ijk} = -6\frac{\hat{x}_i \hat{x}_j \hat{x}_k}{r^5} \qquad (11)$$

where $\hat{\boldsymbol{x}} = \boldsymbol{x} - \boldsymbol{x}_0$ and $r = |\hat{\boldsymbol{x}}|$. Similar equations hold in the presence of solid boundaries and for drop suspensions [2,6].

In contrast to most researchers in this area who employ low-order methods (e.g. see [7–10]), we solve the resulting boundary integral equations employing a (high-order) spectral boundary element method [1,2]. In particular, each boundary is divided into a moderate number of spectral elements as shown in Figure 1. On each element the geometric and physical variables are discretized using Lagrangian interpolation based on the zeros of

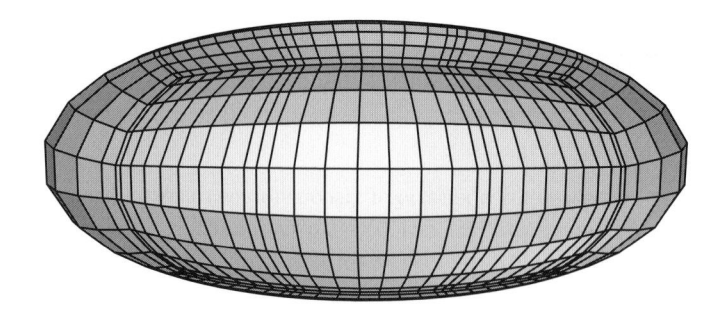

Figure 1. Discretization of a drop surface into $N_E = 14$ spectral elements. The figure illustrates Gauss-Lobatto Legendre distribution of nodal lines with $N_B = 10$ spectral points in each direction.

orthogonal polynomials of Gauss or Gauss-Lobatto type [3]. This is equivalent to an orthogonal polynomial expansion and yields the spectral convergence associated with such expansions. The discretizations are substituted into the appropriate boundary integrals and quadratures evaluated using adaptive Gaussian quadrature.

In order to determine equilibrium fluid interfaces in Stokes flow and/or gravity, we have developed an efficient, Jacobian-free, Newton method based on our spectral boundary element method. This method has proved to be a robust algorithm, of high accuracy and extreme efficiency, valid to study the most complicated three-dimensional problems [2]. To determine transient interfacial shapes, we combine our optimized parallel spectral boundary element algorithm with a time integration of the kinematic condition at the interface, Eq.(7).

The main attraction of our interfacial algorithm is the fact that it exploits all the benefits of the spectral methods, i.e. exponential convergence and numerical stability with the versatility of the finite element method, i.e. the ability to handle the most complicated geometries. In addition, it is not affected by the disadvantage of the spectral methods used in volume discretization; namely, the requirement of dealing with dense systems, because in boundary integral formulations the resulting systems are always dense, independent of the form of the discretization.

We emphasize that our interfacial algorithm shows the exponential convergence of the spectral methodology in any interfacial problem. For example, the exponential convergence in the numerical accuracy as the number of the employed spectral points $N = N_E N_B^2$ increases is clearly evident at the geometric properties of a given shape such as the computed curvature shown in Figure 2. The difference in the accuracy between our spectral algorithm and low-order interpolation methods is dramatic. For example, Zinchenko, Rother and Davis [8] in their Figure 5 reported an error $\approx 4 \times 10^{-2}$ for $N = 5120$; our algorithm shows an error $\approx 1 \times 10^{-10}$ for $N = 3456$.

To be able to access the computational power of supercomputers, our numerical code has been parallelized on shared-memory multiprocessor computers (such as the SGI Origin 2000) by employing OpenMP directives for the calculation of the system matrix, and

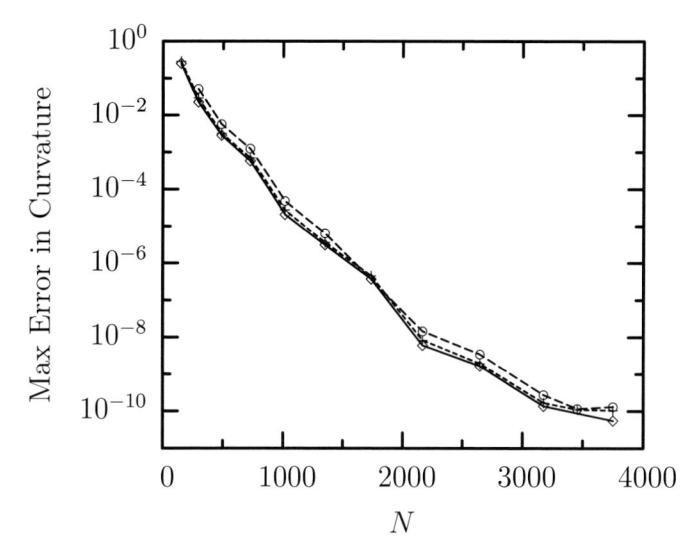

Figure 2. Exponential convergence in the maximum absolute error of the computed curvature as the number of spectral points N increases for different spheroids: ———, $a = b = c = 1$; - - - -, $a = b = 1, c = 0.4$; – – – –, $a = 1, b = c = 0.4$.

highly optimized routines from the LAPACK system library for the solution of the dense system matrix. Multiprocessor runs exploit the parallel nature of calculating the system matrices described by the boundary integral equation, Eq.(10). This results in an overall very good parallel efficiency as shown in Table 1. We emphasize that the size of the problems tested is rather small involving 10–40 spectral elements. Higher efficiency is expected for more complicated problems such as the ones involving the interaction of many droplets.

We emphasize that, to be able to achieve the parallel properties shown in Table 1, it is imperative that the computational load is distributed evenly among the processors. For this, we distribute the load using an "interleave" schedule with a small chunk size, e.g. we set the environmental variable `OMP_SCHEDULE` to `STATIC,1` (see OpenMP API [11]). On the other hand, an uneven load can result in poor parallel performance. For example, a "simple" schedule which divides the load into large chunks based on the number of iterations in the associated `OMP PARALLEL DO` loop produces an efficiency of nearly 75% for the "Integration" on two processors which is much smaller than the optimal efficiency of 99.26% shown in Table 1. Non-optimal load distribution also worsens the efficiency of the systems' solution, i.e. the performance of the parallel LAPACK library.

The parallelization of our algorithm can be performed in two ways: the first way involves all the spectral points on each element while the second one involves all the spectral points on all elements. Considering that each element has $N_B \times N_B$ spectral points (with a typical value of $N_B = 10$) while N_E is the number of the spectral elements (with a typical value of $N_E = 10$–40 for problems with one droplet and much higher values for many-drop

Table 1
Efficiency versus the number of processors N_p for the calculation ("Integration") and the solution ("Solution") of the system matrices resulting from our parallel spectral boundary element algorithm on the shared-memory SGI Origin 2000 provided by the National Center for Supercomputing Applications (NCSA) in Illinois. Note that efficiency denotes the ratio of the wall time for the serial execution T_s to that for the parallel execution T_p multiplied by the number of processors, i.e. efficiency $\equiv T_s/(T_p N_p)$.

N_p	Integration (%)	Solution (%)
1	100.00	100.00
2	99.26	89.62
4	96.72	75.43
8	89.91	72.34

problems), it is easy to realize that by employing the first way of parallelization we may use a moderate number of processors while by employing the second way of parallelization we have the ability to exploit a very high number of processors, if available. For many-drop problems (i.e. study of emulsions and foams), the parallelization can also involve the different drops/bubbles (or teams of them).

To optimize further our algorithm we employ highly-optimized BLAS routines as well as cache optimization. Exploiting the levels n of symmetry of a problem reduces the memory requirements by a factor of 2^n, the computational time for the system matrices by a factor of $2n$ and the time for the direct solution of the linear systems by factor of 2^n.

With this optimized parallel algorithm, we have the ability to study in detail a wide range of problems involving fully three-dimensional interfaces in Stokes flow and/or gravity. For example, in Figure 3 we provide the interfacial dynamics for viscosity ratio $\lambda = 0.2$ in a planar extensional flow near the critical conditions, i.e. near the flow rate at which equilibrium interfacial shapes cease to exist. The critical capillary number (i.e. $Ca \approx 0.155$) is in excellent agreement with experimental findings [12]. Observe that below the critical flow rate, the droplet deformation reaches equilibrium while above it the droplet continues to deform with time.

4. CONCLUSIONS

In this paper, we have presented the efforts of our research group to derive an optimized parallel algorithm so that we are able to efficiently determine interfacial dynamics in Stokes flow and/or gravity. By employing the boundary integral formulation, the problem dimensionality is reduced by one. In addition, our high-order/high-accuracy spectral boundary element approach results in great benefits including exponential convergence, numerical stability and ability to handle the most complicated geometries. The exponential convergence of our spectral methodology results in additional savings in computational time since for a desired accuracy we can use a coarser grid compared to that employed by low-order boundary methods. Our numerical code has been parallelized on shared-memory supercomputers, such as the SGI Origin 2000, using OpenMP directives

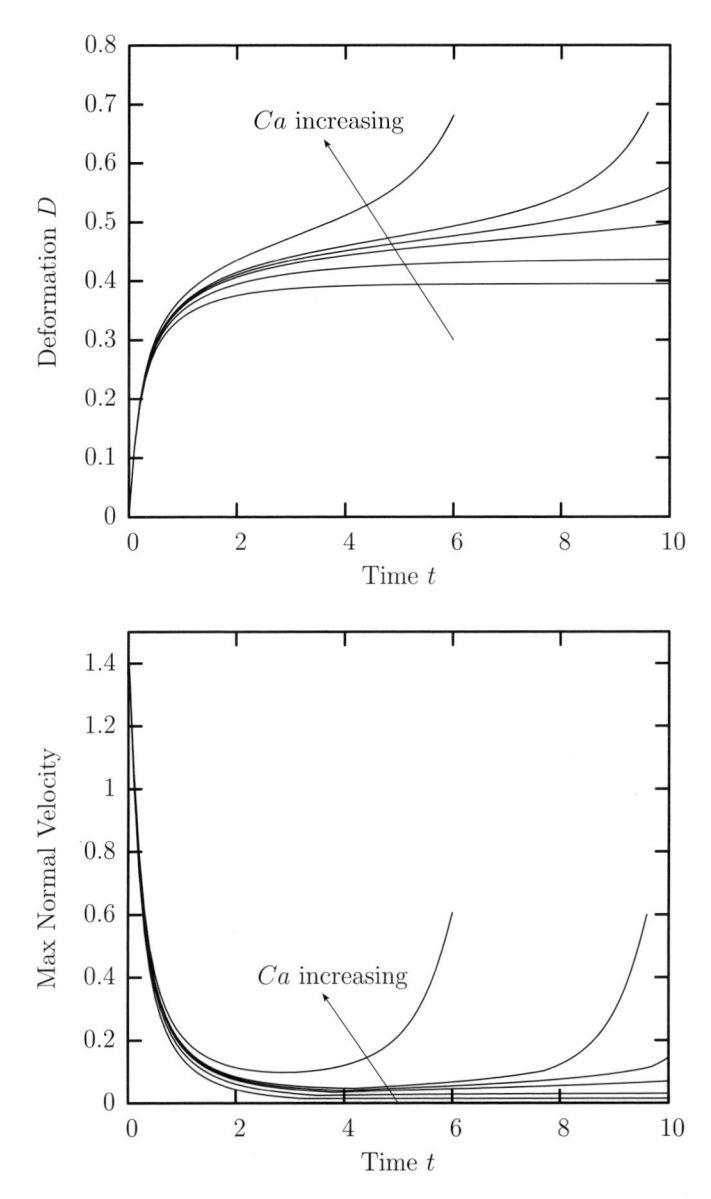

Figure 3. Dynamics near the critical point for a droplet with viscosity ratio $\lambda = 0.2$ in a planar extensional flow $\boldsymbol{u}^\infty = G(x, -y, 0)$. (a) Droplet deformation D versus time t. (b) Maximum normal velocity versus time t. The capillary number is $Ca = 0.15, 0.155, 0.158, 0.159, 0.16, 0.165$. Note that $D \equiv (L - S)/(L + S)$ where L and S are the droplet's length and width, respectively.

for the calculation of the system matrix, and highly optimized routines from the LAPACK system library for the solution of the system matrices. This enables us to utilize the computational power available at supercomputer centers including the National Center for Supercomputing Applications (NCSA) in Illinois. All these properties result in a great computational efficiency which facilitates the study of a wide range of problems involving fully three-dimensional interfaces in Stokes flow and/or gravity.

REFERENCES

1. G. P. Muldowney and J. J. L. Higdon, J. Fluid Mech. 298 (1995) 167.
2. P. Dimitrakopoulos and J. J. L. Higdon, J. Fluid Mech. 377 (1998) 189.
3. C. Canuto, M. Y. Hussaini, A. Quarteroni and T. A. Zang, Spectral Methods in Fluid Dynamics, Springer, 1988.
4. Y. Maday and A. T. Patera, in State of the Art Surveys in Computational Mechanics, A. K. Noor and J. T. Oden (eds.), ASME, 1989.
5. C. Pozrikidis, Boundary Integral and Singularity Methods for Linearized Viscous Flow, Cambridge University Press, 1992.
6. A. Z. Zinchenko and R. H. Davis, J. Fluid Mech. 455 (2002) 21.
7. M. Loewenberg and E. J. Hinch, J. Fluid Mech. 321 (1996) 395.
8. A. Z. Zinchenko, M. A. Rother and R. H. Davis, Phys. Fluids 9 (1997) 1493.
9. C. Pozrikidis, J. Comp. Phys. 169 (2001) 250.
10. I. B. Bazhlekov, P. D. Anderson and H. E. H. Meijer, Phys. Fluids 16 (2004) 1064.
11. OpenMP Architecture Review Board (eds.), OpenMP Application Program Interface, 2005 (available at `http://www.openmp.org`).
12. B. J. Bentley and L. G. Leal, J. Fluid Mech. 167 (1986) 241.

SDLB - Scheduler with Dynamic Load Balancing for Heterogeneous Computers

Stanley Y. Chien*, Lionel Giavelli, Akin Ecer, and Hasan U. Akay

Purdue School of Engineering and Technology
Indiana University-Purdue University Indianapolis
Indianapolis, Indiana 46202 USA
E-mail: schien@iupui.edu

Key words: job scheduler, dynamic load balancing

Abstract: SDLB is a distributed job scheduler that can schedule jobs on multiple clusters of computers with dynamic load balancing ability. This paper describes newly added features to SDLB. SDLB supports computers running in the dedicated mode, the batched mode and multi-user interactive mode. A user can submit single jobs or parallel jobs to the specified machines or let the scheduler to find the best suitable machines for the jobs. Installed in each cluster, an SDLB can negotiate with other SDLBs for acquiring computation resources. The scheduler also has the fault tolerance capability. When there is a failure in part of the computers or computer software that support a particular application job, the scheduler can detect the hanging application job and automatically restart it on other working computers. If the application job can periodically store the checkpoint data, the job will be restarted from the latest checkpoint; otherwise, the job will be restarted from the beginning. SDLB also has an event logger component that keeps the record of all inter-process communication and allows the system developer to selectively read the logged information at different abstraction level for debugging.

1. INTRODUCTION

Many job schedulers support job dispatching on computers running in dedicated mode and batched mode [1-6]. These schedulers perform load balancing from the view of the system and assume that each computer process is independent from other processes. These schedulers do not take parallel applications' information into the consideration of load balancing and hence cannot provide the best load distribution from user's point of view. We have developed a dynamic load balancer that provides load balancing for jobs with closely coupled parallel processes based on information of system load and the parallel application [7, 8] within a computer cluster. We also developed a scheduler on top of the dynamic load balancer for supporting parallel process on multiple computer clusters [9]. This scheduler does load distribution in a two-step process. The first step is to allocate the number of computers needed by the submitted job and the second step makes an optimal load distribution for user's application within the given set of allocated computers.

In this paper, we will describe the new functional features added to SDLB. The enhancement is focused on three main areas. First, a more effective job-scheduling scheme for computers running in batched and interactive modes is added to the scheduler. Secondly, fault tolerance features are added to the scheduler in order to detect and recover faults in the load balancer and in application job execution, if possible. Thirdly, an event logger is added to the scheduler for troubleshooting.

2. SYSTEM ORGANIZATION

The organization of SDLB is shown in Figure 1. The command flow and information flow of SDLB is shown in Figure 2. SDLB consists of four components: a scheduler, dynamic load balancing agents, system agents, and an event logger. The system agent is a demon process that runs continuously on all computers. A DLB agent is a process created and responsible for each user job submitted and is responsible for dynamic load balancing of the job, detecting and recover the fault for all the processes it creates if possible. The DLB agent launches a job agent that is responsible for starting and stopping the job and the fault tolerance of the application job. The event logger can store all the communication between all components of SDLB. A scheduler component is installed on each computer cluster. It is responsible for allocating computers for a job, creating the DLB agent for the job and detecting/recovering the fault if DLB agent cannot do it. Each Scheduler can request computer resources from other schedulers and uses the resources if the request is granted.

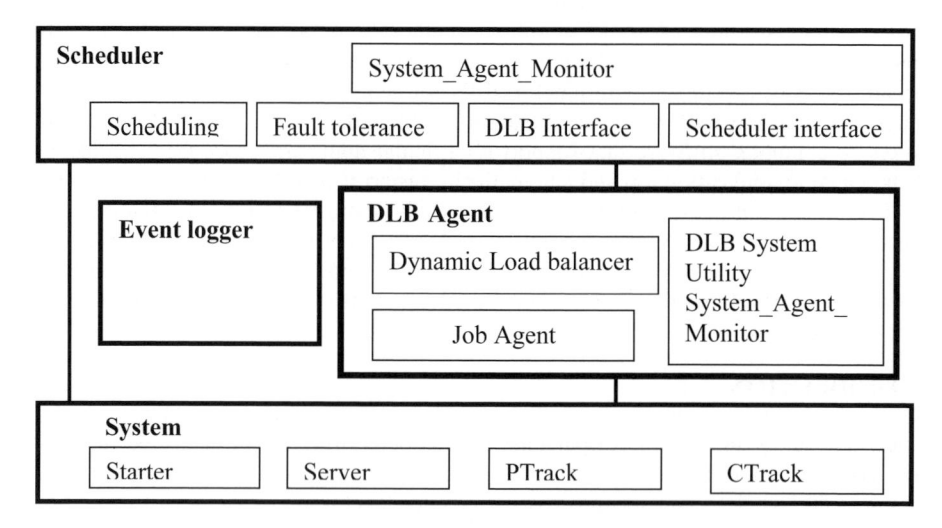

Figure 1. The organization of SDLB

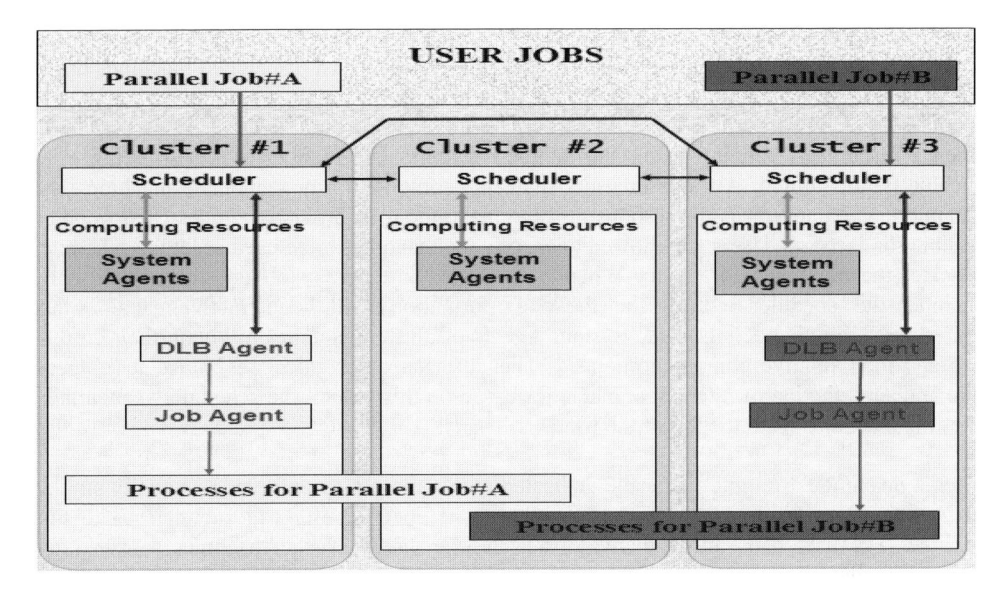

Figure 2. The information flow and commanding chain of SDLB

3. SCHEDULING

Our definition of scheduling refers to the way that the computers are selected from the available ones and attributed to a newly submitted user job. A computer, from user's perspective, is running in one of three operating modes: the batched mode, the interactive (multi-user) mode, and the dedicated mode. Each mode has its own advantages. The system administrator decides the operating mode for each machine. We improved SDLB so that it supports all three operating modes efficiently and user-friendly. The scheduler initially uses the single job and parallel load information provided by the System Agent to do the initial load distribution. Before scheduling a new job, SDLB sorts the available machines according to the effective computation speed that is a function of the computation load and the relative speed of the machines. This step ensures that the best available machines are always attributed to a new job. Moreover, in case of a machine crashing or getting disconnected from the network, the fault tolerance triggered job rescheduling may get some faster machines to the job. Those faster machines may have been highly loaded previously so they were not chosen for the job initially.

In batched mode, only one job can run on a machine at a time and the machine is temporarily dedicated to a job. Jobs get the machines on a first-come first-served basis. SDLB accepts a job-scheduling request only if the requested resource can be satisfied. During job scheduling, if a job cannot get as many machines as required, it will hold the machines and stay in a pending state until all machines are available. In job submission the user can ask for specific machines or just specify the number of machines needed. When a batched job is submitted without specifying the machines, the scheduler refreshes the load information on all the batched machines under the control of the scheduler by connecting to the System Agent to obtain the current load and the relative speed of each machine. Then with the information obtained it calculates the relative effective speed and sorts the list of batched machines based on this value. The scheduler sequentially takes the machines from the top of the list.

When a machine is running under the interactive mode, the number of jobs running on a machine is not limited. All jobs or users share the computation power of the machine. The dynamic load balancer has the responsibility to make the distribution of processes as good as possible for each job. Users are allowed to specify the machines to be used or let the scheduler to select the machines for a job. When a submitted interactive job does not specify the machines, the scheduler first refreshes the load information of all the interactive machines that a job can use by connecting to the System Agents of all the machines to obtain their current load and their relative computation speed: Then the scheduler calculates the relative speed ratio and sorts the list of interactive machines based on this value. The scheduler sequentially takes the machines from the top of the list.

When a machine is running under the dedicated mode, the policy enforced is actually similar to the interactive mode, plus the fact that only one specific user can submit jobs to this machine. The user can run as many jobs as he/she wants on the machines reserved for him/her. Since the user reserves the machines, the scheduler only verifies that the machines are in the right mode and that they are reserved for the current user. The policy is that if a user reserves a given number of machines then it is assumed that he/she wants to use all of them.

4. FAULT TOLERANCE

Fault tolerance is an essential feature for distributed software since the chance of hardware and software failure is much higher in distributed and parallel computing due to the large number of hardware and software components used and long execution time of software. Parallel programs are also sensitive to their environment since they execute many input/output operations. Hence, the parallel applications are more likely to crash in this environment. It is desirable that the failure of one or more components of the system should not cause failure of the job execution. Therefore, distributed software must be able to handle network crashes caused by a hardware problem, a system misconfiguration, traffic jam on the network, software problem or by users. SDLB itself may also crash due to its big number of components distributed on different machines and constantly expecting inputs from each other. It is one of the most important reasons of implementing fault tolerance since a user will not want to use an unreliable tool for their applications. SDLB provides detection and recovery for faults in hardware, and software tools. All serial jobs, loosely coupled parallel jobs, and tightly coupled parallel jobs, are supported by the fault tolerance.

Handling the failures of a system means detecting those failures in the first place. Generally the system will notice a failure when a component is unable to contact another component. There is an evidence for the situation whether the connection might have failed or reached a timeout. In the second case, the failures need to be handled in order to reduce or limit the damages (loss of data or deterioration in the performance of the system). Executing under SDLB, the user will not need to handle the fault and SDLB is able to recover from the fault completely.

SDLB is composed of many components as shown in section 2. In an ideal case, SDLB would provide fault tolerance of all the components, such as machines, network, and the user application. The exhaustive list of all the possible sources of faults in SDLB includes: Scheduler, DLB Agent, Job Agent, System Agent, System Agent Rescuer, Communication Tracker, Process Tracker, the user commands interface program, the user application, and the hardware problem. In SDLB, fault tolerance for all components except for the Scheduler and the user commands is provided.

The DLB agent handles the crash of the user application program on its own. When starting the application job, the DLB agent creates a CheckJobStatusThread object which uses the information collected by the PTrack to make sure that the parallel processes are always running on all the assigned machines. If none of the application processes are detected over a predetermined period, the user application is considered crashed. If none of the application processes are shown as active for a predetermined period, the application process is considered as hung (deadlock). In both cases the DLB agent will stop the job, stop the job agent, and then restart the job agent. This process is entirely local to the DLB agent and the scheduler remains blind to this event.

The fault tolerance for serial jobs is handled in the same way as for parallel jobs except for the fact that a serial job requires only one computer. Serial jobs use the same tools designed for parallel jobs. Each serial job is associated with a DLB Agent and a Job Agent. All the fault tolerance features have been successfully tested for a serial job.

SDLB cannot provide fault tolerance for user commands. Most faults in user commands are due to an improper use of the command, wrong arguments, a wrong configuration file, or the scheduler not running on the machine. Although SDLB can detect but has no mean recover from the errors in these cases. SDLB does not provide fault tolerance in case of a crash of the central SDLB component either. However, if this happens, the jobs being executed currently will not be affected but they will lose some of the fault-tolerance features. The pending jobs will be lost. To recover from this fault the system administrator has to restart the central SDLB component. It is possible to do this automatically using a third party program like the UNIX system's 'crontab' which can be set up to check periodically if a program is running and to start it otherwise.

5. EVENT LOGGING SYSTEM

SDLB is large software. It contains many threads running on many machines at the same time. Each thread has its own behavior and reacts to events diferently. It is not possible to develop a project of this size if one does not have an easy way to debug and trace its execution. Therefore, a Log System that provides convenient tools and methods is needed for the developer.

The following information needs to be logged: (1) All exceptions with a dump of the stack trace; (2) The parts of the code most subjective to crash: network communications, input and output operations, and the essential information before attempting any action. For example, before communication we should log the actors, the IP addresses accessed, the service called, and the nature of the message. This will allow the developer to know the exact context of the crash, so he/she will be able to reproduce it and correct it more easily. (3) The main function calls and events, between modules, and important steps of the process should be logged generally. Those logs would permit an easier understanding and reading of the logs and thus it will be easier to find the origin of a potential error.

It is desirable to have all the threads and different parts of the scheduler to be able to write information to the same log. The log routines must be accessible at all time and be able to handle several requests simultaneously. The logging must be very light and easy to use. There must be a convenient way to disable all loggings. The information contained in theses logs must be exploited to the maximum by a developer in the easiest way as possible. A fast and user-friendly interface should be provided since there will be thousands lines of logs. The developer should be able to view only the logs he/she is interested such as the logs related to a specific job, user, machine, action, and part of the scheduler.

The most convenient way to meet the requirements was to develop an event logger that is independent yet included in SDLB. Being independent, this event logger can also be included in the all component of SDLB if needed. The event logger is composed of two parts: one allows the scheduler to write the information and another one allows the user to retrieve and display the information.

To connect to the database, SDLB used a free java package Java Database Connection Pooling (DBPool) [10]. DBPool provides many advantages such as the possibility of creating and utilizing a pool of connections to the database that stores the logged data. Connection pooling is a very important feature needed by the Event logger because of the number of simultaneous connections to the database. A simple test case was made in order to compare the performance of a simulated system highly demanding on database connections. The results from this test case showed that the use of a pool of connections made the system eight to twenty times faster depending on the kind of requests executed [11]. The database chosen is a MySQL Database Server 4.0.16 which is free to use and distribute for non-commercial applications. MySQL is also known for its reliability and performance. To avoid security weaknesses, two different accounts have been created on the database system for the Diagnostic System. One is only allowed to write and only to a few specific objects, the other is only allowed to read and also only from a few specific objects. Thus, two pools of connections are created and used -- one for each part of the Event logger. This will also prevent one type of request from slowing down the other.

The user interface for the Diagnostic System is a web site implemented with the technology JSP/Servlet. JSP/Servlet is the Java solution to implement applicative websites. We chose to use the Apache HTTP Server 2.0.48 for the web server and Apache Tomcat 5.0.28 for the JSP/Servlet engine. Those two applications are also freely distributed and widely used professionally. The JSP part is a Java code mixed with HTML code which will be precompiled and then interpreted by the web server every time a user requests a page. Only the basic display operations are done in the JSP part. The Servlet part is pure Java code which will be compiled and installed on the web server. The execution of a Servlet code is much faster than the execution of a JSP code; therefore the work is done as much as possible in the Servlet part. The Servlet retrieves and prepares the data, the JSP displays them.

The User Interface of the event logger is shown in Figure 3. This interface allows the user to filter the log according to the level of log, the type of log, the name of the job, or the location (class or function) which generated the message. The log messages displayed are paginated to allow a fast loading of the page and an easy browsing. All those features are implemented in the Servlet part.

6. CONCLUSION

In this project, we improved the existing components and to complete the missing features of the SDLB in order to offer a more coherent and complete package. The scheduling model is made more effective by using information from the System Agents before selecting the machines. The improved features include supporting machines running in dedicated mode and batched mode more effectively, improved fault tolerance ability, and event logger for trouble shooting.

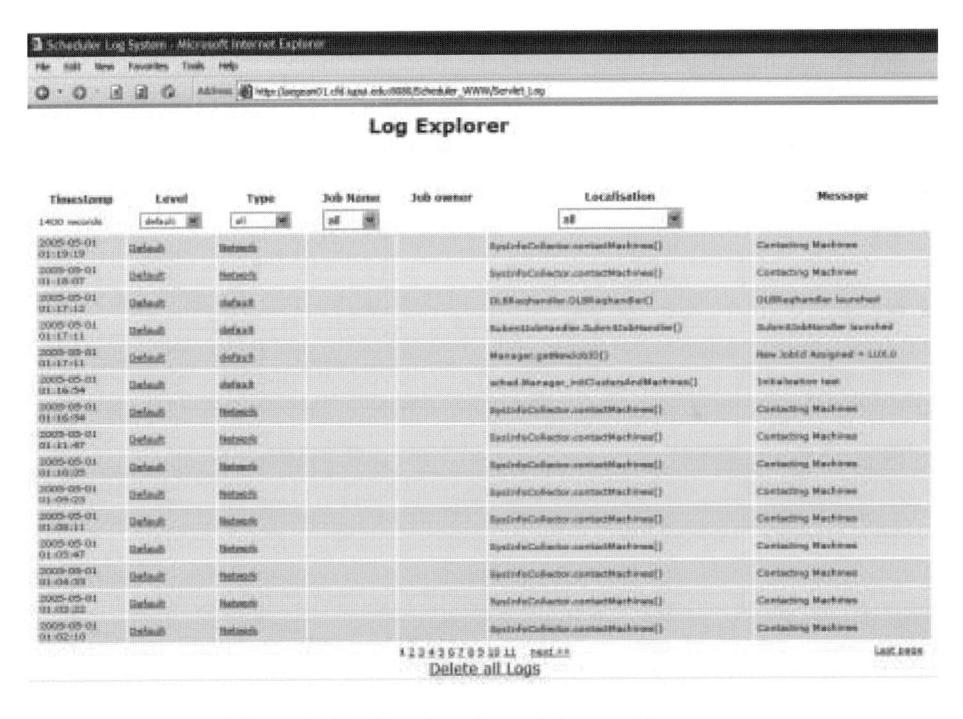

Figure 3. The User Interface of the event logger

REFERENCES

1. Condor – High Throughput Computing, http://www.cs.wisc.edu/condor/
2. An introduction to PORTABLE BATCH SYSTEM, http://hpc.sissa.it/pbs/pbs.html
3. Load Sharing Facility, http://wwwinfo.cern.ch/pdp/bis/services/lsf/
4. "MOAB Cluster Suite™," Cluster Resources Inc., 2005.

5. http://www.clusterresources.com/products/moabclustersuite.shtml
6. The Globus project, http://www.globus.org/toolkit/
7. Y.P. Chien, A. Ecer, H.U. Akay, S. Secer, R. Blech, , "Communication Cost Estimation for Parallel CFD Using Variable Time-Stepping Algorithms," *Computer Methods in Applied Mechanics and Engineering,* Vol. 190, (2000), pp. 1379-1389.
8. S. Chien, J. Zhou, A. Ecer, and H.U. Akay, "Autonomic System for Dynamic Load Balancing of Parallel CFD," *Proceedings of Parallel Computational Fluid Dynamics 2002*, Nara, Japan (May 21-24, 2002), Elsevier Science B.V., Amsterdam, The Netherlands, pp. 185-192, 2003.
9. S. Chien, Y. Wang, A. Ecer, and H.U. Akay, "Grid Scheduler with Dynamic Load Balancing for Parallel CFD," *Proceedings of 2003 Parallel Computational Fluid Dynamics,* May 2003, Moscow, Russia, pp. 259-266.
10. G. Winstanley, "DBPool - Java Database Connection Pooling," Java Consultant, United Kingdom, December 2004. http://homepages.nildram.co.uk/~slink/java/DBPool
11. N. Nordborg, "BASE 2.0 - MySQL Connection from Java," Complex Systems Division Department of Theoretical Physics, Lund University, Sweden, April 2001. http://www.thep.lu.se/~nicklas/base2/prototype/mysql.html

Performance Evaluation of Adaptive Scientific Applications using TAU

Sameer Shende[a]*, Allen D. Malony[a], Alan Morris[a], Steven Parker[b], J. Davison de St. Germain[b]

[a]Performance Research Laboratory, Department of Computer and Information Science, University of Oregon, Eugene, OR 97403, USA

[b]Scientific Computing and Imaging Institute, University of Utah, Salt Lake City, UT 84112, USA

1. Introduction

Fueled by increasing processor speeds and high speed interconnection networks, advances in high performance computer architectures have allowed the development of increasingly complex large scale parallel systems. For computational scientists, programming these systems efficiently is a challenging task. Understanding the performance of their parallel applications is equally daunting. To observe and comprehend the performance of parallel applications that run on these systems, we need performance evaluation tools that can map the performance abstractions to the user's mental models of application execution. For instance, most parallel scientific applications are iterative in nature. In the case of CFD applications, they may also dynamically adapt to changes in the simulation model. A performance measurement and analysis system that can differentiate the phases of each iteration and characterize performance changes as the application adapts will enable developers to better relate performance to their application behavior. In this paper, we present new performance measurement techniques to meet these needs. In section 2, we describe our parallel performance system, TAU. Section 3 discusses how new TAU profiling techniques can be applied to CFD applications with iterative and adaptive characteristics. In section 4, we present a case study featuring the Uintah computational framework and explain how adaptive computational fluid dynamics simulations are observed using TAU. Finally, we conclude with a discussion of how the TAU performance system can be broadly applied to other CFD frameworks and present a few examples of its usage in this field.

2. TAU Performance System

Given the diversity of performance problems, evaluation methods, and types of events and metrics, the instrumentation and measurement mechanisms needed to support performance observation must be flexible, to give maximum opportunity for configuring performance experiments, and portable, to allow consistent cross-platform performance problem solving. The TAU performance system [1,4], is composed of instrumentation, measurement, and analysis

---

*This research was supported by the U.S. Department of Energy, Office of Science, under contracts DE-FG03-01ER25501 and DE-FG02-03ER25561, and University of Utah and LLNL DOE contracts B524196 and 2205056.

parts. It supports both profiling and tracing forms of measurements. TAU implements a flexible instrumentation model that permits a user to insert performance instrumentation hooks into the application at several levels of program compilation and execution. The C, C++, and Fortran languages are supported, as well as standard message passing (e.g., MPI) and multi-threading (e.g., Pthreads) libraries.

For instrumentation we recommend a dual instrumentation approach. Source code is instrumented automatically using a source-to-source translation tool, $tau\_instrumentor$, that acts as a pre-processor prior to compilation. The MPI library is instrumented using TAU's wrapper interposition library that intercepts calls to the MPI calls and internally invokes the TAU timing calls before and after. TAU source instrumentor can take a selective instrumentation file that lists the name of routines or files that should be excluded or included during instrumentation. The instrumented source code is then compiled and linked with the TAU MPI wrapper interposition library to produce an executable.

TAU provides a variety of measurement options that are chosen when TAU is installed. Each configuration of TAU is represented in a set of measurement libraries and a stub makefile to be used in the user application makefile. Profiling and tracing are the two performance evaluation techniques that TAU supports. Profiling presents aggregate statistics of performance metrics for different events and tracing captures performance information in timestamped event logs for analysis. In tracing, we can observe along a global timeline when events take place in different processes. Events tracked by both profiling and tracing include entry and exit from routines, interprocess message communication events, and other user-defined atomic events. Tracing has the advantage of capturing temporal relationships between event records, but at the expense of generating large trace files. The choice to profile trades the loss of temporal information with gains in profile data efficiency.

3. CFD Application Performance Mapping

Observing the behavior of an adaptive CFD application shows us several interesting aspects of its execution. Such applications typically involve a domain decomposition of the simulation model across processors and an interaction of execution phases as the simulation proceeds in time. Each iteration may involve a repartitioning or adaption of the underlying computational structure to better address numerical or load balance properties. For example, a mesh refinement might be done at iteration boundaries and information about convergence or divergence of numerical algorithms is detailed. Also, domain specific information such as the number of cells refined at each stage gives a user valuable feedback on the progress of the computation.

Performance evaluation tools must capture and present key application specific data and corelate this information to performance metrics to provide a useful feedback to the user. Presenting performance information that relates to application specific abstractions is a challenging task. Typically, profilers present performance metrics in the form of a group of tables, one for each MPI task. Each row in a table represents a given routine. Each column specifies a metric such as the exclusive or inclusive time spent in the given routine or the number of calls executed. This information is typically presented for all invocations of the routine. While such information is useful in identifying the routines that contribute most to the overall execution time, it does not explain the performance of the routines with respect to key application phases. To address this shortcoming, we provide several profiling schemes in TAU.

3.1. Static timers

These are commonly used in most profilers where all invocations of a routine are recorded. The name and group registration takes place when the timer is created (typically the first time a routine is entered). A given timer is started and stopped at routine entry and exit points. A user defined timer can also measure the time spent in a group of statements. Timers may be nested but they may not overlap. The performance data generated can typically answer questions such as: *what is the total time spent in $MPI\_Send()$ across all invocations?*

3.2. Dynamic timers

To record the execution of each invocation of a routine, TAU provides dynamic timers where a unique name may be constructed for a dynamic timer for each iteration by embedding the iteration count in it. It uses the start/stop calls around the code to be examined, similar to static timers. The performance data generated can typically answer questions such as: *what is the time spent in the routine $foo()$ in iterations 24, 25, and 40?*

3.3. Static phases

An application typically goes through several phases in its execution. To track the performance of the application based on phases, TAU provides static and dynamic phase profiling. A profile based on phases highlights the context in which a routine is called. An application has a default phase within which other routines and phases are invoked. A phase based profile shows the time spent in a routine when it was in a given phase. So, if a set of instrumented routines are called directly or indirectly by a phase, we'd see the time spent in each of those routines under the given phase. Since phases may be nested, a routine may belong to only one phase. When more than one phase is active for a given routine, the closest ancestor phase of a routine along its callstack is its phase for that invocation. The performance data generated can answer questions such as: *what is the total time spent in $MPI\_Send()$ when it was invoked in all invocations of the IO ($IO => MPI\_Send()$) phase?*

3.4. Dynamic phases

Dynamic phases borrow from dynamic timers and static phases to create performance data for all routines that are invoked in a given invocation of a phase. If we instrument a routine as a dynamic phase, creating a unique name for each of its invocations (by embedding the invocation count in the name), we can examine the time spent in all routines and child phases invoked directly or indirectly from the given phase. The performance data generated can typically answer questions such as: *what is the total time spent in $MPI\_Send()$ when it was invoked directly or indirectly in iteration 24?* Dynamic phases are useful for tracking per-iteration profiles for an adaptive computation where iterations may differ in their execution times.

3.5. Callpaths

In phase-based profiles, we see the relationship between routines and parent phases. Phase profiles [5] do not show the calling structure between different routines as is represented in a callgraph. To do so, TAU provides callpath profiling capabilities where the time spent in a routine along an edge of a callgraph is captured. Callpath profiles present the full flat profiles of routines (or nodes in the callgraph), as well as routines along a callpath. A callpath is represented syntactically as a list of routines separated by a delimiter. The maximum depth of a callpath is controlled by an environment variable.

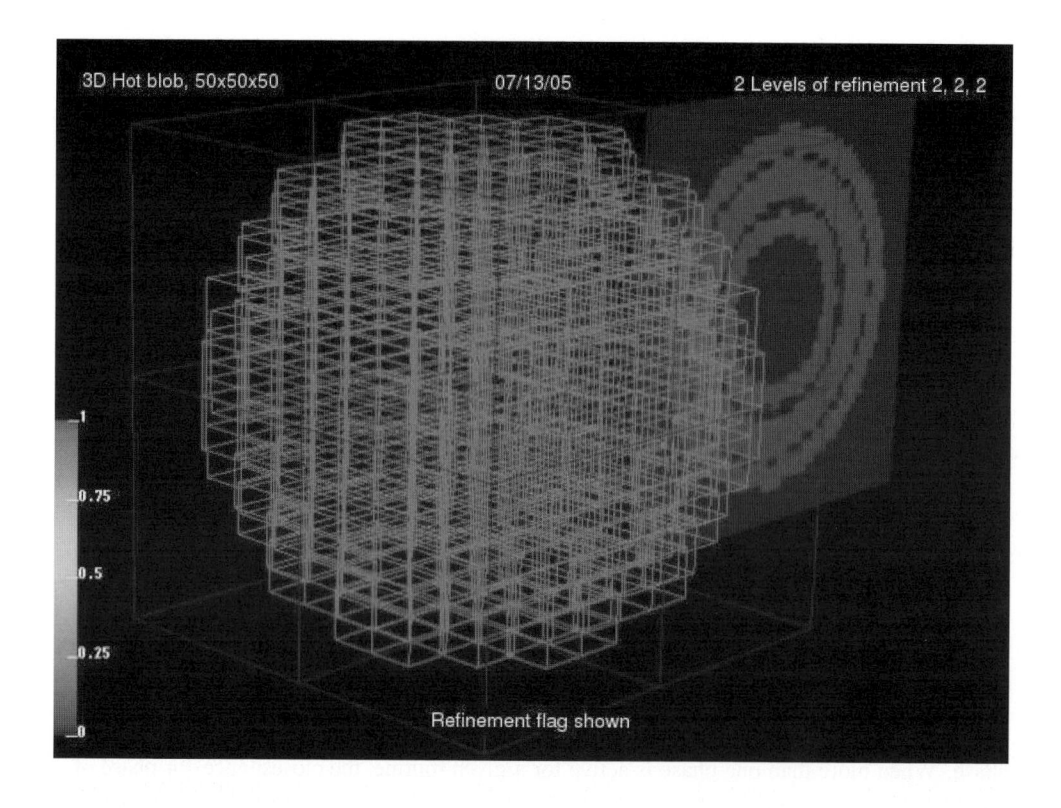

Figure 1. Adaptive Mesh Refinement in a parallel CFD simulation in the Uintah Computational Framework

3.6. User-defined Events

Besides timers and phases that measure the time spent between a pair of start and stop calls in the code, TAU also provides support for user-defined atomic events. After an event is registered with a name, it may be triggered with a value at a given point in the source code. At the application level, we can use user-defined events to track the progress of the simulation by keeping track of application specific parameters that explain program dynamics, for example, the number of iterations required for convergence of a solver at each time step, or the number of cells in each iteration of an adaptive mesh refinement application.

4. Case Study: Uintah

We have applied TAU's phase profiling capabilities to evaluate the performance of the Uintah computational framework (UCF) [2]. The TAU profiling strategy for Uintah is to observe the performance of the framework at the level of patches, the unit of spatial domain partitioning. Thus, we instrument UCF with dynamic phases where the phase name contains the AMR level

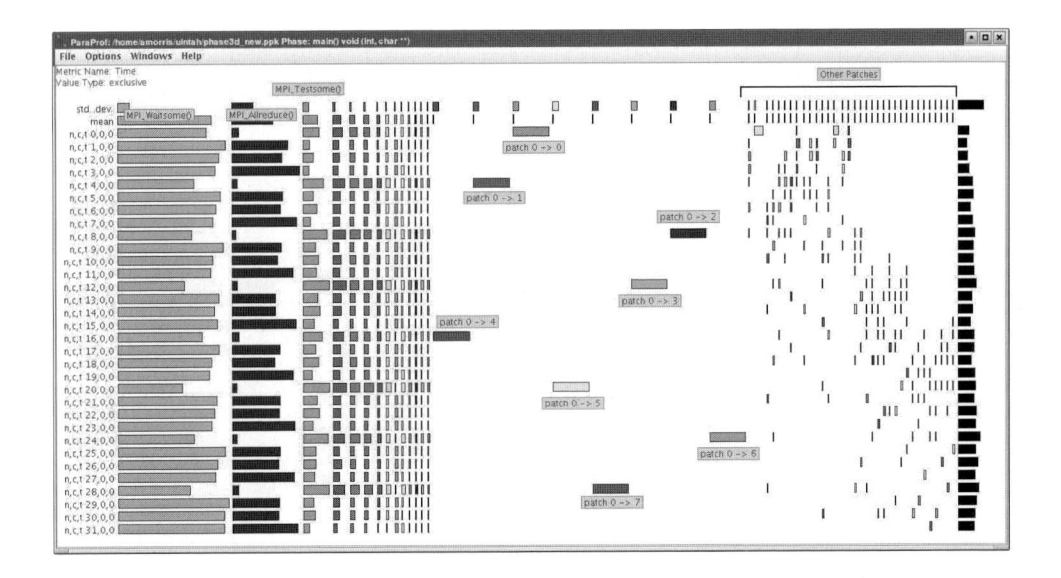

Figure 2. Distribution and time taken in various phases in the Uintah Computational Framework

and patch index. The case study focuses on a 3 dimensional validation problem for a compressible CFD code. Initially, there is a region of gas at the center of the computational domain that is at a high pressure (10atm) and temperature gas (3000K). At time = 0 the gas is allowed to expand forming a spherical shockwave. Eventually, a low pressure region will form in the center and the expanding flow will reverse directions. Figure 1 shows a sample distribution of patches in the domain. The blue outer cubes enclose the eight (2x2x2) level 0 patches. The pink inner cubes cover the "interesting" portions of the domain that have been selected by the AMR subsystem for mesh refinement on level 1.

We instrumented Uintah by creating a phase based on the level index and patch index for a task given to the Uintah scheduler. Phases are given name such as "Patch 0 ->1", which represents the 2nd patch on level 0. All tasks and instrumented functions are then associated by way of phase profiling with the patch on which the computation is done.

Figure 2 shows the performance data obtained from the simulation as displayed by ParaProf. Here we see the eight patches that make up level 0 overwhelm the time taken in other patches. They executed on nodes 0,4,8,12,16,24 and 28. This gave us an immediate result showing that these nodes spent extra time processing the level 0 patches, while the other nodes waited in MPI_Allreduce.

Figure 3 shows the distribution of a given timer, ICE::advectAndAdvanceInTime, across all phases, each representing a patch. We can see that the time spent in this task for the eight level 0 patches is more 9 than seconds, while the time spent for all the other patches is less than 4 seconds each.

Figure 4 shows the profile for the phase "patch 0 ->0" which runs only on node 0. We see the partition of each task that was executed under this phase. If this phase had run on other nodes,

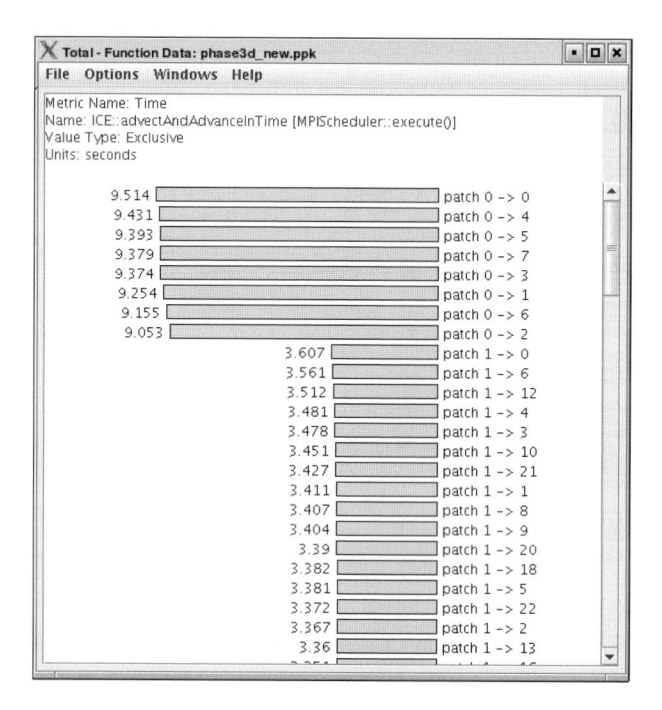

Figure 3. The distrubution of the ICE::advectAndAdvanceInTime task across patches using phase profiling

we would have aggregate statistics as well (mean, std. dev.).

Phase profiling in UCF allows the developers to partition the performance data for tasks and instrumented functions across spatially defined patches with information from the AMR subsystem. This data can be used to identify load balancing issues as well as establish a better understanding of code performance in an AMR simulation.

5. Other frameworks

Besides the Uintah computational framework, TAU has been applied successfully to several frameworks that are used for computational fluid dynamics simulations. These include VTF [9] from Caltech, MFIX [7] from NETL, ESMF [10] coupled flow application from UCAR, NASA and other institutions, SAMRAI [12] from LLNL, Miranda [11] from LLNL, GrACE [13] from Rutgers University, SAGE [6] from SAIC, and Flash [8] from University of Chicago. Our work in performance evaluation of adaptive scientific computations can be broadly applied to other CFD codes. Thus, CFD frameworks can benefit from the integration of portable performance profiling and tracing support using TAU.

```
File  Options  Windows  Help
Phase: patch 0 -> 0
Metric: Time
Value: Exclusive
Units: seconds

12.202                                          reflux_computeCorrectionFluxes [MPIScheduler::execute()]
     9.514                                      ICE::advectAndAdvanceInTime [MPIScheduler::execute()]
            5.501                               ICE::computeEquilibrationPressure [MPIScheduler::execute()]
                  3.01                          ICE::addExchangeToMomentumAndEnergy [MPIScheduler::execute()]
                  2.964                         DataArchiver::output [MPIScheduler::execute()]
                  2.777                         ICE::computeDelPressAndUpdatePressCC [MPIScheduler::execute()]
                  2.599                         ICE::computeLagrangianSpecificVolume [MPIScheduler::execute()]
                  2.512                         AMRICE::errorEstimate [MPIScheduler::execute()]
                  2.399                         ICE::accumulateMomentumSourceSinks [MPIScheduler::execute()]
                  2.24                          ICE::addExchangeContributionToFCVel [MPIScheduler::execute()]
                  1.864                         RegridderCommon::Dilate2 Creation [MPIScheduler::execute()]
                  1.475                         coarsen [MPIScheduler::execute()]
                  1.155                         ICE::computeVel_FC [MPIScheduler::execute()]
                  1.093                         ICE::computeLagrangianValues [MPIScheduler::execute()]
                  0.583                         ICE::computePressFC [MPIScheduler::execute()]
                  0.312                         DataArchiver::checkpoint [MPIScheduler::execute()]
                  0.281                         reflux_applyCorrectionFluxes [MPIScheduler::execute()]
                  0.27                          ICE::accumulateEnergySourceSinks [MPIScheduler::execute()]
                  0.201                         ICE::actuallyComputeStableTimestep [MPIScheduler::execute()]
                  0.119                         ICE::computeTempFC [MPIScheduler::execute()]
                  0.118                         ICE::computeThermoTransportProperties [MPIScheduler::execute()]
                  0.087                         Task execution [MPIScheduler::initiateTask()]
                  0.077                         SchedulerCommon::copyDataToNewGrid [MPIScheduler::execute()]
                  0.065                         ICE::actuallyInitialize [MPIScheduler::execute()]
                  0.048                         HierarchicalRegridder::MarkPatches2 [MPIScheduler::execute()]
                  0.045                         initializeErrorEstimate [MPIScheduler::execute()]
```

Figure 4. Phase profile for patch 0 ->0 shows tasks executing under this phase

6. Conclusions

When studying the performance of scientific applications, especially on large-scale parallel systems, there is a strong preference among developers to view performance information with respect to their "mental" model of the application, formed from the structural, logical, and numerical models used in the program. If the developer can relate performance data measured during execution to what they know about the application, more effective program optimization may be achieved. In this paper, we present portable performance evaluation techniques in the context of the TAU performance system and its application to the Uintah computational framework. We illustrate how phase based profiling may be effectively used to bridge the semantic gap in comprehending the performance of parallel scientific applications using techniques that map program performance to higher level abstractions.

REFERENCES

1. A. D. Malony and S. Shende and R. Bell and K. Li and L. Li and N. Trebon, "Advances in the TAU Performance System," Chapter, "Performance Analysis and Grid Computing," (Eds. V. Getov, et. al.), Kluwer, Norwell, MA, pp. 129-144, 2003.

2. J. D. de St. Germain and J. McCorquodale and S.G. Parker and C.R. Johnson, "Uintah: A Massively Parallel Problem Solving Environment," Ninth IEEE International Symposium on High Performance and Distributed Computing, IEEE, pp. 33–41. 2000.
3. R. Bell and A. D. Malony and S. Shende, "A Portable, Extensible, and Scalable Tool for Parallel Performance Profile Analysis," Proc. EUROPAR 2003 conference, LNCS 2790, Springer, pp. 17-26, 2003.
4. TAU Portable Profiling. URL: http://www.cs.uoregon.edu/research/tau, 2005.
5. A. D. Malony and S. S. Shende and A. Morris, "Phase-Based Parallel Performance Profiling," Proc. ParCo 2005 Conference, Parallel Computing Conferences, 2005.
6. D. Kerbyson and H. Alme and A. Hoisie and F. Petrini and H. Wasserman and M. Gittings, "Predictive Performance and Scalability Modeling of a Large-Scale Application," Proc. SC 2001 Conference, ACM/IEEE, 2001.
7. M. Syamlal and W. Rogers and T. O'Brien, "MFIX Documentation: Theory Guide," Technical Note, DOE/METC-95/1013, 1993.
8. R. Rosner, et. al., "Flash Code: Studying Astrophysical Thermonuclear Flashes," Computing in Science and Engineering, 2:33, 2000.
9. J. Cummings and M. Aivazis and R. Samtaney and R. Radovitzky and S. Mauch and D. Meiron, "A Virtual Test Facility for the Simulation of Dynamic Response in Materials," The Journal of Supercomputing 23(1), pp. 39–50, August 2002.
10. C. Hill and C. DeLuca and V. Balaji and M. Suarez and A. da Silva, "The Architecture of the Earth System Modeling Framework," Computing in Science and Engineering, 6(1), Januaray/February 2004.
11. W. Cabot and A. Cook and C. Crabb, "Large-Scale Simulations with Miranda on BlueGene/L," Presentation from BlueGene/L Workshop, Reno, October 2003.
12. A. Wissinsk and R. Hornung and S. Kohn and S. Smith and N. Elliott, "Large Scale Parallel Structured AMR Calculations using the SAMRAI Framework," Proc. SC'2001 Conference, ACM/IEEE, 2001.
13. Y. Zhang and S. Chandra and S. Hariri and M. Parashar, "Autonomic Proactive Runtime Partitioning Strategies for SAMR Applications," Proc. NSF Next Generation Systems Program Workshop, IEEE/ACM 18th IPDPS Conference, April 2004.

Computational Quality of Service in Parallel CFD [*]

B. Norris,[a] L. McInnes,[a] I. Veljkovic [b]

[a]Mathematics and Computer Science Division, Argonne National Laboratory,
9700 South Cass Avenue, Argonne, IL 60439-4844, *[norris,mcinnes]@mcs.anl.gov*.

[b]Department of Computer Science and Engineering, The Pennsylvania State University,
IST Building, University Park, PA 16802-6106, *veljkovi@cse.psu.edu*.

Abstract. Our computational quality-of-service infrastructure is motivated by large-scale scientific simulations based on partial differential equations, with emphasis on multimethod linear solvers in the context of parallel computational fluid dynamics. We introduce a component infrastructure that supports performance monitoring, analysis, and adaptation of important numerical kernels, such as nonlinear and linear system solvers. We define a simple, flexible interface for the implementation of adaptive nonlinear and linear solver heuristics. We also provide components for monitoring, checkpointing, and gathering of performance data, which are managed through two types of databases. The first is created and destroyed during runtime and stores performance data for code segments of interest, as well as various application-specific performance events in the currently running application instance. The second database is persistent and contains performance data from various applications and different instances of the same application. This database can also contain performance information derived through offline analysis of raw data. We describe a prototype implementation of this infrastructure and illustrate its applicability to adaptive linear solver heuristics used in a driven cavity flow simulation code.

1. INTRODUCTION

Component-based environments provide opportunities to improve the performance, numerical accuracy, and other characteristics of parallel simulations in CFD. Because component-based software engineering combines object-oriented design with the powerful features of well-defined interfaces, programming language interoperability, and dynamic composability, it helps to overcome obstacles that hamper sharing even well-designed traditional numerical libraries. Not only can applications be assembled from components selected to provide good algorithmic performance and scalability, but they can also be changed dynamically during execution to optimize desirable characteristics. We use the term *computational quality of service* (CQoS) [9, 16] to refer to the automatic selection and configuration of components for a particular computational purpose. CQoS embodies the familiar concept of quality of service in networking and

---

[*]This work was supported by the Mathematical, Information, and Computational Sciences Division subprogram of the Office of Advanced Scientific Computing Research, Office of Science, U.S. Department of Energy, under Contract W-31-109-ENG-38.

the ability to specify and manage characteristics of the application in a way that adapts to the changing (computational) environment. The factors affecting performance are closely tied to a component's parallel implementation, its management of memory, the algorithms executed, the algorithmic parameters employed (e.g., the level of overlap in an additive Schwarz preconditioner), and other operational characteristics. CQoS is also concerned with functional qualities, such as the level of accuracy achieved for a particular algorithm.

This paper presents an overview of new software infrastructure for automated performance gathering and analysis of high-performance components, a key facet of our CQoS research, with emphasis on using these capabilities in parallel CFD simulations, such as flow in a driven cavity and compressible Euler flow. The remainder of this paper is organized as follows. Section 2 discusses parallel CFD applications and algorithms that motivate this infrastructure. Section 3 introduces the new framework for enabling CQoS in parallel nonlinear PDE-based applications. Section 4 illustrates the performance of a simple adaptive algorithm strategy. Section 5 concludes with a summary and discussion of future work.

2. MOTIVATING APPLICATIONS AND ALGORITHMS

Flow in a Driven Cavity. The first parallel application that motivates and validates this work is driven cavity flow, which combines lid-driven flow and buoyancy-driven flow in a two-dimensional rectangular cavity. The lid moves with a steady and spatially uniform velocity and thus sets a principal vortex and subsidiary corner vortices. The differentially heated lateral walls of the cavity induce a buoyant vortex flow, opposing the principal lid-driven vortex. We use a velocity-vorticity formulation of the Navier-Stokes and energy equations, which we discretize using a standard finite-difference scheme with a five-point stencil for each component on a uniform Cartesian mesh; see [7] for a detailed problem description.

Compressible Euler Flow. Another motivating application is PETSc-FUN3D [2], which solves the compressible and incompressible Navier-Stokes equations in parallel; the sequential model was originally developed by W. K. Anderson [1]. The code uses a finite-volume discretization with a variable-order Roe scheme on a tetrahedral, vertex-centered unstructured mesh. The variant of the code under consideration here uses the compressible Euler equations to model transonic flow over an ONERA M6 wing, a common test problem that exhibits the development of a shock on the wing surface. Initially a first-order discretization is used; but once the shock position has settled down, a second-order discretization is applied. This change in discretization affects the nature of the resulting linear systems.

Newton-Krylov Algorithms. Both applications use inexact Newton methods (see, e.g., [14]) to solve nonlinear systems of the form $f(u) = 0$. We use parallel preconditioned Krylov methods to (approximately) solve the Newton correction equation $f'(u^{\ell-1})\, \delta u^\ell = -f(u^{\ell-1})$, and then update the iterate via $u^\ell = u^{\ell-1} + \alpha \cdot \delta u^\ell$, where α is a scalar determined by a line search technique such that $0 < \alpha \leq 1$. We terminate the Newton iterates when the relative reduction in the residual norm falls below a specified tolerance. Our implementations use the Portable, Extensible Toolkit for Scientific computation (PETSc) [3], a suite of data structures and routines for the scalable solution of scientific applications modeled by PDEs. PETSc integrates a hierarchy of libraries that range from low-level distributed data structures for vectors and matrices through high-level linear, nonlinear, and time-stepping solvers.

Pseudo-Transient Continuation. For problems with strong nonlinearities, Newton's method often struggles unless some form of continuation is employed. Hence, we incorporate pseudo-transient continuation [11], a globalization technique that solves a sequence of problems derived from the model $\frac{\partial u}{\partial t} = -f(u)$, namely,

$$g_\ell(u) \equiv \frac{1}{\tau^\ell}(u - u^{\ell-1}) + f(u) = 0, \ \ell = 1, 2, \ldots, \tag{1}$$

where τ^ℓ is a pseudo time step. At each iteration in time, we apply Newton's method to Equation (1). As discussed by Kelley and Keyes [11], during the initial phase of pseudo-transient algorithms, τ^ℓ remains relatively small, and the Jacobians associated with Equation (1) are well conditioned. During the second phase, the pseudo time step τ^ℓ advances to moderate values, and in the final phase τ^ℓ transitions toward infinity, so that the iterate u^ℓ approaches the root of $f(u) = 0$.

Adaptive Solvers. In both applications the linearized Newton systems become progressively more difficult to solve as the simulation advances due to the use of pseudo-transient continuation [11]. Consequently both are good candidates for the use of *adaptive* linear solvers [4,5,13], where the goal is to improve overall performance by combining more robust (but more costly) methods when needed in a particularly challenging phase of solution with faster (though less powerful) methods in other phases. Parallel adaptive solvers are designed with the goal of reducing the overall execution time of the simulation by dynamically selecting the most appropriate method to match the characteristics of the current linear system.

A key facet of developing adaptive methods is the ability to consistently collect and access both runtime and historical performance data. Our preliminary research in adaptive methods [4,5,13], which employed ad hoc techniques to collect, store, and analyze data, has clearly motivated the need for a framework to analyze performance and help to manage algorithmic adaptivity.

3. COMPUTATIONAL QUALITY OF SERVICE FOR PARALLEL CFD

For a given parallel fluids problem, the availability of multiple solution methods, as well as multiple configurations of the same method, presents both a challenge and an opportunity. On the one hand, an algorithm can be chosen to better match the application's requirements. On the other hand, manually selecting a method in order to achieve good performance and reliable results is often difficult or impossible. Component-based design enables us to automate, at least partially, the task of selecting and configuring algorithms based on performance models, both for purposes of initial application assembly and for runtime adaptivity.

The Common Component Architecture (CCA) specification [6] defines a component model that specifically targets high-performance scientific applications, such as parallel CFD. Briefly, CCA components are units of encapsulation that can be composed to form applications; *ports* are the entry points to a component and represent public interfaces through which components interact; *provides* ports are interfaces that a component implements, and *uses* ports are interfaces through which a component invokes methods implemented by other components. A runtime *framework* provides some standard services to all CCA components, including instantiation of components and connection of *uses* and *provides* ports. At runtime, components can be instantiated/destroyed and port connections made/broken, thereby allowing dynamic adaptivity of CCA component applications and enabling the implementation of the adaptive linear solver methods introduced above.

In this paper, we present a CCA component infrastructure that allows researchers to monitor and adapt a simulation dynamically based on two main criteria: the runtime information about performance parameters and the information extracted from metadata from previous instances (executions) of a component application. This infrastructure includes components for performance information gathering, analysis, and interactions with off-line databases. Figure 1 (a) shows a typical set of components involved in nonlinear PDE-based applications; no explicit performance monitoring or adaptive method support is shown here. Figure 1 (b) shows the same application with the new performance infrastructure components. This design makes development of adaptive algorithms easier and less error-prone by separating as much as possible unrelated concerns from the adaptive strategy itself. In contrast, because our initial adaptive linear solver implementations were tightly interleaved and accessible only through a narrow PETSc interface intended for simple user-defined monitoring functions, the software [5, 13] became difficult to understand and maintain (mixed code for multiple heuristics) and extend (e.g., when adding new adaptive heuristics). This situation motivated us to define a simple interface that is flexible enough to enable the implementation of a wide range of adaptive heuristics, with an initial focus on adaptive linear solvers. We have reproduced some of our original results using this new infrastructure, incurring only the expected minimal fixed overhead of component interactions, for example as shown in [15].

We briefly introduce the terminology used in our CQoS infrastructure. We collectively refer to performance-relevant attributes of a unit of computation, such as a component, as *performance metadata*, or just *metadata*. These attributes include algorithm or application parameters, such as problem size and physical constants; compiler optimization options; and execution information, such as hardware and operating system information. Performance metrics, also referred to as CQoS metrics, are part of the metadata, for example, execution time and convergence history of iterative methods. Ideally, for each application execution, the metadata should provide enough information to duplicate the run; in practice, not all parameters that affect the performance are known or can be obtained, but the most significant ones are usually represented in the metadata we consider. We collectively refer to such metadata as an application instance, or *experiment*.

The design of our infrastructure is guided by the following goals: (1) low overhead during the application's execution: since all the time spent in performance monitoring and analysis/adaptation is overhead, the impact on overall performance must be minimized; (2) minimal code changes to existing application components in order to encourage use of this performance infrastructure by as many CCA component developers as possible; and (3) ease of implementation of performance analysis algorithms and new adaptive strategies, to enable and encourage the development and testing of new heuristics or algorithms for multimethod components.

Within our framework we differentiate between tasks that have to be completed during runtime and tasks that are performed when the experiment is finished (because of stringent overhead constraints during an application's execution). Consequently, we have two databases that serve significantly different purposes. The first is created and destroyed during runtime and stores performance data for code segments of interest and application-specific performance events for the running experiment. The second database is persistent and contains data about various applications and experiments within one application. The second database also contains metadata derived by performance analysis of raw performance results. At the conclusion of an experiment, the persistent database is updated with the information from the runtime database.

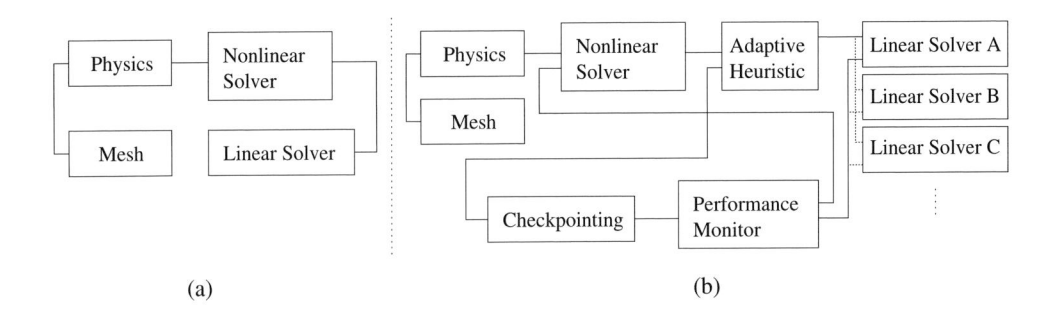

Figure 1. Some of the components and port connections in a typical PDE application: (a) in a traditional nonadaptive setting, and (b) augmented with performance monitoring and adaptive linear solver components.

Our initial implementation of this infrastructure relies on the Tuning and Analysis Utilities (TAU) toolkit [8] and the Parallel Performance Data Management Framework (PerfDMF) [10]. We now briefly describe the principal components involved in collecting and managing performance metadata and runtime adaptation.

The **Adaptive Heuristic** component implements a simple *AdaptiveAlgorithm* interface, whose single method, *adapt*, takes an argument containing application-specific metadata needed to implement a particular adaptive heuristic and to store the results. Specific implementations of the *AdaptiveContext* interface contain performance metadata used by adaptive heuristics, as well as references to the objects that provide the performance metadata contained in the context.

The **TAU Measurement** component collects runtime data from hardware counters, timing, and user-defined application-specific events. This component was provided by the developers of TAU; complete implementation details can be found in [12]. The **Performance Monitor** component monitors the application, including the selection of algorithms and parameters based on runtime performance data and stored metadata.

The **Checkpoint** component collects and stores metadata into a runtime database that can be queried efficiently during execution for the purpose of runtime performance monitoring and adaptation. When we started our implementation, the TAU profiling API could give only either callpath-based or cumulative performance information about an instrumented object (from the time execution started). Hence, we have introduced the Checkpoint component to enable us to store and retrieve data for the instrumented object during the application's execution (e.g., number of cache misses for every three calls of a particular function). The period for checkpointing can be variable; the component can also be used by any other component in the application to collect and query context-dependent and high-level performance information. For example, a linear solver component can query the checkpointing component for performance metadata of the nonlinear solver (the linear solver itself has no direct access to the nonlinear solver that invoked it). We can therefore always get the latest performance data for the given instrumented object from the database constructed during runtime.

The **Metadata Extractor** component retrieves metadata from the database at runtime. After running several experiments, analyzing the performance data, and finding a common perfor-

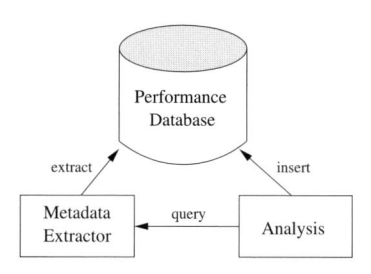

Figure 2. Components for offline query, management, and analysis of CQoS metadata.

mance behavior with some parameter values, we store data summarizing this behavior in the database. An example of derived metadata is the rate of convergence of a nonlinear or a linear solver. During runtime, these data are used in adapting our parameter and algorithm selection, and the Metadata Extractor component can retrieve compact metadata from the database efficiently.

Offline Analysis Support. The portions of the infrastructure that are not used at runtime are illustrated in Figure 2. They include a performance data extractor for retrieving data from the performance database, which is used by the offline analysis algorithm components. At present, the extractor also produces output in Matlab-like format, which is convenient for plotting some performance results; this output can be enhanced to interface with tools that provide more advanced visualization capabilities, such as an extension of ParaProf (part of the TAU suite of tools).

Many analyses can be applied offline to extract performance characteristics from the raw execution data or the results of previous analyses—in fact, facilitating the development of such analyses was one of our main motivations for developing this performance infrastructure. Initially we are focusing on simple analyses that allow us to replicate results in constructing adaptive, polyalgorithmic linear solvers from performance statistics of base linear solver experiments [4, 5, 13]. For the longer term, we plan to use this infrastructure for rapid development of new performance analyses and adaptive heuristics.

4. APPLICATION EXAMPLE

We illustrate the use of our performance infrastructure in the parallel driven cavity application briefly described in Section 2. As mentioned earlier, the use of pseudo-transient continuation affects the conditioning of the linearized Newton systems; thus, the resulting linear systems are initially well-conditioned and easy to solve, while later in the simulation they become progressively harder to solve.

In this parallel application, metadata describing the performance of the nonlinear solution, as well as each linear solution method, can be used to determine when to change or reconfigure linear solvers [5, 13]. Figure 3 shows some performance results comparing the use of a simple adaptive heuristic with the traditional single solution method approach. Even this simple automated adaptive strategy performs better than most base methods, and almost as well as the best base method (whose performance, of course, is not known a priori). Recent work on more

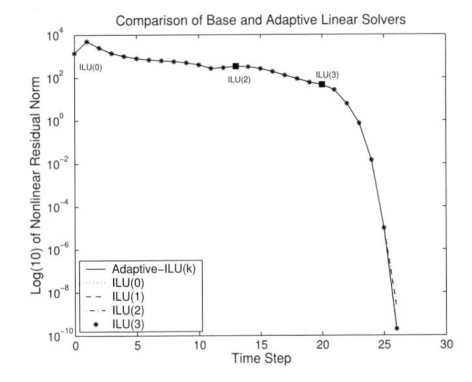

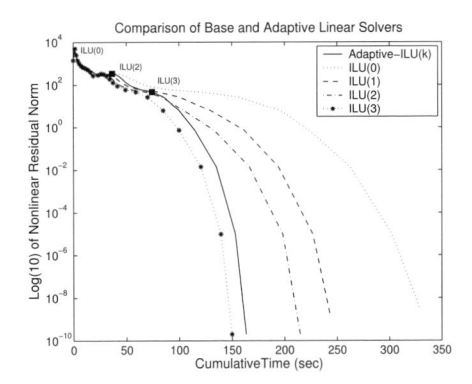

Figure 3. Comparison of single-method linear solvers and an adaptive scheme. We plot the nonlinear convergence rate (in terms of residual norm) versus both time step (left-hand graph) and cumulative execution time (right-hand graph).

sophisticated adaptive heuristics has achieved parallel performance that is consistently better than any single solution method [4].

Another use of our performance infrastructure is selection of application parameters based on performance information with the goal of maximizing performance. For example, in the application considered here, the initial CFL value is essential for determining the time step for the pseudo-transient Newton solver, which in turn affects the overall rate of convergence of the problem. One can query the database to determine the best initial CFL value from the experimental data available. Similarly, the values of other parameters that affect performance can be determined based on past performance data and offline analysis.

5. CONCLUSIONS AND FUTURE WORK

This work has introduced new infrastructure for performance analysis and adaptivity of parallel PDE-based applications, with a focus on computational fluids dynamic simulations. We are currently completing the migration of our existing adaptive heuristics to the new component infrastructure. New heuristics for adaptive method selection will also be investigated, including components for offline analyses of performance information. In addition to runtime adaptation, our performance infrastructure can potentially support initial application assembly and is being integrated with existing CCA component infrastructure that uses component performance models for automated application assembly. The accuracy of such models can be enhanced by using historical performance data to select the best initial set of components.

While our current adaptive linear solver implementations are based on PETSc solvers, the component infrastructure is not limited to one particular library. We plan to experiment with adaptive algorithms based on other libraries as well as to evaluate their effectiveness on new applications. Another topic of future research concerns defining and using CQoS metrics that reflect differences in algorithmic scalability in large-scale parallel environments. Such metrics would rely at least partly on performance models of the scalability of various algorithms.

REFERENCES

1. W. K. Anderson and D. Bonhaus. An implicit upwind algorithm for computing turbulent flows on unstructured grids. *Computers and Fluids*, 23(1):1–21, 1994.
2. W. K. Anderson, W. D. Gropp, D. K. Kaushik, D. E. Keyes, and B. F. Smith. Achieving high sustained performance in an unstructured mesh CFD application. In *Proceedings of Supercomputing 1999*. IEEE Computer Society, 1999. Gordon Bell Prize Award Paper in Special Category.
3. S. Balay, K. Buschelman, W. Gropp, D. Kaushik, M. Knepley, L. McInnes, Barry F. Smith, and H. Zhang. PETSc users manual. Technical Report ANL-95/11 - Revision 2.2.1, Argonne National Laboratory, 2004. http://www.mcs.anl.gov/petsc/ .
4. S. Bhowmick, D. Kaushik, L. McInnes, B. Norris, and P. Raghavan. Parallel adaptive solvers in compressible PETSc-FUN3D simulations. Argonne National Laboratory preprint ANL/MCS-P1279-0805, submitted to *Proc. of the 17th International Conference on Parallel CFD*, Aug 2005.
5. S. Bhowmick, L. C. McInnes, B. Norris, and P. Raghavan. The role of multi-method linear solvers in PDE-based simulations. *Lecture Notes in Computer Science, Computational Science and its Applications-ICCSA 2003*, 2667:828–839, 2003.
6. CCA Forum homepage. http://www.cca-forum.org/ , 2005.
7. T. S. Coffey, C.T. Kelley, and D.E. Keyes. Pseudo-transient continuation and differential algebraic equations. *SIAM J. Sci. Comp*, 25:553–569, 2003.
8. Department of Computer and Information Science, University of Oregon, Los Alamos National Laboratory, and Research Centre Julich, ZAM, Germany. *TAU User's Guide (Version 2.13)*, 2004.
9. P. Hovland, K. Keahey, L. C. McInnes, B. Norris, L. F. Diachin, and P. Raghavan. A quality of service approach for high-performance numerical components. In *Proceedings of Workshop on QoS in Component-Based Software Engineering, Software Technologies Conference*, Toulouse, France, 20 June 2003.
10. K. Huck, A. Malony, R. Bell, L. Li, and A. Morris. PerfDMF: Design and implementation of a parallel performance data management framework. In *Proc. International Conference on Parallel Processing (ICPP 2005)*. IEEE Computer Society, 2005.
11. C. T. Kelley and D. E. Keyes. Convergence analysis of pseudo-transient continuation. *SIAM Journal on Numerical Analysis*, 35:508–523, 1998.
12. A. Malony, S. Shende, N. Trebon, J. Ray, R. Armstrong, C. Rasmussen, and M. Sottile. Performance technology for parallel and distributed component software. *Concurrency and Computation: Practice and Experience*, 17:117–141, Feb 2005.
13. L. McInnes, B. Norris, S. Bhowmick, and P. Raghavan. Adaptive sparse linear solvers for implicit CFD using Newton-Krylov algorithms. In *Proceedings of the Second MIT Conference on Computational Fluid and Solid Mechanics*, Boston, USA, June 2003. Massachusetts Institute of Technology.
14. J. Nocedal and S. J. Wright. *Numerical Optimization*. Springer-Verlag, 1999.
15. B. Norris, S. Balay, S. Benson, L. Freitag, P. Hovland, L. McInnes, and B. Smith. Parallel components for PDEs and optimization: Some issues and experiences. *Parallel Computing*, 28(12):1811–1831, 2002.
16. B. Norris, J. Ray, R. Armstrong, L. McInnes, Bernholdt, W. Elwasif, A. Malony, and S. Shende. Computational quality of service for scientific components. In *Proceedings of the International Symposium on Component-Based Software Engineering, (CBSE7)*, Edinburgh, Scotland, 2004.

Parallel Numerical Simulation of Flame Extinction and Flame Lift-Off

G. Eggenspieler, S. Menon [a]

[a]Computational Combustion Laboratory,
School of Aerospace Engineering,
Georgia Institute of Technology,
Atlanta, Georgia 30332

Large Eddy Simulation of premixed flame propagation in the thin reaction zone and broken reaction zone is studied using the Linear-Eddy Mixing model as the subgrid model for species mixing and molecular diffusion. This modeling approach is proven to capture all length scales relevant to flame propagation in highly turbulent flow. Flame extinction and flame lift-off are observed when heat losses are simulated. The computational cost of the method described herein is more expensive than more classical flamelet modeling approaches but is possible using the parallel CFD technique.

1. Introduction

State of the art gas turbines engines using Lean-Premix-Prevaporized (LPP) technology are operated at equivalence ratio slightly higher than the lean extinction limit. Although it is often assumed that the turbulent combustion processes taking place in these gas turbines are described by the flamelet regime, where the flame thickness (δ_F) is smaller than the Kolmogorov scale (η) and the chemical time scales (τ_C) is smaller than the characteristic turbulent flow time scales (τ_F), other turbulent combustion regimes are also locally present. In regions of high turbulence, the smallest eddy can be smaller than the flame thickness. In this case, eddies penetrate the preheat zone increasing heat and species transport. As a results the flame thickness increases. Furthermore, experiments [1] as well as theoretical studies and numerical simulations [2] have shown that very high level of turbulence can result in flame quenching. Experimental studies [3] showed that stabilized flame can exhibit local quenching (quenching of a very small portion of the flame) without global quenching (quenching of a large portion in the flame creating a hole in the flame where unburnt reactants penetrate into the product region). Both local and global flame quenching are not fully understood. In theory, flame quenching is a direct result of the action of turbulent structures small and/or powerful enough to break-up the structure of the flame, but this was never formally demonstrated neither by experiments nor by numerical simulations. Other factors such as reactants equivalence ratio distribution, combustion chamber geometry, heat losses, etc. have to be considered when flame extinction is studied.

This numerical study uses the Large-Eddy-Simulation (LES) technique to solve the compressible Navier-Stokes equations (conservation of mass and momentum) and the

Linear-Eddy-Model (*LEM*) [4–7] to solve the species and energy conservations equations. The solver is denoted as *LEMLES* [8,9].

The first section focuses on the description of the *LEMLES* implementation and its ability to resolve all turbulent and chemical scales involved in the turbulent propagation of a premixed flame. The second section deals with the cost and technique used to compute the chemical reaction rates. The computational cost of classical techniques is prohibitive and a simple, cost effective and accurate method for the computation of the reaction rates is described. The next section presents the computational cost of the different *LEMLES* simulations. The last section focuses on flame extinction via the introduction of heat losses.

2. *LEMLES* Formulation and Implementation

For brevity, the *LES* model is not described here. Details can be found elsewhere [10]. While the *LES* methodology solves the flow scales up to the cut-off size determined by the grid size and models the effects and characteristics of the flow structures smaller than the grid size, the *LEM* technique solves the species and energy equations up to the Kolmogorov scale. This is made possible by the use of a one dimensional grid (also called *LEM* domain) embedded in each *LES* cells. Species molecular diffusion, chemical reactions and turbulent convection from structures smaller than the grid size are simulated on this 1-D computational domain while the *LES* information are used to convect the species and energy through the *LES* cells. Details of the *LEMLES* models can be found elsewhere [8].

The choice of the number of *LEM* cells per *LES* cell is a compromise between the need to resolve all turbulent and chemical length scales and the computational cost. The *LEM* resolution is chosen as 18 *LEM* cells per *LEM* line in this study. Fig. 1 (a) shows the typical distribution of the *LES* subgrid velocity fluctuations (u') in the combustion chamber. For each level of turbulence (i.e. for each u'), a *PDF* of the eddy size can be computed [11]. Using this eddy size *PDF* distribution, the expected eddy size ($\bar{L}$) is computed for all values of u' and is expressed as the number of *LEM* cells needed to resolve it. A minimum of six *LEM* cells is needed to resolve an eddy. All eddies that are larger than six *LEM* cells are fully resolved. Fig. 1 shows that 97 % of subgrid eddies present in the combustion chamber are resolved (Domain (A)).

It is also of great interest to evaluate the resolution of the LEM line with regards to the resolution of the flame/eddy interactions. Fig. 2 shows the *CDF* distribution of the eddy size and the Karlovitz number ($Ka=(\delta_F/\eta)^2$) as a function of u' and for $\Phi=0.5$. Fig. 2 shows that both the flamelet and the *TRZ* regime are fully resolved (domain (I)). 6 % of the domain has a level of turbulence corresponding to the *BRZ* regime (domain (II) and (III)). In domain (II), all scales relevant to the flame/eddy interactions are resolved at the *LEM* level whereas these interactions are under-resolved in domain (III). It is important to note Fig. 2 shows that the flame does not propagate in the flamelet regime ($Ka<1$). This is due to the fact that only *LES* subgrid turbulent scales are considered in Figs. 1 and 2. For low level of u', the frequency of occurrence of subgrid eddies is negligible and only *LES* resolved eddies affect the flame front. *LES* resolved eddies are larger than the flame front. Therefore, for low level of turbulence, the flame propagates in the flamelet regime.

The flame front thickness is 0.15 mm and 1.2 mm for $\Phi=1.0$ and $\Phi=0.45$ (equivalence ratio used for the study of flame propagation. See section 5), respectively. In the flame region, the typical LES resolution is 0.45 mm. Thus, on average, with 18 LEM cells per LES cell, the flame is resolved using 5 LEM for $\Phi=1.0$ and 40 LEM cells for $\Phi=0.45$.

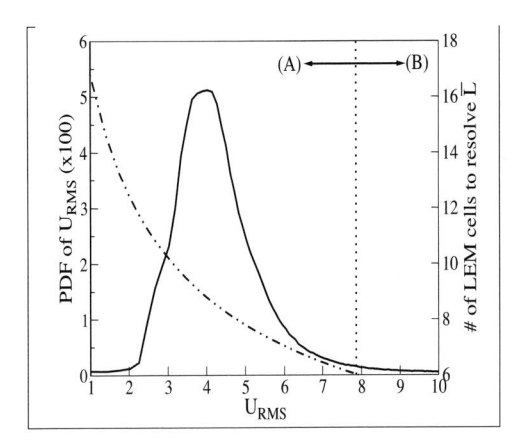

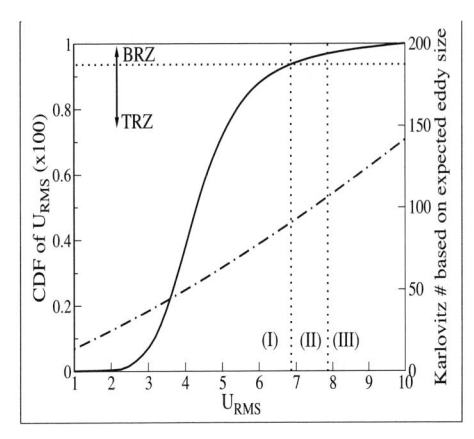

Figure 1. PDF distribution of the subgrid velocity fluctuations u' $(-)$ and expected eddy size $\overline{L}$ $(-\cdot\cdot-)$ as a function of u'. Domain (A) and Domain (B) are defined in sec. 2.

Figure 2. CDF distribution of the subgrid velocity fluctuations u' $(-)$ and Karlovitz number $(-\cdot\cdot-)$ associated with the expected eddy size $\overline{L}$. Domain (I), (II) and (III) are defined in sec. 2.

3. Efficient Computation of the Chemical Rates

The mass reaction rate is computed using either a 5-species, 1-step [12] chemical mechanism or a 8-species, 4-steps [13] chemical mechanism.

Direct integration is performed using the $DVODE$ (Double precision and Variable coefficient ODE) solver and is proved to have a prohibitive cost when the computational times of $LEMLES$ are compared to the computational cost of LES simulations using the $G-equation$ model [14] (see Table 1). $LEMLES$ simulations are 28 and 80 times more expensive when the 5-species and 8-species mechanism are used, respectively. This cost is divided by a factor of 10 when the reaction rates are not computed. The use of an In-Situ-Adaptive-Tabulation ($ISAT$) method was proved to not speed-up the chemical rates computation. The idea behind the $ISAT$ [15] method is to store reaction rates computed via direct integration as a function of the initial species mass fraction and temperature. As the computation progresses, more and more reaction rates are obtained by retrieving information from the table rather than re-computing them via direct integration. As long as the table is not too large, the computational time needed to retrieve information is

Table 1
Comparison of the computational cost of different approaches for the 5- and 8-species mechanisms. The simulation using a level set approach like the $G - equation$ model ($GLES$) is used as a reference.

Model	Time (compared to $GLES$)
$LEMLES$ - 5-species - No reaction	3
$LEMLES$ - 5-species - Reaction	7
$LEMLES$ - 5-species - Reaction (DVODE)	28
$LEMLES$ - 8-species - No reaction	6
$LEMLES$ - 8-species - Reaction	28
$LEMLES$ - 8-species - Reaction (DVODE)	80

smaller than the time required to perform the direct integration. In other studies [7], this speed-up was found to be of the order of 30 for a 16-species mechanism. The ratio of the $ISAT$ table size (N_{ISAT}) to the total number of computational points (N_{PTS}) is a critical parameter. In the previous study [7], the number of mixture states (species mass fraction and temperature) that could be stored in the table was of the order of 10,000 while the number of computational points was of the order of 250 ($N_{ISAT}/N_{PTS} = 40$). In the computations presented here, N_{ISAT} is chosen to be 20,000 while N_{PTS} is 320,000. In this case, $N_{ISAT}/N_{PTS} = 0.0625$. When $N_{ISAT}/N_{PTS} >> 1$, the table is never filled and quasi-irrelevant information (i.e. mixture states that have a low probability of occurrence) are stored without impacting the retrieval speed. However, when $N_{ISAT}/N_{PTS} << 1$, the $ISAT$ table is saturated after only a limited number of computational steps and this is not acceptable. Thus, in the computation performed in this study, the limiting factor for the use of $ISAT$ is the fast rate of information storage in the table. One solution to achieve a slower growth of the $ISAT$ table is to reduce the accuracy of the retrieval of information in the table. For obvious reasons, this solution can not be considered.

A simple, cost effective and accurate method for the computation of the reaction rates was designed. This method is valid for both chemical mechanisms. When the $DVODE$ solver is used, the computation of the reaction rate per LES time step is performed by dividing the actual LES computational time step (Δt_{LES}) and performing a direct computation of the reactions rate over this smaller time step (Δt_{DVODE}), which is computed by the $DVODE$ solver. It was found that the time step computed by the $DVODE$ solver is fairly constant ($\Delta t_{DVODE} = \Delta t_{LES}/K$) and K can be determined by studying the chemical mechanism on a simple $1D$ problem. K is an input to the LES program. Replacing the use of $DVODE$ to compute the local Δt_{DVODE} by a fixed Δt_{DVODE} speeds-up the $LEMLES$ computation by a factor of 4 (see Table 1). When compared to the computation of the reaction rate when the $DVODE$ solver is used, the error is less than 1% for both the 5- and 8-species mechanism.

4. Computational Performances

The total computational times of one Flow-Through-Time (FTT), or 50,000 computational time-steps, are $3.8*10^{-9}$, $2.6*10^{-8}$ and $1.06*10^{-7}$ CPU seconds per grid point and per time step for the level set approach, the $LEMLES$ approach with 5-species and the $LEMLES$ approach with 8-species, respectively. The level set approach model corresponds to the LES simulation where a flamelet model like the $G-equation$ model is used to model the flame propagation (this approach is denoted $GLES$). Computations time are estimated on a Compaq SC45. This computer has 128 nodes connected by a 64-port, single-rail Quadrics high-speed interconnect switch. Each node contains four 1,000 MHz Alpha EV 68 processors and 4 gigabytes of RAM.

A full computation includes the generation of a stable initial flow and scalars fields used as initial conditions for $LEMLES$. This initial flow is obtained using the fast $GLES$ solver. As a rule of thumb, this requires 3 FTT. Data are collected with the $LEMLES$ over 3 FTT after an initial FTT considered as a transition between the $GLES$ and $LEMLES$ computations. Therefore, a complete simulation requires 4610 and 17090 hours, for the 5- and 8-species mechanism, respectively. In this study, 50 processors are used and the wall-clock time are 98 (5 days) and 341 (2 weeks) hours, for the 5- and 8-species mechanism, respectively.

In conclusion, the 8-species mechanism is more accurate but this accuracy has a cost (this mechanism is 3.5 times more computationally expensive than the 5-species mechanism). The computational cost of a $LEMLES$ simulation with detailed chemistry (i.e. more than 5 species) is the biggest disadvantage of the model. In conclusion, the choice of the chemical mechanism must be done after a careful cost-benefits analysis.

5. Flame Extinction

Premixed flame propagation is studied in a real case scenario: combustion inside a full scale industrial dump combustion chamber. The total combustion chamber length is 0.21m. The length of the inflow pipe is 0.015m. The radius of the combustion chamber and the inflow pipe are 0.045m and 0.017m, respectively. The centerline region is meshed using a Cartesian grid and the rest of the domain is meshed using a cylindrical grid. Both grids are continuous. For the cylindrical grid, the resolution is 140x75x81 grid points in the axial, radial and circumferential directions, respectively. For the Cartesian grid, the resolution is 140x21x21 grid points in the axial, horizontal and vertical directions, respectively. The inflow pressure is 6 bar, the inflow temperature is 644 K and the reacting mixture is a premixed mixture of methane and air. The reactant enter the combustion chamber in a swirling manner. The swirl number is 1.1.

A first simulation does not exhibit any flame extinction. Hence, perturbations to the flame propagation are implemented in the form of local heat losses. This technique is described in the next section.

Heat losses occur at the walls of the combustion chamber. The walls materials cannot withstand the high product temperature and walls are usually cooled or cool air is injected via holes in the combustion chamber to create a protective colder layer of gases. Considering wall related heat losses is not practical in LES because the LES resolution does not allow to resolve the wall thermal boundary layer. However, one can simulate heat losses by directly including them locally. Following the study of [16], heat losses (dT_{HL}/dt) are included in the energy balance in the form of a linear term $\frac{dT_{HL}}{dt} = \frac{h}{\rho C_P}(T - T_1)$, where T_1 is the reactant temperature. The coefficient h is defined as: $h = \lambda Sc^2(\frac{S_L}{\nu})^2 \frac{c}{\beta}$. $\lambda = \mu C_P/P_r$,

β is the reduced activation energy [17] and c is a dimensionless heat-loss coefficient. Heat losses are taken into account only in the post-flame region and are a direct function of the product temperature and $c = f(r)c^*$ where $c^*=10^{-4}$ and $f(r)=max((1-r/0.015),0)$ with r (expressed in meters) is the distance from the considered point to the edge of the inflow pipe.

The sum of all local fuel reaction rates (in the region from the dump plane to 5.0 mm downstream of the dump plane) is computed as a function of time and is plotted in Fig. 3. The maximum absolute value of the sum is used to normalize the data. The first mode of the oscillations is extracted. The flame is lifted-off when the normalized sum of the reaction rate is close to 0 (see Fig. 4). The flame is attached to the inflow pipe lips when the normalized sum of the reaction rate is close to -1. The first mode of the oscillations has a frequency of approximately 1,000 Hz. The flame extinction frequency matches the longitudinal half quarter mode of the combustion chamber.

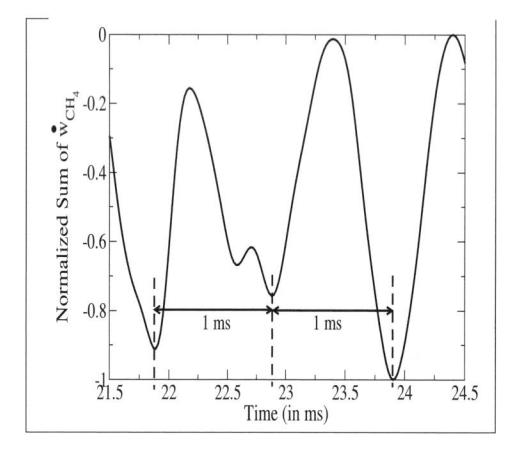

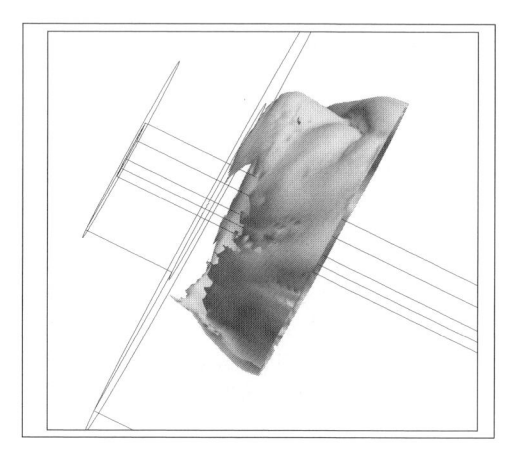

Figure 3. Normalized sum of the reaction rate in the region close to the dump plane as a function of time. The data is smoothed in order to extract the first frequency mode.

Figure 4. Instantaneous snapshot of a reaction rate iso-surface ($\dot{w}_{CH_4}$=-5 s^{-1}). Lines delimiting the combustion chamber and inflow pipe are shown. The flame does not propagate, thus lifts-off, in the region close to the combustion chamber dump plane. Only a portion of the reaction rate isosurface is shown for clarity.

6. Conclusion

Premixed flame extinction and lift-off was simulated in a *LES* computation. All subgrid length scales relevant to turbulent combustion as well as all relevant combustion regimes are resolved through the use of the Linear-Eddy Mixing Model. A fast and accurate method to compute the reaction rates was developed and tested. Even though still computationally expensive, this study demonstrates the capability of the *LEMLES* to simulate flame propagation in full scale combustion chamber. Method like Artificial Neural Networks may greatly speed-up the computation of the reaction rate. Such a method is under development. Even if the numerical implementation is more complex than flamelet models, it is valid for premixed, partially-premixed and non-premixed combustion and is considered as an alternative to other techniques, like the *PdF* models.

REFERENCES

1. A. Buschmann, F. and Dinkelacker, Measurement of the Instantaneous Detailed Flame Structure in Turbulent Premixed Combustion, Twenty-Sixth Symposium (International) on Combustion, The Combustion Institute, (1996) 437-445.
2. M.S. Mansour, Y.C. Chen and N. Peters, The Reaction Zone Structure of Turbulent Premixed Methane-Helium-Air Flames Near Extinction, Twenty-Fourth Symposium (International) on Combustion, The Combustion Institute, (1992) 461-468.
3. D. Most, V. Hoeller, A. and Soika, F. and Dinkelacker and A. Leipertz, Ninth International Symposium on Application of Laser Techniques to Fluid Mechanics, (1998) 25.6.1-7.
4. A. R. Kerstein, Linear-Eddy Model of Turbulent Transport II, Combustion and Flame, 75 (1989) 397-413.
5. A. R. Kerstein, Linear-Eddy Model of Turbulent Transport III, Journal of Fluid Mechanics, 216 (1990) 411-435.
6. V. K. Chakravarthy and S. Menon, Modeling of turbulent premixed flames in the flamelet regime, Proceedings of first International Symposium on Turbulent and Shear Flow Phenomena, Begel House, (1999) 189-194.
7. V. Sankaran and S. Menon, The Structure of Premixed Flame in the Thin-Reaction-Zones Regime, Proceedings of the Combustion Institute, 28 (2000) 203-210.
8. G. Eggenspieler and S. Menon, Modeling of Pollutant Formation near Lean Blow Out in Gas Turbine Engines, Direct and Large Eddy Simulation V, Kluwer Press, (2003).
9. V. Sankaran, I. Porumbel and S. Menon, Large-Eddy Simulation of a Single-Cup Gas Turbine Combustor, (2003) AIAA-2003-5083.
10. W. W. Kim, S. Menon and H.C. Mongia, Large-Eddy Simulation of a Gas Turbine Combustor Flow, Combustion Science and Technology, 143 (1999) 25-62.
11. S. Menon, P. A. McMurtry and A. R. Kerstein, A Linear Eddy Mixing Model for Large Eddy Simulation of Turbulent Combustion, LES of Complex Engineering and Geophysical Flows, Galperin, B. and Orszag, S., Cambridge University Press, (1993).
12. C. K. Westbrook and F.L. Dryer, Simplified Reaction Mechanisms for the Oxidation of Hydrocarbon Fuels in Flames year, Combustion Science and Technology, 27 (1981), 31-43.
13. J. M. Card, J. H. Chen, M. Day and S. Mahalingam, Direct Numerical Simulation of Turbulent Non-Premixed Methane-Air Flames Modeled with Reduced Kinetics, Studying turbulence using Numerical Simulation Databases - V, Center for Turbulence Research, (1994) 41-54.

14. N. Peters, Turbulent Combustion, Cambridge Monographs on Mechanics, (2000).
15. S. B. Pope, Computationally Efficient implementation of combustion chemistry using in situ adaptive tabulation, Combustion Theory Modelling, 1 (1997) 41-63.
16. T. Poinsot, D. Veynante and S. Candel, Quenching processes and premixed turbulent combustion diagrams, Journal of Fluid Mechanics, 228 (1991) 561-606.
17. F. A. Williams, Combustion Theory, Second Edition, The Benjamin/Cummings Publishing Company, Inc., (1985).

Parallel computation of the flow- and acoustic fields in a gas turbine combustion chamber

Szasz, R.Z.[a], Mihaescu, M.[a], Fuchs, L.[a]

[a]Dep. of Heat and Power Engineering, Lund Institute of Technology,
P.O. Box 118, 221 00 Lund, Sweden

A parallel solver of a splitting approach for Computational Aero-Acoustics (CAA) is presented. The solver is based on a hybrid approach, where the flow field and acoustic fields are computed as separate but coupled processes. The Large Eddy Simulation (LES) based flow solver is parallelized using a domain decomposition approach. The performance of the solver has been evaluated and good parallel performance has been obtained. The acoustical solver utilizes the LES data for computing the acoustical sources. The combined solver has been successfully applied for the computation of the aeroacoustic field in a model gas turbine combustion chamber.

1. INTRODUCTION

A major issue in current gas turbine combustor design is the appearance of thermo-acoustic instabilities. Small acoustic fluctuations generated by the turbulent flow field or by spatial or temporal variations in heat release rate may be amplified through a resonance with the eigen-modes of the combustion chamber. Such thermo-acoustic instabilities may lead to structural failure or in mild cases to shorter life-time of the combustor.

Here, we focus on the generation and propagation of the acoustic waves generated by the turbulent flow field. One way to deal with the acoustic field is to solve the compressible Navier-Stokes equations. This method, however, is inefficient for low-Mach number flows. Generally speaking solving a full 3-D compressible flow problem requires the solution of five PDEs together with the equation of state. The corresponding semi-compressible problem is simpler and easier to solve since the stability limits are not determined by the speed of sound (which is large relative to the flow speed at low Mach numbers). Additionally, for external CAA problems, one is not interested in the far-field flow and hence the computation of the flow field there becomes redundant. Under these conditions it is computationally more efficient to split the dependent variables into a semi-compressible component and a correction accounting for the error in the equation of state. Such a splitting leads to the semi-compressible Navier-Stokes equations for the flow field and the non-acoustic pressure and temperature. These variables can be used to compute the acoustical source and use a lower order PDE (inhomogeneous wave equation or linearized Euler equations) for the acoustical component.

In the following we use such a decomposition. The approach allows one to use two completely different PDEs and solvers for the flow and acoustic fields. The flow solver will determine the flow field and will provide the acoustic source terms which have to be transferred to the acoustic solver. In this way only one equation has to be solved for the acoustic field. Another advantage is that the grid used in the acoustic solver does not necessarily has to be the same as the one used in the flow computations. This advantage is especially pronounced when the acoustic sources are located in a small area of the computational domain.

The flow and acoustic fields in the same geometry have been computed before using a serial version of the solver [1]. There is a need for a relatively fine grid in order to capture the complex geometry of the inlet and to solve the turbulence scales in the inertial subrange. The increased number of computational cells, however, demands the parallelization of the software to maintain the computational effort within reasonable limits. The flow solver used in the present computations has been parallelized previously and is characterized by a good speed-up performance, the communication time being of the order of 4% of the total computational time[2].

The separation of the flow solver and the acoustic solver gives an inherent possibility to parallelize the solution procedure, by solving the two solvers as separate processes. However, one has to examine if the gain obtained by parallelization will overcome the computational time lost by transferring the acoustic sources from the flow solver to the acoustic one.

Here, we present a parallel solver which is based on both domain decomposition and equation decomposition. The primary objective of the paper is to evaluate the performance of the solver. Furthermore, the solver will be used to evaluate the aero-acoustic field in a model gas turbine burner geometry.

2. NUMERICAL METHODS

In the present paper we are using the same method as the one used by Mihaescu et al.[3]. The incompressible Navier-Stokes equations (Equations 1, 2) are solved on a Cartesian staggered grid, with the possibility of having local refinements. The energy equation is written in terms of a non-dimensionalized temperature (Equation 3). The solver is based on the Finite Differences method. Third and fourth order schemes are used for the discretization of the convective and diffusive parts, respectively. The high-order accuracy is achieved using a defect-correction approach. Convergence is enhanced by using multi-grid technique.

$$\rho_{,t} + (\rho u_i)_{,i} = 0 \qquad (1)$$

$$(\rho u_i)_{,t} + (\rho u_i u_j)_{,j} = -p_{,i} + 1/Re(u_i)_{,jj} \qquad (2)$$

$$T_{,t} + u_i T_{,i} = 1/(RePr) T_{,ii} \qquad (3)$$

In order to accurately capture the sound sources, one has to use appropriate turbulence models. Of course, the most exact solution would be provided by Direct Numerical Simulation (DNS). However, DNS is computationally expensive (the computational effort is almost proportional to Re^3), and is limited for relatively low Reynolds number cases. Models based on Reynolds-averaging of the Navier-Stokes equations are not suitable for acoustic computations, since they are based on a time averaging procedure which filters out the fluctuations of the flow field. A good compromise is offered by Large Eddy Simulation (LES) which is computationally less expensive than DNS ($\approx Re^2$), and in the same time resolves the large, energy containing eddies. Rembold et al.[4] computed the acoustic field generated by a rectangular low Reynolds number jet using both DNS and LES. Their results showed that the low frequency part of the radiated noise was well captured by LES, and corrections for the subgrid scale contributions to the acoustic spectrum did not change the LES prediction. Here, turbulence is modeled by LES, without an explicit SGS model.

The acoustic sources are computed each time-step and transferred to the acoustic solver. The acoustic solver resolves an inhomogeneous wave equation (Equation 4) which is discretized as well with finite differences. The source term is given by equation 5 and takes into account the noise generated due to both velocity and entropy fluctuations. Since the present case is isotherm, the second term on the RHS in equation 5 is zero. In the acoustic solver, second order schemes are used for the discretization of all terms.

$$(\rho')_{,t} - 1/(M^2)(\rho')_{,ii} = T_{ij,ij} + aH_{,t} \tag{4}$$

$$T_{ij} = \rho u_i u_j + \delta_{ij}(p - 1/M^2 \rho) \tag{5}$$

Here H is the rate of heat-release (and s is the appropriate coefficient). Thus, the acoustic sources are attributed to variations in the spatial distribution of the $u_i u_j$ tensor, the compressibility effect and the last term accounts for the acoustic source due to heat-release. The parallelization of the solver is based on the MPI libraries. The acoustic solver is not expected to be computationally expensive (as compared to the flow solver), thus it is solved on a single processor. On the other hand, the flow computations are solved on the remaining nodes, the computational effort being distributed among the processors by using domain decomposition (Figure 1). The acoustic sources are computed by the flow solvers after each time-step and sent to the acoustic solver. The interpolation of the acoustic sources onto the acoustic grid is carried out by the acoustic solver.

The communication time between the flow solvers and the acoustic solver is expected to be larger than the time needed for communication among the flow solvers. First, the flow solvers exchange only boundary data, while the acoustic solver has to receive data from the whole volume. Second, the flow solvers communicate with at most two other flow solvers (provided that the domain is decomposed in a single direction), while the acoustic solver receives data from each flow solver. However, the fact that only the source term has to be transferred (along with information related to the position of the acoustic sources, but this information is small for cartesian grids) suggests that the communication time between the acoustic solver and flow solvers might be shorter than the communication time between the flow solvers, especially if the number of processes is small.

Since the computational domains of the flow solvers are overlapping, special care has to be taken to not impose twice the acoustic sources in the overlapping region. Here, this is done by using a flag which marks the cells which received acoustic sources from previously treated flow processes.

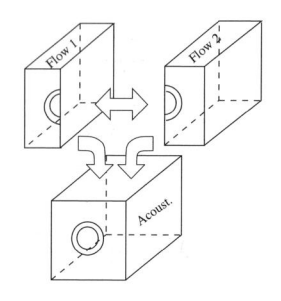

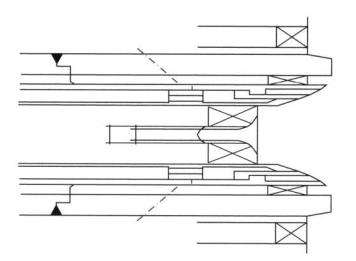

Figure 1. Distribution of the computational domain among the parallel processes

Figure 2. The geometry of the nozzle

2.0.1. PROBLEM SET-UP

Three co-annular swirling jets with different swirl angles and mass flows are entering a model gas turbine combustion chamber (see Figure 2 for a sketch of the nozzle). All dimensions are non-dimensionalized with the outer diameter of the outer swirler. The combustion chamber has a rectangular cross section of 3x3 diameters and a length of 8 diameters.

Constant top-hat velocity profiles are imposed in each of the swirler inlets, while at the combustion chamber outlet flux-conserving zero gradient boundary conditions are used. The walls are treated as non-slip. For the acoustic solver all boundaries are treated as reflecting, except the outlet boundary where non-reflecting boundary conditions are imposed. The Reynolds number was set to 20000, while the Mach number was 0.15. Totally 80000 time-steps were computed, but only the last 70000 were used to compute the statistics.

The computations have been carried out on a Linux cluster. The computational nodes of the cluster are equipped with AMD Opteron 148 2.2 GHz processors and 1 GB RAM. The system interconnect consists of a GigaBit Ethernet network connected to a stacked D-Link switch. The solvers were compiled with the Portland Group Fortran compiler v.5.2. For parallel running MPICH 1.2.5 has been used.

3. RESULTS

3.1. Solver performance

To evaluate the performance of the solver two set of tests have been carried out. First, the computational and communication times have been measured for different grid sizes (both in the flow and acoustic solvers). For these tests, the domain was split in two for the flow computations. Second, keeping a constant grid size in the acoustic solver the communication/computational times have been measured for increasing number of processes (up to ten) for the flow solver. In practical applications, one is interested to use the maximum number of grid points (allowed by the available memory or by the required computational effort). Thus, in the second series of tests, we chose to keep a constant number of computational cells per flow process (instead of dividing the same grid in more and more parts).

Figure 3 shows the time spent with communication among flow processes, as percentage of the total time. Three flow grid sizes (with approximately 0.4, 1.5 and 3.4 million cells) and three acoustic grid sizes (roughly 1.8, 4.6 and 8.7 million cells) are considered. As it was expected the acoustic grid size has no effect on the communication time between the flow processes. Increasing the flow grid size leads to a decrease in the relative time spent for communication. Even for the coarsest considered grid the communication time is only about 4%.

The time spent for communication between the flow processes and the acoustic process is shown in Figure 4. As one can observe, The acoustic grid size has again negligible influence on the timing results. Large grid sizes used in the flow solvers, on the other hand, lead to a better performance, the relative time spent for communication decreases as the grid size increases. One can also observe, that the time spent by flow solvers for communication with the acoustic process is about half of the time spent with communication with the other flow processes (compare Figures 3 and 4).

With respect to the acoustic solver, the time spent with communication dominates over the computational time (see Figure 5). This is due to the fact, that the acoustic computations are much faster than the flow computations and the acoustic process has to wait for the flow processes to receive data. This is why larger flow grid sizes lead to longer time spent for communication. Even if the relative time spent for communication decreases with increasing acoustic grid size, the time spent for waiting the source terms is still significant. This suggests that parallelization of the acoustic solver is not needed. Additionally, the available extra time can be used e.g. by implementing more complex schemes or carrying out other tasks.

One advantage of solving the acoustic problem as a separate process is that the interpolation of the acoustic sources from the flow grid(s) to the acoustic grid can be done by the acoustic solver. As one can see in Figure 6, the time needed for interpolation is significantly longer than the time needed to transfer the acoustic sources from the flow processes to the acoustic process (Figure 4), thus the parallelization of the solvers is definitely worth.

To assess the speed-up of the solver a series of tests have been run with increasing number of processes (2 - 10) for the flow solver. The load of the flow solvers was kept constant (approximately 1.5 million cells). As it is shown in Figure 7, the total time per time-step increased with approximately 17% when the number of flow processes has been increased from 2 to 10. Thus, the speed-up performance of the code is considered to be good. No tests have been carried out for larger number of processors since ten is already towards the upper limit of the amount of processors which a user can get on a cluster.

Figure 8 shows the relative time spent for communication among flow processes (circles) and the flow processes and the acoustic process (triangles). One can observe that the increase of the computational time when the number of processes is increased is mostly due to the increase of the time needed for the communication among the flow processes.

3.2. Flow field computations

The parallel code described in the previous sections has been applied to compute the flow and acoustic fields in a model gas turbine combustion chamber. The flow field has

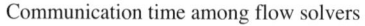

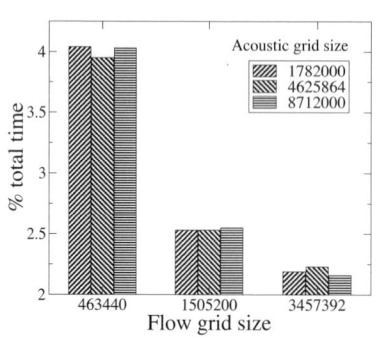

Figure 3. Relative communication time among the flow processes

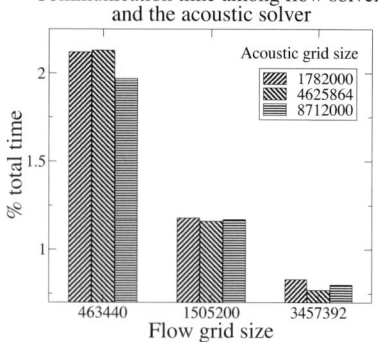

Figure 4. Relative communication time among the flow processes and the acoustic process

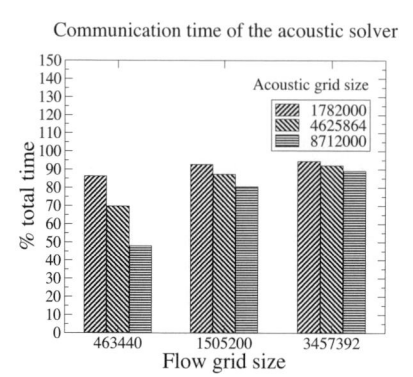

Figure 5. Relative time spent for communication by the acoustic solver

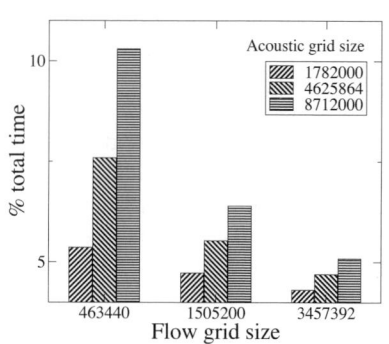

Figure 6. Relative time spent for interpolating the acoustic sources

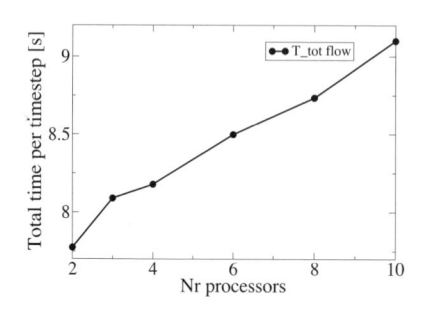

Figure 7. Total time per time-step for increasing number of flow processes

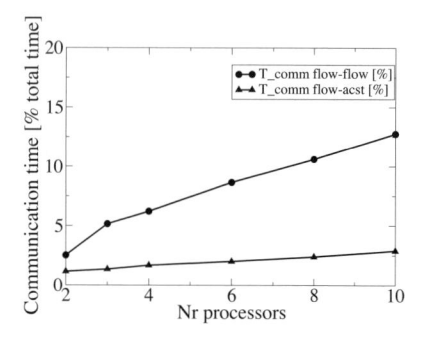

Figure 8. The speed-up of the parallel solver

been computed on two processes, while the acoustic field has been solved on a third processor.

Figure 9 presents the isocontours of the average axial velocity field in a longitudinal cross section. One can observe that the three co-axial swirling jets are merged close to the combustion chamber inlet. The jets attach to the combustion chamber wall around one diameter downstream of the inlet and an elongated inner recirculation zone is formed. Additionally, outer recirculation zones can be seen in the corners of the combustion chamber.

Due to the presence of multiple shear layers strong turbulence is generated at the inlet. The high turbulence levels, on the other hand, are expected to generate strong noise. This is confirmed by Figure 10 where the average acoustic sources are shown. One can observe that the magnitude of the acoustic sources decay rapidly with increasing distance from the inlet region. As it is seen in Figure 11 the largest acoustic density fluctuations are found in the regions with the strongest acoustic sources. The magnitude of the acoustic density fluctuations decay as well rapidly towards the exit of the combustion chamber.

To evaluate the radiated sound field the frequency spectrum of the acoustic density fluctuation has been computed in monitoring points situated on the symmetry axis at 5 and 7 diameters downstream from the nozzle. Two additional monitoring points were located at the same axial positions, but close to the combustion chamber walls, at 1.22 diameters from the symmetry axis. Figure 12 shows the frequency spectra obtained at the monitoring points situated on the symmetry axis. As one can observe, in both points there is a dominant frequency at around $St = 1$. The two spectra have peaks at the same frequencies, however, the amplitude of the higher frequencies is lower in the point situated further downstream. The frequency spectra at two monitoring points located at the same axial but different radial positions (not shown here) are practically identical.

4. CONCLUSIONS

A hybrid approach has been presented for solving problems in Computational Aero-Acoustic (CAA). The parallelized solver has been applied to the computation of the LES and the generated acoustical fields in a model gas turbine combustion chamber. The acoustic field has been computed parallel to the flow field. The flow field computations being computationally more expensive than the acoustic computations have been parallelized using a domain decomposition approach. The timing results have shown that the solver has good speed-up performance. The computation of the acoustic field as a separate process turned out to be efficient, the gain by solving the acoustic field on a different processor overcame the loss of efficiency due to the time needed to transfer the source terms. The available extra time to the acoustic solver makes possible the implementation of more complex methods to solve the acoustic field or to carry out additional tasks without loss of computational efficiency. The solver can be easily improved to handle e.g. multiple identical noise source regions or moving noise sources.

5. Acknowledgements

The authors would like to acknowledge the financial support of the Swedish Research Council (VR). The computational time offered at the Center for Scientific and Technical

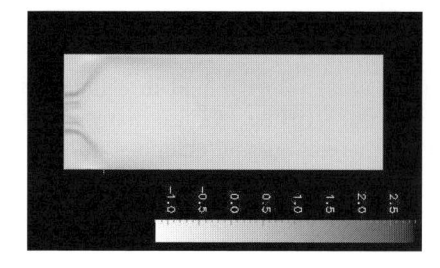

Figure 9. Average axial velocity

Figure 10. Average acoustic sources

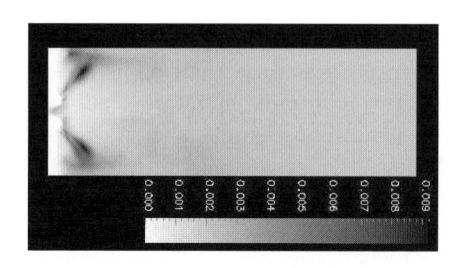

Figure 11. RMS of the acoustic density fluctuation

Figure 12. Frequency spectrum of the acoustic density fluctuation at monitoring points on the symmetry axis

Computing at Lund University (LUNARC) and within the allocation program SNAC is highly appreciated.

REFERENCES

1. R. Szasz, Numerical modeling of flows related to gas turbine combustors, Ph.D. thesis, Lund Institute of Technology (2004).
2. R. Szasz, L. Fuchs, Numerical modeling of flow and mixing of single and interacting swirling co-annular jets, in: N. Kasagi, J. Eaton, R. Friedrich, J. Humphrey, M. Leschziner, T. Miyauchi (Eds.), Proc. of Third International Symposium on Turbulence and Shear Flow Phenomena, Vol. 2, 2003, pp. 663–668.
3. M. Mihaescu, R. Szasz, L. Fuchs, E. Gutmark, Numerical investigations of the acoustics of a coaxial nozzle, AIAA paper 2005-0420.
4. B. Rembold, J. Freund, M. Wang, An evaluation of LES for jet noise prediction, in: Proc. of Summer Program, Center for Turbulence Research, Stanford University, 2002, pp. 5–14.

Performance of Lattice Boltzmann Codes for Navier-Stokes and MHD Turbulence on High-End Computer Architectures

George Vahala,[a] Jonathan Carter,[b] Min Soe,[c] Jeffrey Yepez,[d] Linda Vahala,[e] and Angus Macnab[f]

[a]Department of Physics, William & Mary, Williamsburg, VA 23187, USA

[b]Lawrence Berkeley National Laboratory, Berkeley, CA 94720, USA

[c]Rogers State University, OK 74017, USA

[d]Air Force Research Laboratory, Hanscom AFB, MA 02139, USA

[e]Old Dominion University, Norfolk, VA 23529, USA

[f]CSCAMM, University of Maryland, MD 20742, USA

Keywords: Lattice Boltzmann; parallelization and vectorization; turbulence; vortex-merging; vector and scalar processors; entropic lattice Boltzmann; discrete H-theorem

1. LATTICE BOLTZMANN ALGORITHM

The accurate resolution of the nonlinear convective derivatives in the incompressible Navier-Stokes equations causes much concern – especially for turbulent flows in nontrivial geometry. Lattice Boltzmann (LB) algorithms [1] embed the fluid problem into a higher dimensional (kinetic) phase space within which the solution trajectory is simpler, and hence easier to compute accurately, than that for the nonlinear fluid turbulent flow. Moreover, the LB kinetic equation can be discretized on a specific phase velocity lattice with just linear advection and a minimal number of discrete velocities required so that the nonlinear macroscopic equations can be recovered in the Chapman-Enskog limit [1]. The nonlinear convective derivatives of the Navier-Stokes equations can be recovered from the LB collision operator by a simple polynomial expansion in the mean velocity. The discrete LB scheme (second order accurate in space-time) for the distribution functions in 3D is (in LB units)

$$f_\alpha(\mathbf{x}+\mathbf{c}_\alpha,t+1) = f_\alpha(\mathbf{x},t) - \tau^{-1}\left[f_\alpha(\mathbf{x},t) - f_\alpha^{eq}(\mathbf{x},t)\right], \quad \alpha = 1....27 \qquad (1)$$

$\mathbf{c}_\alpha, \alpha = 1..27$ are the 27 discrete velocities on a body-centered cube with speeds: 1 (the 6 face centers), $\sqrt{2}$ (the 12 dihedral centers), $\sqrt{3}$ (the 8 corners) and a rest speed. τ is the relaxation time for collisions to drive $f_\alpha \to f_\alpha^{eq}$ where f_α^{eq} is an appropriately chosen relaxation distribution function. The viscosity $\mu = (2\tau-1)/6$ so that the Reynolds number $\to \infty$ as $\tau \to 0.5_+$. In the standard LB approach, a polynomial representation of f_α^{eq} in terms of the mean velocity $\mathbf{u}$ is sufficient to recover the Navier-Stokes incompressible equations to $O(Ma^3) = O(u^3/c_s^3)$, where c_s is the local sound speed. MHD can be modeled by LB if one further introduces a vector distribution function $\mathbf{g}$ for the magnetic field $\mathbf{B}$, with the LB-kinetic equations coupled through the corresponding the relaxation distribution functions $f_\alpha^{eq} = f_\alpha^{eq}(\mathbf{u},\mathbf{B})$, $\mathbf{g}_\alpha^{eq} = \mathbf{g}_\alpha^{eq}(\mathbf{u},\mathbf{B})$ through polynomial expansions in the moments $\mathbf{u}$ and $\mathbf{B}$ [2-4]. The viscosity and resistivity are controlled by the corresponding relaxation rates in the kinetic equations.

Unlike spectral methods, these explicit LB schemes are applicable to arbitrary geometries, using second order accurate bounce-back rules for spatial nodes near/on the boundary [1, 5]. However nonlinear numerical instabilities arise as the transport coefficients are reduced to achieve higher and higher Reynolds and magnetic Reynolds numbers. These instabilities arise in the straightforward LB scheme because there are no realizeability constraints imposed that enforce the non-negativity of the discrete distribution functions. As the transport coefficients are sufficiently reduced, the distribution functions become negative in certain regions of the domain. However, unconditionally stable explicit LB schemes have now been developed [6-8] for Navier-Stokes turbulence. These so-called Entropic Lattice Boltzmann (ELB) schemes enforce a discrete H-theorem constraint that ensures positive-definiteness of the distribution functions at arbitrary high Reynolds numbers at every grid point and at each time step. In ELB, a two-parameter generalized collision operator is now introduced

$$f_\alpha(\mathbf{x}+\mathbf{c}_\alpha,t+1) = f_\alpha(\mathbf{x},t) - \beta\,\gamma(\mathbf{x},t)\left[f_\alpha(\mathbf{x},t) - f_\alpha^{eq}(\mathbf{x},t)\right] \qquad (2)$$

where $0 < \beta < 1$ is a fixed (arbitrary) parameter, playing a role similar to the relaxation time τ in the standard LB scheme. The function $\gamma(\mathbf{x},t)$ is introduced so that the generalized collision process obeys a discrete H-theorem, with the discrete H-function taking a Boltzmann-like form

$$H[\mathbf{f}] = \sum f_\alpha \ln[f_\alpha / w_\alpha] \qquad (3)$$

where the weights w_α are dependent on the chosen geometric lattice and particle speed. On minimizing the H-function subject to the local conservation of mass and momentum one obtains an explicit algebraic (but non-polynomial) expression for the equilibrium distribution function f_α^{eq} [8]:

$$f_\alpha^{eq} = \rho\, w_\alpha \prod_{j=1}^{3} \left(2 - \sqrt{1 + 3u_j^2}\right) \left(\frac{2u_j + \sqrt{1 + 3u_j^2}}{1 - u_j}\right)^{c_{\alpha j}} \qquad (4)$$

where $c_{\alpha j}$ is the jth Cartesian component of the streaming velocity $\mathbf{c}_\alpha$. Using a Newton-Raphson algorithm, the function $\gamma(\mathbf{x},t)$ is determined at each grid point and time step so that

$$H[\mathbf{f}] = H[\mathbf{f} - \gamma(\mathbf{f} - \mathbf{f}^{eq})] \quad , \quad \gamma \neq 0 \qquad (5)$$

This ensures that the collisional term in the entropic LB equation always increases the entropy everywhere and thus maintains the positive-definiteness of f_α throughout the (discrete) simulation. Local equilibrium occurs when $\gamma(\mathbf{x},t) = 2$. The effective viscosity of the entropic LB simulation and the (bare) molecular viscosities are given by

$$\mu_{eff}(\mathbf{x},t) = \frac{1}{6}\left[\frac{2}{\gamma(\mathbf{x},t)\,\beta} - 1\right] \quad , \quad \mu_{bare} = \frac{1}{6}\left[\frac{1}{\beta} - 1\right] \qquad (6)$$

In Sec. 2, we shall consider the parallelization and vectorization performance of LB on various platforms and then briefly discuss the expected changes for entropic LB. In Sec. 3, we discuss some results from entropic LB to interlocking multi-vortex layers in 2D turbulence. Results on the Earth Simulator for our 3D MHD LB-code [a two hour wallclock run on 4800 PEs for an 1440^3 grid yielding 26.25 Tflops, see Ref. 4] will be presented at the Gordon Bell session at SC|05.

2. HEC Platforms, Computational Implementation and Performance of LB

We consider five parallel HEC architectures, noting that the vector machines have higher peak performance and better system balance than the superscalar platforms. Additionally, they have high memory bandwidth (as measured by the STREAM benchmark) relative to peak CPU speed (bytes/flop), allowing them to more effectively feed the arithmetic units. Finally, the vector platforms utilize interconnects that are tightly integrated to the processing units, with high performance network buses and low communication software overhead.

Two superscalar platforms are considered. The IBM Power3 380-node *Seaborg* running AIX 5.2 (xlf compiler 8.1.1) with each SMP node consists of 16 375-MHz processors (1.5 Gflops/peak) connected to the main memory via the Colony switch using an omega-type topology. The Intel Itanium runs were performed on the *Thunder* system consisting of 1024 nodes, each containing 4 1.4-GHz Itanium2 processors (5.6 Gflops/s peak) and tunning Linux Chaos 2.0 (Fortran version ifort 8.1). The system is interconnected using Quadrics Elan4 in a fat-tree configuration.

The 3 state-of-the-art parallel vector systems considered here: the Cray X1, the Earth Simulator (*ES*), and the newly-released NEC SX8. The Cray X1 is designed to combine traditional vector strengths with the generality and scalability features of modern superscalar cache-based parallel systems. The computational core, called the single-streaming processor (SSP), contains 2 vector pipes running at 800 MHz. Each SSP operates at 3.2 Gflops/peak for 64-bit data. 4 SSP can be combined into a logical computational unit called the multi-streaming processor (MSP) with a peak of 12.8 Gflops/s – the X1 system can operate in either SSP or MSP modes and we present performance results using both approaches. The 4 SSPs share a 2-way set associate 2 MB data Ecache, a unique feature for vector architectures that allows extremely high bandwidth (25-51 GB/s) for computations with temporal data locality. The X1 node consists of 4 MSPs sharing a flat memory, and large system configurations are networked through a modified 2D torus interconnect. All reported X1 experiments were performed on the 512-MSP system running UNICOS/mp 2.5.33 (5.3 programming environment). The *ES* uses a traditional vector architecture, with no data cache and extremely high bandwidth to main memory. The 1000 MHz *ES* vector processor was a precursor to the NEC SX6, containing a 4-way replicated vector pipe with a peak performance of 8.0 Gflops/s per CPU. The *ES* contains 640 8-way nodes connected through a custom single-stage IN crossbar. The 5120-processor *ES* runs Super-UX, a 64-bit Unix operating system based on System V-R3 with BSD4.2 communication features. The newly-released NEC SX8 is the world's most powerful vector processor. The SX8 architecture operates at 2 GHz, and contains 4 replicated vector pipes for a peak performance of 16 Gflops/s per processor. The SX8 architecture has several enhancements compared with the ES/SX6 predecessor, including improved divide performance, hardware square root functionality, and in-memory caching for reducing bank conflict overheads. However, the SX8 in our study uses commodity DDR-SDRAM; thus, we expect higher memory overhead for irregular accesses when compared with the specialized high-speed FPLRAM (Full Pipelined RAM) of the ES. Both the *ES* and SX8 processors contain 72 vector registers each holding 256 doubles, and utilize scalar units operating at the half the peak of their vector counterparts. All reported SX8 results were run on the 36 node system (soon to be upgraded to 72 nodes).

Table 1 Processor characteristics of architectures

Platform	CPU/node	Clock (MHz)	Peak (Gflop/s)	Stream BW (GB/s)	Peak (Byte/Flop)
Power3	16	375	1.5	0.4	0.26
Itanium2	4	1400	5.6	1.1	0.19
X1	4	800	12.8	14.9	1.16
ES	8	1000	8.0	26.3	3.29
SX8	8	2000	16.0	41.0	2.56

This HLRS SX8 is interconnected with the NEC Custom IXS network and runs Super-UX (Fortran Version 2.0 Rev.313)

Table 2 Network characteristics of architectures

Platform	Network	MPI Latency (μs)	MPI BW (GB/s/proc)	Bisect BW (GB/flop)	Network Topology
Power3	Colony	16.3	0.13	0.09	Fat-tree
Itanium2	Quadrics	3.0	0.25	0.04	Fat-tree
X1	Custom	7.3	6.3	0.09	4D Hypercube
ES	Custom (IN)	5.6	1.5	0.19	Crossbar
SX8	IXS	5.0	2.0	0.13	Crossbar

Computational Implementation for 3D LB-MHD : The 3D LB-MHD algorithm requires two multi-dimensional arrays for the distribution functions (f, g) to be stored at times t and $t+1$. This leads to quite large memory requirements: ~ 1 KB/grid point. After initialization, the algorithm cycles through the two stages: *COLLISION* and *STREAM* to update (f, g) to time $t+1$. In the *COLLISION* stage, the standard (discrete) moments of (f, g) are computed to determine the macroscopic variables $(\rho, \mathbf{u}, \mathbf{B})$ - and from these variables we can determine the equilibrium functions (f^{eq}, g^{eq}) and hence the r.h.s of Eq. (1). In the *STREAM* stage, the r.h.s for each α are then streamed to the appropriate neighboring cell, according to the vector $\mathbf{c}_\alpha$. The *COLLISION* step is computationally intensive, but requires only data local to the grid point. The *STREAM* step is a set of shift operations, moving data from grid point to grid point according to $\mathbf{c}_\alpha$. For most architectures [9], the optimal layout for the (f, g) requires the leading dimensions being the spatial coordinates, followed by an index representing the streaming vector :

f(x,y,z, 27) , *g(x,y,z, 13:27, 3)* since only 15 velocities are needed to model **B** [3,4].

Fortran array syntax is assumed with x varying fastest when stepping contiguously through memory. For the parallel implementation each array is partitioned onto a 3D Cartesian processor grid, and MPI is used for communication. As in most simulations of this nature, ghost cells are used to hold copies of the planes of data from neighboring processors. In the shift algorithm during the *STREAM* stage, after the first exchange is completed in one direction, we have only partially populated ghost cells. The next exchange includes this data, further populating the ghost cells. This procedure [10] has the advantage of reducing the number of neighbors included in message passing from 26 to 6 – a beneficial optimization considering the MPI latency on most architectures is reasonably high compared with bandwidth. Because different lattice vectors contribute to different spatial directions, the data to be exchanged are not contiguous: e.g., 12 of the 26 non-zero lattice vectors have a component in the +x-direction, and must be sent in this direction, but are not contiguous in the arrays *(f, g)*. The data is packed into a single buffer, resulting in 6 message exchanges/time step. The MPI performance was essentially constant using *mpi_isend/mpi_irecv/mpi_wait* or *mpi_sendrecv* implementations since little computational work (small amount of coupling to the send buffers) could be overlapped with communication.

Performance Results : In Table 3 we summarize the performance of our 3D LB-MHD code, noting that the vector architectures clearly outperform the scalar systems. Across these architectures, the LB algorithm exhibits an average vector length (AVL) very close to the maximum and a very high vector operation ratio (VOR).

Table 3 Performance across architectures in Gflop/s/PE. Percent of peak is shown in parenthesis.

PE	Grid	Power3	Itanium2	X1 (MSP)	X1 (SSP)	ES	SX8
16	256^3	0.14 *(9)*	0.26 *(5)*	5.19 *(41)*		5.50 *(69)*	7.89 *(49)*
64	256^3	0.15 *(10)*	0.35 *(6)*	5.24 *(41)*		5.25 *(66)*	8.10 *(51)*
256	512^3	0.14 *(9)*	0.32 *(6)*	5.26 *(41)*	1.34 *(42)*	5.45 *(68)*	9.66 *(60)*
512	512^3	0.14 *(9)*	0.35 *(6)*		1.34 *(42)*	5.21 *(65)*	

In absolute terms, the SX8 is the leader by a wide margin, achieving the highest per processor performance to date for our 3D LB-MHD code. The *ES*, however, sustains the highest fraction of peak across all architectures. Examining the X1 behavior, we see that in the MSP mode absolute performance is similar to the *ES*, while X1(SSP) and X1(MSP) achieve similar percentages of peak. The high performance of the X1 is gratifying since we noted several warnings concerning vector register spilling during the optimization of the collision routine. Because the X1 has fewer vector registers than the *ES*/SX8 (32 vs. 72), vectoring these complex loops will exhaust the hardware limits and force spilling to memory. The fact that we see no performance penalty is probably due to the spilled registers being effectively cached.

Turning to the superscalar architectures, although the Itanium2 outperforms the Power3 by more than a factor of 2, the percentage of peak achieved is considerably lower. We attribute this to the lower relative memory bandwidth available in this architecture. Although the SX8 achieves the highest absolute performance, the percentage of peak is somewhat lower than that of *ES*. This, we think, is related to the memory subsystem and use of DDR-SDRAM.

In considering the performance for the same grid size by different number of processors, there are 2 competing effects: (i) smaller messages and a higher surface to volume ratio

Table 4 Performance on ES at high processor counts

PEs	Grid	% MPI comm.	Avg. Msg. Size (MB)	Perf./proc. (Gflop/s)	VOR	AVL
1024	1024^3	5.1	2.3	5.44	99.62	254.5
2048	1024^3	8.6	2.1	5.36	99.71	254.5
4096	1024^3	-	1.1	5.16	99.58	253.3

for an increased number of processors, and (ii) the benefit of fitting more of the problem into cache (at least for those architectures that have it). Sometimes the performance increases with increasing processor counts, sometimes decreases. We performed several additional experiments on the *ES*, explicitly measuring the MPI overhead, VOR and AVL, shown in Table 4. For fixed grid size, the performance per processor drops slightly with increasing concurrency. This is mainly the effect of communication overhead increasing, due to both the cost of communication and the increasing ratio of communication to computation. Both of these effects can be seen in the column listing the percentage of time spent in MPI communication. The VOR and AVL values show that the performance of the computational kernel is hardly affected by the scaling up of the problem. The results clearly indicate that LB simulations benefit greatly from the vector architectures, owing to very high memory bandwidth requirement and their data parallel nature.

3. ELB simulations for 2D Navier-Stokes Turbulence

The ELB contains a new Newton-Raphson iteration step in *COLLISION*. However the gridpoint loop is so long and an appreciable number of gridpoints will remain at each iteration, so we see no large impediments to vectorizing almost as efficiently as the MHD algorithm discussed in Sec. 2. Here we consider 2D simulations of multi interlacing vortex layers (15 x 15) at sufficiently high Reynolds numbers (grid 1024^2) that the standard LB algorithm fails because of (severe) numerically instability. There is a broken symmetry at the bottom and right edge, as can be seen in the earlier stage breakdown of the vortex layers into isolated vortices. These isolated vortices are confined to well structured mini-cells, Fig. 1(a), which are both meso- and macro-symmetric about the main diagonal. The yellow vortices rotate counterclockwise while the blue vortices rotate clockwise. Rapidly, the mini-cell confinement is broken, and

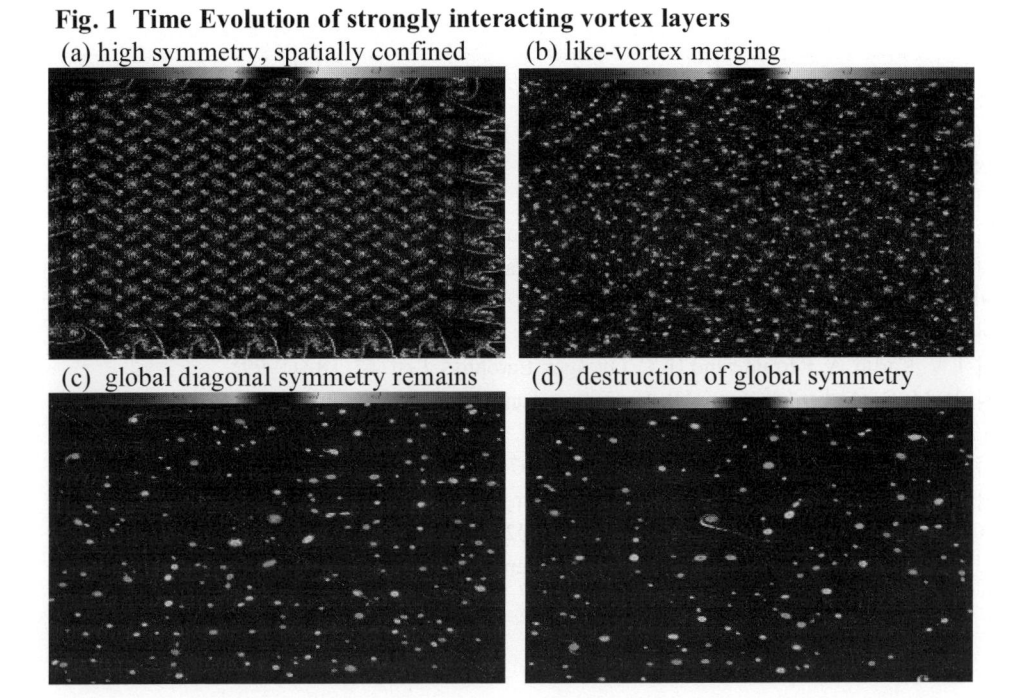

Fig. 1 Time Evolution of strongly interacting vortex layers
(a) high symmetry, spatially confined   (b) like-vortex merging
(c) global diagonal symmetry remains   (d) destruction of global symmetry

like-vortex merging occurs, Fig. 1(b). The global diagonal symmetry in (c) is being destroyed by time frame (d) where a co- and counter-rotating vortex near the center are seen to tear each other apart. This somewhat unusual sequence is shown below in Fig. 2 (the white faint lines in the plots are the streamlines).

Fig. 2 Tearing apart of a co- and counter-rotating vortex

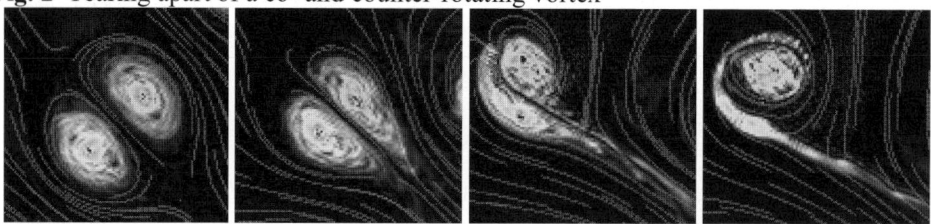

Acknowledgments

The authors would like to thank the staff of the Earth Simulator Center, especially T. Sato, S. Kitawaki and Y. Tsuda, for their assistance during our visit (J.C). We are also grateful for the early SX8 system access provided by HLRS, Germany. The other architectures used were the IBM Power3 at NERSC, LBNL; Intel Itanium at LLNL; Cray X1 at ORNL. The entropic LB runs were performed on the IBM Power4 at NAVO MSRC. The authors wish to express their indebtedness to Sean Ziegeler at NAVO MSRC VADIC for the visualization.

The authors are grateful for support from U.S Department of Energy (J.C under DEAC02-05CH11231) and Air Force Research Laboratory.

References

[1] S. Succi, The Lattice Boltzmann Equation for Fluid Dynamics and Beyond, Clarendon Press, Oxford 2001 – and references therein.
[2] P. J. Dellar, J. Comput. Phys. 179, 95 (2002)
[3] A. Macnab, G. Vahala and L. Vahala, Prog. Comput. Fluid Dyn, **5**, 37 (2005)
[4] J. Carter, M. Soe, L. Oliker, Y. Tsuda, G. Vahala, L. Vahala and A. Macnab, SC|05 (to be published)
[5] H. Chen, Phys. Rev. **E58**, 3955 (1998)
[6] I. V. Karlin, A. Ferrante and H. C. Ottinger, Europhys. Lett. **47**, 182 (1999)
[7] B. M. Boghosian, J. Yepez, P. V. Coveney and A. J. Wagner, Proc. Roy. Soc. Lond. **A457**, 717 (2001)
[8] S. Ansumali and I. V. Karlin, J. Stat. Phys. **107**, 291 (2002)
[9] G. Wellein, T. Zeiser, S. Donath and G. Hager, Computers & Fluids, (2005, to be published)
[10] B. Palmer and J. Neiplocha, Proc. PDCS **2002**, 192 (2002)
[11] S. Ziegeler, G. Vahala, L. Vahala and J. Yepez, NAVO MSRC Navigator, Spring 2005, p. 19

An Incompressible Navier-Stokes with Particles Algorithm and Parallel Implementation

Dan Martin[a], Phil Colella[a], and Noel Keen[a*].

[a]Applied Numerical Algorithms Group, Lawrence Berkeley National Laboratory,
1 Cyclotron Road, Berkeley, CA 94720

We present a variation of an adaptive projection method for computing solutions to the incompressible Navier-Stokes equations with suspended particles. To compute the divergence-free component of the momentum forcing due to the particle drag, we employ an approach which exploits the locality and smoothness of the Laplacian of the projection operator applied to the discretized particle drag force. We present convergence and performance results to demonstrate the effectiveness of this approach.

1. Introduction

Projection methods enable computation of incompressible and low Mach number flows with computational timesteps dictated by advective timescales rather than the more restrictive acoustic timescales. [1] The projection operator $\mathcal{P}$ projects a vector field onto the space of divergence-free vectors through use of the Hodge-Helmholtz decomposition. Given a vector field u, there exists a vector field u_d and a scalar ϕ such that:

$$u = u_d + \nabla \phi \tag{1}$$
$$\nabla \cdot u_d = 0. \tag{2}$$

Then, the projection operator may be written in the form:

$$\mathcal{P}u = u_d \tag{3}$$
$$\mathcal{P}u = \left(I - \nabla(\Delta^{-1})\nabla\cdot\right)u \tag{4}$$

By refining the computational mesh in regions of the domain where greater accuracy is desired, adaptive mesh refinement (AMR) allows greater computational efficiency, focusing computational resources where they are most needed. We use block-structured local refinement of a Cartesian mesh [1], which enables parallelization in a straightforward way by distributing logically rectangular patches among processors. [2]

---

*This work supported by the NASA Earth and Space Sciences Computational Technologies Program and by the U.S. Department of Energy: Director, Office of Science, Office of Advanced Scientific Computing, Mathematical Information, and Computing Sciences Division under Contract DE-AC03-76SF00098.

2. Problem Description

We wish to solve the incompressible Navier-Stokes equations with suspended particles. The particles exert drag on the fluid, while particle motion is induced by the particle-fluid drag forces on the particles, so the particles may move with a velocity different from that of the local fluid. The momentum equation and divergence constraint are:

$$\frac{\partial \boldsymbol{u}}{\partial t} + (\boldsymbol{u} \cdot \nabla)\boldsymbol{u} = -\frac{1}{\rho}\nabla p + \nu \Delta \boldsymbol{u} + \boldsymbol{f} \tag{5}$$

$$\nabla \cdot \boldsymbol{u} = 0, \tag{6}$$

where $\boldsymbol{u}$ is the velocity field, p is the pressure, ν is the kinematic viscosity, and $\boldsymbol{f}$ is the sum of the drag force exerted by the particles on the fluid:

$$\boldsymbol{f}(\boldsymbol{x},t) = \sum_{k=1}^{N} \boldsymbol{f}^{(k)}(t) \delta(\boldsymbol{x} - \boldsymbol{x}^{(k)}(t)). \tag{7}$$

N is the number of particles. We denote quantities associated with the kth particle by the superscript (k) (other quantities are assumed to be associated with the fluid); $\boldsymbol{f}^{(k)}$ is the drag force exerted by the kth particle on the fluid, and $\boldsymbol{x}^{(k)}$ is the location of the kth particle. We treat the particle drag force as a point source, spread to the computational mesh using $\delta_\epsilon(\boldsymbol{x})$, a smoothed numerical approximation to the Dirac delta function $\delta(r)$.

The fluid-particle drag force is given by a simple drag law with drag coefficient k_{drag}:

$$\boldsymbol{f}^{(k)}(t) = k_{drag}\left(\boldsymbol{u}^{(k)} - \boldsymbol{u}(\boldsymbol{x}^{(k)}(t))\right). \tag{8}$$

The motion of a particle with mass $m^{(k)}$ is due to the equal and opposite force on the particle from the fluid, along with a gravitational acceleration $\boldsymbol{g}$:

$$\frac{\partial \boldsymbol{u}^{(k)}}{\partial t} = -\frac{\boldsymbol{f}^{(k)}}{m^{(k)}} + \boldsymbol{g} \tag{9}$$

$$\frac{\partial \boldsymbol{x}^{(k)}}{\partial t} = \boldsymbol{u}^{(k)}. \tag{10}$$

We approximate $\delta(r)$ numerically by a discrete delta function δ_ϵ with the properties:

$$\delta_\epsilon = \frac{1}{2^{D-1}\pi\epsilon^D} g\left(\frac{r}{\epsilon}\right) \tag{11}$$

$g(r) \geq 0$

$g(r) = 0 \quad \text{for } r > 1$

$$\int_0^1 g \, r^{D-1} dr = 1$$

The parameter ϵ is the particle spreading radius. We use a quadratic function for $g(r)$.

3. Projecting the Particle Force

Computing the update to the momentum equation requires the divergence-free contribution of the particle-induced forces. There are several options.

The simplest approach is to add the forcing due to particle drag directly to the momentum equation, and then project the resulting velocity field. Unfortunately, this forcing tends to be singular, so taking the derivatives necessary for a projection method is problematic from an accuracy standpoint. [3]

A second approach is to analytically determine the projection of the discrete delta function used to spread the particle force onto the mesh. [3] If the projection operator is $(\boldsymbol{I} - grad(\Delta^{-1})div)$, then we can define $\boldsymbol{K}_\epsilon = \{K_{ij}\}$ such that $\mathcal{P}(\boldsymbol{f}) = \boldsymbol{K}_\epsilon \boldsymbol{f}$:

$$\boldsymbol{K}_\epsilon(\boldsymbol{x}) = (\boldsymbol{I} - grad(\Delta^{-1})div)\delta_\epsilon \tag{12}$$

In an infinite domain, the operators can commute:

$$\boldsymbol{K}_\epsilon(\boldsymbol{x}) = \delta_\epsilon \boldsymbol{I} - grad\ div(\Delta^{-1})\delta_\epsilon \tag{13}$$

Note that $\Delta^{-1}\delta_\epsilon$ may be evaluated analytically with the proper choice of δ_ϵ.

Then, the projection of the forces on the grid may be computed:

$$\mathcal{P}\boldsymbol{f} = \sum_k \boldsymbol{f}^{(k)} \boldsymbol{K}_\epsilon(\boldsymbol{x} - \boldsymbol{x}^{(k)}) \tag{14}$$

While this approach avoids the accuracy issues of the first approach, it is expensive. Since $\boldsymbol{K}_\epsilon$ does not have compact support, the cost of this approach is $O(N_p N_g)$, where N_p is the number of particles, and N_g is the number of grid points.

We surmount this with the realization that while the projection of the drag force does not have compact support, the Laplacian of the projected drag force does. Taking the Laplacian of (13), again using the commutability of the operators in an infinite domain,

$$\Delta \boldsymbol{K}_\epsilon = \Delta \delta_\epsilon - grad\ div(\delta_\epsilon \boldsymbol{I}) \tag{15}$$

Note that this *does* have compact support, since $\delta_\epsilon = 0$ for $r > \epsilon$.

Now define a discrete approximation to (15) at a grid location indexed by $\boldsymbol{i}$:

$$\boldsymbol{D}_{\boldsymbol{i}}^{(k)} = \Delta^h \boldsymbol{f}^{(k)} \boldsymbol{K}_\epsilon(\cdot - \boldsymbol{x}^{(k)}) \tag{16}$$

where Δ^h is the discrete Laplacian operator with grid spacing h, and the $(\cdot - \boldsymbol{x}^{(k)})$ signifies evaluation at grid points, i.e. $(\boldsymbol{i}h - \boldsymbol{x}^{(k)})$. Then,

$$(\Delta^h)^{-1} \boldsymbol{D}^{(k)} = \boldsymbol{f}^{(k)} \boldsymbol{K}_\epsilon(\cdot - \boldsymbol{x}^{(k)}) \tag{17}$$

Using the compact support of δ_ϵ. we may evaluate $\boldsymbol{D}$ in the local neighborhood of the particle:

$$\boldsymbol{D}_{\boldsymbol{i}}^{(k)} = \Delta^h \boldsymbol{f}^{(k)} \boldsymbol{K}_\epsilon(\cdot - \boldsymbol{x}^{(k)}) \text{ for } |\boldsymbol{i}h - \boldsymbol{x}^{(k)}| < (\epsilon + Ch) \tag{18}$$

where C is a safety factor. Then,

$$D_i = \sum_k D_i^{(k)} \tag{19}$$

$$\mathcal{P}_I f(ih) \approx (\Delta^h)^{-1} D \tag{20}$$

We solve (20) with infinite-domain boundary conditions on $\mathcal{P}_I f(x)$ (the subscript I indicates the use of infinite-domain boundary conditions as opposed to the standard projection operator $\mathcal{P}(u)$, which includes physical boundary conditions on the velocity).

To better approximate the no-normal-flow boundary condition at physical walls, we also use image particles for all particles near the wall. For each particle within $(Ch + \epsilon)$ of the wall, we add an image particle on the other side of the wall with the opposite velocity field normal to the wall. This provides a first-order approximation to the no-flow boundary condition, taking account of the particles near the wall. The boundary condition will be strictly enforced by the projection step of the fluid update.

4. Discretization of Advance

The particle drag force projection outlined above is combined with the incompressible AMR Navier-Stokes algorithm in [1] to produce an AMR incompressible Navier-Stokes with suspended particles solver. For algorithmic simplicity, this implementation does not refine in time.

We begin with the discrete solution on a locally refined Cartesian mesh at time t^n. The velocity field u and pressure p are cell-centered. Each particle's position $x^{(k),n}$ and velocity $u^{(k),n}$ are also known. To advance the solution from time t^n to time $t^{n+1} = t^n + \Delta t$, we proceed as follows.

We first compute the drag force at time t^n, $f^{(k),n}$ on each particle using (8). Fluid velocities are computed at particle locations using quadratic interpolation of the cell-centered velocity u^n.

Using the approach outlined above, we compute the projected force $\mathcal{P}_I(f^n)$ using infinite-domain boundary conditions.

Then, we compute a provisional update u^* in much the same way as in [1]:

$$u^* = u^n + \Delta t \left(-[(u \cdot \nabla)u]^{n+\frac{1}{2}} - grad(p^{n-\frac{1}{2}}) + [\nu \Delta u] + \mathcal{P}_I(f^n) \right) \tag{21}$$

where the nonlinear advective term $[(u \cdot \nabla)u]^{n+\frac{1}{2}}$ is computed using the second-order upwind scheme outlined in [1], including $\mathcal{P}_I(f^n)$ as a forcing term in the predictor step, and $[\nu \Delta u]$ is computed using a second-order L_0-stable Runge-Kutta scheme [5].

We now update the particle velocities and positions using the analytic solutions for (9-10) (for compactness of notation, $u^{(k),n}$ refers to the particle velocity at time t^n, while u^n refers to the fluid velocity at time t^n interpolated to the particle position $x^{(k),n}$):

$$u^{(k),n+1} = \left(u^{(k),n} - u^n - \frac{m^{(k)}g}{k_{drag}}\right) e^{-\frac{\Delta t k_{drag}}{m^{(k)}}} + u^n + \frac{m^{(k)}g}{k_{drag}} \tag{22}$$

$$x^{(k),n+1} = x^{(k),n} + \frac{m^{(k)}}{k_{drag}}\left(u^{(k),n} - \frac{m^{(k)}g}{k_{drag}} - u^n\right)\left(1 - e^{-\frac{\Delta t k_{drag}}{m^{(k)}}}\right) + \Delta t\left(u^n + \frac{m^{(k)}g}{k_{drag}}\right) \tag{23}$$

We then use $u^{(k),n+1}$ and $x^{(k),n+1}$ to compute the projected drag force (again using infinite-domain boundary conditions) $\mathcal{P}_I(f^*)$. Then, we modify u^* to make the update second-order in time, and project to complete the update:

$$u^{n+1} = \mathcal{P}\left(u^* + \Delta t\left[grad(p^{n-\frac{1}{2}}) - \frac{1}{2}\mathcal{P}_I(f^n) + \frac{1}{2}\mathcal{P}_I(f^*)\right]\right) \quad (24)$$

$$grad(p^{n+\frac{1}{2}}) = (\mathbf{I} - \mathcal{P})\left(u^* + \Delta t\left[grad(p^{n-\frac{1}{2}}) - \frac{1}{2}\mathcal{P}_I(f^n) + \frac{1}{2}\mathcal{P}_I(f^*)\right]\right) \quad (25)$$

Note that the projection is also applied to $\mathcal{P}_I(f)$; to enforce the physical boundary conditions, since $\mathcal{P}_I(f)$ was computed using infinite-domain boundary conditions. We use the approximate cell-centered projection described in [1].

5. Evaluating $\mathcal{P}_I(f)$

We want to approximate the divergence-free contribution of the drag force $\mathcal{P}_I(f)$. In indicial notation,

$$(\mathcal{P}_I f(x))_i = \sum_k f_j^{(k)}(\delta_{ij}\Delta - \partial_i\partial_j)(\Delta^{-1})\delta_\epsilon(x - x^{(k)}) \quad (26)$$

Define $K_{ij}^{(k)}(x) = (\delta_{ij}\Delta - \partial_i\partial_j)(\Delta^{-1})\delta_\epsilon(x - x^{(k)})$. Then,

$$\mathcal{P}_I f(x)_i = \sum_k f_j^{(k)} K_{ij}^{(k)} \quad (27)$$

Approximate $\mathbf{D} = \Delta \mathbf{K}$ by $\tilde{\mathbf{D}} = \{\tilde{D}_{ij}\}$:

$$\tilde{D}_{ij}^{(k)}(x) = \begin{cases} (\Delta^h K_{ij}^{(k)})(x) & \text{if } r < \epsilon + Ch \\ 0 & \text{otherwise} \end{cases} \quad (28)$$

Then,

$$\mathcal{P}_I f(x)_i = \sum_k f_j^{(k)} K_{ij}^{(k)}(x) \quad (29)$$

$$\Delta^h \mathcal{P}_I f(x)_i = \sum_k f_j^{(k)} \Delta^h K_{ij}^{(k)}(x) \quad (30)$$

$$\approx \sum_k f_j^{(k)} \tilde{D}_{ij}^{(k)}(x) \quad (31)$$

$$\mathcal{P}_I f(x)_i \approx (\Delta^h)^{-1} \sum_k f_j^{(k)} \tilde{D}_{ij}^{(k)}(x). \quad (32)$$

We solve (32) with infinite-domain boundary conditions on $\mathcal{P}_I f(x)$. [4].

6. AMR implementation

By focusing computational effort on the neighborhood of the particles, AMR can increase efficiency and improve parallel performance. The particles are distributed onto the same block-structured meshes as the fluid solver, but with a different processor distribution to balance the particle loads independently from the fluid solver workload. This results in a better overall balance, at the cost of some added communication between the two processor distributions. The code was implemented using the Chombo framework [6].

7. Convergence – Single Particle Settling

We use a simple test problem to demonstrate the accuracy and effectiveness of this approach. A single particle with mass 0.001g starts at rest in a fluid in a 1 m^3 cubic domain. As the particle accelerates downward due to gravity, a velocity is induced in the fluid due to the particle drag. We use $k_{drag} = 0.04$, $\epsilon = 6.25cm$, and $\nu = 0.004 \frac{cm^2}{s}$.

Because there is no analytic solution available for this problem, we compute a solution on a uniform 256^3 fine mesh and treat this as the "exact" solution against which we compare other computed solutions. The convergence of the $x-$velocity in the L_2 norm (other velocity components and norms are similar) is shown in Figure 1 for uniform mesh and for a single level of refinement with refinement ratios of 2 and 4. Because the timestep and the cell spacing are reduced simultaneously (the timestep is halved when the cell spacing is halved), the second order convergence in these plots demonstrates convergence in both time and space. Also, if AMR is effective, the errors of the adaptive computations should approach those of the uniform mesh computation with the equivalent resolution (i.e. a 64^3 computation with one level of refinement with a refinement ratio of 2 should have the same error as a 128^3 uniform mesh computation. This is borne out in Figure 1.

Figure 2 shows the serial CPU times on a 2 GHz Opteron processor and the total number of cells advanced for the 256^3 uniform-mesh computation and for the equivalent-resolution adaptive cases with a single refinement level with refinement ratios of 2 and 4. The values are normalized by the uniform-mesh values, to make it easier to evaluate. Even for this simple test case, the use of AMR results in significant savings, both in the number of cells advanced (a crude indicator of memory use), and in CPU time. The space between the two lines represents the overhead due to adaptivity.

8. Parallel Performance – particle cloud with a vortex ring

To demonstrate the parallel performance of this algorithm, we use a different problem which has enough particles to distribute effectively, and which has more complicated fluid dynamics beyond the drag-induced flow. We compute a three-dimensional vortex ring in a 1-meter cube domain with 32,768 particles arranged in a $32 \times 32 \times 32$ array, spanning $15cm \leq x, y \leq 85cm$, and $25cm \leq z \leq 75cm$. For this problem, the vorticity distribution is specified, from which the initial velocity is computed. The vortex ring is specified by a location of the center of the vortex ring (x_0, y_0, z_0), the radius of the center of the local cross-section of the ring from the center of the vortex ring r, and the strength Γ.

The cross-sectional vorticity distribution in the vortex ring is given by $\omega(\rho) = \frac{\Gamma}{a\sigma^2}e^{(\frac{\rho}{\sigma})^3}$. ρ is the local distance from the center of the ring cross-section, $a = 2268.85$, and $\sigma = 2.75$.

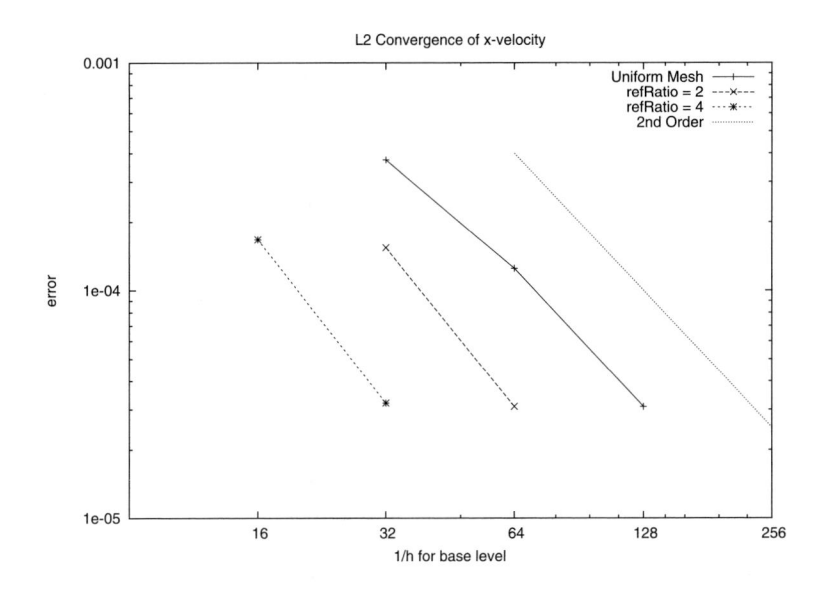

Figure 1. Convergence for the single particle settling problem. $x-$ axis is $\frac{1}{h_0}$, while the $y-$ axis is the L_2 error.

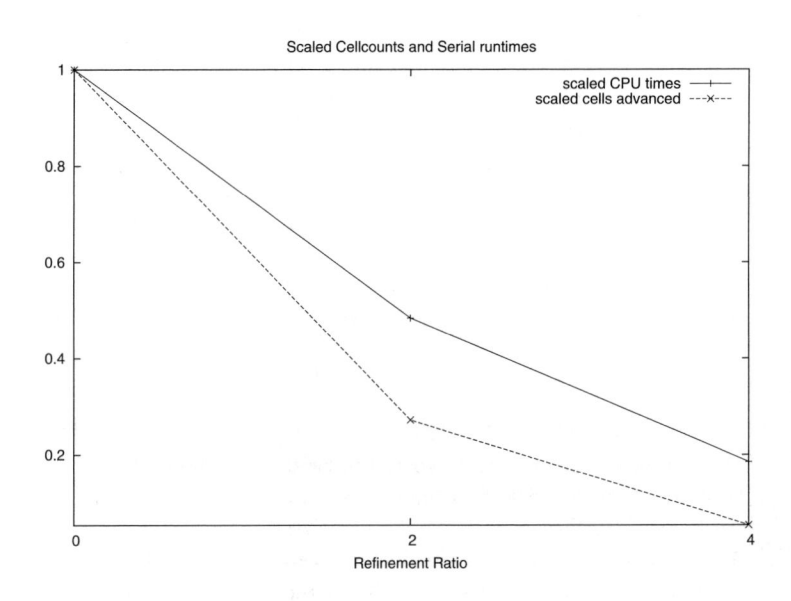

Figure 2. Scaled run times and cell counts for the single particle settling problem.

Prob size	Num Procs	Max Mem (MB)	AMR Run (sec)	Particle update (sec)
32x32x32	8	72.5	100.4	0.59
64x64x64	16	115.7	178.4	1.2
64x64x64	64	75.8	103.4	0.38
128x128x128	32	317.8	597.3	3.73
128x128x128	64	175.1	352.4	2.24
128x128x128	128	117.3	220.8	1.41

Table 1
Parallel performance for vortex-ring problem with 32,768 particles.

Base Problem Size	Num Procs	Large Problem Size	Large num processors	Scaled Efficiency
32x32x32	8	64x64x64	64	0.97
64x64x64	16	128x128x128	128	0.81

Table 2
Scaled Efficiencies computed from Table 1.

The vortex ring is centered at $(50cm, 50cm, 40cm)$, with a radius of 2cm and $\Gamma = 1.5 \times 10^5$.

The number of particles is held fixed while we decrease the mesh spacing. In three dimensions, as we halve the cell spacing while holding the number of particles constant, the asymptotic computational size (both in CPU time and memory) of the problem increases by a factor of 8. The particle spreading radius ϵ is also held fixed at $6.25cm$ as the mesh spacing is decreased because the particles in this problem represent physical particles, rather than point charges. Therefore, the work involved in the particle-fluid drag force projection should also increase by a factor of 8 as the mesh spacing is halved.

Run times and maximum memory usage for this problem on a Compaq AlphaServer ("halem.gsfc.nasa.gov") are shown in Table 1. We compute a scaled efficiency by comparing the CPU times between two runs which differ by a factor of two in base grid size and a factor of 8 in number of processors. The resulting efficiencies are shown in Table 2.

REFERENCES

1. D Martin and P. Colella, J Comp Phys. No. (2000).
2. C A Rendleman et al, Computing and Visualization in Science 3 (2000), 147.
3. R Cortez and M Minion, J Comp Phys 161 (2000) 428.
4. R A James, J Comp Phys 25 (1977).
5. E H Twizell, A B Gumel, and M A Arigu, Advances in Comput. Math. 6 (1996) 333.
6. P Colella et al, "Chombo Software Package for AMR Applications", available at http://seesar.lbl.gov/ANAG/software.html.

Determination of Lubrication Characteristics of Bearings using the Lattice Boltzmann Method

G. Brenner[a], A. Al-Zoubi,[a] H. Schwarze[b], S. Swoboda[b]

[a]Institute of Applied Mechanics, Clausthal University of Technology, D-38678 Clausthal-Zellerfeld, Adolph-Roemer Str. 2A, Leibnizstr. 32Germany

[b]Institute of Tribology, Clausthal University of Technology, D-38678 Clausthal-Zellerfeld, Leibnizstr. 32, Germany

The effect of surface textures and roughness on shear- and pressure forces in tribological applications is analyzed by means of full numerical simulations taking into account the geometry of real surface elements. Topographic data of representative surface structures are obtained with high spatial resolution using an optical interference technique. The three dimensional velocity field is obtained using the lattice-Boltzmann method for laminar flows of Newtonian fluids. Subsequently, pressure and shear flow factors are obtained by evaluating the velocity field according to the extended Reynolds equation. The approach allows determining efficiently the hydrodynamic characteristics of micro structured surfaces in lubrication. Especially the influence of anisotropies of surface textures due to the manufacturing process on the hydrodynamic load capacity and friction is determined. The results obtained indicate, that full numerical simulations should be used to compute accurately and efficiently the characteristic properties of tribological film flows and therefore may contribute to a better understanding and prediction of EHL problems.

1. INTRODUCTION

In elasto-hydrodynamic lubrication (EHL) there is an increasing interest in studying laminar flows past rough surfaces. This is due to the fact, that the conditions under which lubricated machine elements such as bearings have to operate become more and more severe. As a result, the nominal thickness of lubrication films has decreased to a level where surface roughness effects in laminar flows become significant [1]. Here, the understanding and quantification of surface roughness effects can significantly contribute to the improvement and performance of machine devices, i.e. reducing friction and wear [2]. The goal of the present paper is to contribute to the hydrodynamic characterization of shear flows past rough surfaces in lubrication regime. The results are obtained using the lattice-Boltzmann method.

The present topic is subject of various research activities, where the focus is mainly on the modelling of rough surface contacts [3] using statistical approaches [4,5] or the application of mathematical methods such as the homogenization technique [6,7] or the local volume averaging method [8,9] for simplified geometries such as sinusoidal waviness.

Jacob et al. [1] studied the effect of longitudinal surface textures on friction in EHL contacts based on a sinusoidal waviness of the surface. Only few publications concentrate on the determination of the micro-hydrodynamics of real, measured surfaces using direct methods. In [2] a FEM method is used to investigate the flow in a representative volume within the lubricant film driven by pressure or shear forces. Besides synthetically generated surface geometries, also measured surfaces haven been evaluated. Notable, Xu et al. [10] studied thermal Newtonian EHL contact using measured roughness distributions as an input.

The assessment of real, measured surface roughness distributions is hampered by two circumstances. The application of classical CFD methods such as finite Element or finite volume techniques usually requires substantial effort in pre-processing the measured data and generating suitable meshes for the flow calculation. Additionally, the requirements regarding the spatial resolution are very high and the resulting problems are time consuming to solve. The newly developed lattice-Boltzmann method (LBM) may circumvent these problems since it allows a quasi automatic mesh generation. Besides that, this technique has been shown to be very efficient, in particular when using high performance parallel computers.

The present paper is organized as follows: After this introduction, the micro-hydrodynamic model based on Reynolds theory is summarized. Subsequently, the numerical method is briefly described. Finally, the focus in section 4 is on the presentation and discussion of results.

2. PROBLEM DESCRIPTION

Instationary, elastohydrodynamic simulations are usually based on Reynolds equation to account for friction due to shear in the flow. In its original form, the Reynolds equation is valid to describe the flow between two smooth surfaces with small film thickness and slowly varying cross section. The flow is driven by the pressure gradient or a difference of wall velocities. In the following, x and y denote the directions parallel to the walls and z the direction normal to the walls. For simplicity, it is assumed, that the wall velocities U_1 and U_2 are in the x-direction, see Fig. 1. Integrating the Navier-Stokes equation for a stationary flow and no inertia effects yields the mean volumetric flux in the x and y-direction as

$$q_x = -\frac{h_T^3}{12\mu}\frac{\partial p}{\partial x} + \frac{U_1 + U_2}{2}h_T \quad \text{and} \quad q_y = -\frac{h_T^3}{12\mu}\frac{\partial p}{\partial y}. \tag{1}$$

Here, h_T denotes the local film thickness, p the static pressure and μ the dynamic viscosity. Defining average values of the fluxes over a volume element with rough walls yields

$$\bar{q}_x = -\phi_x\frac{\overline{h}_T^3}{12\mu}\frac{\partial \overline{p}}{\partial x} + \frac{U_1 + U_2}{2}\overline{h}_T + \frac{U_1 - U_2}{2}\sigma\phi_x^s \quad \text{and} \quad \bar{q}_y = -\phi_y\frac{\overline{h}_T^3}{12\mu}\frac{\partial \overline{p}}{\partial y}. \tag{2}$$

where $\overline{h}_T$ denotes the average gap width, $\partial\overline{p}/\partial x$ the mean pressure gradient, σ the variance of the surface roughness and μ the dynamic viscosity. The pressure flow factors

ϕ_x and ϕ_y relate the volumetric flux or mean velocity due to a prescribed pressure gradient in a rough bearing to that between smooth surfaces. They may be obtained from a numerically computed velocity field with imposed constant pressure gradient $\partial \overline{p}/\partial x$ or $\partial \overline{p}/\partial y$ and with fixed walls (i.e. $U_1 = U_2 = 0$) by integrating the mean flux according to

$$\phi_x = \overline{q}_x / \frac{\overline{h}_T^3}{12\mu} \frac{\partial \overline{p}}{\partial x} \quad \text{and} \quad \phi_y = \overline{q}_y / \frac{\overline{h}_T^3}{12\mu} \frac{\partial \overline{p}}{\partial y}. \qquad (3)$$

The shear flow factors ϕ_x^s and ϕ_y^s quantify additional flow transport due to sliding in a rough bearing. To determine shear flow factors the flow is driven by a relative movement of the rough walls without pressure gradient. Since only the velocity difference between the surfaces is relevant, we may set $U_2 = 0$. Making use of the definition above we obtain for the shear flow factor, e.g. in the x-direction

$$\phi_x^s = \frac{\overline{h}_T}{\sigma} \left(\frac{2}{U_1} \frac{\overline{q}_x}{\overline{h}_T} - 1 \right). \qquad (4)$$

Again, the mean velocity is obtained directly from the computed velocity field. For isotropic surface roughness there is no directional dependence of the flow factors. If the film thickness is large compared to the surface roughness, the pressure flow factor should approach unity. In the same limit, the shear flow factors should approach zero. In the present analysis, the mean gap width and the surface roughness represented by the rms value plays a crucial role, especially if the ratio $\overline{h}_T/\sigma$ is of the order of 3 to 10. For lower ratios asperities start interacting with each other and the friction is influenced also by non fluid dynamical effects. Thus, this regime is precluded in the present analysis.

3. NUMERICAL METHOD

The motion of an incompressible Newtonian fluid may be described by the conservation equations of mass and momentum. Usually, the solution is based on approximations such as the finite volume method. In the present work an alternative approach, the lattice Boltzmann method, is utilized. Though quite new, this technique is in the meantime well documented in the literature [11,12]. Here, we present only the basic steps of the LB method and some aspects of grid generation and formulation of boundary conditions. For further details of the present implementation we refer to [13]. The basic idea of the LBM is to solve the Boltzmann kinetic equation describing the rate of change of the single particle distribution function under the premise of molecular chaos, binary collisions and negligible external forces. In order to obtain a formulation that is amenable to a numerical solution the collision term of the Boltzmann equation is reduced to a simple relaxation term. This reflects the fact that the state of the fluid, described by the distribution function, relaxes towards an equilibrium value characterized by the Maxwellian distribution with some time constant. A further simplification is to discretice the velocity space by a small number of degrees of freedom resulting in a set of discrete velocity Boltzmann equations for the discrete velocity distribution functions. Subsequently, a first order discretization in space and time results in the lattice Boltzmann equation

$$f_\alpha(\vec{x} + \vec{c}_\alpha \Delta t, t + \Delta t) = f_\alpha(\vec{x}, t) - \frac{\Delta t}{\tau}(f_\alpha(\vec{x}, t) - f_\alpha^{eq}(\vec{x}, t)). \qquad (5)$$

Eq. 5 may be regarded as a finite difference approximation of a system of first order differential equations in diagonal form. From the moments of the discrete velocity distribution functions the hydrodynamic properties such as fluid density, mass flux and the momentum flux tensor may be obtained from the moments of the distribution function. The present implementation is the D2Q9 model (2 dimensional, 9 velocity) and D3Q19 model (3 dimensional, 19 velocity) according to the classification of [12]. The kinematic viscosity of the fluid is controlled by the relaxation parameter. It has be shown [14] that the flow field obtained using the above approach obeys the Navier-Stokes equations in the limit of small Mach and Knudsen numbers. The algorithm described above may be understood as an explicit time stepping consisting of an advection of distribution functions in the direction of the discrete velocities to the next neighboring nodes followed by a relaxation towards an equilibrium state. This algorithmic simplicity allows a very efficient implementation on computers and an efficient solution of flow problems with high spatial resolution. This is advantageous in particular when complicated boundaries, as in the present application, have to be considered. At the boundaries the distribution functions entering the computational domain have to be prescribed in order to satisfy the no-slip condition at solid walls and the periodicity assumption elsewhere. The latter is simply realized by reintroducing the leaving distribution functions at the opposing boundaries. The no slip condition is realized employing the bounce back approach. This is motivated by the gas kinetical picture of particles being reflected at a solid surface. In the lattice Boltzmann context, the zero velocity or no slip condition is realized by reflecting the distributions reaching a wall node into their original direction and position. In case of a moving wall, a source term is added to the distribution functions to enforce a velocity parallel to the boundary with prescribed magnitude. Details concerning the formulation of boundary conditions may be found in [16,17,15]. The initial values of the distribution functions are the respective equilibrium functions.

In the following, some more practical aspects are explained. An important feature of the lattice-Boltzmann method is the fact that the flow domain is mapped on a Cartesian, equidistant computational grid. At the nodes of this grid, the flow variables such as velocity and pressure are obtained from the distribution functions. Boundaries are treated as immersed solids in the MAC (marker and cell) fashion, i.e. at a solid-fluid interface, cells are marked in a suitable way ("bitmap") and the bounce back scheme is imposed to obtain locally a zero velocity. The bitmaps representing the geometry of the surfaces are obtained by an interference topography method (Veeco Instruments WYKO NT 1100) with a lateral resolution of about 100nm and 3nm in the direction normal to the surfaces. This results finally in a three dimensional, Cartesian domain with a length of typically 0.07 mm (L_x, L_y) by 0.01 mm (L_z). The computational mesh typically consists of up to one million grid nodes.

4. RESULTS

In the present study, the hydrodynamic characteristics of two different bearings are investigated. The first probe originates from the central crank shaft of a diesel engine (configuration A). The second example stems from the bearing shell (configuration B). The topography of the surfaces was scanned using an interferometer with a spatial resolution

$0.167 \mu m$ in the lateral direction (pixel size). Scanned surface elements of each probe are shown in Fig. 2. Due to the manufacturing of the crank shaft surface, regular textures with a wavelength of about $3-5\mu m$ are observed. In contrast to that, the second probe shows a very isotropic roughness. For the pressure flow simulations, the following boundary conditions and flow parameter are employed. The computational domain consists of a box with a length of about $8.35\mu m$ by $8.35\mu m$ in x- and y-direction. With the given pixel size of the scanning device, this corresponds to 50 by 50 grid points. The lower surface represents the rough wall while the upper one is assumed to be smooth. No-slip conditions are prescribed at all solid surfaces. In the x- and y-direction periodic boundary conditions are prescribed for the velocity components. The flow is driven by a constant volume force in either x- or y-direction. The Reynolds number based on the mean velocity in the gap and the mean gap width is kept constant at unity in all simulations. To illustrate the influence of the flow direction on the flow patterns iso-surfaces of the streamwise velocity component, i.e. the x-component of velocity for the flow driven by a pressure gradient in x-direction and vice versa are shown in Fig. 3. In case that the flow is aligned with the orientation of the surface textures a channelling effect is observed. If the mean flow crosses the textures the periodically changing deformation which a fluid element encounters during its passage the corresponding viscous loss of momentum is higher. Consequently, the pressure flow factor should be smaller than in the case of aligned flow. This is confirmed in Fig. 4, where the pressure flow factor is presented for both flow configurations and both surface geometries. In case of isotropic surface roughness the flow factor is independent of the orientation of the flow. For large values of $\overline{h}_T/\sigma$ the flow factors approach unity. For the anisotropic surface, the flow factors differ significantly depending on the flow orientation, if $\overline{h}_T/\sigma$ is small. If the flow is aligned with the textures, the flow factor becomes larger than unity. Thus, in this case the effect of surface roughness supports the flow transport compared to smooth walls due to the channelling effect. For the shear flow simulation, in principle the same set up as before is used. Instead of a pressure force, the flow is driven by the movement of the upper smooth lid. The parameters are chosen such that the Reynolds number, based on the wall velocity and the mean height in the gap, is unity. The results for both configurations and both flow directions are presented in Fig. 5. The results indicate that for low $\overline{h}_T/\sigma$ the transport of fluid due to the roughness decreases. However, in case of the anisotropic surface texture the effect is less pronounced. For high $\overline{h}_T/\sigma$ the shear flow factor approaches zero and the effect of roughness is not important.

5. SUMMARY AND CONCLUSION

In the present paper the lattice Boltzmann method is utilized to investigate the influence of surface roughness in lubricated machine elements on friction in laminar flow regimes. The results indicate that LB method can be efficiently employed to quantify exactly tribological parameters that are needed for elasto-hydrodynamic lubrication (EHL) computations.

REFERENCES

1. B. Jacob, C.H. Venner, P.P. Lugt, Influence of Longitudinal Roughness on Friction in EHL Contacts, ASME J. Tribol, 126 (2004) pp. 473-481.
2. G. Knoll, V. Lagemann, Simulationsverfahren zur triologischen Kennwertbildung rauer Oberflächen, Teil 1- Einfluss der bearbeitungsbedingten Oberfläche auf die hydrodynamische Tragfähigkeit geschmierter Kontakte, Tribologie und Schmierungstechnik, 49 (2002), pp. 12-15.
3. R. F. Salant, A. H. Rocke, Hydrodynamic Analysis of the Flow in a Rotary Lip Seal Using Flow Factors, ASME J. Tribol, 126 (2004), pp. 156-161.
4. N. Patir, H. S. Cheng, An Average Flow Model for Determining Efects of Three-Dimensional Roughness on Partial Hydrodynamic Lubrication, ASME J. Tribol, 100 (1978) pp. 12-17.
5. N. Patir, H. S. Cheng, Application of Average Flow Model to Lubrication Between Rough Sliding Surfaces, ASME J. Tribol, 101 (1979), pp. 220-230.
6. M. Kane, B. Bou-Said, Comparison of Homogenization and Direct Techniques for the Treatment of Roughness in Incompressibile Lubrication, ASME J. Tribol, 126 (2004), pp. 733-737.
7. G. C. Buscalglia, M. Jai, Homogenization of the Generalized Reynolds Equation for Ultra-Thin Gas Films and its Resolution by FEM, ASME J. Tribol, 126 (2004), pp. 547-552.
8. N. Letalleur, F. Plouraboue, M. Prat, Average Flow Model of Rough Surface Lubrication: Flow Factors for Sinusoidal Surfaces, ASME J. Tribol, 124 2002), pp. 539-545.
9. M. Prat, F. Plouraboue, N. Letalleur, Averaged Reynolds Equation for Flow Between Rough Surfaces in Sliding Motion, Transport in Porous Media, 48 (200), pp 291-313.
10. G. Xu, G., F. Sadeghi, Thermal Analysis of Circular Contacts with Measured Roughness, ASME J. Tribol, 113 (1996), pp. 473-483.
11. X. He, L.S. Luo, Theory of the lattice Boltzmann method: From the Boltzmann equation to the lattice Boltzmann equation,Phys. Rev. E, 56(6), 1997, pp. 6811-6817.
12. S. Chen, Z. Wang, X. Shan, G. D. Doolen, Lattice Boltzmann computational fluid dynamics in three dimensions, J. Stat. Phys., 68 (1992), pp. 379-400.
13. G. Brenner, Numerische Simulation komplexer fluider Transportvorgänge in der Verfahrenstechnik, Habilitation, University of Erlangen-Nuremberg, 2002.
14. S. Chen, G. D. Doolen, Lattice Boltzmann Method for Fluid Flows, Ann. Rev. Fluid Mech ., 30(1998), pp. 329-364.
15. D. Yu, R. Mei, L. S. Luo, W. Shyy, Viscous Flow Computation with the Method of Lattice Boltzmann Equation, Prog. Aero. Sci., 39 (2003), pp. 329-367.
16. Q. Zhou, X. He, On Pressure and Velocity Boundary Conditions for the lattice Boltzmann BGK Model, Nonlinear Science e-Print Archive, 1995, comp-gas/958001.
17. H. He, Q. Zhou, L.-S. Luo, M. Dembo, Analytic Solutions of Simple Flows and Analysis of Nonslip Boundary Conditions for the lattice Boltzmann BGK Model, J. Stat. Phys., 87(1/2), 1997, pp. 115-136.

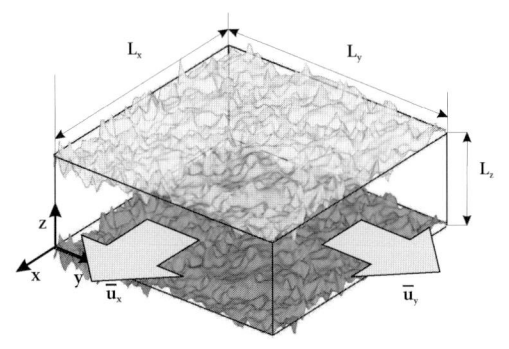

Figure 1. Sketch of the computational domain for the calculation of pressure and shear flow factors.

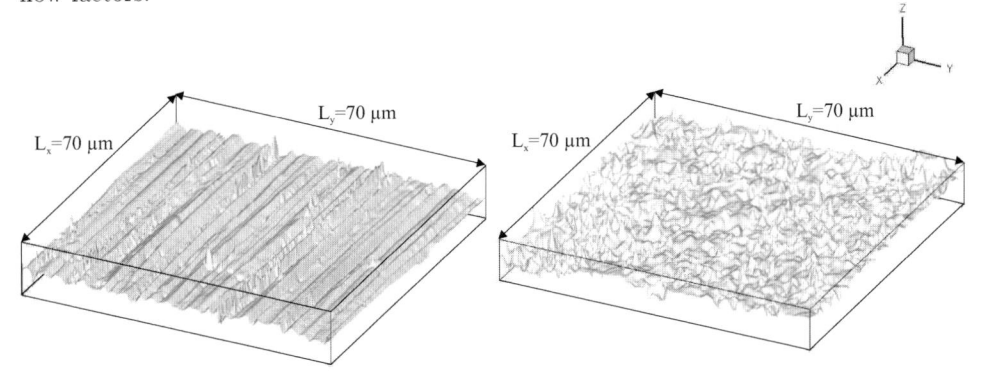

Figure 2. Scanned surface elements of the crankshaft (left: configuration A) and the bearing shell (right, configuration B).

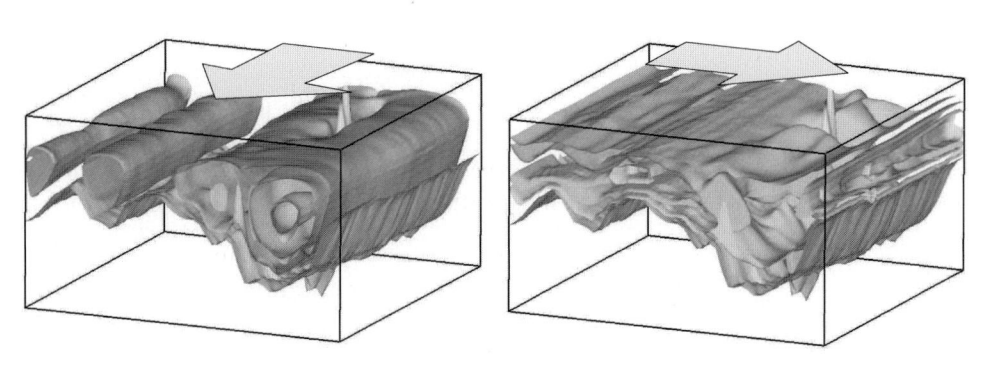

Figure 3. Iso-surfaces of the streamwise velocity component for the pressure driven flow past the crankshaft element with anisotropic surface roughness. Left: flow in x-direction, right: flow in y-direction.

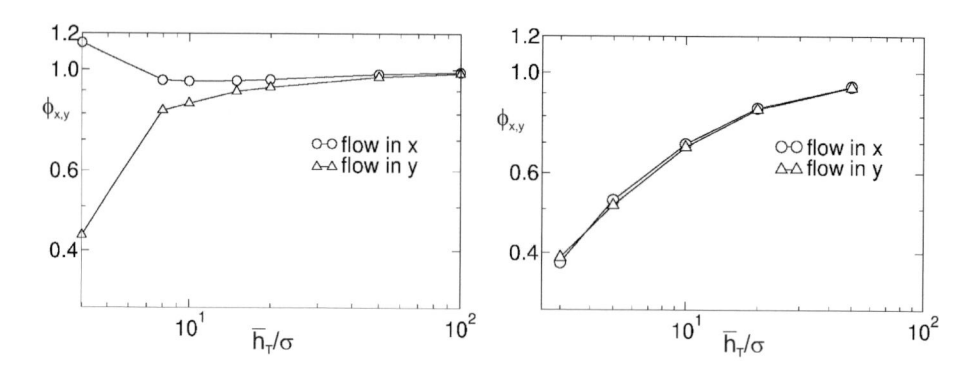

Figure 4. Simulated pressure flow factors for the flow past configuration A and B and for both flow directions as function of the $\bar{h}_T/\sigma$ ratio.

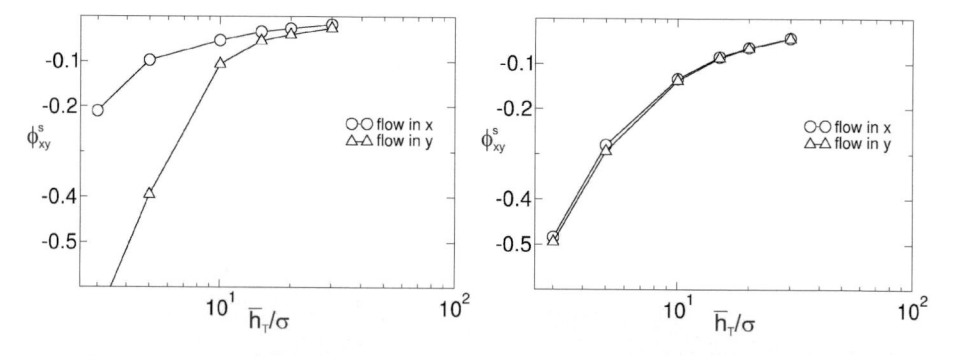

Figure 5. Simulated shear flow factors for the flow past configuration A and B and for both flow directions as function of the $\bar{h}_T/\sigma$ ratio.

MPI-OpenMP hybrid parallel computation in continuous-velocity lattice-gas model

Masaaki Terai[a] and Teruo Matsuzawa[b]

[a]School of Information science, Japan Advanced Institute Science and Technology, 1-1 Asahidai, Nomi, Ishikawa 923-1292, Japan

[b]Center for Information science, Japan Advanced Institute Science and Technology, 1-1 Asahidai, Nomi, Ishikawa 923-1292, Japan

The particle based models are popular scheme for fluid flow analysis based on micro-kinetic scale features. To take into account the features, these models can analyze various complicated fluid in a relatively straightforward way. The continuous-velocity lattice-gas model is one of the particle based models. By using this model, the macroscopic fluid behavior are estimated from the microscopic condition. This model is featured with two physical objects as a particle and a site. A particle is relevant to the microscopic condition, and a site is relevant to the macroscopic condition. These models have computational advantages which are very suitable for parallelization. In parallel computation, by use a particle decomposition, a load balancing is accomplished. It is implemented simply. However a particle decomposition was shown not good efficiency than domain decomposition in previous results. In that case, particles are distributed equally, without thermal difference. On the other hand, particles are not distributed equally with differecne in temperature. By use domain decomposition, difference in particle arises difference in load on each processor. In present study, the authors introduced hybrid MPI-OpenMP parallel computation to keep simple code and to reduce calculation time in the condition. As a result, a hybrid MPI-OpenMP method had spent fewer elapsed time than other parallel method, in larger processors, with difference in temperature. This parallel method is good candidate to treat more complicated fluids with thermal.

1. Introduction

In the industry, the fluids were mostly complicated flow with thermal and gravity. A natural convection study is especially providing experimental data suitable for testing and validating computer codes to wide variety of practical problems. In these case, a finite difference method or a finite element method based on Navier-Stokes equation was applied to problems, with Boussinesq approximation.

On the other hand, the particle based model such as lattice gas cellular automata(LGCA), lattice Boltzmann model(LBM) are popular scheme for fluid flow analysis based on micro-kinetic scale features. To take into account the features, these models can analyze various complicated fluids in a relatively straightforward way. One of these investigation is change of phase, boilling and evapolation. In these complicated fluids, thermal will play important role.

One of the particle based model, the continuous-velocity lattice gas model[1][2] was introduced by Malevanets and Kapral. This model is featured with two physical objects as particle and site. A site resembles a lattice-point in LGCA. On the other hand, a particle position and velocity is treated as continuous variables. Therefore, this new model can analyze thermal fluids without to solve the energy equation.

Generally in particle based model, more detailed simulation requires more large number of particles or large spaces, and produces increase of the calculation time. Therefore, working hardwares and software resources had improved in the efficiency of computation for some decades. A computational advantage of the particle based models are very suitable for parallelization in both shared and distribution memory paradigms[3][4].

In driven cavity flow benchmark test by using the continuous-velocity lattice gas model, the authors obtained high speed up ratio with parallel computation in previous study[6]. In that case, the domain decomposition method obtained good efficiency. However the number of particle is nearly uniform in throughout the cavity. A difference in number of particles between processors cause a difference in load, since a load of the computer is depend on number of particles at each processor. And the calculation time will be increased. In this study, this new model was applied to thermal cavity to make a difference in number of particles and to a estimate benchmark test with a parallel computation.

2. Continuous-velocity lattice gas model

The continuous-velocity lattice gas model was proposed by Malevants and Kapral[1]. This new model is particle based model to analyze fluid flow by using micro-kinetic features. This model resembles previous lattice gas model. Characteristics of introduced new model can be summarized as follows.

Firstly, in initial condition, we gave N particles which have velocity of Maxwellian distribution to the system. These particles are treated as two-dimensional ideal gas.

In collision process, Particle dynamics are based on microscopic description of fluids from the viewpoint of micro-kinetic scale dynamics. A collision between particles uses a site, which resembles lattice-point in LGCA. The essential properties to ensure the conservation of mass, momentum and energy are preserved without compromise. This collision rule use of a stochastic rotational matrix $\boldsymbol{\sigma}$ for particle interactions. It is similar to that in Direct Simulation Monte Carlo, although the scheme in the continuous-velocity lattice gas model includes the multi-particle collision. Collision rule as follow :

$$c_i' = \boldsymbol{V} + \boldsymbol{\sigma}\left(c_i - \boldsymbol{V}\right) + \boldsymbol{g} \tag{1}$$

Where $\boldsymbol{V}$ is velocity of the center of mass, $\boldsymbol{\sigma}$ is rotation matrix. A i-th particle position is $\boldsymbol{x}_i$, velocity is c_i. A post value is denoted by prime simbol.

In streaming process, positions and velocities of particles are treated as continuous variables. Thus, particle dynamics evolve in synchronous discrete time evolution and in continuous space. A next particle position is decided by adding present position to velocity simply. Streaming rule as follow :

$$\boldsymbol{x}_i' = \boldsymbol{x}_i + c_i \tag{2}$$

By use of the collision rule, coefficient of dynamic viscosity ν as follow :

$$\nu = \frac{1}{12} + T_M \frac{1 + \rho_0 - e^{-\rho_0}}{2(\rho_0 - 1 + e^{-\rho_0})} \tag{3}$$

Where ρ_0 is density on a site, in initial condition. T_M is median temperature between heating wall temperature T_H and cooling temperature T_C. By using above equation, coefficient of viscosity $\eta = \rho_0 \nu$, thermal conductance $\lambda = C_v \eta$ are obtained. C_v is specific heat at constant volume.

In boundary condition, adiabatic boundaries were imposed nonslip model. Isothermal boundaries were imposed diffuse reflection model. Where n means normal vector of boundary surface. t means tangential vector of boundary surface. T_W is a wall boundary temperature. Particle velocities follow obey distribution $f_{h,n}$.

$$f_h = \frac{1}{\sqrt{2\pi T_W}} \exp\left(-\frac{c_h^2}{2T_W}\right) \tag{4}$$

$$f_n = \frac{c_n}{T_W} \exp\left(-\frac{c_n^2}{2T_W}\right) \tag{5}$$

A distance between sites is fixed as 1. Each site placed at regular intervals. Fluid flow properties were observed on these site by spatial and temporal average.

3. Parallel method

3.1. Particle decomposition

When particle decomposition algorithm is chosen, particles are globally numbered, partitioned, and allocated to processors. Each processor has particles equally. Therefore a load balancing accomplish. A communication between processors arises in collision process to estimate velocity of the center of mass on a site.

3.2. Domain decomposition(flat MPI)

In the case of the domain decomposition algorithm, the simulation space is divided into sub space and each processor is responsible for the calculation of collision, positions, and velocities of all particles within the sub space. As calculation progresses, difference in number of particles between sites increases. Therefore a load balancing does not accomplish. A communication between processors arises in streaming process to translate particles from own sub space to another sub space. Fig.1(a) show chart of domain decomposition in MPI flat version.

3.3. Domain decomposition(hybrid MPI-OpenMP)

MPI processes run their piece of the problem on different processors usually having their own distributed memory. This piece of the problem is generally space in a system. Therefore MPI should be used in a manner that minimizes communications, typically by decomposition the problem. That most of work is done independently. On the other hand, OpenMP codes have directives that describe how parallel sections or loop iterations are to be split up among threads. The team of threads share a single memory space. Since

OpenMP forks a team of threads at the beginning of a parallel region and joins them at the end of the parallel region, OpenMP parallelism should, in general, be exploited at as high a level as possible in the parallel code.

In present study, we introduced MPI and OpenMP hybrid parallelization to reduce communication, keep implementation simply. Firstly, MPI processes was created. Secondly, each MPI processes create OpenMP processes(threads). Each MPI and OpenMP processes were applied to each processors separately. Fig.1(b) show chart of domain decomposition in hybrid MPI-OpenMP version.

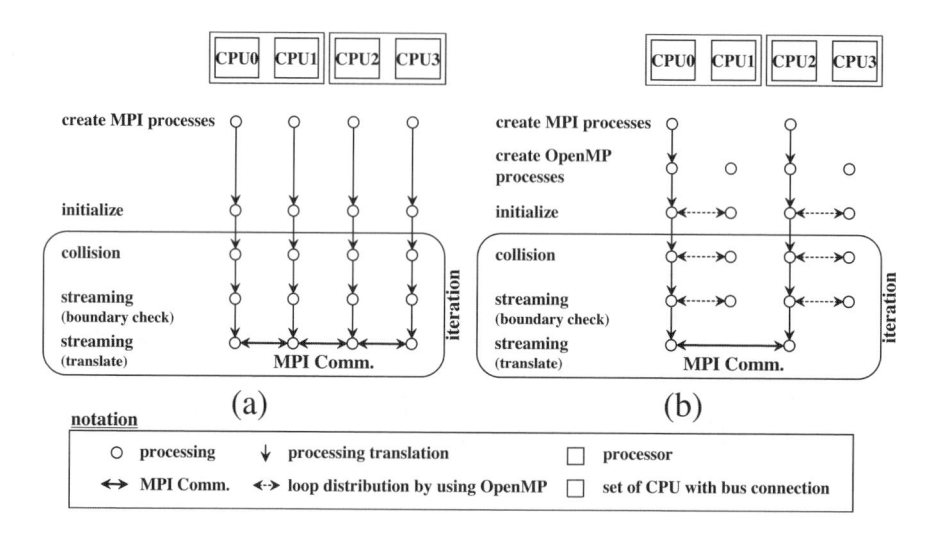

Figure 1. chart of domain decomposition. (a) flat MPI. (b) hybrid MPI-OpenMP .

4. Results

4.1. Environment

This benchmark test platform is SGI Altix 3700[5]. The SGI Altix 3700 is ccNUMA multiprocessor computer which can access global shared memory across multiple nodes. Each node consists of two Intel Itanium 2 microprocessors which can access same local memory through a control SHUB. The developed parallel code is used for the performance evaluation with the number of processors used from 1 to 64. OMP_NUM_THREADS was fixed as 2.

4.2. Natrural convection

Only a few thermal study in continuous-velocity lattice-gas model have reported. Therefore a natural cavity is good candidate to estimate thermal study and code confirmation, in continuous-velocity lattice-gas model.

In initial condition, the left wall is a heating. The right wall is a cooling. Other walls are adiabatic walls. All variables are dimensionless. We gave a heating wall temperature $T_H = 2.0$, a cooling temperature $T_C = 1.0$, $\Delta T = T_H - T_C = 1.0$, density in a site $\rho_0 = 16$, $g = 0.001$ as gravitational acceleration $\boldsymbol{g} = (0, -g)$.

Rayleigh number Ra is introduced in order to compare qualitatively with previous result. Ra is given by production of Grashof number Gr=$g\beta\Delta TL^3/\nu^2$ and Prandtl number Pr= 2 in two-dimensional ideal gas. Where a coefficient of volume expansion $\beta = 0.66$. It estimated from preliminary experiment. A representative length L is variable in order to increase Ra. Ra is 10^3, 10^4, 10^5 in $L = 88, 191, 412$, respectively. We calculated 30000 steps, applied spatial and temporal average. In continuous-velocity lattice-gas model, we gave thermal difference and gravity to system. A natural convection arised without Boussinesq approximation. As a result, we compared with Daivs[7], obtained qualitative agreement.

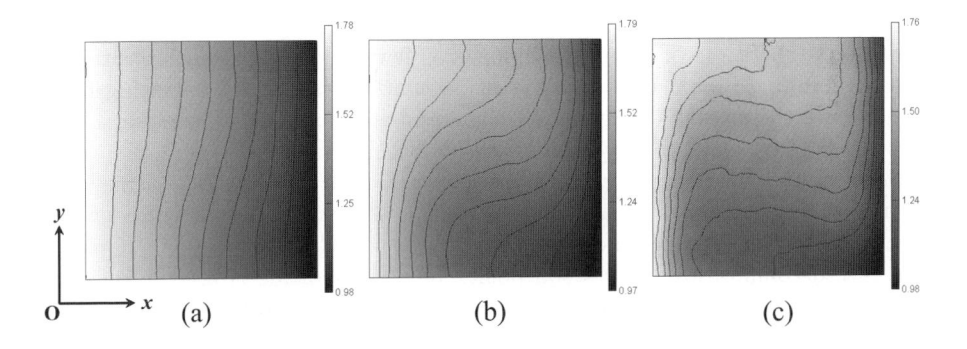

Figure 2. contour of temperature. (a) Ra=10^3. (b) Ra=10^4. (c) Ra=10^5.

4.3. Parallel method comparision

At the beginning of the simulation, sites were divided along x-axis into several domains with equal number of sites and each domain was assigned to a different process. The elapsed time was defined as computation time per 100 steps. A difference in temperature condition measured from 5000 steps since density in around initial step was not equilibrium. A problem size were $L = 64, 128, 256$ in size S, M, L, respectively. The gravity was not imposed on this system in order to simplified benchmark model. A difference in number of particles arose only a differece in temperature without gravity. Other conditions were same as section 4.2.

The results with no difference in temperature were shown as Fig.4 (a)(c)(e). A domain decomposition of flat MPI version was faster than a particle decomposition of flat MPI version. A particle decomposition used reduction operation of MPI in collision process to calculate a velocity of center of mass. This collective communication time was bigger than a point to point communication time. A domain decomposition used mostly point

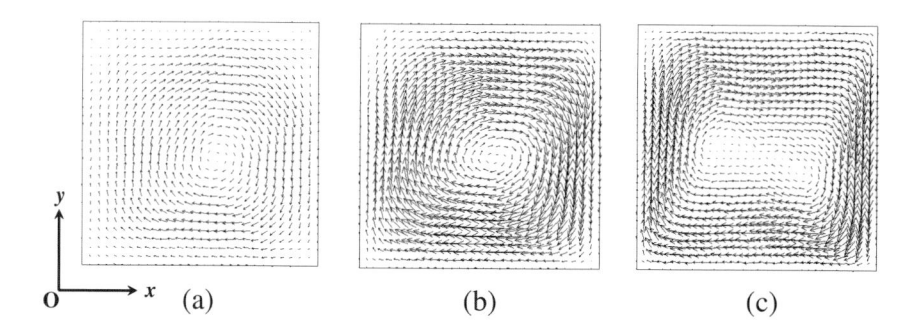

Figure 3. vector of velocity (downsize per each L). (a) Ra=10^3(3). (b) Ra=10^4(6). (c) Ra=10^5(12).

to point. As a result, a communication time was fewer in domain decomposition than in particle decomposition. And the load of calculation was nearly balancing on each processor, since these results were density equally in whole system. Therefore, a domain decomposition of flat MPI accomplished good performance.

In previous study[6], a domain decomposition of flat MPI version was achieved high speed up ratio on CRAY T3E-1200E, even though this code was not considered load balancing. We obtained same tendency of the results on SGI Altix 3700.

A particle decomposition had a good point. It load balancing is accomplished in all cases, since elapsed time was not affected by difference in temperature. However, in larger processors, elapsed time of a particle decomposition was bigger than domain decomposition. A particle decomposition was not suitable for a large scale computation.

On the other hand, a hybrid MPI-OpenMP version had spent more elapsed time than elapsed time of other parallel methods. Since each OpenMP process made loop distribution, it added small overhead. However other parallel method had problem when applied thermal. Since difference in temperature arose difference in number of particles on each processor. Therefore an unbalancing of load arose increasing calculation time.

The results with a difference in temperature were shown as Fig.4 (b)(d)(f). It made to set difference in density, due to set difference in temperature between walls. A hybrid MPI-OpenMP had spent fewer elapsed time than others, in larger processors. Since hybrid MPI-OpenMP version was divided a space coarsely than flat MPI version. Particles on each partition space were distributed to OpneMP threads dynamically. A hybrid MPI-OpenMP parallel method is good candidate for thermal fluid flow analysis in continuous-velocity lattice-gas model.

All results tended to spend more elapsed time for difference in temperature than no difference in temperature, and bigger size problem than smaller size problem. A thermal calculation is heavy task. A parallel computation is important to treat more complicated fluids with thermal.

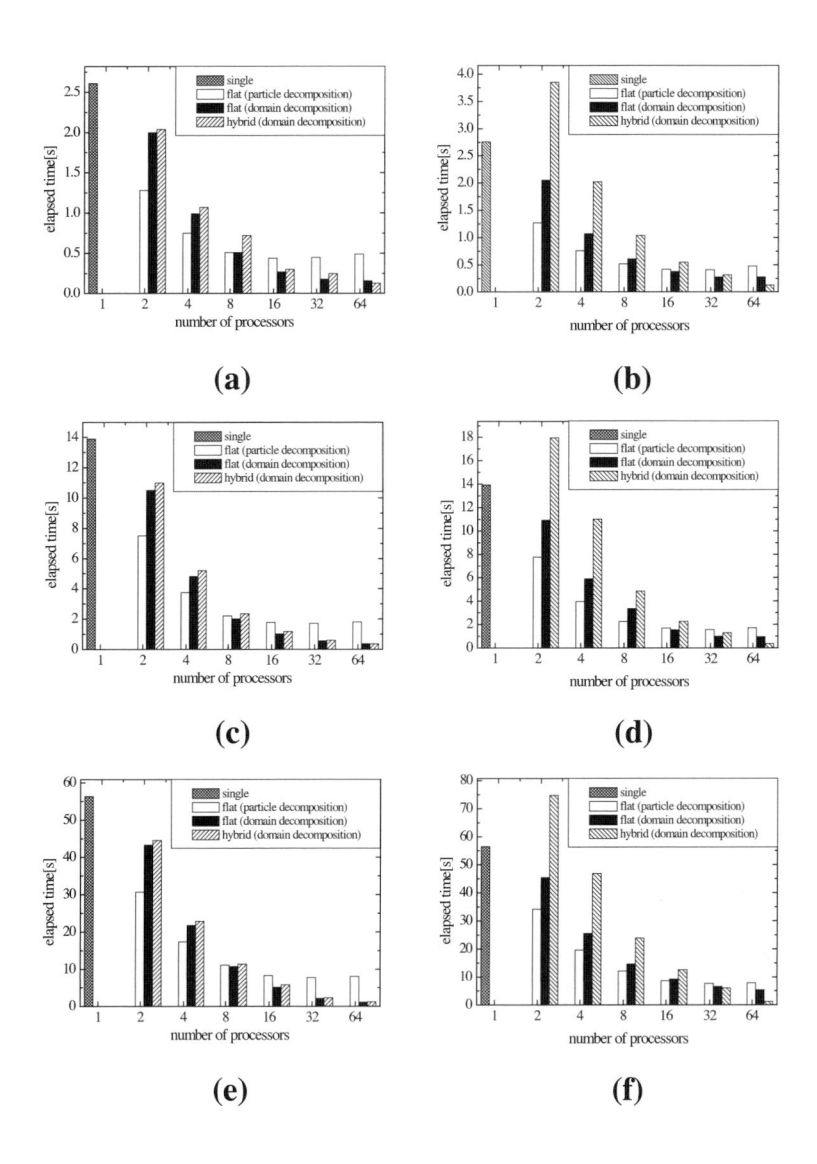

Figure 4. elapsed time. A no difference in temperature are (a)(c)(e). A difference in temperature are (b)(d)(f). A size S problem are (a)(b). A size M problem are (c)(d). A size L problem are (e)(f). A flat means flat MPI. A hybrid means hybrid MPI-OpenMP

5. Concluding Remarks

We imposed parallel computation with flat MPI and hybrid MPI-OpenMP in continuous-velocity lattice-gas model. The benchmark experiment was estimated on SGI Altix 3700. The MPI-OpenMP hybrid parallelization accomplished to reduce calculation time with increasing processors, with difference in temperature. This way provide relatively easier implementation than only flat MPI load balancing. And a natural convection is qualitative agreement with previous results.

REFERENCES

1. A. Malevanets, R.Kapral. Continuous-velocity lattice-gas model for fluid flow, Europhys. Lett., Vol.44, No.5, 1998, pp. 552-558.
2. Y.Hashimoto, Y.Chen, H.Ohashi, Immiscible real-coded lattice gas, Computer Phys. Comm., 129, 2000, pp. 56-62.
3. G. Bella, S. Filippone, N. Rossi, S. Ubertini, Using OpenMP on a Hydrodynamic Lattice-Boltzmann Code, EWOMP2002, 2002.
4. C.Pan, J.F.Prins, C.T.Miller, A high-performance lattice Boltzmann implementation to model flow in porous media, Computer Phys. Comm., 158, 2004, pp.89-105.
5. http://www.sgi.com/products/servers/altix/
6. M. Terai, T. Matsuzawa, Data-structure parallel computation for continuous-velocity lattice-gas model(Japanese), IPSJ, Vol.45, No.SIG06(ACS6), 2004, pp. 151-160.
7. G. de. Vahl Davis, Natural convection of air in a square cavity: a bench mark numerical solution, International Journal for Numerical in Fluids, 3, 1983, pp. 249-264.

Parallel Atmospheric Modeling with High-Order Continuous and Discontinuous Galerkin methods

Amik St-Cyr[a] and Stephen J. Thomas[a]

[a]National Center for Atmospheric Research,
1850 Table Mesa Drive, Boulder, 80305 CO, USA

High-order finite element methods for the atmospheric shallow water equations are reviewed. The accuracy and efficiency of nodal continuous and discontinuous Galerkin spectral elements are evaluated using the standard test problems proposed by Williamson et al (1992). The relative merits of strong-stability preserving (SSP) explicit Runge-Kutta and multistep time discretizations are discussed. Distributed memory MPI implementations are compared on the basis of the total computation time required, sustained performance and parallel scalability. Because a discontinuous Galerkin method permits the overlap of computation and communication, higher sustained execution rates are possible at large processor counts.

1. Introduction

High-order finite element methods are well-suited to atmospheric modeling due to their desirable numerical properties and inherent parallelism. A spectral element atmospheric model received an honorable mention in the 2001 Gordon Bell award competition, Loft et al (2001). Discontinuous Galerkin approximations are an extension of low order finite-volume techniques for compressible flows with shocks (Cockburn et al 2000). Either nodal or modal basis functions can be employed in high-order finite elements and the methods are spectrally accurate for smooth solutions. To avoid excessive memory requirements, global assembly of finite element matrices is avoided in the continuous Galerkin method by applying a direct-stiffness summation, Deville et al (2002). Computations within an element are based on tensor-product summations, taking the form of dense matrix-matrix multiplications. These are naturally cache-blocked and can be unrolled to expose instruction level parallelism to processors containing multiple floating point units.

The shallow water equations are a prototype for atmospheric general circulation models. The parallel performance of a 3D model can be estimated by solving identical 2D shallow water problems on multiple layers. To evaluate the efficiency of various time integrators, schemes of equivalent order will be compared on the basis of the total wall-clock time required to solve a given initial value problem (i.e. time to solution). The amount of computation required by a method depends on the Courant number or equivalently the time step size. Moreover, the parallel performance depends on the number of right-hand side evaluations per time step and the associated parallel communication. In the case of the discontinuous Galerkin method, communication of conserved variables can be overlapped with the weak divergence and source term computations. The continuous Galerkin spectral element model developed by Loft et al (2002) will serve as the parallel performance baseline for our simulations.

2. Shallow Water Equations

The shallow water equations contain the essential wave propagation mechanisms found in atmospheric general circulation models. These are the fast-moving gravity waves and nonlinear Rossby waves. The latter are important for correctly capturing nonlinear atmospheric dynamics. The flux form shallow-water equations in curvilinear coordinates are described in Sadourny (1972).

$$\frac{\partial u_1}{\partial t} + \frac{\partial}{\partial x^1} E = \sqrt{G}\, u^2 (f + \zeta),$$

$$\frac{\partial u_2}{\partial t} + \frac{\partial}{\partial x^2} E = -\sqrt{G}\, u^1 (f + \zeta),$$

$$\frac{\partial}{\partial t}(\sqrt{G}\,\Phi) + \frac{\partial}{\partial x^1}(\sqrt{G}\, u^1 \Phi) + \frac{\partial}{\partial x^2}(\sqrt{G}\, u^2 \Phi) = 0,$$

where

$$E = \Phi + \frac{1}{2}(u_1 u^1 + u_2 u^2), \quad \zeta = \frac{1}{\sqrt{G}}\left[\frac{\partial u_2}{\partial x^1} - \frac{\partial u_1}{\partial x^2}\right]$$

h is the height above sea level. u^i and u_j are the contravariant and covariant velocities. $\Phi = gh$ the geopotential height. f is the Coriolis parameter. The metric tensor is G_{ij} and $G = \det(G_{ij})$.

3. Space Discretization

The computational domain Ω is partitioned into finite elements Ω_k. An approximate solution u_h belongs to the finite dimensional space $\mathcal{V}_h(\Omega)$. u_h is expanded in terms of a tensor-product of the Lagrange basis functions defined at the Gauss-Lobatto-Legendre points

$$u_h^k = \sum_{i=0}^{N}\sum_{j=0}^{N} u_{ij} h_i(x) h_j(y)$$

A weak Galerkin variational problem is obtained by integrating the equations with respect to a test function $\varphi_h \in \mathcal{V}_h$. In the continuous Galerkin spectral element method, integrals are directly evaluated using Gauss-Lobatto quadrature and continuity is enforced at the element boundaries.

To illustrate the discontinuous Galerkin approach, consider a scalar hyperbolic equation in flux form, By applying the Gauss divergence theorem, the weak form becomes

$$\frac{d}{dt}\int_{\Omega_k} \varphi_h u_h\, d\Omega = \int_{\Omega_k} \varphi_h S\, d\Omega + \int_{\Omega_k} \mathcal{F}\cdot\nabla\varphi_h\, d\Omega - \int_{\partial\Omega_k} \varphi_h \mathcal{F}\cdot\hat{n}\, ds$$

The jump discontinuity at an element boundary requires the solution of a Riemann problem where the flux function $\mathcal{F} \cdot \hat{n}$ is approximated by a Lax-Friedrichs numerical flux. The resulting semi-discrete equation is given by

$$\frac{du_h}{dt} = L(u_h). \tag{1}$$

4. Time Discretization

Strong-stability preserving (SSP) time discretization methods were developed for semi-discrete method of lines approximation of hyperbolic PDE's in conservative form, Gottlieb et al (2001). Strong stability is a monotonicity property for the internal stages and the numerical solution. A general m-stage SSP Runge-Kutta method is given by

$$u^{(0)} = u^n$$
$$u^{(i)} = \sum_{k=0}^{i-1} \alpha_{ik} u^{(k)} + \Delta t \beta_{ik} L(u^{(k)}), \quad i = 1, \ldots, m, \tag{2}$$
$$u^{n+1} = u^{(m)}.$$

To compute $u^{(i)}$ for each stage requires up to m evaluations of the right-hand side $L(u^{(k)})$. Thus, higher-order SSP Runge-Kutta methods can be expensive, in terms of the number of floating point operations, memory to store intermediate stages and parallel communication overhead per time step. Linear multistep methods (LMM) substitute time levels for stages.

Higueras (2004) discovered second order SSP Runge-Kutta methods with three stages and having Courant number $C = 2$. Therefore, the most efficient second order explicit integrator for the discontinuous Galerkin approximation would appear to be the three stage SSP Runge-Kutta scheme with an efficiency factor of $C/3 = 2/3$. Indeed, our numerical experiments confirmed this to be the case. In fact it was found in practice that this scheme integrates twice as fast as the two stage method. Moreover, the maximum time step for the three stage SSP RK2-3 matches that of the second order leap frog integrator employed in the continuous Galerkin spectral element model of Loft et al (2001).

5. Numerical Experiments

Our numerical experiments are based on the shallow water test suite of Williamson et al (1992). Test case 5 is a zonal flow impinging on an isolated mountain. The center of the mountain is located at $(3\pi/2, \pi/6)$ with height $h_s = 2000\,(1 - r/R)$ meters, where $R = \pi/9$ and $r^2 = \min[R^2, (\lambda - 3\pi/2)^2 + (\theta - \pi/6)^2]$. The initial wind and height fields take the same form as test case 2 with $\alpha_0 = 0$, $gh_0 = 5960$ m^2/s^2 and $u_0 = 20$ m/s. A total of 150 spectral elements containing 8×8 Gauss-Lobatto-Legendre points are employed. The explicit time step was $\Delta t = 90$ sec. A spatial filter was not applied during this integration. Figure 1 contains a plot of the geopotential height field after 15 days of integration using the discontinuous Galerkin approximation. These results compare favorably with the continuous spectral element model.

6. Parallel Performance Results

Both the continuous and discontinuous Galerkin spectral element models are implemented within a unified software framework. A hybrid MPI/OpenMP programming model is supported where the entire time step is threaded according to an SPMD shared-memory approach. MPI message passing calls are serialized in hybrid mode. The cubed-sphere computational domain is partitioned across the compute nodes of a distributed-memory machine using either Metis or space-filling curves (Dennis 2003). The latter algorithm is applied in the case where the number of elements along a cube face edge is divisible by $2^n 3^m$. Unlike continuous Galerkin spectral elements, almost all the computations within a discontinuous Galerkin finite element do not require any information from neighboring elements. Both the weak divergence and source terms can be computed independently on each element and then the local contribution of the boundary integrals can be added later. By employing non-blocking MPI communication, these computations can be performed while the exchange of conserved variables between elements proceeds. Single processor optimizations are based on data structures designed for extensive re-use and stride-1 memory access to minimize cache misses. Finite elements are represented as Fortran 90 derived types which are allocated statically on each processor at run-time. Spherical and cartesian coordinates along with the metric tensor are defined in the element type. Extensive loop unrolling is applied to expose instruction level parallelism to the processor. An effective technique for fast computation of the maximum eigenvalue α of the flux Jacobian is to invoke vector intrinsics and the use of square roots is minimized.

An experiment was designed to test if overlapping communication and computation in a pure MPI code has any measurable effect on an IBM p690 cluster with a Colony switch. The machine consists of 32-way SMP nodes containing 1.3 Ghz Power4 processors capable of four flops per clock cycle. A node can be configured as either a single 32-way or four 8-way logical partitions. Both the continuous and discontinuous Galerkin models were compared using test case 5, but integrated for only five days. For this test, the problem is replicated on 40 independent vertical levels. The discontinuous Galerkin code was integrated using the three step linear multistep method (LMM2-3) with $\Delta t = 15$ sec and three stage SSP Runge-Kutta (RK2-3) scheme with $\Delta t = 60$ sec. The continuous Galerkin model uses a second order leap frog integrator and $\Delta t = 60$ sec. $K = 384$ elements were employed with 10×10 Gauss-Lobatto-Legendre points. The LMM2-3 and RK2-3 single processor execution rates are 667 MFlops and 528 MFlops, respectively. The leap frog code sustains 724 MFlops. The wait time is defined as the average time for the MPI non-blocking communication to complete, summed over all time steps and all processors. We find that the wait time for 8-way is less than for 32-way partitions (see figure 2). Figure 3 is a plot of the integration times using 8-way partitions.

7. Conclusions

The three stage RK2-3 integrator discovered by Higueras (2004) was the most efficient SSP method examined. The discontinuous Galerkin method converges exponentially for smooth solutions and standard error metrics compare favorably with a continuous Galerkin spectral element model. Overlapping communication with computation was suggested by Baggag et al (1999). However, we only communicate u_h and compute fluxes

$\mathcal{F}(u_h)$ locally. Both non-overlapping and overlapping implementations were compared. The latter was found to be clearly beneficial on SMP clusters such as the IBM p690 and leads to improved scalability in the strong sense for a fixed problem size. The performance within a 32-way node is higher but 8-way nodes scale better. Our experience with a nodal Galerkin method indicates that a filter is required for long integrations to stabilize the scheme, thereby extending the results of Nair et al (2004).

Acknowledgements.

NCAR is supported by the National Science Foundation. This work was partially supported by an NSF collaborations in mathematics and the geosciences grant (0222282) and the DOE climate change prediction program (CCPP).

REFERENCES

1. Baggag, A., H. Atkins, D. Keyes, 1999: Parallel implementation of the discontinuous Galerkin method. ICASE Report 99-35, NASA/CR-1999-209546.
2. Cockburn, B., G. E. Karniadakis, and C. W. Shu, 2000: Discontinuous Galerkin Methods. Springer-Verlag, New York, 470 pp.
3. Dennis, J. M., 2003: Partitioning with space-filling curves on the cubed-sphere. Proceedings of Workshop on Massively Parallel Processing at IPDPS'03. Nice, France, April 2003.
4. Deville, M. O., P. F. Fischer, and E. H. Mund, 2002: High-Order Methods for Incompressible Fluid Flow. Cambridge University Press, 499 pp.
5. Fischer, P. F., and J. S. Mullen, 2001: Filter-Based stabilization of spectral element methods. *Comptes Rendus de l'Académie des sciences Paris*, t. 332, Série I - Analyse numérique, 265–270.
6. Giraldo, F. X., J. S. Hesthaven, and T. Warburton, 2003: Nodal high-order discontinuous Galerkin methods for spherical shallow water equations. *J. Comput. Phys.*, **181**, 499-525.
7. Gottlieb, S., C. W. Shu, and E. Tadmor, 2001: Strong stability preserving high-order time discretization methods. *SIAM Review*, **43**, 89–112.
8. Higueras, I., 2004: On strong stability preserving time discretization methods. *J. Sci. Comput.*, **21**, 193-223.
9. Loft, R. L., S. J. Thomas, and J. M. Dennis, 2001: Terascale spectral element dynamical core for atmospheric general circulation models. Proceedings of Supercomputing 01, IEEE/ACM.
10. Nair, R. D., S. J. Thomas, and R. D. Loft, 2004: A discontinuous Galerkin global shallow water model. *Mon. Wea. Rev.*, to appear.
11. Sadourny, R., 1972: Conservative finite-difference approximations of the primitive equations on quasi-uniform spherical grids. *Mon. Wea. Rev.*, **100**, 136–144.
12. Thomas, S. J., and R. D. Loft, 2002: Semi-implicit spectral element atmospheric model. *J. Sci. Comp.*, **17**, 339–350.
13. Williamson, D. L., J. B. Drake, J. J. Hack, R. Jakob, P. N. Swarztrauber, 1992: A standard test set for numerical approximations to the shallow water equations in spherical geometry *J. Comp. Phys.*, **102**, 211–224.

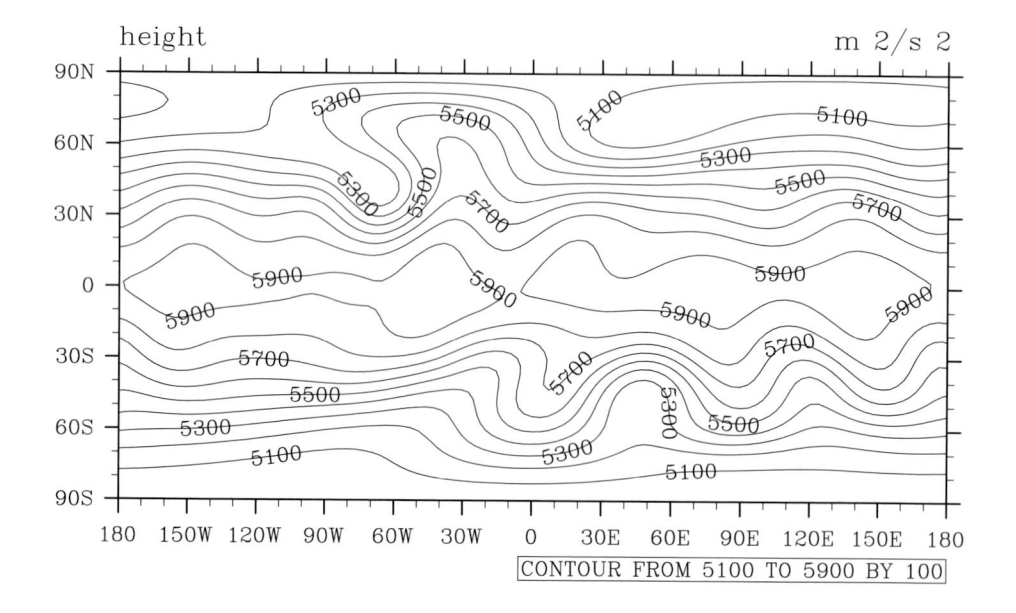

Figure 1. Shallow water test case 5: Flow impinging on a mountain. 150 spectral elements, 8 × 8 Gauss-Legendre Lobatto points per element. Geopotential height field h at fifteen days produced by discontinuous Galerkin method.

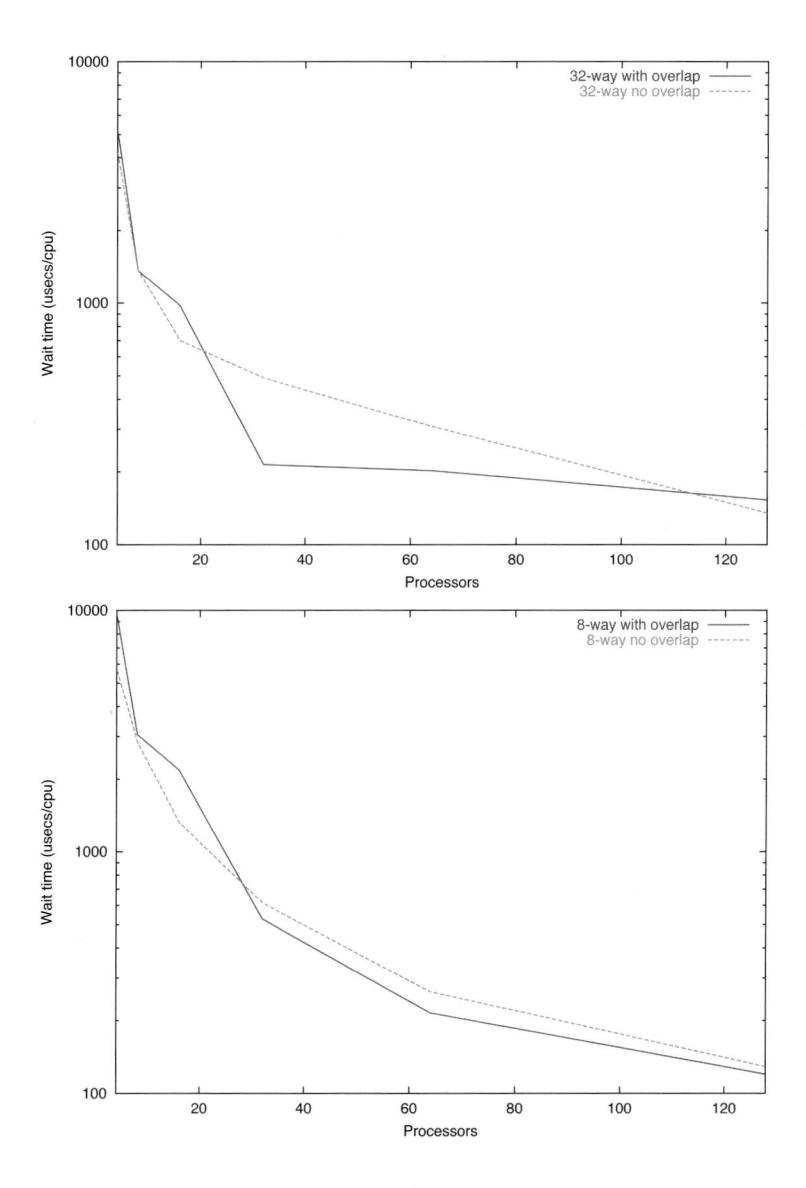

Figure 2. Shallow water test case 2: Steady state zonal geostrophic flow. 150 spectral elements, 40 levels, 10 × 10 Gauss-Legendre Lobatto points per element. Top panel: Average wait time in microseconds for IBM p690 32-way partitions. Bottom panel: Average wait time in microseconds for IBM p690 8-way partitions.

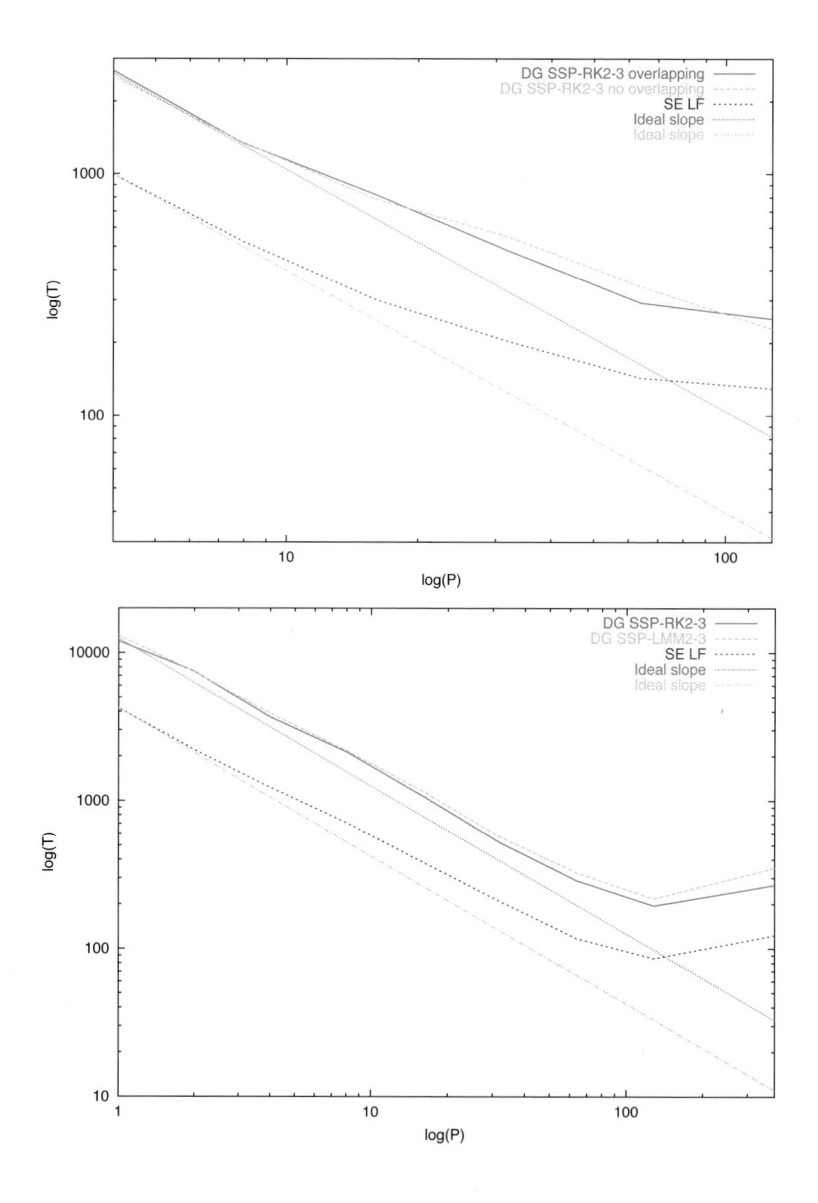

Figure 3. Shallow water test case 5: Flow impinging on a mountain. 384 spectral elements, 40 levels, 10×10 Gauss-Legendre Lobatto points per element. Top panel: Integration time in seconds for IBM p690 32-way partitions. Bottom panel: Integration time in seconds for IBM p690 8-way partitions.

A new formulation of NUDFT applied to Aitken-Schwarz DDM on nonuniform meshes

A. Frullone [*] and D. Tromeur-Dervout [†] [a]

[a]CDCSP/UMR5208, University Lyon 1, 15 Bd Latarjet, 69622 Villeurbanne, France

Keywords: Non Uniform Discrete Fourier Transform, Aitken acceleration of Convergence, Schwarz Domain Decomposition, Parallel Computation, Non Uniform Grid.
MSC : 65T50, 65B99, 65N55, 65Y05, 65Y10.

1. Motivations

Aitken-Schwarz method [9,8] is well designed to solve linear and nonlinear problems in a metacomputing framework. The method is highly tolerant to low bandwidth and high latency and its performances had been tested on 1256 processors of 3 Cray T3E connected with a standard 5Mb/s ethernet network in the experiments related in [2], showing its competitivity in the case of $O(10)$ subdomains.

Some Extension of this DDM methodology has been done on heterogeneous DDM with FEM and spectral Chebychev collocation method [3] using a regular third grid for interface and also on irregular cartesian meshes [1] and in the finite volume framework [7]. In the two later references, approximations of the main eigenvalues and associated eigenvectors V of the trace of error operator P are searched in the physical space associated with the Euclidian scalar product. The advantage is then to have an accurate acceleration of the interface solution. Nevertheless this approach claims to build the trace of error operator in the physical space and to compute the eigenpairs. The cost to compute the full P is $N + 1$ Schwarz iterates and the computational cost to compute several main eigenvalues and associated eigenvectors with a sufficient accuracy is not a priori known (the cost for all eigenpairs for a system of size N is $O(N^3)$). A band approximation of P can be perform in order to reduce the number of Schwarz iterates to be performed but in this case the impact of this approximation in the computing of the right eigenpairs is not a priori known.

The present approach opts for the use of the Fourier transform of the iterated solution's traces at the artificial interfaces. More generally speaking, the principle consists in defining an orthogonal basis $V_{[[.,.]]}$ with respect to an hermitian $[[.,.]]$ form associated to the interface mesh M. Then we can write a trace operator $P_{[[.,.]]}$ with respect to the basis $V_{[[.,.]]}$ corresponding on how the components of the error at artificial interfaces in the basis $V_{[[.,.]]}$ are reduced by the Schwarz iterate. We then have access to a posteriori

---
[*]This work is backward to the Région Rhône-Alpes thru the project: "Développement de méthodologies mathématiques pour le calcul scientifique sur grille".
[†]This author was partially supported thru the GDR MOMAS:"solveur multidomaines"

error estimates on the approximation of eigenvectors associated to $P_{[[.,.]]}$ and also we can adaptively construct an approximation of $P_{[[.,.]]}$ in order to reduce the number of Schwarz iterates based on a posteriori estimates.

The plan of this paper is as follows. Section 2 presents the construction of the hermitian form in 1D and the non uniform discrete Fourier transform associated. Section 3 shows results on the NUDFT approximation accuracy in 1D and 2D. Section 4 applied this approximation in the Aitken-Schwarz domain decomposition context with a discussion on the adaptive computing of the trace operator $P_{[[.,.]]}$ while section 5 concludes and gives the perspectives.

2. Non Uniform Discrete Fourier Transform

The nonuniform discrete Fourier transform (NUDFT) arises in many practical situations where the input data are not equally spaced, as for example in image reconstruction and in spectral methods on adaptive grids in CFD problems. The problem of efficiently computing NUDFT has been addressed by several authors in recent years see for example [10] and its references. Since now, as long as we could check, the problem of computing efficiently NUDFT has been solved by interpolation techniques associated with FFT on uniform meshes and gridding techniques, as for example in [4][5][10].

The new formulation for NUDFT proposed here, is based on an appropriate choice of the Fourier basis functions $(\phi_l(x))_{0 \leq l \leq N}$, strictly related to the nonuniform mesh. The main property of this formulation is that the basis functions $\Phi_l = (\phi_l(x_j))_{0 \leq j \leq N}$ are orthogonal with respect to a sesquilinear form $[[.,.]]$, i.e $[[\phi_l, \phi_k]] = 0$, if $l \neq k$, which is not the case for the classical hermitian discrete scalar product $(x, y) = y^* x$.

Mathematically the approximation problem can be formulated as follows:

definition 2.1 *Let $(x_i)_{0 \leq i \leq N}$ and $(z_i)_{0 \leq i \leq N}$ with $z_i = \frac{2\pi i}{N}$ be a set of points defined on $[0, 2\pi[$ such that $x_i = z_i + \epsilon_i$ with $(\epsilon)_{0 \leq i \leq N}$ defining the irregularity of the mesh. Set $\psi_l(x) = \exp(ilx)_{0 \leq l \leq N/2}$, then ψ_l verify $\psi_{N-l}(x) = D^N \overline{\psi_l(x)}$, with $D = diag(\epsilon_i)_{0 \leq i \leq N}$. We define the functions*

$$\phi_l(x) = \begin{cases} \psi_l(x), & 0 \leq l \leq N/2, \\ D^{-N}\psi_{N-l}(x), & N/2 + 1 \leq l \leq N \end{cases} \quad (1)$$

definition 2.2 *We consider the sesquilinear form on $S_N = \text{span}\{\phi_l(x), 0 \leq l \leq N\}$:*

$$[[f, g]] = \sum_{l=0}^{N} \gamma_l f(x_l)\overline{g(x_l)} + \sum_{l=0}^{N} \beta_l (f'(x_l)\overline{g(x_l)} + f(x_l)\overline{g'(x_l)}) \quad (2)$$

This sesquilinear form is none but the Hermite quadrature formula for $\int_0^{2\pi} f\bar{g}dx$.

proposition 2.1 *According to definitions 2.1-2.2, the orthogonality of the basis functions, expressed by the equation $[[\phi_l, \phi_k]] = \delta_{lk}$, where δ_{lk} is the Kronecker symbol, is guaranteed by solving the following linear system of $2(N + 1)$ equations:*

$$\{\sum_l \gamma_l \exp(ikx_l) + \sum_l \beta_l ik \exp(ikx_l) = \delta_{0k}, k = -N, \cdots, N, \text{ and } \sum_{l=0}^{N} \beta_l = 0 \quad (3)$$

Now, since the matrix $H = ([[\phi_l, \phi_k]])$ is equal to the identity matrix, the sesquilinear form $[[.,.]]$ is hermitian.

definition 2.3 *The discrete Fourier coefficients of f are given by:*

$$\tilde{f}_k = [[f, \Phi_k]], \quad k = -N/2, ..., N/2 \tag{4}$$

or, in an algebraic framework:

$$\tilde{f} = M_1 f + M_2 f', \quad M_1, M_2 \in \mathcal{M}_{N+1}(\mathbb{C}), \quad M_1(k,l) = \gamma_l \overline{\phi_k(x_l)} + \beta_l \overline{\phi'_k(x_l)}, \quad M_2(k,l) = \beta_l \overline{\phi_k(x_l)}$$

We are now in the position to prove :

proposition 2.2 *With the notations of definition 2.1-2.3, then the trigonometrical approximation $\Pi_N^F(f(x)) = \sum_{l=0}^{N} \alpha_l \phi_l(x)$ where $\alpha_l = [[f, \Phi_l]]/[[\Phi_l, \Phi_l]]$ with $f = (f(x_i))_{0 \leq i \leq N} \in \mathbb{C}^{N+1}$, is exact for all $f \in \mathbb{T}^{N/2}([0, 2\pi[)$ the space of the trigonometrical polynomials of degree less or equal to $N/2$.*

Proof Since $f \in \mathbb{T}^{N/2}([0, 2\pi[)$, we can write it as a linear combination of the basis functions:

$$f(z) = \sum_{k=-N/2}^{N/2} \eta_k \exp(ikz) = \sum_{m=0}^{N} \eta_m \Phi_m(z). \tag{5}$$

Now, since $[[\cdot, \cdot]]$ is hermitian

$$\tilde{f}_k = [[f, \Phi_k]] = \sum_{m=0}^{N} \eta_m [[\Phi_m, \Phi_k]] = \sum_{m=0}^{N} \eta_m \delta_{mk} = \eta_k \tag{6}$$

Thus the trigonometrical approximation of f becomes:

$$\Pi_N^F(f(z)) = \sum_{l=0}^{N} \tilde{f}_k \phi_k(z) = \sum_{k=0}^{N} \eta_k \Phi_k(z) = f(z), \tag{7}$$

which proves the proposition. $\square$

The main advantage of this approach is, once the hermitian form defined, the interpolation has an a priori known cost (of $O(N^2)$ operations). In the applications one is given the vector f which represents the values of a function $f(x)$ on the points $(x_i)_{0 \leq i \leq N}$. No information is given on the vector f' which is needed in definition 2.3. To satisfy this requirement, we determine the vector f' implicitly by imposing

$$\frac{d}{dx}(\Pi_N^F(f(x)))_{|x=x_l} = f'(x_l), \quad l = 0, ..., N \tag{8}$$

In an algebraic form, if Φ' is the matrix whose elements are:

$$\Phi'(l, k) = \phi'_k(x_l) \tag{9}$$

then the vector f' is obtained by solving the algebraic system:

$$(id_{N+1} - \Phi' M_2) f' = \Phi M_1 f \tag{10}$$

where id_N is the identity matrix in $\mathcal{M}_{N+1}(\mathbb{C})$.

We can extend the formulation introduced above to the two dimensional case considering cartesian product of of **x** and **y** grids. The basis functions write:

definition 2.4

$$\phi_{lm}(x,y) = \begin{cases} \psi_l^x(x)\psi_m^y(y),\ 0 \le l,m \le N/2 \\ D_x^{-N}\psi_{N-l}^x(x)\psi_m^y(y),\ N/2+1 \le l \le N, \\ \qquad\qquad\qquad\qquad 0 \le m \le N/2 \\ \psi_l^x(x)D_y^{-N}\psi_{N-m}^y(y),\ 0 \le l \le N/2, \\ \qquad\qquad\qquad\qquad N/2+1 \le m \le N \\ D_x^{-N}\psi_{N-l}^x(x)D_y^{-N}\psi_{N-m}^y(y), \\ \qquad\qquad\qquad\qquad N/2+1 \le l,m \le N \end{cases} \quad (11)$$

3. Results on NUDFT approximation

This section shows some numerical results on the approximation accuracy.

3.1. Numerical results 1D

First we check on an analytical function the approximation property with respect to the irregularity of the 1D mesh and the condition number of the matrix involved in the definition of the hermitian form. Table 2 shows the approximation error with respect to

N	$\varepsilon = h_u/2$	$\varepsilon = h_u$	$\varepsilon = 2h_u$	$\varepsilon = 4h_u$	$h_u = 2\pi/N$
2^6	6.22E-4	1.23E-3	3.64E-3	8.48E-3	$\|f - \Pi_N^F(f)\|_\infty$
	6.93E-3	1.34E-2	2.75E-2	5.03E-2	$\frac{\|h-h_u\|_2}{Nh_u}$
	0.45E+0	0.86E+0	1.78E+0	3.26E+0	$\frac{\|h-h_u\|_\infty}{h_u}$
	1.35E+3	5.53E+4	2.55E+7	5.38E+10	$cond_2([[.,.]])$
	2.15E+0	2.92E+2	4.40E+5	1.05E+10	$cond_2(eq.(10))$
2^7	1.85E-12	5.92E-12	1.43E-10	3.22E-3	$\|f - \Pi_N^F(f)\|_\infty$
	3.68E-3	7.26E-3	1.17E-2	2.66E-2	$\frac{\|h-h_u\|_2}{Nh_u}$
	0.47E+0	0.93E+0	1.51E+0	3.43E+0	$\frac{\|h-h_u\|_\infty}{h_u}$
	5.75E+3	3.41E+5	4.48E+9	7.53E+12	$cond_2([[.,.]])$
	2.50E+0	2.45E+2	6.65E+5	1.25E+11	$cond_2(eq.(10))$
2^8	6.32E-14	2.52E-12	3.52E-8	5.43E-2	$\|f - \Pi_N^F(f)\|_\infty$
	1.83E-3	3.82E-3	7.65E-3	1.28E-2	$\frac{\|h-h_u\|_2}{Nh_u}$
	0.47E+0	0.98E+0	1.96E+0	3.28E+0	$\frac{\|h-h_u\|_\infty}{h_u}$
	2.58E+4	9.98E+5	8.82E+10	9.75E+12	$cond_2([[.,.]])$
	2.58E+0	6.12E+2	3.05E+7	3.67E+11	$cond_2(eq.(10))$

Table 1
$\|f - \Pi_N^F(f)\|_\infty$ as a function of the number of discretisation points N and the maximum distance between two points ε, for $f(x) = \exp(-40(x - (2\pi/3))^2)$.

the number of discretisation points and the maximum distance between two points (ϵ), for an analytical function $f(x) = \exp(-40(x-(2\pi/3))^2)$. h_u represents the step size on an uniform mesh, $\frac{\|h-h_u\|_2}{Nh_u}$ the distance of the non uniform mesh to a uniform mesh, $\frac{\|h-h_u\|_\infty}{h_u}$

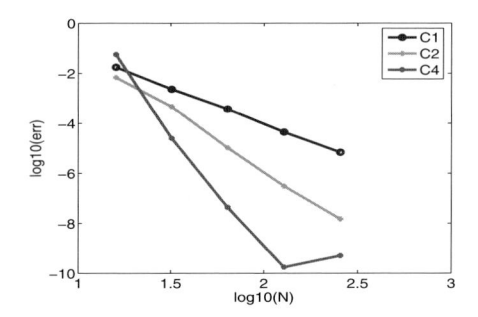

Figure 1. $\|f - \Pi_N^F(f)\|_\infty$ as a function of the number of discretisation points and the regularity of the test function, for $\varepsilon = h_u$, $f(x) = (1 - \cos(2x))^{\eta+1/2}$, $\eta = 1, 2, 4$.

the relative factor of the non uniformity compared to an uniform mesh. The results exhibit that good accuracy are obtained until the condition numbers for computing the weight of the hermitian form and for solving equation 10 are still reasonable.

Figure 1 shows the quality of the approximation with respect to the function regularity. Results exhibit that the convergence behaves in $O(1/N^{2\eta})$ where η is the regularity of the approximate function.

3.2. Numerical results 2D

The results obtained for the $2D$ case are consistent with the $1D$ ones, as proved in the following table:

4. Application in the Aitken-Schwarz DDM to solve Poisson problem

We have applied this new formulation of NUDFT to the solution of 2D Poisson's equation with Aitken-Schwarz DDM. Our approach provides a framework where the trace of the iterate solutions on the irregular mesh interfaces are decomposed in a Fourier orthogonal basis. Then we can accelerate the Fourier modes through the Aitken technique as in [8]. We are going to describe briefly the numerical ideas behind the Aitken-Schwarz

N	$\epsilon = h_u/2$	$\epsilon = h_u$	$\epsilon = 2h_u$	$\epsilon = 4h_u$	$h_u = 2\pi/N$
2^6	1.1E-13	3.7E-13	9.5E-7	2.09E+3	$\|f - \Pi_N^F(f)\|_\infty$
	1.5E+3	8E+3	2.5E+6	2.2E+12	$cond_2([[.,.]])$
2^7	2.62E-13	1.48E-10	8E-4	3E+6	$\|f - \Pi_N^F(f)\|_\infty$
	6E+3	5E+5	1.7E+10	1E+14	$cond_2([[.,.]])$

Table 2
$\|f - \Pi_N^F(f)\|_\infty$ and $cond_2([[.,.]])$ as a function of the number of discretisation points in each direction N and the maximum distance between two points ϵ, for $f(x,y) = \cos^2(x)\cos(y)$, with $h_u = 2\pi/N$.

method. We refer to [8] for more details. Let us consider the linear differential problem
$$L[U] = f \text{ in } \Omega, U|\partial\Omega = 0 \tag{12}$$
For simplicity, we restrict ourselves to the two overlapping subdomains case and the additive Schwarz algorithm:
$$L[u_1^{n+1}] = f \text{ in } \Omega_1, u_{1|\Gamma_1}^{n+1} = u_{2|\Gamma_1}^n, \text{ and }, L[u_2^{n+1}] = f \text{ in } \Omega_2, u_{2|\Gamma_2}^{n+1} = u_{1|\Gamma_2}^n. \tag{13}$$
If U_{Γ_i} denotes the exact solution on interface Γ_i then the operator T,
$$u_{i|\Gamma_i}^n - U_{\Gamma_i} \to u_{i|\Gamma_i}^{n+2} - U_{\Gamma_i} \quad i = 1, 2 \tag{14}$$
is linear. If $\tilde{u}_i^n$ denotes the NUDFT coefficients of $u_{i|\Gamma_i}^n$, then in a matrix form, we have
$$\tilde{u}_i^{n+2} - U_{\Gamma_i} = P_{[[.,.]]}(\tilde{u}_i^n - U_{\Gamma_i}) \quad i = 1, 2. \tag{15}$$
In the general case, the interface operator P relies every mode and every interface. If the operator is represented by means of a $m \times m$ matrix, we can construct it numerically by using m Schwarz iterations, then accelerate the solution using the formula:
$$\tilde{u}^\infty = (Id - P)^{-1}(\tilde{u}^{n+1} - P\tilde{u}^n). \tag{16}$$
The advantage of the nonuniform discrete Fourier decomposition is that we can check the modes which converge faster, thus reducing the complexity of the problem by solving only q Schwarz iterations (with $q < m$). Thus we can reduce the number of communications between the suddomains and also the amount of data to be exchange. We can also adapt the communication following the mode convergence. If the error vanish quickly, it is not necessary to exchange data between far sudomains but only between neighboring subdomains. In the case of the Laplacian operator, indeed, we can approximate P with a block diagonal matrix (taking into a account the uncoupling of the Fourier modes). The special version of Aitken-method introduced in [9] is **Algorithm 1:**

1- given traces of the interfaces, make 3 steps of Schwarz method

2- take the NUDFT of the last 4 traces

3- apply the one dimension Aitken acceleration formula to each mode of these transformed traces

4- recompose the physical traces from the result of step 3

5- from these traces, make one step of Schwarz method.

Figure 2 exhibits the convergence Fourier modes of the iterate solution at the interfaces and the global convergence of the method for the 2D Poisson problem on two overlapping subdomains, showing the gain obtained using our NUDFT in comparison with a FFT on the nonuniform mesh (showing that the mesh is far from an uniform mesh). We consider that the NUFDT coefficients are totally decoupled in equation (16) which is not actually the case. Nevertheless, with this approximation we obtained quite reasonable convergence.

Figure 3 exhibits results in the case of nonuniform non matching grids: in this situation, a different space discretisation in the y-coordinate is managed by a projection of the NUDFT Fourier polynomial of the interface for a subdomain to the points of the neighbor subdomain.

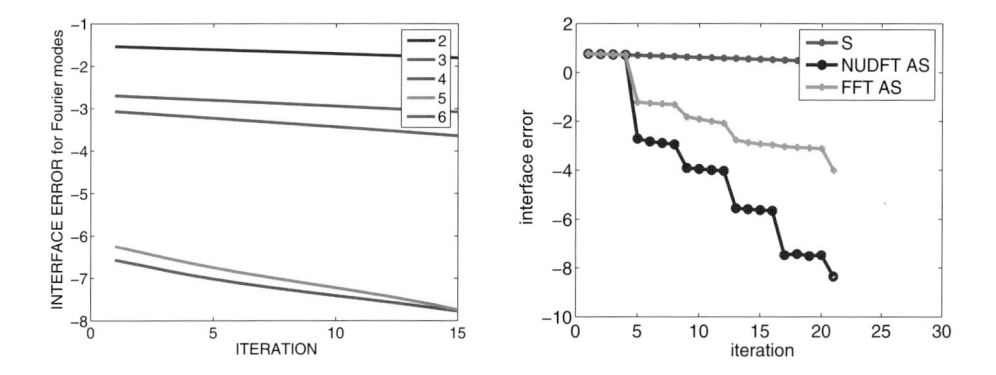

Figure 2. convergence of first Fourier modes (left) of the Schwarz iterate interface solutions, and effect of the Aitken acceleration on the overall Schwarz algorithm without considering the coupling between Fourier coefficients for acceleration (right).

4.1. Discussion about the adaptivity in the approximation of $P_{[[.,.]]}$

The main advantage of this approach to compute the $P_{[[.,.]]}$ in the Fourier space instead than in the physical domain space, is to introduce adaptivity in its approximation. The adaptation is based on a posteriori estimates on the Fourier mode of the error between two Schwarz iterates. Considering few Schwarz iterates, based on the value of the non uniform Fourier component of the error at the artificial interfaces between the last two Schwarz iterates performed, one can select the m_τ modes which are greater in module than a tolerance τ. The others modes component less in module than τ are considered to be converged. We then can iterate again until to reach $m_\tau + 1$ Schwarz iterates and we can construct an approximation $\tilde{P}_{[[.,.]]}$ of $P_{[[.,.]]}$ which corresponds to the trace operator error of the non converged modes. The value of τ gives the accuracy expected as the non accelerated modes continue to converge during the remaining Schwarz iterates. Results using this approach will be given in an extended paper [6].

5. Perspectives and conclusions

A new formulation of the non uniform Fourier transform strictly related to the value of the function at the non equidistant points have been proposed. Theoretical results on approximation properties have been confirmed numerically in the 1D and 2D cases. This formulation needs to solve a linear system of size $2N$ with sufficiently accuracy to compute the weights of the sesquilinear form. Works are under progress to use the Tikinov regularizing procedure to solve this linear system when is badly conditioned.

This approximation allows the extensions of the Aitken-Schwarz domain decomposition methodology to non uniform and non matching meshes with good accelerations. It provides a framework to obtain an efficient parallel solver for elliptic non separable operators that arise in CFD problems with non structured grids.

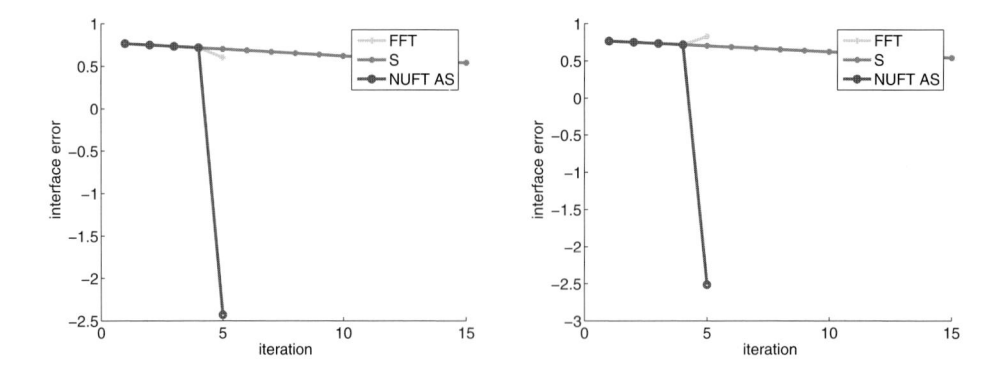

Figure 3. Effect of the Aitken acceleration on the overall Schwarz algorithm, using the classical FFT and the NUDFT for two different nonuniform non matching grids .

REFERENCES

1. J. Baranger, M. Garbey, F. Oudin-Dardun, Recent development on Aitken-Schwarz method, *int. conf DD13 on decomposition methods in science and engineering* pp. 289–296, 2000.
2. N. Barberou, M. Garbey, M. Hess, M. Resch, T. Rossi, J. Toivanen and D. Tromeur-Dervout, Efficient metacomputing of elliptic linear and non-linear problems, *Journal of Parallel and Distributed Computing*, 63(5), pp. 564–577, 2003.
3. I. Boursier, D. Tromeur-Dervout, Y. Vassilevski, Solution of Mixed Finite Element / Spectral Element systems for convection-diffusion equation on non matching grids, *Technical Report* CDCSP-03-00, 2004.
4. A. Dutt, V. Rokhlin, Fast Fourier transforms for nonequispaced data, *SIAM J. Sci. Comput.* 14(6), pp. 1368–1393, 1993.
5. A. Dutt, V. Rokhlin, Fast Fourier transforms for nonequispaced data. II, *Appl. Comput. Harmon. Anal.* 2(1),pp. 85–100, 1995.
6. A. Frullone and D. Tromeur-Dervout, Adaptive acceleration of the Aitken-Schwarz Domain Decomposition on non uniform non matching grids, *in preparation*.
7. M. Garbey, Aitken-Schwarz method for CFD solvers, *Parallel computational fluid dynamics PCFD03*, pp. 267–275, Elsevier, Amsterdam, 2004
8. M. Garbey, D. Tromeur-Dervout, On some Aitken-like acceleration of the Schwarz method, *Internat. J. Numer. Methods Fluids* 40 (12), pp. 1493–1513, 2002.
9. M. Garbey, D. Tromeur-Dervout, Two level Domain Decomposition for Multiclusters,*Domain Decomposition in Sciences and Engineering*, T.Chan & Al editors, published by DDM.org, pp. 325–339, 2001.
10. L. Greengard, J.-Y. Lee, Accelerating the Nonuniform Fast Fourier Transform, *SIAM Review*, 46(3), pp. 443-454, 2004.

Unstructured Mesh Processing in Parallel CFD Project GIMM.[*]

B. Chetverushkin[a], V. Gasilov[a], M. Iakobovski[a], S. Polyakov[a], E. Kartasheva[a], A. Boldarev[a], I. Abalakin[a] and A. Minkin[a].

[a]*Institute for Mathematical Modelling, Russian Ac.Sci., 4-A, Miusskaya Sq., 125047, Moscow, Russia*

Key words: Computational fluid dynamics, three-dimensional evolutionary viscid flow, parallel algorithms, distributed computations.

1. Introduction

Rapid development of networks caused the growth of interest to heterogeneous systems applications in remote computations resulting thereby in the progress of special software. In this paper we present a CFD project GIMM carried out in the Institute for Mathematical Modeling, RAS. The project takes its grounds in multidisciplinary studies carried out by the IMM RAS team. The development of advanced computer models is of paramount importance in effective exploitation of new parallel systems. It's essential that the design of numerical methods should be in close correlation with the development of algorithms for parallel computing. GIMM pursues the goal of joining up new results in numerical analysis with latest achievements in creation of network facilities. Special attention is paid to the reasonable management of multiprocessor systems (MPS) with distributed memory. Design of new parallel algorithms and proper programming is considered as having paramount importance in effective exploitation of parallel systems. Design of new numerical methods is implemented in close correlation with the development of algorithms for parallel computing. The use of high-dimension irregular meshes permits to approximate accurately realistic 3D geometry of the simulated objects. Processing of irregular data structures is a complicated and very time-consuming numerical work. New effective algorithms of domain decomposition over processors taking into account numerical expenses in each point are developed that provides the good load balancing and minimize data exchange.

---

[*] The study is done under the auspices of Russian Ac. Sci. (The State contract № 10002-251/OMH-03/026-023/240603-806 and RFBR (Project № 04-01-08024-OFI-A).

2. General description.

The current version of GIMM components includes:
- CAD compatible data structures,
- Surface/solid grid generation tools,
- Libraries of a numerical package kernel,
- Problem-oriented application software for PC and parallel systems,
- Server tools for data storage, visualization, pre- and postprocessor tools for PC and parallel systems,
- Client tools for data storage, visualization, pre- and postprocessor tools for PC,
- Server control tools for processing the user tasks by parallel system,
- Client interface for PC.

The work of software components is supervised by the GIMM program manager. The first program manager version works under Windows and ensures the interactive user work with remote computing resources and computer systems possessing both distributed and shared memory. The server modules of the GIMM package are developed as Linux applications. The MPI is used for task start and interprocessor communications during the run. In case when the mixed type parallel systems are used for calculations, e.g. the two- or four-processor nodes are incorporated into the computational net, MPI and OpenMP tools are combined. To this end the work at the processors' level is done via algorithms based on the common memory using while at the nodes' level the appropriate algorithms use MPI utilities and distributed memory.

The program manager supports operations usual for CFD studies:
1. Preprocessing I, i.e. creation of a geometry model and setting of physical data as well as initial/ boundary conditions.
2. Preprocessing II, or mesh generation and creation of a computational model equipped by problem attributes.
3. Formation of algebraic equations approximating the governing system of differential fluid mechanics equations.
4. Support of the task starting operations and run control functions.
5. Postprocessing including interactive data analysis and visualization tools for both the user PC and remote distributed computing system.

3. Numerical methods.

At present in GIMM the following models are available:
Model 1: 3D Navier-Stokes system for compressible heat-conductive flow
Model 2: 3D Navier-Stokes system for incompressible flow
Model 3: 3D single-phase nonlinear flow in porous media.

The governing systems are approximated by means of unstructured tetrahedral meshes and conjoint systems of finite volumes. The mixed finite volume (FV) - finite element (FE) approximations provide sufficiently high accuracy: the approximations to convective terms are done in terms of FV technique, and dissipative terms are approximated by simplex FE representations of dependent variables.

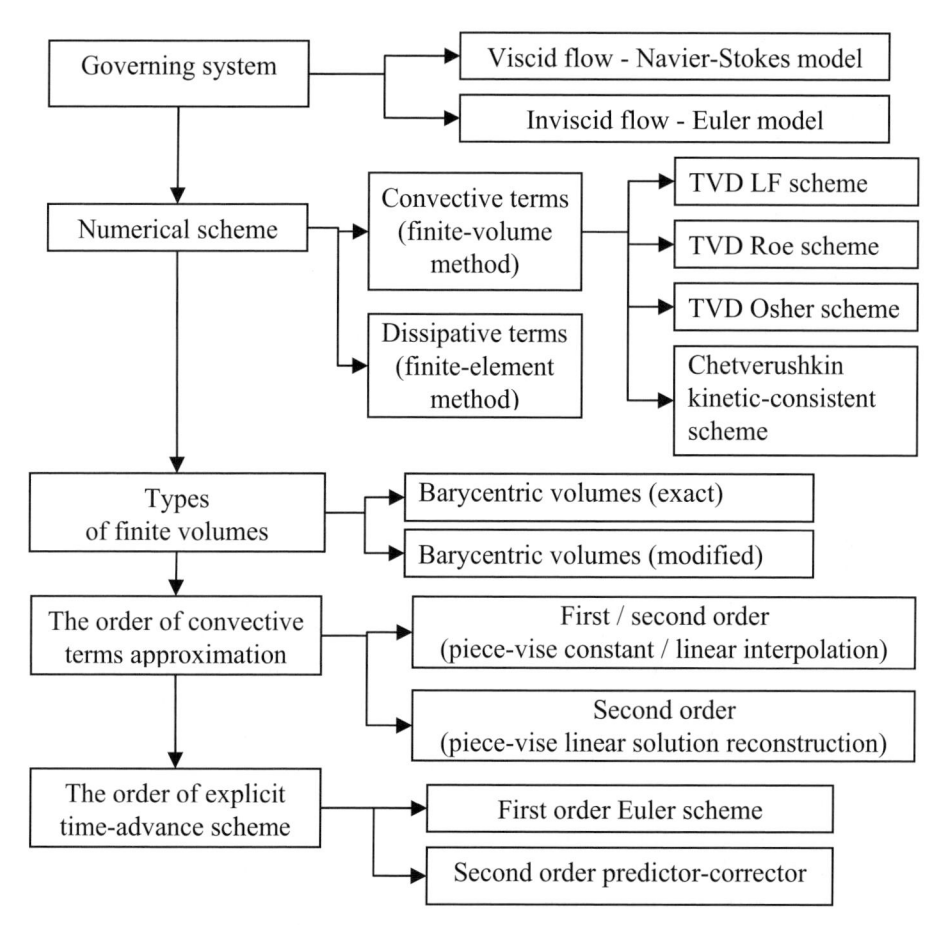

```
┌─────────────────┐      ┌──────────────────────────────────┐
│ Governing system├─────▶│ Viscid flow - Navier-Stokes model│
└────────┬────────┘   │  └──────────────────────────────────┘
         │            │  ┌──────────────────────────────────┐
         │            └─▶│ Inviscid flow - Euler model      │
         ▼               └──────────────────────────────────┘
```

Fig.1 GIMM numerical basis.

For treatment of convective fluxes we use high-resolution TVD schemes of Roe, Osher and Van-Leer [1]. We also use the Chetverushkin kinetically-consistent method [2] and TVD Lax-Friedrichs schemes with intermediate solution reconstruction which we extended for the case of 3D unstructured meshes [3]. The time-marching algorithm is implemented as an explicit 2-nd order predictor-corrector. The diagram at Fig.1 (next page) presents an overview of GIMM numerical techniques.

4. Geometric models and meshing.

Preprocessing related to geometric properties of a studied object usually is implemented by means of some CAD system. A set of universal program tools is developed for input and acquisition of geometrical and physical data related to a 3D problem with mixed initial and boundary conditions. Simultaneously these tools are used for storage and treatment of a discrete computational model.

Our data model takes its origin in finite – element technique and provides a description of geometric and topological properties for computational domains as well as for numerical meshes. The geometry of any computational model is formed of parametric surfaces expressed in the terms of rational B-splines (NURBS). Thus a boundary representation (B-REP) is obtained. Any 3D geometric complex is represented via sets of elements: 0-order elements = "nodes", 1-order elements = "edges", 2-order elements = "faces", 3-order elements = "cells".

A topology complex describes the relations between numbered data sets irrelative of their geometric nature. A typical 3D mesh consists of a large number of cells. Therefore only a few incidence relations are stored permanently and other relations are calculated every time they are needed.

The main topological relation between pairs of cells in a cellular complex is "to be a face". It is also named a boundary relation. Other relations can be defined on the base of this one. The incidence relations serve for elements of different dimensions, and adjacency relations serve for elements having the same dimension.

We suppose that computational mesh can change in the process of calculations. Data assigned to mesh elements are not predefined and can vary from one application to another so we introduced the conception of a numerated set. We use such sets for representation of collection of mesh elements. We imply that all mesh elements are enumerated and each of them has its unique number. So, having the number of an element, it is possible to find all the data assigned to this element. Numeration allows implementing changes in all data structures when some elements are added or deleted. Fig. 2 shows the geometric data techniques applied in GIMM.

Evidently realistic results in 3D simulations of objects with complicated shapes like aircrafts, engines, cars etc. can be achieved with meshes including at least $10^6 - 10^7$ nodes. For such meshes the proper amount of processed information is so large that it is necessary to develop parallel mesh generation algorithms.

For applications in parallel computing we developed a technique based on domain decomposition with the following mesh generation by means of combined octree and modified advancing front algorithms [4] permitting to generate cells of prescribed shapes and sizes. The technique of multicolouring allows implementing parallel

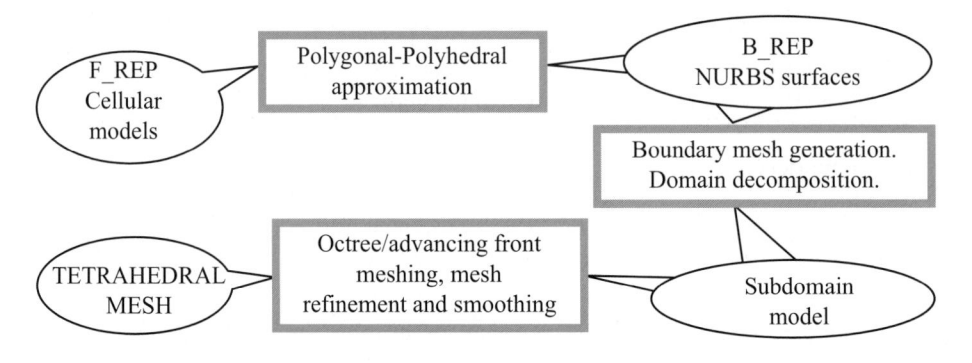

Fig. 2. Mesh generation for functionally represented (F-REP) solids.

meshing in subdomains in such sequence that avoids the data exchange between adjacent regions and ensures fully independent meshing of subdomains. The preliminary operations i.e. geometric data input, formation of a computational domain in the form of a cellular geometric complex and setting of necessary physical attributes can be accomplished in a single- processor regime. The octree technique is applied for mesh generation in the interior of a subdomain while the advancing front algorithm works in the near-boundary regions. The initial front is formed by external surface triangulation and used for every subdomain adjacent to the external boundary. For a given subdomain the triangulations along interfaces with its adjacent subdomains are taken from the previous steps of the general procedure or are newly generated in the case they are not yet constructed to the current step. Some optimizing procedures are developed for getting meshes of the higher quality.

The meshing algorithms are implemented in two programs. The first one serves for triangulation of closed surfaces described in terms of NURBS-segments. The second performs the tetrahedral meshing starting from a given surface triangulation.

Two versions of ANSI C/C++ software were developed – a single-processor program operating under OS Windows and a multiple-processor one under OS Linux. In the latter case interprocessor communication library *MPICh 1.2* is used and the programs are evoked via the standard script *mpirun*.

5. The data treatment for the high-dimension meshes.

Let's consider the main aspects of computational technologies concerning the large volume data treatment created under the GIMM project. The one is the correct load balancing whish is a very important factor strongly affecting a rate of parallel computations. Another problem providing the evident difficulties in large-scale 3D simulations is the lack of acceptable visualization tools.

The load balancing is done by means of the rational decomposition of the computational mesh via the constructing of a multilevel graph system. Due to this method a mesh is represented as an ensemble of coherent subdomains. The initial graph is formed by pairs of neighbor mesh vertices. In turn, these pairs are combined into the higher level pairs. Thus the multilevel graph decomposition is formed successively, while every new graph possesses a simpler structure than the previous one. As the result of this work the computational mesh is represented as a set of connected subdomains. This procedure repeats until the graph of a proper size is obtained which can be easily divided into the required number of macrodomains. At the farther stage of the run these macrodomains are distributed over the set of processors. The macrodomains serve also for estimations of the numerical work. During the macrodomain creation performance it is possible to take into account not only the mesh topology, but an a priori estimated numerical complexity of calculations per every mesh node. Thus the resulted load distribution can be made very close to the optimum.

The data storage in the GIMM code is organized via the data distribution over the multiprocessor system itself. This technology is supported by specially developed hierarchical distributed file system (HDFS). It is based on the client-server technology in the multiprocessor variant. In accordance with this technology some number of specially appointed processors are used for communications through the entire disc space of the massive parallel system (MPS) including both local (processor self memory) and external (RAID server) devices. The other processors implementing the run are the clients connected to definite servers. The read/write operations corresponding to the distributed files are performed via the system of demands addressed to the HDFS servers. The cash memory using leads to somewhat information doubling which is almost inevitable. Therefore the main idea of organizing the HDFS system is to minimize the doubling of the information stored on the individual discs and simultaneously to maintain the high exchange rate and information integrity. In addition the HDFS is used for data compressing necessary for data storage and further treatment at the postprocessing stage.

The convenient tools for data visualization and postprocessing are of primary importance in 3D simulations of complex FD problems. For a typical mesh consisting of 10^6-10^7 nodes the whole amount of 3D results data exceeds the memory resources of a personal computer. Besides that, the data exchange through the net connecting PC with MPS is restricted by other operations related to multiprocessor computations.

To mitigate this problem we developed the distributed visualization system which is especially suitable for treatment of large data volumes. The GIMM visualization system RV – "Remote Viewer" is constructed according to a Client/Server model. It works in close interaction with HDFS. This construction allows to implement the most part of visualization process by supercomputer and then to transfer the compressed information to a user.

The final image is formed at the user's workplace and it is possible to use modern multimedia hardware (helmets, stereo glasses, three-dimensional manipulators etc.) for better image perception.

6. Numerical results

At the current stage of the GIMM development we performed hydrodynamic test studies pursuing the two main goals:
- approbation of the algorithms and numerical techniques (accuracy, efficiency, etc.)
- study of the effectiveness of parallel processing with different distributed computing systems architecture.

The package working ability was examined through a number of test studies. Here we present a brief description of results pertinent to such famous benchmark as the problem of viscid low-Mach (M~0.1) and low-Reynolds (Re~25) flow around a sphere. A numerical grid was constructed so to condense in the sphere vicinity and the volumetric ratio of largest cells to smallest ones was ~ 100. The total mesh consisted of 2,356,196 nodes and 14,018,176 tetrahedrons.

The steady-state calculations were done by the time-marching procedure. For the test studies we used the two parallel systems operating in the Joint Supercomputer Center (Russian Ac. Sci.): MVS BETA (equipped by Dec Alpha-21167 processors) and BC1 (equipped by IBM Power P4 processors). The initial configuration included 4

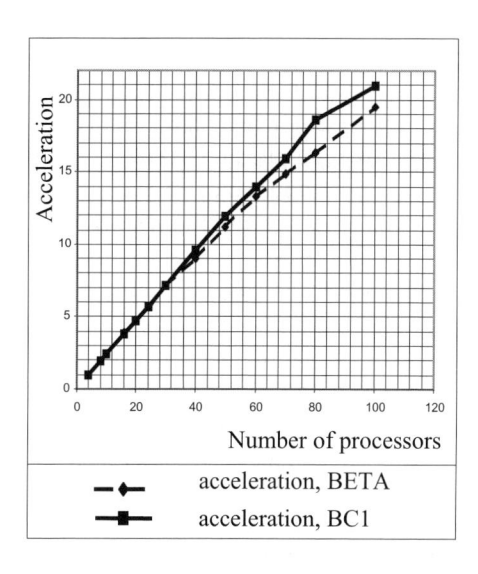

Fig. 3. Acceleration rates for BETA and BC1 parallel systems.

Fig. 4. Mesh partition.

processors. Thus the acceleration for increased number of the processors in use was calculated as $S_N = T_4 / T_N$, where N is the number of processors, T_N is the computing time for these processors. The effectiveness was estimated by the formula $E_N = (S_N \cdot 4/N) \cdot 100\% = (S_N /(N/4)) \cdot 100\%$. One may see at the Fig. 3 that maximum effectiveness for BETA was 98.3 % when $N=10$, and the minimum effectiveness was 77.8% when $N=100$. Fig.4 shows an example of mesh partition (4 subdomains) and Table 1 includes the load balance effectiveness factors for two numerical methods.

Number of subdomains	Percentage of cut edges	Finite volume method effectiveness	Finite element method effectiveness
4	1.54%	0.96	0.94
10	4.50%	0.92	0.88
40	10.18%	0.87	0.78

Table 1. Processor load balance.

The numerical experiments demonstrated that the algorithm is efficient even when using strongly irregular tetrahedral meshes.

7. Conclusion.

Dealing with any standard industrial software the user may encounter the restrictions of his activity caused by the prescribed program structure and by the set of standard models and methods. Being reasonable in CAD-CAE project design, such restrictions fall into contradiction with the concept of a research code which is often aimed at studies of new models and algorithms. Considering the application aspects of high-performance calculations one may conclude that the industrial codes yet can not use perfectly all the resources which can be given by modern massive computational systems. Keeping in mind these points the GIMM team concentrated efforts on the development of open code with predominant applications in the area of parallel and distributive computations. The main goal pursued at the first stage of the project development was the development of versatile algorithms suitable for various applications in CFD studies. This was the principal motivation for the development of numerical algorithms using unstructured meshes and their adaptation to calculations with various multiprocessor systems.

The modern version of the GIMM code is aimed at numerical simulation of evolutionary multistage 3D hydrodynamic phenomena. It combines as traditional as new numerical technologies which were incorporated into the code with the primary aim to have a universal simulation tool. GIMM provides a possibility of massive parallel computations with 10^8-10^9 mesh nodes and by the order of 10^3 processors involved. The main feature of GIMM package is the comprehensive use of parallel programming at all stages of a problem solution, i.e. from geometry modeling till postprocessing. Some numerical tools which are already incorporated into GIMM were developed 3-5 years ago and passed through comprehensive approbation in IMM RAS. The created code is highly effective compared to the traditional PC-oriented tools and allows 10 to 100 times reduction of the period necessary for the numerical study of any fundamental or applied problems.

REFERENCES

1. D. Kroner. Numerical schemes for conservation laws. – B.G.Teubner Publ., Stutgart, Leipzig, 2000.
2. B.N.Chetverushkin. Kinetic schemes and quasigasdynamic system. – MAKS Press, Moscow, 2004.
3. V.A.Gasilov and S.V.D'yachenko. Quasimonotonous 2D MHD scheme for unstructured meshes. Mathematical Modeling: modern methods and applications. Moscow, Janus-K, 2004, pp.108-125.3
4. P.J. Frey and P.L. George. Mesh generation. - Hermes Sci. Publ., Oxford, UK, 2000.
5. M. Yakobovski, S. Boldyrev, and S. Sukov. Big Unstructured mesh Processing on Multiprocessor Computer Systems.In: Parallel Computational Fluid Dynamics – Advanced numerical methods, Software and Applications//B.Chetverushkin, A.Ecer, J.Periaux, M. Satofuka and P.Fox (Editors). – Elsevier B.V., 2004, p. 73-80.

Investigation of the parallel performance of the unstructured DLR-TAU-Code on distributed computing systems

Thomas Alrutz [a] *

[a] DLR Institute of Aerodynamics and Flow Technology, 37073 Göttingen, Germany

High Performance Computing is widely used in the Computational Fluid Dynamics (CFD) community to satisfy the increasing needs for a faster prediction of simulated flows. Despite the advances in computer manufacturing and chip design over the last years, there is still a lack of computational resources required to solve CFD problems in real time. Moreover, for most research centers it is difficult to buy a supercomputer like the NEC-SX 6 based Earth Simulator out of their budgets. To encounter this gap, there is an increasing installation of supercomputers built out of commodity components like PC's or workstations connected via high speed networks. These cluster systems often lack tight integration between the processor, network and main memory. This often results in inefficiencies in the communication subsystem, such as high software overheads and/or message latencies. In this paper we investigate the parallel performance of the unstructured DLR TAU-Code on a set of up-to-date Linux-Clusters. The TAU-Code flow solver is parallelized by domain decomposition and uses MPI-1 for the inter domain communication. The influence of the different cluster configurations on the parallel performance is investigated with two real life applications, which are typical for the day-to-day work of a CFD-engineer or researcher. The parallel efficiency of the TAU-Code is investigated with two different versions of cache optimization techniques, which will have a significant impact on the scalability of the code. The potential of the installed high speed network to handle the enormous amount of data transfer caused by a parallel CFD calculation will be evaluated to determine the most promising platform for this type of CFD-Code.

1. Introduction

The improvements in high speed networks, and the exponential increase in processor performance over the last years, have made commodity clusters a growing alternative to classical supercomputers (vector computers or shared memory systems). On the other hand most of the systems built out of commodity components lack tight integration between processor, memory and network. Along with these drawbacks another problem of the commodity systems is that the used hardware is not designed to support parallel computation. Therefore the latency of the messages sent with a MPI [8] implementation is poor compared to classical supercomputers [2]. Thus the benefit of a commodity cluster installation depends highly on the applications that will run on such a system. For

---
*Research Scientist, Numerical Methods.

common CFD codes the requirements on the processor main memory and the network interconnection are high [10], especially if the code is an unstructured code which uses multigrid as a convergence accelerator [7]. In this paper we evaluate the parallel efficiency of the unstructured DLR TAU-code on a set of Linux-clusters with two typical aerodynamic configurations. For the investigation of the limits of the used high speed network, we apply two different versions of cache optimization integrated in the code. Furthermore we provide a guideline how to test the currently available hardware to select the most promising platform.

2. Performance evaluation method

2.1. The TAU-Code

The DLR TAU-Code is a finite-volume Euler/Navier-Stokes solver working on unstructured hybrid grids. The code is composed of independent modules: Grid partitioner, preprocessing module, flow solver and grid adaptation module. The grid partitioner and the preprocessing are decoupled from the solver in order to allow grid partitioning and calculation of metrics on a different platform than used by the solver. The flow solver is a three-dimensional finite volume scheme for solving the Unsteady Reynolds-Averaged Navier-Stokes equations. Further details of the TAU-Code may be found in [5].

2.2. Flow solver settings

For the performance evaluation a RANS calculation for both of the aerodynamic configurations is selected. The following settings for the flow solver reflects a standard parameter setup for most of the calculations computed with TAU in production:

- Turbulence model: Spalart Allmaras
- Solver type: central scheme
- Number of Runge-Kutta steps: 3
- Multigrid: 4w cycle
- Number of iterations: 30 (only for benchmark)

Due to the way the solver is parallelized, there is an exchange of messages between the domains on every grid level. Thus rather small messages are exchanged on the coarsest grid level and larger messages on the finer grid levels. Keeping this in mind, we can investigate the influence of latency and bandwith with respect to message passing on the parallel efficiency of the code with the selected multigrid cycle.

2.3. Aerodynamic applications

In order to simulate the requirements on a CFD code by the typical day-to-day work of a CFD-engineer/researcher, we selected the DLR F6 configuration presented in [6] and a A380 configuration obtained from Airbus industries. The DLR F6 (figure 1) is a wing/body/pylon/nacelle configuration from which a hybrid unstructured grid with 2 million grid points and 5 million volume elements (tetra, prisms and pyramids) was generated. The A380 (figure 2) is a very complex configuration from which a grid with 9 million

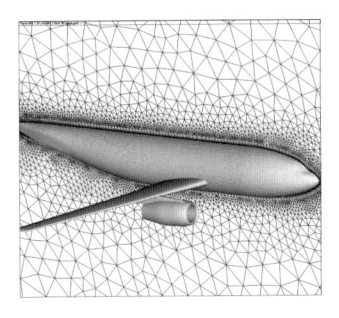

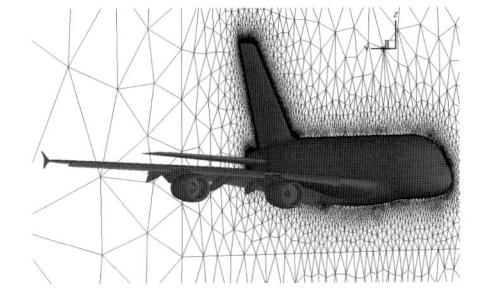

Figure 1. DLR F6 configuration

Figure 2. A380 configuration

grid points and 25 million volume elements was generated. The memory requirements for the A380 on 1 CPU exceeds the resources of most of the tested systems, such that we had to start the benchmarks for this testcase on 8 CPUs.

2.4. Partitioning

The parallelization of the flow solver is based on a domain decomposition of the computational grid. Due to the explicit time stepping scheme, each domain can be treated as a complete grid. First of all we had to split the grids into the number of desired domains for every parallel run of the flow solver. The partitioning is done by a sequential run of the primary grid partitioner, which uses a recursive bisection algorithm to compute the desired number of domains. Figure 3 gives an overview of the maximal dimensions (elements and points) of the partitioned grids for each configuration.

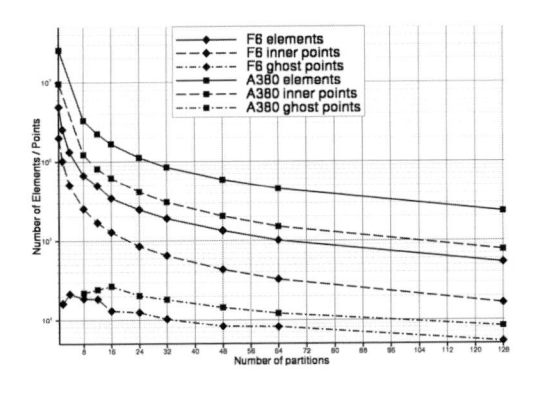

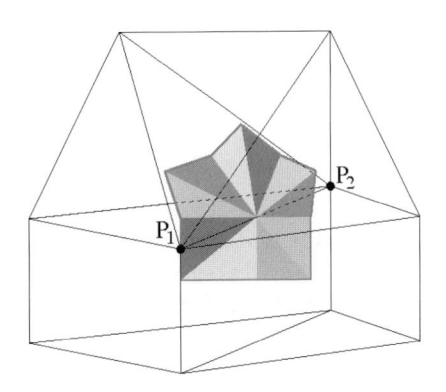

Figure 3. Maximal grid dimensions

Figure 4. Dual mesh face

2.5. Cache optimization

After the partitioning the parallel preprocessor is started to compute the dual grids out of the partitioned grids. Furthermore the preprocessor is responsible for the cache or vector optimization. A major part of the calculations in the flow solver is performed by considering the grid faces one by one, while the computed values (e.g. fluxes over the face/edge) are adjoined to the respective points on both sides of the face (figure 4). Therefore, the cache optimization for scalar machines is achieved by a sorting of faces/edges in the dual grid. The main difference between the two implemented cache optimization techniques, with respect to performance issues, is the strategy to sort the faces/edges. In the basic version (released 2002) the faces/edges are sorted into different colors depending on a treshold which the user can define as an input parameter. The aim is to prevent cache misses while looping over the edges/faces within the same color. In the enhanced version (released 2004) the strategy is to minimize the disadvantage of the indirect adressing. In this version the edges and the adjoined points are sorted into colors to achieve direct strided data access in each color.

2.6. Computing platforms

For our benchmarks we have selected a couple of up-to-date cluster systems, but we were not able to run the benchmarks on all architectures currently present in the TOP 500 [1]. However, we were able to test a broad variety of cluster interconnects along with a major set of currently used processors in the High Performance Computing field. The relevant technical details of the tested systems are listed in table 1:

Table 1
Technical details of the tested systems (MPI latency and bandwidth are taken from [3])

System	Chip	GHz	Ram	Network	Latency	Bandwidth	CPU/Nodes
p690	Power4+	1.70	333 MHz	HPS[4]	8 μs	0.72 GB/s	160/5
Cluster	Xeon	3.06	133 MHz	Myrinet[9]	22 μs	0.03 GB/s	64/32
Cluster	Opteron	1.40	155 MHz	Gigabit	70 μs	0.02 GB/s	32/16
Cluster	Opteron	2.40	166 MHz	Infiniband[12]	10 μs	0.15 GB/s	256/128
XD1[11]	Opteron	2.20	200 MHz	RapidArray	2 μs	0.23 GB/s	72/36

2.7. Performed benchmarks

Normally a CFD-engineer/researcher is interested in the turn around time for a complete computation of an (un)steady state problem. Unfortunately the number of iterations the flow solver needs until the solution of the flow field is converged depends highly on the type of flow problem and on the aerodynamic configuration. Thus we will focus on the wall clock or real time (t_w) that the solver needs to perform 1 iteration of the selected multigrid cycle. For the analysis of the parallel performance on the tested systems we calculate the relative speedup

$$s_n = \frac{t_{w_k}}{t_{w_n}}, \tag{1}$$

with $k \in \{1, 8\}$ and $n \in \{1, 2, 4, 8, 12, 16, 24, 32, 48, 64, 96, 128\}$ for each configuration and cache optimization strategy. Thus we have to perform 4 parallel benchmark runs with a different number of domains (n) on each of the systems in table 1 in order to get a significant data basis for our discussion. If we have the relative speedup s_n we are able to conclude about the parallel efficiency of the code

$$e_n = \frac{q_n}{s_n}, \text{ with } q_n = \frac{n}{k}. \tag{2}$$

3. Results of the benchmarks

3.1. Analysis of the sequential benchmarks with the DLR F6

In [6] we have presented a quite good parallel efficiency (95 % on 12 CPUs) for the code on a NEC SX-5 vector machine with the DLR F6 configuration. In order to get an idea of how fast the tested cluster systems are compared to the well known vector computers, we have done a reference calculation on the old NEC SX-5 on 1 CPU. The obtained wall clock time for 1 iteration with the settings from 2.2 on the NEC SX-5 was

$$\bar{t}_{w_1} = 33.03s \approx 926 \text{ MFlops}. \tag{3}$$

We take the time from (3) as a reference value and calculate the relative performance

$$p^i = \frac{\bar{t}_{w_1}}{t^i_{w_1}} \tag{4}$$

for a comparison of all of the tested cluster nodes in the following table 2:

Table 2
The relative performance of cluster nodes scaled to NEC SX-5

Computing platform				Basic optimization			Enhanced optimization		
Chip	GHz	GB/s	L2 MB	t_{w_1}	p	MFlops	t_{w_1}	p	MFlops
Power4+	1.70	6.8	0.7	70.62 s	0.47	433	38.36 s	0.86	797
Xeon	3.06	4.2	0.5	65.64 s	0.50	466	39.16 s	0.84	781
Opteron	1.40	4.8	1.0	85.83 s	0.43	356	47.05 s	0.70	650
Opteron	2.40	5.2	1.0	48.26 s	0.67	634	32.03 s	1.03	955
Opteron	2.20	6.4	1.0	49.40 s	0.67	619	32.96 s	1.00	928

The enhanced cache optimization speeds up the computation by a minor factor of 1.51 and a maximal factor of 1.83 for the tested nodes. The performance gain of the enhanced cache optimization is higher on the systems which have a smaller memory bandwidth (Xeon) or a third level cache (Power4+). Furthermore it can be expected that the benchmark with the enhanced cache optimization will stress the installed high speed network of the cluster far more than the one with the basic cache optimization. Another observation is that the 2.4 GHz Opteron which has a smaller memory bandwidth than the one with 2.2 GHz is slightly faster with the enhanced cache optimization (2.3 % versus 2.9 %). The results in table 2 leads to the acceptance, that an unstructured CFD code will benefit more from a higher memory bandwidth than from a higher CPU clock.

3.2. Analysis of the Parallel benchmarks

In order to analyse the results of the parallel benchmarks the mainloop time for 1 iteration and the relativ speedup to 1/8 CPUs over the number of used processors for each of the tested system is plotted in the following diagrams. The acceptance that the

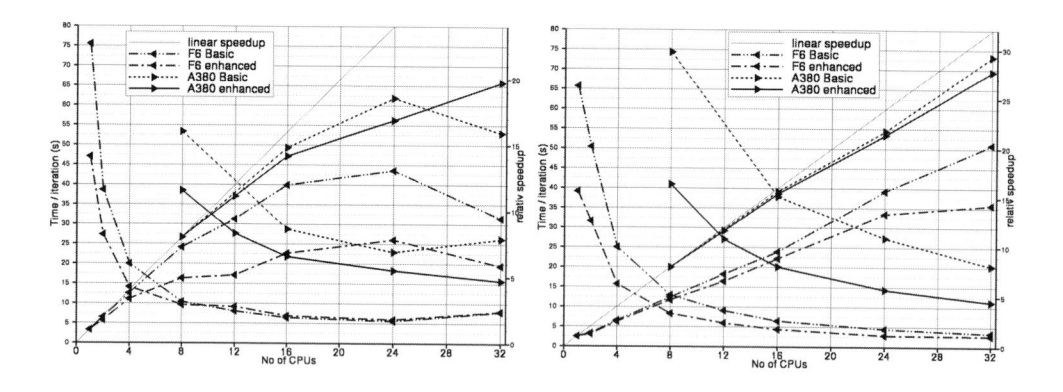

Figure 5. Opteron 1.4 GHz with Gigabit

Figure 6. Xeon 3.06 GHz with Myrinet

enhanced cache optimization has a significant impact on the parallel efficiency of the code is correct for all of the systems with one exception. Indeed the gigabit opteron cluster shows a better parallel performance for the A380 configuration with the enhanced cache optimization turned on. This is not expected, but will have no influence of the overall bad scalability of the code on this system (figure 5). The parallel performance degrades for

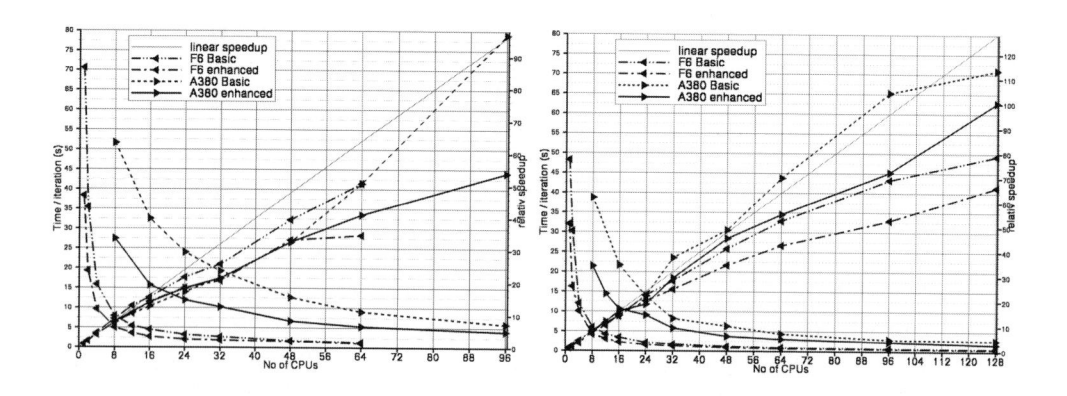

Figure 7. Power4+ 1.7 GHz with HPS

Figure 8. Opteron 2.4 GHz with Infiniband

both configurations at 24 CPUs. Here it is quite obvious that the latency of the gigabit network is the bottleneck, which can be seen by a closer view on the bad speedup curve for the F6 with the enhanced cache optimization turned on.

The Myrinet system on the other hand performs quite well up to 32 CPUs, but with the F6 configuration the sytem shows its limitations (figure 6). The Myrinet system is not capable to scale-up as good as the XD1 or the Infiniband system. This could be explained by the bigger latency of the Myrinet network and the smaller memory bandwidth of the Xeon bus topology when two CPUs share the memory controller.

The p690 system shows a different behavior (figure 7). This system is the only one with a third level cache and the effect on the parallel performance can be seen with A380 configuration. With the basic cache optimization the speed-up curve increases when 48 CPUs or more are used. For the F6 the cache effect starts at 32 CPUs. When the enhanced cache optimization is turned on, the cache effect is not so strong (A380 at 32 CPUs) and the speed-up curve is reasonable for both configurations. The limitations of the network can be seen by a closer look at the degradation of the parallel performance at 64 CPUs for the F6.

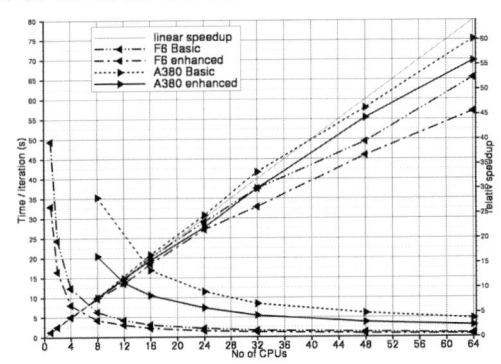

Figure 9: Opteron 2.2 GHz with RapidArray

The Infiniband system seems to have enough resources to provide a sustainable performance for both configurations (figure 8). There is no significant degradation in the parallel performance and the speed-up curves are reasonable with the exeption of the A380 with basic cache optimization. A possible explanation for the super linear speed-up could be the bad performance on 8 CPUs (38.77 s), if we compare the data with the XD1 (35.25 s). Furthermore we can see a cache effect at 128 CPUs for the A380 with enhanced cache optimization, which was also observed for the p690 at 48 CPUs.

The XD1 (figure 9) is the only system which shows a nice speed-up curve for all configurations. No degradation in the parallel performance is observed and it seems that the system is not affected by a bad performance in some combinations of testcase and number of CPUs. We can see a super linear speedup for the A380 with basic cache optimization for up to 32 CPUs, but no other effect is observed.

4. Conclusion

The parallel efficiency of the TAU-code is evaluated on different distributed memory systems. For the parallel performance and scalability of the code, the high performance network has the strongest influence. The memory bandwith of the cluster nodes has a major influence as well, but is found to be not so important when the enhanced cache optimization is used. The Gigabit cluster gives acceptable performance for up to 16 CPUs with an efficiency of 43%[2]. The Myrinet cluster performs quite well up to 32 CPUs

---
[2] F6 with enhanced cache optimization

(54% effiency[3]) and it can be expected that the network will provide enough resources for configurations like the A380 to calculate with up to 128 CPUs (87% efficiency[3] on 32 CPUs). The p690 system shows a good performance up to 64 CPUs (54% effiency[3]) and it can be expected to give the same performance like the Myrinet system for larger configurations. The Infiniband cluster shows the potential of the attached high speed network and performs very good up to 128 CPUs (52%[3] and 78%[4] effiency). On the XD1 the code shows the best scalability. The parallel efficiency is about 71%[3] and 87%[4] at 64 CPUs and therefore higher as on all other systems. Due to the good balance between memory bandwidth, high speed network and processor, the XD1 is the most promising platform for an unstrucutured code like TAU.

Acknowledgments

We wish to thank Cray Inc and Dr. Monika Wierse for the use of the XD1 system and the provided benchmark results.

REFERENCES

1. TOP 500. $http://www.top500.org/lists/2005/06$, 2005.
2. Christian Bell, Dan Bonachea, Yannick Cote, Jason Duell, Paul Hargrove, Parry Husbands, Costin Iancu, Michael Welcome, and Katherine Yelick. An evaluation of current high-performance networks. Paper lbnl-52103, Lawrence Berkeley National Laboratory, January 2003.
3. HPC Challenge. $http://icl.cs.utk.edu/hpcc/hpcc_results.cgi$, 2005.
4. IBM. $http://www-03.ibm.com/servers/eserver/pseries$, 2005.
5. N. Kroll and J. K. Fassbender, editors. *MEGAFLOW — Numerical Flow Simulation for Aircraft Design Results of the second phase of the German CFD initiative MEGAFLOW presented during its closing symposium at DLR, Braunschweig, Germany, December 10th and 11th 2002*, volume 89 of *Notes on Numerical Fluid Mechanics and Multidisciplinary Design*, Berlin, 2005. Springer Verlag.
6. Norbert Kroll, Thomas Gerhold, Stefan Melber, Ralf Heinrich, Thorsten Schwarz, and Britta Schöning. Parallel Large Scale Computations for Aerodynamic Aircraft Design with the German CFD System MEGAFLOW. In *Proceedings of Parallel CFD 2001*, 2001. Egmond aan Zee, The Netherlands, May 21-23.
7. D.J. Mavripilis. Parallel Performance Investigation of an Ustructured Mesh Navier-Stokes Solver. *The International Journal of High Performance Computing*, 2(16):395–407, 2002.
8. MPICH. $http://www-unix.mcs.anl.gov/mpi/mpich$, 2005.
9. Myrinet. $http://www.myri.com/myrinet/overview/index.html$, 2005.
10. M. Sillen. Evaluation of Parallel Performance of an unstructured CFD-Code on PC-Clusters. paper 2005-4629, AIAA, Toronto, Ontario, Canada, 2005.
11. Cray Inc. Seattle USA. Cray XD1 Technical Specifications, $http://www.cray.com/products/xd1/specifications.html$, 2004.
12. Voltaire. $http://www.voltaire.com/hpc_solutions.htm$, 2005.

---

[3] A380 with enhanced cache optimization

Parallel visualization of CFD data on distributed systems

Marina A. Kornilina, Mikhail V. Iakobovski, Peter S. Krinov, Sergey V. Muravyov, Ivan A. Nesterov, Sergey A. Sukov[a]

[a]*Institute for Mathematical Modelling RAS, mary@imamod.ru*

Keywords: parallel algorithms, interactive visualization, data compression, unstructured meshes.

The application of up-to-date computer systems allows solving extremely difficult problems arising both in numerical modeling and in processing of datasets acquired from scientific experiments. The size of calculation mesh used in simulations reached $10^6 \div 10^8$ nodes and it grows roughly one order of magnitude each 10 years. Obtained results (usually three-dimensional scalar datasets and vector fields) may occupy rather a large volume of a computer disk space. The increase of amount of calculated data makes extremely important the problem of developing efficient methods of data visualization and analysis. It has led to the deep gap between the capabilities of parallel computer systems and availability of powerful up-to-date interactive tools for data visualization and analysis.

The approach for building parallel interactive system of CFD data visualization is discussed. New algorithms and program **RemoteViewer** are developed for data processing defined on 2D and 3D unstructured meshes. The program permits intelligent partitioning of unstructured meshes, compression of triangulated surfaces and visualization of 3D data on distributed systems.

This work is supported by the Russian Foundation for Basic Research (grant 05-01-00750-a, 04-01-08034-a).

1. INTERACTIVE VIZUALIZATION

Visualization tools for detailed analysis of CFD results should allow to represent both the general picture of gas dynamic flow and the detailed picture in some particular zones. The efficient system for parallel processing and visualization of large amounts of data obtained from CFD simulations should take the following aspects into consideration:

- individual processor node of parallel computer system can't perform analysis of the full processed data in acceptable time for interactive visualization;
- the RAM of individual processor is generally not enough for full data storing;
- the majority of users have remote access to supercomputers via slow channels of global networks, which do not allow to transfer full data for reasonable time.

For the purpose of visualization of numerical experiments we suggest the distributed system for interactive visualization **RemoteViewer** [3-5] based on client-server technology. The **RemoteViewer** uses multiprocessor system or metacomputer for mesh data processing and the graphic system, displays and other graphical devices at the users workplace for image representation. Thus visualization server performs the crucial part of data processing while the client can displays the data received from server.

It is clear, that typical graphical output devices such as monitors, projectors and etc. can represent only small volume of data in a recognizable form. Furthermore the improvements and new possibilities of modern monitors are incomparably slight in comparison with steady grow of CPU performance. Therefore the necessary component of visualization server is a module of data compression allowing compression ratios of hundreds, thousands and even more times. The discussed compression ratios are impossible without data coarsening. Yet, a certain controlled loss of data accuracy is by no means significant for visualization purposes. The main aim of visualization is to represent the characteristic features of the object. The details are hardly seen when we study the object overview, but that is the case when we have a certain loss of accuracy. In this approach the smaller the visual fragment is, the less loss of accuracy we have. Potentially, the considered approach allows us to perform the analysis of mesh data of any type with a desired resolution.

Nowadays there are two main principle approaches to large volume distributed data representation. The main difference is in data communication between client and multiprocessor cluster. The client can receive either 3D scene or 2D bitmap for final representation. Both approaches have their advantages. While both of them can produce similar graphical outputs, first mentioned approach can better meet scientific needs from the practical standpoint. 3D scene, even coarsened, instead of 2D image provides better interactive possibilities such as rotating and scaling of the loaded 3D object in real-time mode. This approach also allows to benefit from modern graphical hardware presented most likely on the client side rather than on the server side.

From a broad spectrum of methods for 3D mesh data visualization the following methods have been chosen: the method of isosurfaces - for scalar fields, trajectories of massless test particles - for vector fields. Both methods can dramatically decrease the number of vertices and graphical primitives needed to represent the input data. That is why these approaches together with distributed data processing can handle and permit interactive analysis of data, which exceeds a computer's RAM tens and hundreds times.

The system can perform the following operations on the data files that are in the direct access only on the parallel visualization server:
1. Setting of boundaries of the visual area (zoom) and resolution.
2. Setting of basic image characteristics (number of isosurfaces, represented on the screen, corresponding function values, number of represented trajectories of test particles and corresponding initial coordinates).
3. Data processing, compression and transfer of compressed image by visualization server to the client part at the users workplace.
4. The 3D image is displayed on the client computer screen and can be studied using rotation and zooming without referring back to the server.
5. If the closer examination of a smaller object fragment is required, the demand for image of this fragment is sent to the server. The new image can approximate the object with the higher accuracy due to the reduction in data size.

The application of up-to-date computer systems allows solving extremely difficult problems arising both in numerical modeling and in processing of datasets acquired from scientific experiments. The large data volume presents essential problems in many fields of their application: visualization, data transmission via narrow Internet channels, real-time data investigation, etc. Mostly the visualization of large datasets in their full with the accuracy of simulation experiment is of no use or simply impossible. Profusely dense spatial datasets acquired with the modern computer sistems are firstly too large in volume and secondly redundant and over-detailed for visualization. So, the problem domain arises: it's necessary to process the datasets with some compression algorithm in order to diminish the volume and redundancy but to keep visually qualitative appearance of the datasets. The lossless compression doesn't solve the problem of visualization of a large data volume since it doesn't reduce their true size but only reduces a volume needed for the data storage. The lossless compression is used in case of necessity of initial data storage for further full exact restoration. It doesn't solve the redundancy problem at all. So, for visualization, one must use an algorithm of lossy compression. It must be based on some simplification method.

2. SCALAR FIELDS VISUALIZATION

We considered compression algorithms for datasets of two basic types. The first type is an arbitrary tree-dimensional (3D) surfaces. The exploration of surfaces might be important for example in investigation of a scalar physical parameter given in a 3D volume. The examples of the physical parameters being investigated might be pressure, temperature, density, etc. The second data type considered is 3D scalar datasets defined on a tetrahedral mesh. Such datasets may represent results of mathematical modeling of various 3D physical problems, results of natural sciences, etc.

The size of data that specifies the isosurface can be comparable to the one that specifies the initial tetrahedral mesh and it can not be visualized directly at the client computer. The data compression is required which allows approximation of the initial surface with given accuracy. There are a number of approaches to the lossy compression based on the object synthesis and coarsening. The new methods created on this basis are used in the developed visualization system **RemoteViewer**. The most attractive methods for practical use are the coarsening ones, because they allow control not only over the compression ratio, but also over the accuracy of data representation. Actually, some loss of accuracy may occur for complex surface parts. But this drawback is not significant for visualization because only proper representation of object topology is primarily important. The fragments which do not have the desired order of approximation can be studied with zooming and consequently with a lower loss of accuracy. Note that isosurface, which represents the points having the same constant value of data, generally is not the "real surface" or any surface fragment. The isosurface may have cavities, self-intersection, be nonsmooth or disconnected, be composed of surface fragments, or fill some 3D areas completely. For meshes isosurface is a set of triangles in 3D space.

2.1. SINGLE-PROCESSOR ALGORITHM FOR SURFACE SIMPLIFICATION

The widespread and convenient way of surface representation is a triangulation of points in 3D space. Therefore, it is this type of surface representation that our algorithm of lossy compression is intended for. The coarse work-scheme of the algorithm includes the following steps: an initial surface (input data) is processed to the algorithm's data representation, then it is simplified and finally a new surface (output data) of possibly smaller size is created. This new surface approximation occupies an appropriate for visualization volume and corresponds to the required accuracy. The algorithm is meant for any type of surfaces: they may represent a result of a multi-valued function, possess holes, be unsmooth, disconnected, non-manifold or more complicated.

The basic idea of the algorithm consists in iterative gradual simplifications of various small surface areas (mini-areas) within the limits of the global (relative to the initial surface) user-specified error tolerance or in other words with the demanded accuracy preservation. While doing it, there is also a control of the obtained mini-area not to fold over and not to change a topology type for the new surface. The user-specified error tolerance is the absolute approximation error for the output surface with respect to the initial surface. This error is calculated in terms of L_2 metric in R^3. The basic simplification operation of the algorithm is a contraction of selective (primarily shorter) edges. After each mini-area simplification, the number of vertices (and triangles) of this area decreases that gives rise to the final compression. The iterative simplification of various surface mini-areas (relative redundancy removal) is performed as long as the condition of the required accuracy preservation can remain inviolate.

The task to achieve a desired volume with required accuracy preservation may be unrealizable in general case. In order to obtain some visualizable result in any case, the desired volume achievement has a higher priority in the algorithm. If the appropriate volume cannot be achieved within the limits of the required accuracy, then the parameter of maximum tolerable error begins to increase gradually.

The testing of algorithm showed that the best compression level is achieved when the surface consists of numerous subregions with low curvature and in case of high data redundancy (point density overage). The Fig.1 shows the result of some test surface compression with different accuracy levels.

Before compression   After compression with different accuracy levels

Figure 1. Surface compression with different accuracy levels.

In practice, single-processor simplification of large 3D datasets (contrary to surface processing) is a very time-consumining operation. This problem is another one reason for utilizing multiprocessor computer systems (MCS).

2.2. PARALLEL ALGORITHM FOR SURFACE SIMPLIFICATION

The processing of large volume surfaces is more effective when performed with the MCS. The proposed algorithm can be easily parallelized, as data is processed locally and different surface fragments can be handled simultaneously. Actually, the domain decomposition is used for parallelization: initial surface is divided into parts, with each part being processed by a single processor. It allows to get the most scalability and performance improvements due to benefits of geometric parallelism. No inter-processor interaction occurs during the principal simplification process. In order to provide further connectivity of subdomains, all shared vertices of these domains are marked preliminarily as undeletable and remain unhandled during the whole process.After simplification, all surface parts are gathered from processors and "sewed" together with the marked vertices. On the final stage of the algorithm, all seams are processed with the uniprocessor algorithm with a decreased valuue of the maximum allowed error tolerance. Testing of algorithm showed that the visual quality of the seam-areas of the surface obtained is nearly as good as after simplification of these areas with the single-processor algorithm. The Fig.2 demonstrates the result of the simplification of a test surface given on a rectangular region. The compression was performed with three processors, with high accuracy.

Side view   View from above

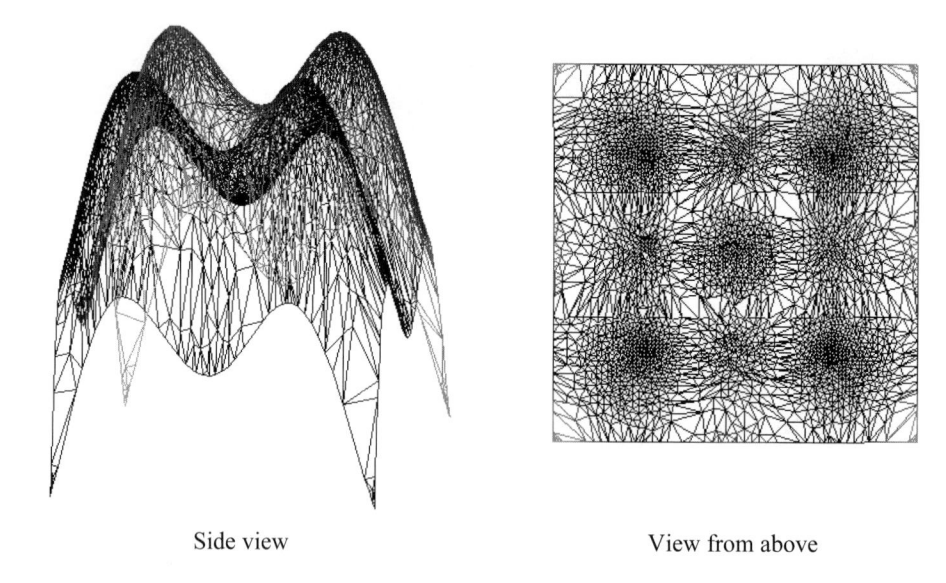

Figure 2. Parallel surface simplification done with 3 processors.

3. PARALLEL ALGORITHM FOR 3D SCALAR DATASETS SIMPLIFICATION

Based on the algorithm for surface compression the algorithm is extended for 3D data defined on tetrahedral meshes. Considering that 3D scalar data has more complicated structure and takes much more processor memory, the compression time in this case can be hundred times as much as the compression time for isosurfaces. The main idea of the compression algorithm for the data defined on tetrahedral meshes is to represent the data as a set of points in a 4-dimensional space and make consequent simplifications for small data fragments similarly to the surface simplification. In this approach, physical values of a mesh function are handled as a new geometrical coordinate. For the correct working of this method with the classical L_2-distance between points in 4D Euclidean space it's necessary beforehand to normalize (rescale) values of a mesh function according to the geometrical size of the dataset. In general the algorithm for processing 3D scalar data differs from the algorithm for surfaces only in the space dimension. The few differences mainly concern the procedure of monitoring the topology preservation.

Visualization of flow around the plane and the sphere is taken as an example. The gas flows are described by Navier-Stokes equations. The tetrahedral mesh is used. The triangles which make up the external and internal surfaces of this mesh are represented at the Fig. 3.

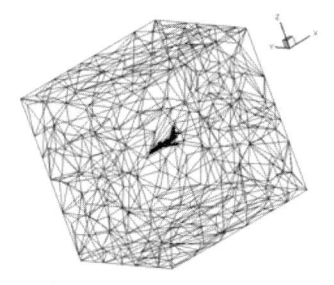

Figure 3. Calculation area for simulation of flow around plane.

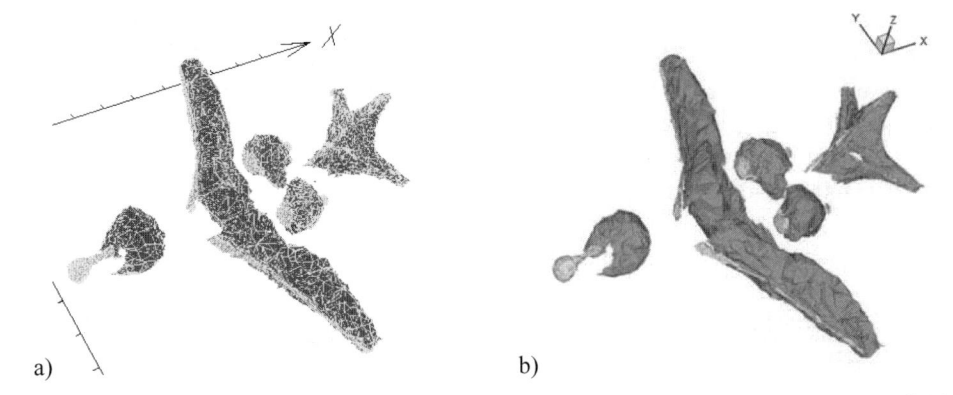

a)   b)

Figure 4. Isosurfaces of density field, created by **RemoteViewer** (a). Iso-surfaces of density field, created by Tecplot (b).

Visualization system **RemoteViewer** was compared to the commonly used system Tecplot [7]. Visual images of the same field prepared by these systems are demonstrated in the Fig. 4(a), 4(b). In spite the fact that **RemoteViewer** performed a lossy compression using 10 processors the figures show that there is no any considerable difference between them. Visual coincidence of both images verifies that **RemoteViewer** is a competitive visualizer in the software field. The algorithm showed good results during visual comparisons between isosurfaces of initial and compressed datasets. However, there still remain various aspects of the algorithm that could be improved upon.

The general algorithms of multi-processor simplification outlined here are able to effectively compress initial datasets with any accuracy required. The application of these methods is quite topical and urgent since the compression of large datasets is a necessary key to the possibility of their visualization. The speed of the algorithms can be roughly represented as $O(n)$, where n is an initial number of vertices. The gradualism of the methods allows their application in multiresolution modeling. These methods are rather universal as they are able to process the wide class of initial datasets. The algorithms showed their efficiency on a large number of numerical modeling results for many physical problems and for various test datasets.

4. VECTOR FIELDS VISUALIZATION

Static and dynamic visualization of 3D vector fields is very essential for CFD applications. For unsteady flows a set of vector fields at different moments of time can be interactively visualized. Integration of a particle trajectory over all vector fields allows to reveal vortex zones, shock waves and other peculiarities, inherent to unsteady flows. Vector fields visualization permits to compare numerical results with experimental data, obtained using coloring or dust injection.

To obtain perfect load balancing we used asynchronous message passing, dynamic task queue and current data allocation table for each processor node. This allowed to significantly reduce the dependence of the speed of calculation from the speed of disk input/output subsystem. This approach ensured the effective localization of initial particle positions on unstructured meshes as well as as transfer of trajectories for calculation between processor nodes.

5. INPUT/OUTPUT AND STORING OF MESH DATA

We need to parallelize not only analysis and visualization operations, but also data input and output. The two-level method [1,2] of storing large meshes is used. Microdomains [5,6] (small compact mesh fragments) and macrograph of connections between microdomains are stored separately. This method makes the parallel data input efficient. Microdomains can be read on each processor simultaneously and independently. The corresponding programs are designed as a library for decomposition and distributed input/output of 3D meshes and mesh functions. This library allows to store the distributed mesh on several disks of file servers or processor nodes [6].

References
1. B. Hendrickson, R. Leland. Multilevel Algorithm for Partitioning Graphs. Supercomputing '95 Proceedings. San Diego, CA, 1995.
2. G. Karypis, V. Kumar. Multilevel Graph Partitioning Schemes. ICPP, 3 (1995) 113-122.
3. M.V.Iakobovski, D.E.Karasev, P.S.Krinov, S.V.Polyakov. Visualisation of grand challenge data on distributed systems. In L.A. Uvarova, ed. Mathematical Models of Non-Linear Excitations, etc. Proc. of a Symp. Kluwer Academic Publishers. New York. 2001. 71-78.
4. Krinov P.S., Iakobovski M.V., Muravyov S.V. Large Data Volume Visualization on Distributed Multiprocessor Systems. In B.Chetverushkin et al., eds. Proc. of the Parallel CFD 2003 Conference. Elsevier B.V., Amsterdam. 2004. 433-438.
5. M.V.Iakobovski. VANT. Ser.: Mat. Mod. Fiz. Proc., 2 (2004). 40-53. (in Russian)
6. Iakobovski M.V., Boldyrev S.N., Sukov S.A. Big Unstructured Mesh Processing on Multiprocessor Computer Systems. In B.Chetverushkin et al., eds. Proc. of the Parallel CFD 2003 Conference. Elsevier B.V., Amsterdam. 2004. 73-79.
7. http://www.tecplot.com/